DATA ANALYSIS
and STATISTICS
for Geography, Environmental Science, and Engineering

DATA ANALYSIS
and STATISTICS
for Geography, Environmental Science, and Engineering

MIGUEL F. ACEVEDO

CRC Press
Taylor & Francis Group
Boca Raton London New York

CRC Press is an imprint of the
Taylor & Francis Group, an **informa** business

CRC Press
Taylor & Francis Group
6000 Broken Sound Parkway NW, Suite 300
Boca Raton, FL 33487-2742

First issued in paperback 2019

© 2013 by Taylor & Francis Group, LLC
CRC Press is an imprint of Taylor & Francis Group, an Informa business

No claim to original U.S. Government works

ISBN-13: 978-0-4398-8501-7 (hbk)
ISBN-13: 978-0-367-86679-2 (pbk)

Library of Congress Cataloging-in-Publication Data

Acevedo, Miguel F.
    Data analysis and statistics for geography, environmental science, and engineering / Miguel F. Acevedo.
        p. cm.
    Includes bibliographical references and index.
    ISBN 978-1-4398-8501-7 (hardcover : alk. paper)
    1. Geography--Data processing. 2. Geography--Statistical methods. 3. Environmental sciences--Data processing. 4. Environmental sciences--Statistical methods. 5. Engineering--Data processing. 6. Engineering--Statistical methods. I. Title.

G70.2.A26 2012
519.5--dc23   2012032357

Visit the Taylor & Francis Web site at
http://www.taylorandfrancis.com

and the CRC Press Web site at
http://www.crcpress.com

Contents

PART II *Matrices, Tempral and Spatial Autoregressive Processes, and Multivariate Analysis*

Preface

This book evolved from lecture notes and laboratory manuals that I have written over many years to teach data analysis and statistics to first-year graduate and fourth-year undergraduate students. I have developed this material during 15 years while teaching a first-year graduate course in quantitative techniques for the Applied Geography and the Environmental Sciences program at the University of North Texas (UNT). In that course, we focus on data analysis methods for problem solving in geographical and environmental sciences, emphasizing hands-on experience. Quantitative methods applied in these sciences share many attributes; of these, we emphasize the capabilities to analyze multiple factors that vary both spatially and temporally.

Statistical and probabilistic methods are the same in a broad range of disciplines in science and engineering, and so are the computational tools that we can use. Methods may vary by discipline either because of academic tradition or because of the priority given to certain problems. However, methods not traditionally employed in a discipline sometimes become part of its arsenal as priorities shift and methods are "imported" from other fields where they have shown to be effective.

Some of the principles inspiring this book are that educating twenty-first-century scientists and engineers in statistical and probabilistic analysis requires a unified presentation of methods, the inclusion of how to treat data that vary in space and time, as well as multiple dimensions, and a practical training of how to perform analysis using computers. Furthermore, given the importance of interdisciplinary work in sustainability, this book attempts to bring together methods applicable across a variety of science and engineering disciplines dealing with earth systems, the environment, ecology, and human–nature interactions. Therefore, this book contributes to undergraduate and graduate education in geography and earth science, biology, environmental science, social sciences, and engineering.

OVERVIEW

I have divided the book into two parts:

- Part I, Chapters 1 through 8: Probability, random variables and inferential statistics, applications of regression, time series analysis, and analysis of spatial point patterns
- Part II, Chapters 9 through 15: Matrices, multiple regression, dependent random processes and autoregressive time series, spatial analysis using geostatistics and spatial regression, discriminant analysis, and a variety of multivariate analyses based on eigenvector methods

The main divide between the two parts is the use of matrix algebra in Part II to address multidimensional problems, with Chapter 9 providing a review of matrices.

Although this organization may seem unconventional, it allows flexibility in using the book in various countries, various types of curricula, and various levels of student progress into the curriculum. In the United States, for example, most undergraduate students in the sciences do not take a linear algebra course, and in some engineering programs, linear algebra is not required until the third year (juniors) upon completion of a second calculus class. In other countries and many U.S. engineering programs, colleges expose their undergraduate students to matrix algebra earlier in the curriculum; for example, engineering students in China typically take linear algebra in their second year of college. Therefore, I have left the multidimensional material for last after a substantial review of matrix algebra, allowing the students to become familiar with time series and spatial analysis at an earlier stage.

USE OF THE BOOK

There are several ways to use this book. For example, a junior-level third-year course for under-graduate students can cover The eight chapters of Part I at a rhythm of about two chapters per week during a typical 15-week semester. A more challenging or honors section could include the review of matrices (Chapter 9) and a couple of chapters from Part II. A senior-level combined with first-year graduate course can be based on the entire book. Depending on the students' background, the course could cover the material of Part I as a review (except spatial analysis and time series) in order to spend the majority of time on the topics presented in Part II. My experience has been that last-year (senior) undergraduate and first-year graduate students in the sciences are unfamiliar with matrix algebra and would need Chapter 9. However, a senior-level undergraduate or graduate engineering course may not need coverage of this chapter, except for the computer session. Many graduate students would be familiar with the material in the first two chapters of Part I, and they could read it rapidly to refresh the concepts. For other students, this material may be new and may require additional reading beyond the basics provided here.

PEDAGOGY

Each chapter starts with conceptual and theoretical material covered with enough mathematical detail to serve as a firm foundation to understand how the methods work. Over the many years that I have used this material, I have confirmed my belief that students rise to the challenge of understanding the mathematical concepts and develop a good understanding of basic statistical analysis that facilitates their future learning of more advanced and specialized methods needed in their profession or research area. To facilitate learning these concepts, I have included examples that illustrate the applications and how to go from concepts to problem solving. The conceptual and theoretical section ends with exercises similar to the examples.

In each chapter, a hands-on computer session follows the theoretical foundations, which helps the student to "learn by doing." In addition, this computer session allows the reader to grasp the practical implications of the theoretical background. This book is not really a software manual, but the computer examples are developed with sufficient detail that the students can follow and perform themselves either in an instructor-supervised computer classroom or lab environment or unassisted at their own pace. This design gives maximum flexibility to the instructor and the student. In a similar fashion to the theoretical section, the computer session ends with exercises similar to the examples.

COMPUTER EXAMPLES AND EXERCISES

I have organized the computer examples using the R system, which is open source. This is very simple to download, install, and run. As some authors put it, R has evolved into the *lingua franca* of statistical computing (Everitt and Hothorn, 2010). R competes with major systems of scientific computing, yet because it is open source, it is free of commercial license cost while having access to thousands of packages to perform a tremendous variety of analysis. At the time of this writing, there are 3398 packages available. Even students with no prior knowledge of programming are quickly acquainted with the basics of programming in R. For those users who still prefer a graphical user interface (GUI), there is diversity of GUIs also available as open source.

R is a GNU project system (GNU stands for Gnu's Not Unix). The GNU project includes free software and general public license. R is available from the comprehensive R archive network (CRAN), the major repository mirrored all over the world. The simplest approach is to download the precompiled binary distribution for your operating system (Linux, Mac, or Windows). In this book, we will assume a Windows installation because it is a very common situation in university environments, but the computer exercises given here would work under all platforms.

In addition to the R GUI, there are several other GUIs available to simplify or extend R. For example, (1) a web GUI to enter commands over a web browser, which can be used from smart phones and pads with web access, and (2) the R Commander GUI and its several plug-in packages to simplify entering commands.

As mentioned earlier, students can execute the computer examples and exercises in the classroom environment or at their own pace using their computers. Over the years, I have tested this material in both modes. I conduct a weekly instructor-supervised session in the computer classroom, where I run demonstrations from the instructor machine equipped with a projector or systematic instructions followed by students in their assigned computer or simply letting the students follow the instructions given in the book and asking for help as needed. Students can go to the computer lab to work on their assigned exercises or complete the assignments off-campus by installing R on their computers or running R from a browser.

HOW TO USE THE BIBLIOGRAPHY

In each chapter, I provide suggestions for supplementary reading. These items link to several textbooks that cover the topics at similar levels and have been written for different audiences. This can help students read tutorial explanations from other authors. Often, reading the same thing in different words or looking at different figures helps students to understand a topic better. In addition, the supplementary readings point to several introductory texts that can serve to review. I have also included references that provide entry points or hooks to specialized books and articles on some topics, thus helping advanced students access the bibliography for their research work.

SUPPLEMENTARY MATERIAL

Packages seeg and RcmdrPlugin.seeg available from CRAN provide all data files and scripts employed here. These are also available via links provided at the Texas Environmental Observatory (TEO) website www.teo.unt.edu and the author's website, which is reachable from his departmental affiliation website www.ee.unt.edu. The publisher also offers supplementary materials available with qualifying course adoption. These include a solutions manual and PowerPoint® slides with figures and equations to help in preparing lectures.

Miguel F. Acevedo
Denton, Texas

Acknowledgments

I am very grateful to the many students who took classes with me and used preliminary versions of this material. Their questions and comments helped shape the contents and improve the presentation. My sincere thanks to several students who have worked as teaching assistants for classes taught with this material and helped improve successive drafts over the years, in particular H. Goetz and K. Anderle who kindly made many suggestions. I would also like to thank the students whom I have guided on how to process and analyze their thesis and dissertation research data. Working with them provided insight about the type of methods that would be useful to cover in this textbook.

Many colleagues have been inspirational, to name just a few: T.W. Waller and K.L. Dickson, of the UNT Environmental Science program (now emeritus faculty), M.A. Harwell (Harwell Gentile & Associates, LC), D.L. Urban (Duke University), M. Ataroff and M. Ablan (Universidad de Los Andes), and J. Raventós (Universidad de Alicante), and S. García-Iturbe (Edelca).

Many thanks to Irma Shagla-Britton, editor for environmental science and engineering at CRC Press, who was enthusiastic from day one, and Laurie Schlags, project coordinator, who helped immensely in the production process. Several reviewers provided excellent feedback that shaped the final version and approach of the manuscript.

Special thanks to my family and friends, who were so supportive and willing to postpone many important things until I completed this project. Last, but not least, I would like to say special thanks to the open source community for making R such a wonderful tool for research and education.

Author

Miguel F. Acevedo has 38 years of academic experience, the last 20 of these as faculty member of the University of North Texas (UNT). His career has been interdisciplinary, especially at the interface of science and engineering. He has served at UNT in the Department of Geography, the Graduate Program in Environmental Sciences of the Department of Biology, and more recently in the Department of Electrical Engineering.

Dr. Acevedo received his PhD in biophysics from the University of California, Berkeley (1980) and his MS and ME in computer science and electrical engineering from the University of Texas at Austin (1972) and from Berkeley (1978), respectively. Before joining UNT, he was at the Universidad de Los Andes, Merida, Venezuela, where he taught since 1973 in the School of Systems Engineering, the Graduate Program in Tropical Ecology, and the Center for Simulation and Modeling (CESIMO).

Dr. Acevedo has served on the Science Advisory Board of the U.S. Environmental Protection Agency and on many review panels of the U.S. National Science Foundation. He has received numerous research grants and has written many journal and proceeding articles as well as book chapters. UNT has recognized him with the Regents Professor rank, the Citation for Distinguished Service to International Education, and the Regent's Faculty Lectureship.

Part I

Introduction to Probability, Statistics, Time Series, and Spatial Analysis

1 Introduction

In this introductory chapter, we start with a brief historical perspective of statistics and probability, not to exactly account for its development, but to give the reader a sense of why statistics, while firmly grounded in mathematics, has an empirical and applied flavor.

1.1 BRIEF HISTORY OF STATISTICAL AND PROBABILISTIC ANALYSIS

In the western world, **statistics** started in the seventeenth century a little over 300 years ago attributed by many to John Graunt who attempted to relate data of mortality to public health by constructing life tables (Glass, 1964). Thus, the birth of statistics relates to solving problems in demography and epidemiology, subjects very much at the heart of geographical science and ecology. It is interesting to note that Graunt's 1662 book *Natural and Political Observations Made upon the Bills of Mortality* took an interdisciplinary approach linking counts of human mortality at various ages to public health. In addition, the word "observations" in the title of his book illustrates statistics' **empirical** heritage, i.e., collecting and organizing data and performing analysis driven by these data.

Subsequently, and for more than a century, statistics helped governments, or **stat**es, analyze demographical and economical issues. That is how statistics got its name; from the word "Statistik" used in Germany and the Italian word "statista" for public leader or official.

There has been some historical interest in determining whether statistics was born much earlier because of the Greek's achievements in mathematics and science, the Egyptian's censuses conducted to build the pyramids, and evidence of tabular form of data in China. However, the Greeks did not build an axiomatic apparatus as the one they had for geometry (Hald, 2003). Some attribute early use of data in tables in China to Yu the Great (~2000 years BC), founder of the Xia dynasty, who also developed flood control, a human–nature interaction of relevance to sustainability even today in many parts of the world. It seems that early use of data compilation was associated with geographical description of the Chinese state including keeping data in tabular form on number of households, in the various provinces and economic production (Bréard, 2006). Summation and calculation of means is present in the seventh-century compilation *Ten Books of Mathematical Classics*. Officials used means to calculate grain consumed per person and tax payment required per household (Bréard, 2006).

Back to Europe, also during the seventeenth century, Pierre de Fermat and Blaise Pascal, motivated by attempts to analyze games of chance, established the mathematical basis of **probability theory**. Chevelier de Mere piqued mathematical interest to solve a famous game of chance problem that was around for a century. Jacob Bernoulli and Abraham de Moivre continued to develop probability theory during the eighteenth century. A legacy of this period is the cornerstone discovery that averaging the outcomes for a **large number** of trials yields results approaching those **expected** by theory. Also from the eighteenth century, Thomas Bayes contributed the concept of conditional probabilities, and Pierre-Simon Laplace contributed the central limit theorem, inspired in calculating the distribution of meteor elevation angles. Laplace's paper on the central limit theorem appeared in the first years of the nineteenth century and he continued to be very active well into that century.

Major thrusts of the nineteenth century included how to use probability to deal with uncertainty in the natural sciences. This quest furthered the link between theory and observations, as the theory of errors was developed to solve problems in geodesy and astronomy. Inspired by this problem,

Carl Friedrich Gauss contributed a jewel for prediction models, the **least-squares** method. Then, as that century ended, statistical mechanics became a milestone in physics employing probability theory to explain macroscopic behavior (e.g., temperature of a gas) from microscopic behavior (random motion of a large number of particles).

During the ending years of the nineteenth century and first part of the twentieth century, statistical approaches and probability theory were further integrated. In the first quarter of the twentieth century, R.A. Fisher introduced the concepts of inductive inference, level of significance, and parametric analysis. These concepts served as the basis for estimation and hypothesis testing and were further developed by J. Neyman and E.S. Pearson. Together with this enhanced linking of the empirical approaches of statistics and theoretical approaches of probability, there was an increased use in social sciences (e.g., economics, psychology, sociology), natural sciences (biology, physics, meteorology), as well as in industry and engineering (e.g., ballistics, telephone systems, computer systems, quality control). Major methods in factor analysis and time series were developed for a variety of applications. Such an integration of empirical and theoretical approaches saw an increase through the twentieth century in specialized fields of science and engineering.

In the latest part of the twentieth century at AT&T Bell Laboratories (now Lucent Technology), major development included theoretical contributions by J. Tukey and several others, and the language S, predecessor of R used in this book (Becker et al., 1988; Chambers and Hastie, 1993; Crawley, 2002).

This brief historical account helps us understand how statistics came to be firmly grounded on mathematics, and at the same time be rather unique in its emphasis due to its empirical heritage and applicability in many fields. Thus, statistics and probability play a central role in applied mathematics. Interested students can read more on prospects for directions of statistics in the twenty-first century in several references (Gordon and Gordon, 1992; Raftery et al., 2002).

1.2   COMPUTERS

Computers and networks of computers became readily accessible in the latest part of the twentieth century, having two major kinds of impacts on statistics and probabilistic analysis: (1) we can now easily perform complicated calculations and visualize the results, and (2) we now have an amazing availability of data in an increasingly networked cyber-infrastructure.

Therefore, in addition to understanding the theory, it is important to acquire the computer skills to be a successful practitioner of statistical and probabilistic analysis in any modern profession. In particular, the integration of theoretical and computational aspects facilitates statistical analysis as applied in many areas of social and natural sciences, as well as medicine and engineering. Moreover, the computational aspects are not limited to the ability to perform a calculation but to manage data files and visualize results.

1.3   APPLICATIONS

Many examples in this book are from ecology, geosciences, environmental science, and engineering, not only because of my experience in these areas, but because this book attempts to bring together methods that can be applied in sustainability science and engineering. The complexity of environmental problem solving requires professionals skilled in computer-aided data and statistical analysis.

1.4   TYPES OF VARIABLES

A **variable** represents a characteristic of an object or a system that we intend to measure or to assign values, and of course, that varies (Sprinthall, 1990). Take for example time t and water level of a stream h. A variable can take one of several or many values; for example, time $t = 2$ days denotes

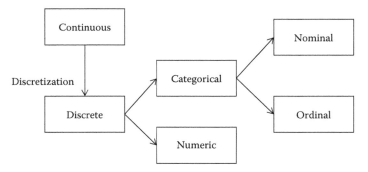

FIGURE 1.1 Simple classification of variables according to the nature of the values.

that t takes the value 2 days, and $h = 1.1\,$m that water level takes the value 1.1 m. It is common to classify variables in several types (Figure 1.1).

1.4.1 DISCRETE

There are variables that can take values from a **discrete** set of values, say the integers from 0 to 10, or colors red, green, and blue. The values can be numbers and the variable is **numeric**, or just categories (as colors in the example) and the variable would be **categorical** (Figure 1.1). We can use numbers 1, 2, 3 to code the categories, but the variable is still categorical. Numeric discrete variables often result from counting; for example, number of red balls and number of individuals in a classroom.

Moreover, we can distinguish **nominal** vs. **ordinal** categorical variables. When the categories have no underlying order, the variable is **nominal**, but if there is a meaningful order, the variable is **ordinal** (Figure 1.1). For an example of a nominal variable, think of three values given by colors red, blue, and green; or of two values head and tails from a coin toss. For an example of ordinal variables, think of qualitative scores low, medium, and high, or a scale used in polling or assessment 1 to 5, where 1 denotes "strongly disagree" and 5 the other extreme "strongly agree," with 3 denoting "agree" in the middle. In this last case, the ordinal values are coded with numbers but the variable is still categorical, not numeric.

1.4.2 CONTINUOUS

Other variables can take **real** values and therefore can theoretically have infinite number of values. For example, $h = $ water level in a stream could take any value in the interval from zero to a maximum. We refer to this type of variable as **continuous**. Many measurements of physical, chemical, and biological properties yield continuous variables. For example, air temperature, wind speed, salinity, and height of individuals.

1.4.3 DISCRETIZATION

We can convert continuous variables into discrete variables by **discretization**, i.e., dividing the range in intervals, and using these to define the values (Figure 1.1). Counting how many times the measurements fall in each interval yields a discrete numeric variable. For example, divide air temperature T in three intervals—cold $= T < 0°C$, medium $= 0°C \leq T < 10°C$, and hot $= 10°C \leq T$—and we now have a discrete variable with values cold, medium, and hot. Note that we need to make sure that boundaries are clear, for example, cold extends to less than zero but does not include zero, which is included in the medium interval. Counting the number of days such that the maximum temperature of the day falls in these intervals yields a numeric variable.

1.4.4 Independent vs. Dependent Variables

In addition, two major classes of variables are **independent** and **dependent** variables. We assume that the independent forces or drives the dependent, as in cause–effect, or factor–consequence. In some cases, this distinction is clear and, in other cases, it is an arbitrary choice to establish the predictor of a dependent variable Y based on the independent variable X. As the popular cautionary statement warns us, we could draw wrong conclusions about cause and effect when the quantitative method employed can only tell about the existence of a relationship.

1.5 PROBABILITY THEORY AND RANDOM VARIABLES

A good grasp of **probability** concepts is essential to understand basic statistical analysis, and in turn learning basic statistics is essential to understand advanced methods. These include regression, multivariate analysis, spatial analysis, and time series. Defining events and using probability trees are basic and very useful concepts, as well as using conditional probabilities and Bayes' theorem.

Random variables are those with values associated to a function defining the probability of their occurrence. **Distribution** is a general word used for this correspondence between the values and their probabilities. When the variable is discrete, this function is a **probability mass function (pmf)**, and when the variable is continuous, the function is a **probability density function (pdf)**. Accumulation of probability along the possible values of the variable leads the definition of **cumulative** functions. Then, calculations on the distributions are **moments**, such as the **mean** and **variance**. As we will see in forthcoming chapters, the law of large numbers and the central limit theorem are crucial to linking theoretical distributions and their moments to **samples**. Using computers, we can generate numbers that look random and seemingly drawn from an assumed distribution.

1.6 METHODOLOGY

Terms and methods will become clear as we go through the different chapters. When tackling quantitative problems, we can use the following general methodological framework.

1. Problem definition
   a. Define the questions to be answered by the analysis
   b. Define possible assumptions that could simplify the questions
   c. Identify components and their relationships aided by graphical block diagrams and concept maps
   d. Identify domains and scales in time and space
   e. Identify data required and sources
   f. Design experiments for data collection (in this case would need to go to step 4 and return to step 2)
2. Data collection and organization
   a. Measurements and experiments
   b. Collecting available data
   c. Organize data files and metadata
3. Data exploration
   a. Variables, units
   b. Independent and dependent variables
   c. Exploratory data analysis
   d. Data correlations
4. Identification of methods, hypotheses, and tests
   a. Identify hypotheses
   b. Identify methods and their validity given the data

5. Analysis and interpretation
   a. Perform calculations
   b. Answer to the questions that motivated the analysis
   c. Describe limits of these answers given the assumptions
   d. Next steps and new hypotheses
6. Based on the results, return to one of steps 1–4 as needed and repeat

1.7 DESCRIPTIVE STATISTICS

Step 3 of the methodological process just outlined includes applying **descriptive statistics**. By descriptive we mean that we are not attempting to make inferences or predictions but mainly to characterize the data, typically a **sample** of the theoretical population defined by the distribution. In simpler words, we want to tell what the data look like. Here, we use the term **statistic** to refer to a calculation on the sample, such as the sample mean and variance that we will discuss in Section 1.14. **Exploratory Data Analysis** refers to performing descriptive statistics, and using a collection of visual and numerical tools such as quantile–quantile plots, boxplots, and autocorrelation.

1.8 INFERENTIAL STATISTICS

Once we require answers to specific questions about the samples, we enter the realm of **inferential statistics**. The following are examples of questions we typically pose. Is a sample drawn from a **normal** distribution? Is it drawn from the same distribution as this other sample? Is there a trend? The question is often posed as a **hypothesis** that can be tested to be falsified. We must learn two important classes of methods: **parametric** (e.g., t and F tests) and **nonparametric** (e.g., Wilcoxon, Spearman). The first type is applied when the assumed distribution complies with certain conditions, and thus more conclusive, whereas the second type is less restrictive in assumptions but less conclusive.

Similarities between distributions are studied by goodness of fit (GOF) methods, which can be parametric (χ^2) and nonparametric (Kolmogorov–Smirnov) methods. Some simple and useful inferential methods are based on counts and proportions, and others such as contingency tables allow unveiling associations between categorical variables.

Providing sound **design of experiments** to test hypotheses has been an important mission of statistics in science and engineering. Analysis of variance (ANOVA) is a well-known method as its nonparametric counterpart (Kruskal–Wallis and Friedman). The mathematical formulation of ANOVA share basis with prediction, making their joint study helpful.

1.9 PREDICTORS, MODELS, AND REGRESSION

Prediction is at the core of building empirical models. We assume that there are drivers and effects. In other words, we want to build predictors of dependent variables Y based on the independent variables X. Regression techniques are the basis for many prediction methods. There are many types defined according to a variety of criteria and uses. The mathematical nature or structure of the predictor determines the type of method, such as linear regression vs. nonlinear regression; the number of variables determines the dimensionality, simple regression vs. multiple regression; the nature of the variables and their explicit variation with time and space determines specific methods, such as spatial autoregressive vs. autoregressive time series.

Varying with time and space is so pervasive in geographical and environmental systems that their study becomes essential even at an introductory level. Traditionally, the geography student is familiar with spatial analysis, and the engineering student with time series analysis. However, it is one of main tenets of the book that it is important to understand both spatial analysis and time series analysis.

1.10   TIME SERIES

A **random or stochastic process** is a collection of random variables with a distribution at each time t. When the distribution, or being less restrictive its moments, do not vary with time, we have the special case of **stationary** process that facilities analysis. A **time series** is a realization or sample of a random process. In Chapter 7, we cover independent random processes or processes such that the value at a given time is independent of past values. Then, after covering matrices in Chapter 9, we will study dependent random processes in Chapter 11. In these processes, the value at a given time is dependent of past values; we will study these focusing on **Markov** and **autoregressive** models.

Many methods in time series deal with finding out correlations between values at different times, and building predictors based on autoregressive and moving average models. Periodic process is amenable to spectral or periodical analysis, where we find the differences in the various or many periods or frequencies contained in the data.

1.11   SPATIAL DATA ANALYSIS

Throughout the book, we will look at two main types of spatial data: (1) **point patterns** and (2) **lattice** arrangements. A point pattern is a collection of points placed (often irregularly) over a spatial domain. We may have values of variables at each point. Lattice data correspond to values given to regions. These can be regularly arranged (as in a grid) or irregularly arranged (as in polygons). In Chapter 8, we will cover the analysis of point patterns, but we will wait until after Chapter 9 to cover lattice data because it requires matrix algebra. We will restrict ourselves to cover only the fundamentals of spatial data analysis since nowadays many of these methods are available in Geographic Information Systems (GIS) software.

A frequently encountered problem is examining the spatial distribution of points, for example, to check whether points are **clustered** or **uniformly** distributed, as well as its spatial variability. We can address this question by two alternative procedures. One procedure is dividing the spatial domain in **quadrats** and examining how many points fall in each quadrat and how it compares to the expected number of points. Another procedure is to calculate the distance to **nearest-neighbors**, examining the rhythm of change of this distance and comparing it to the one expected if it were to be uniform.

Once we consider that each point is **marked** or that there is a value of some variable associated to the point, we can also calculate the relationships between these values in the form of covariance. This leads to the concepts of variograms and semivariance that constitute the basis of **geostatistics**. Central to this collection of methods that emerged from engineering is the **kriging** method to predict values of variables in non-sampled points using a collection of sampled points.

Departing from point patterns are polygon patterns. **Spatial regression** helps predict values of variables in polygons from values at the neighboring polygons. Crucial to this method is defining the neighbor structure. Traditional methods include spatial autocorrelation (Moran and Geary) and several types of spatial regression models.

1.12   MATRICES AND MULTIPLE DIMENSIONS

Matrices are fundamental to understand the analysis of multiple variables and methods related to regression. All professionals in science and engineering benefit from understanding **matrix algebra**. It is important to know how to perform algebra operations with matrices, including the concepts of determinant and inverse. Formulating and solving linear equations using matrices, covariance matrices of sets of variables, and eigenvalues and eigenvectors, and singular value decomposition are crucial to all **multidimensional analysis** methods.

Many methods relate to inference such as multiple analysis of variance (MANOVA) and discriminant analysis. Others are related to prediction of a dependent variable based on a set of independent variables, as in multiple regression. We will cover extensions of some of the simple methods, particularly multiple linear regression, in order to build a model to predict Y from several independent variables X_i.

In addition, we will study methods to analyze a set of multiple variables to uncover relationships among all the variables and ways of reducing the dimensionality of the dataset. In some cases, we have basis to assume that we have several independent variables X_i, influencing several dependent or response variable Y_j. There are many multivariate analysis methods, but we will concentrate on those based on eigenvectors; such as principal components, factor analysis, correspondence analysis, multiple discriminant functions, canonical correlation, multidimensional scaling, and cluster analysis.

1.13 OTHER APPROACHES: PROCESS-BASED MODELS

A model is a simplified representation of reality, based on concepts, hypotheses, and theories of how the real system works. This book focuses on **empirical** models that build a quantitative relationship between variables based on data without an explicit consideration of the process yielding that relation. For example, using regression we can derive a predictor of tree height as a function of tree diameter based on measured data from 20 trees of different heights and diameters.

In contrast, some models are formulated by a set of mathematical **equations** based on the **processes** or **mechanisms** at work. For example, a differential equation representing tree growth over time based on increment of its diameter. For this purpose, we use the concept that diameter increases faster when the tree is smaller and that growth decreases when the tree is large. I have written a related book emphasizing **process-based** or **mechanistic** models, as opposed to **empirical** models (Acevedo, 2012). Simulation modeling is often the subject of a different course and taken at the senior undergraduate and first-year graduate level.

Both approaches complement each other. Empirical models help estimate parameters of the process-based models based on data from field and laboratory experiments. For example, we can use a mechanistic model to calculate flow of a stream using water velocity and cross-sectional area, but estimate velocity using an empirical relation of velocity to water depth. In addition, we will use empirical models to convert output variables of process-based models to other variables. For example, we can predict tree diameter increase from a process-based model of tree growth and then convert diameter to height using an empirical relation of height vs. diameter.

1.14 BABY STEPS: CALCULATIONS AND GRAPHS

1.14.1 MEAN, VARIANCE, AND STANDARD DEVIATION OF A SAMPLE

In this chapter, we introduce three simple concepts: the mean, variance, and standard deviation of a set of values (Carr, 1995; Davis, 2002). Most likely, you know these concepts but let us review them for the sake of common terminology and notation. A **statistic** known as the **"sample mean"**, which is the arithmetic average of n data values x_i comprising a sample

$$\bar{X} = \frac{1}{n} \sum_{i=1}^{n} x_i \tag{1.1}$$

The sample mean or average is denoted with a bar on top of X, i.e., $\bar{X}$.

Second, the **statistic** known as the **"sample variance"**, which is the variability measured relative to the arithmetic average of n data values x_i comprising a sample

$$s_X{}^2 = \frac{1}{n} \sum_{i=1}^{n} (x_i - \bar{X})^2 \tag{1.2}$$

This is the average of the square of the deviations from the sample mean. Alternatively

$$s_X{}^2 = \frac{1}{n-1} \sum_{i=1}^{n} (x_i - \bar{X})^2 \tag{1.3}$$

where $n - 1$ is used to account for the fact that the sample mean was already estimated from the n values. We can convert this equation to a more practical one by using Equation 1.1 in Equation 1.3 and doing algebra to obtain

$$s_X{}^2 = \frac{1}{n-1} \left[\sum_{i=1}^{n} x_i{}^2 - \frac{1}{n} \left(\sum_{i=1}^{n} x_i \right)^2 \right] \tag{1.4}$$

This is easier to calculate because we can sum the squares of x_i and subtract the square of the sum of the x_i. Now, a third well-known concept is the standard deviation calculated as the square root of the variance

$$sd(X) = s_X = \sqrt{s_X{}^2}. \tag{1.5}$$

1.14.2   Simple Graphs as Text: Stem-and-Leaf Plots

The easiest visual display of data is a **"stem-and-leaf"** plot. It is a way of displaying a tally of the numbers and the shape of the distribution. In a stem-and-leaf plot, each data value is split into a "stem" and a "leaf." The "leaf" is usually the last digit of the number and the other digits to the left of the "leaf" form the "stem." For example, the number 25 would be split as stem = 2, leaf 5, shortened as 2|5.

First, list the data values in numerical ascending order. For example,

```
23, 25, 26, 30, 33, 33, 34, 35, 35, 37, 40, 40, 41
```

Then separate each number into stem|leaf. Since these are two-digit numbers, the tens digit is the stem and the units digit is the leaf. Group the numbers with the same stems in numerical order

```
2|356
3|0334557
4|001
```

A stem-and-leaf plot shows the shape and distribution of data. It can be clearly seen in the diagram that the data cluster around the row with a stem of 3 and has equal spread above and below 3.

1.14.3 Histograms

A histogram is a graphical display of the distribution of the data; it graphs the frequency with which you obtain a value in a sample (if discrete numbers) or values falling in intervals ("bins") of the range of the variable (if continuous). Given a large enough sample, a histogram can help to characterize the distribution of a variable (Carr, 1995; Davis, 2002).

1.15 EXERCISES

Exercise 1.1
Use a variable X to denote human population on Earth. Explain why it varies in time and space and give examples of a value at a particular location or region and time.

Exercise 1.2
Suppose you build a model of light transmission through a forest canopy using measured light (treated as dependent variable) at various heights (treated as independent variable) and use it to predict light at those heights where it is not measured. Would this be a process-based model or an empirical model?

Exercise 1.3
Extend Exercise 1.2 to use the concept from physics that light is attenuated as it goes through a medium. Propose that attenuation is proportional to the density of foliage at various heights, and then propose a model based on an equation before you collect data. Would this be a process-based model or an empirical model?

1.16 COMPUTER SESSION: INTRODUCTION TO R

1.16.1 Working Directory

The computer sessions of this book assume that you have access to write and read files in a **working directory** or **working folder** typically located in a local hard disk **c:** or a network home drive **h:** or in a removable drive ("flash drive") say **e:**. For the purpose of following the examples in this book, a convenient way to manage files is to create a working directory, for example, **c:\labs**, to store all the files to be used with this book. Then a folder or directory for each computer session will be created within **c:\labs** working directory. For example, for session 1, it would be the folder given by path **c:\labs\lab1**. In each folder, you will store your data files and programs for that session.

1.16.2 Installing R

Download R from the Comprehensive R Archive Network (CRAN) repository http://cran.us.r-project.org/ by looking for the **precompiled binary distribution** for your operating system (Linux, Mac, or Windows). In this book, we will assume a Windows installation. Thus, for Windows, select the **base** and then the executable download for the current release; for example at the time this chapter was last updated, the release was R-2.14.0. Save in your disk and run this program, following installation steps. It takes just a few minutes. During installation, it is convenient to choose the option to install manuals in PDF.

1.16.3 Personalize the R GUI Shortcut

You can establish your working directory as default by including it in the **Start In** option of the shortcut and avoid changing directory every time. It takes setup time but it will be worthwhile because it saves time in the end.

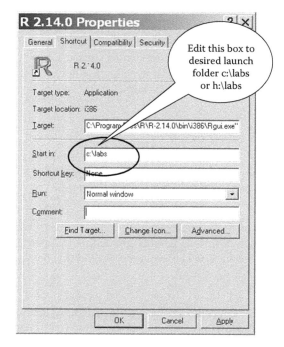

FIGURE 1.2   Modifying Start in folder of R shortcut properties.

To do this, Right click on shortcut, go to **Properties**, then type for example **c:\labs** or **h:\labs** as your launch or work directory (Figure 1.2). Double click to Run. R will start from this folder named labs. This remains valid unless edited. You need to create this folder beforehand if it does not exist yet. Note that now when you select Change dir, the working folder will be the one selected as Start In for the shortcut and therefore there is no need to reset.

When working on machines that are shared by many individuals (e.g., a university computer lab), a more permanent solution is to create a new personalized Shortcut in your working directory. Find the Shortcut on the desktop; right click, select **Create Shortcut**, and browse to the desired location (folder **labs**) as shown in Figure 1.3. Then right click on this new shortcut and select properties, edit **Start In** as shown before. Thereafter run R by a double click on this new personalized shortcut.

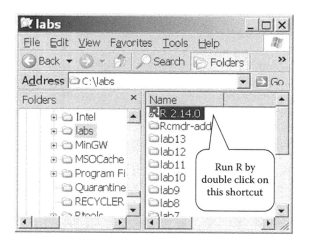

FIGURE 1.3   New R shortcut to reside in your working folder h:\labs.

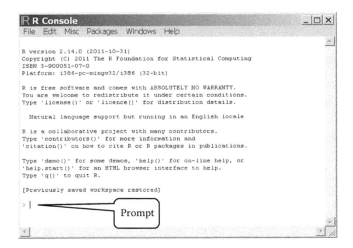

FIGURE 1.4   Start of R GUI and the R Console.

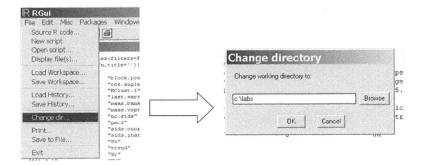

FIGURE 1.5   "Change dir" option in File menu to change working directory.

1.16.4   RUNNING R

You can just double click on the shortcut created during installation. Once the R system loads we get the R Graphical User Interface (**Rgui**) that opens up with the ">" prompt on the **R Console** (Figure 1.4). Another practical rule: make sure you are in the desired directory or folder. Use the **File|Change Dir** option in the RGui menu, and type the path or browse to your working directory.

In other words, under the **File** menu you can find **Change dir**, to select the working directory (Figure 1.5). You may have to repeat this **Change dir** operation every time you start the R system. However, once you have a **workspace** file .**Rdata** created in a folder, you can double click it to launch the program from that folder. We will explain more on this later when we describe the workspace.

1.16.5   BASIC R SKILLS

This session is a very brief tutorial guide to use **R** interactively from the GUI under windows. There is a lot more to **R** than summarized in these notes; see supplementary readings at the end of the chapter. These indications illustrate only how to get started, input data from files, simple graphics and programming loops, and are intended only as a starting point. We will study more details on R in later sessions.

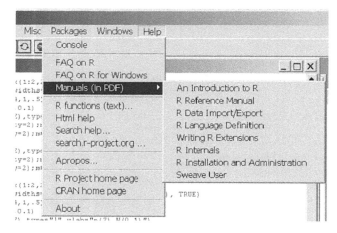

FIGURE 1.6   Finding R manuals from help menu item.

R Manuals are available online in PDF and in HTML formats via the help menu item. PDF (**P**ortable **D**ocument **F**iles) can be viewed and printed using the ***Acrobat Reader***. HTML (**H**yper**T**ext **M**arkup **L**anguage) can be viewed using a web browser. For example, *An introduction to R* in PDF (Figure 1.6). To obtain help in HTML format use the **Help** menu item, select **Html help** (Figure 1.7). This will run a browser with the help files and manuals. Just follow the links.

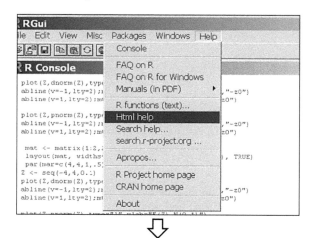

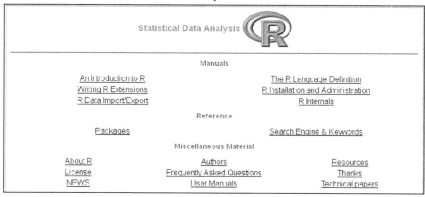

FIGURE 1.7   Help in HTML format and browser.

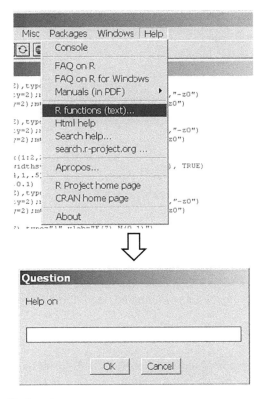

FIGURE 1.8   Help on specific functions.

You can obtain help on specific functions from Help menu, select R functions (text) as shown in Figure 1.8, then type the function name on the dialog box (Figure 1.8). From the R console, you can obtain help on specific functions. One way is to type **help**(*name of function*) at the prompt in the console; for example, **help(plot)** to get help on function **plot**. Also typing question mark followed by the function name would work, for example, **?plot**. You can also launch the help with **help.start** function. The following doc may help as simple reference to R at the CRAN website: http://cran. us.r-project.org/doc/contrib/Verzani-SimpleR.pdf.

1.16.6   R CONSOLE

This is where you type commands upon the > prompt and where you receive text output (Figure 1.4). We can also enter scripts or programs in a separate text editor, and then copy and paste to the R console for execution. Better you can use the **Script** facility of the Rgui described in Section 1.16.7.

1.16.7   SCRIPTS

From the **File** menu, select **New Script** to type a new set of commands. This will generate an editor window where you can type your set of commands. In this window, you can type and edit commands, and then right click to run one line or a selection of lines by highlighting a section (Figure 1.9). Then save the script by using **File|Save as** followed by selecting a folder to save in, say lab1, and a name. Scripts are saved as files with name ***.R** that is with extension "**.R**". Later the script can be recalled by **File|Open Script**, browse to your folder, select the file, and edit the script. Further edits can be saved using **File|Save**.

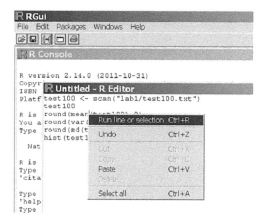

FIGURE 1.9   Script Editor: right click to run a line or lines of the script.

An alternative way of running your script (if you do not need to edit it before running it) is to use **File|Source R Code**, browse to folder, select the script. This is equivalent of using the function **source**("myscript.R") to execute the set of commands stored in file **myscript.R**.

1.16.8   GRAPHICS DEVICE

This is where you receive graphical output (Figure 1.10). To print hard copy of a graph, you can just select **File|Print** from menu while the graphics window is selected. To save in a variety of graphics formats use **File|Save as**. These formats include Metafile, Postscript, Bmp, PDF, and Jpeg. To capture on the clipboard and then paste on another application, you could simply use Copy and Paste from graphics window. To do this you can also use **File|Copy to clipboard** and then paste clipboard contents on selected cursor position of the file being edited in the application. Notice that one can work with windows as usual, go to **Windows** in the **menu** bar; here you can **Tile** the windows or can **Cascade** the windows, etc. In addition, this is where you can open the **Console** window.

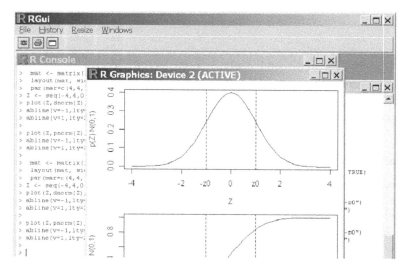

FIGURE 1.10   R graphics windows.

1.16.9 Downloading Data Files

Datasets and functions for the book are available as **archive** from the book website. After downloading, you can **unzip** and this will include several folders each containing several files. Each computer session corresponds to a folder. Download the archive **labs.zip** save to **labs** and unzip. This can be done under Windows using the Win Explorer and assuming you have a compression program available (e.g., WinZip and WinRAR). Then use extract button and select options in dialog box. Then folders **lab1, lab2, etc.**, will be created within **labs**. We can examine the contents of each folder with the file explorer. It is convenient to configure the Explorer to show all the file extensions. Otherwise, you will not see the **.txt** part of the filename.

To keep files organized, I recommend that you store the files of each computer lab session in separate folders. To do this, you would use the subdirectory (e.g., **lab1**) in your working directory in your home drive **c:\labs**. In this directory, you will store your data files and functions (mostly ASCII text files), figures and results.

1.16.10 Read a Simple Text Data File

The first thing to do before you import a data file is to look at your data and understand the contents of the file. This is a practical rule, which applies regardless of what software you use. For example, let us look at file **test100.txt** in **lab1**. As we can see, this is a text file and it opens using the notepad. It does not have a header specifying the variable name and that it is just one field per line with no separator between fields (Figure 1.11). This is a straightforward way of entering a single variable.

A more convenient text editor is **Vim**, also an open-source program that you can download from the internet (www.vim.org). I recommend that use this editor instead of the notepad or wordpad to work with text files and scripts. Some nice features are that you get line numbers and position within a line (Figure 1.12), a more effective find/replace, and tool and color codes for your script.

Now, going back to file **lab1\test100.txt**, it contains 100 numbers starting like this

```
48
38
44
41
56
...
```

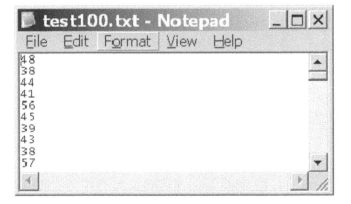

FIGURE 1.11 Example of a file with a single variable, one field per line. Viewed with the notepad text editor.

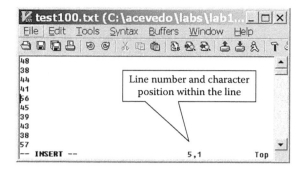

FIGURE 1.12   Same file as in previous figure viewed with the Vim editor.

Make sure you **File|Change Dir** to the working folder **labs** so that the path to the file is relative to this folder. If you have personalized the R shortcut to start in folder **labs**, then the path to the file is relative to this folder, for example, in this case, **lab1\ test100.txt**. Therefore, you could use this name to **scan** the file. Use forward slash "/" to separate folder and filename.

```
> scan("lab1/test100.txt")
```

On the console, we receive the response

```
Read 100 items
 [1] 48 38 44 41 56 45 39 43 38 57 42 31 40 56 42 56 42 46 35 40 30
     49 36 28 55
[26] 29 40 53 49 45 32 35 38 38 26 38 26 49 45 30 40 38 38 36 45 41
     42 35 35 25
[51] 44 39 42 23 44 42 52 55 46 44 36 26 42 31 44 49 32 39 42 41 45
     50 39 55 48
[76] 49 26 50 46 56 31 54 26 29 32 34 40 53 37 27 45 37 34 32 33 35
     50 37 74 44
```

Alternatively, you could also use two backward slashes > `scan("lab1\\test100.txt")`. It has the same effect as one forward slash. We will use forward slashes for simplicity. Next, create an object by scanning an input data file

```
> x100 <- scan("lab1/test100.txt")
```

object `x100` is assigned the results scanned from the file. The operator "**<-**" is used for **assignment**. Equivalently you can write the same using the equal sign "=". However, the equal sign is used for other purposes, such as giving values to arguments of functions.

Double check that you have the newly created object by **Misc|List objects** or using `ls()`.

```
> ls()
[1] "x100"
```

The object x100 is stored in the workspace **labs\.Rdata** but file **test100.txt** resides in **labs\lab1.** Double check the object contents by typing its name

```
> x100
 [1]  48 38 44 41 56 45 39 43 38 57 42 31 40 56 42 56 42 46 35 40 30
      49 36 28 55
[26]  29 40 53 49 45 32 35 38 38 26 38 26 49 45 30 40 38 38 36 45 41
      42 35 35 25
[51]  44 39 42 23 44 42 52 55 46 44 36 26 42 31 44 49 32 39 42 41 45
      50 39 55 48
[76]  49 26 50 46 56 31 54 26 29 32 34 40 53 37 27 45 37 34 32 33 35
      50 37 74 44
```

We can see that this object is a one-dimensional array. The number given in brackets on the left-hand side is the position of the entry first listed in that row. For example, entry in position 26 is 29. Entry in position 51 is 44.

Since this object is a one-dimensional array, we can check the size of this object by using function length()

```
> length(x100)
[1] 100
```

Important tip: when entering commands at the console, you can recall previously typed commands using the up arrow key. For example, after you type

```
> x100
```

you can use the up arrow key and edit the line to add length

```
> length(x100)
```

1.16.11 SIMPLE STATISTICS

Now we can calculate sample mean, variance, and standard deviation

```
> mean(x100)
[1] 40.86
> var(x100)
[1] 81.61657
> sd(x100)
[1] 9.034189
>
```

It is good practice to round the results, for example, to zero decimals

```
> round(mean(x100),0)
[1] 41
> round(var(x100),0)
[1] 82
> round(sd(x100),0)
[1] 9
>
```

We can concatenate commands in a single line by using the semicolon ";" character. Thus, for example, we can round the above to two decimals

```
> mean(x100); round(var(x100),2); round(sd(x100),2)
[1] 40.86
[1] 81.62
[1] 9.03
```

1.16.12 SIMPLE GRAPHS AS TEXT: STEM-AND-LEAF PLOTS

We can do stem-and-leaf plots with a simple function `stem` on the Rconsole. For example,

```
> stem(x100)

The decimal point is 1 digit(s) to the right of the |

2 | 35666667899
3 | 001112222344555556667778888889999
4 | 000001112222222234444445555556668899999
5 | 000233455566667
6 |
7 | 4
```

1.16.13 SIMPLE GRAPHS TO A GRAPHICS WINDOW

When applying a graphics command, a new graph window will open by default if none is opened; otherwise, the graph is sent to the active graph window. Plot a histogram by using function `hist` applied to a single variable or one-dimensional array object,

```
>hist(x100)
```

to obtain the graph shown in Figure 1.13. Bar heights are counts of how many measurements fall in the bin indicated in the horizontal axis.

1.16.14 ADDRESSING ENTRIES OF AN ARRAY

We can refer to specific entries of an array using brackets or square braces. For example, entry in position 26 of array `x100` above

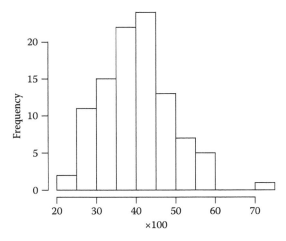

FIGURE 1.13 Histogram of x100.

```
> x100[26]
[1] 29
```

That is to say `x100[26]` = `29`. A colon ":" is used to declare a sequence of entries. For example, the first 10 positions of `x100`

```
> x100[1:10]
[1] 48 38 44 41 56 45 39 43 38 57
```

Entries can be removed, for example, `x100[-1]` removes the first entry

```
> x100[-1]
[1] 38 44 41 56 45 39 43 38 57 42 31 40 56 42 56 42 46 35 40 30 49
  36 28 55 29
> length(x100[-1])
[1] 99
```

Using a blank or no character in the bracket means that all entries are used

```
> x100[]
[1] 48 38 44 41 56 45 39 43 38 57 42 31 40 56 42 56 42 46 35 40 30
  49 36 28 55
[26] 29 40 53 49 45 32 35 38 38 26 38 26 49 45 30 40 38 38 36 45 41
  42 35 35 25
[51] 44 39 42 23 44 42 52 55 46 44 36 26 42 31 44 49 32 39 42 41 45
  50 39 55 48
[76] 49 26 50 46 56 31 54 26 29 32 34 40 53 37 27 45 37 34 32 33 35
  50 37 74 44
> length(x100[])
[1] 100
```

1.16.15 EXAMPLE: SALINITY

Next, we work with an example of data collected in the field. Salinity is an environmental variable of great ecological and engineering importance. It conditions the type of plant and animals that can live in a body of water and impacts the quality of water and the potential use of saline water. At the interface between rivers and sea, such as estuaries, salinity experiences spatial and temporal gradients. It is traditional to express salinity in parts per thousand ‰ instead of percentage % because it is the same as approximately grams of salt per kilogram of solution. **Freshwater**'s salinity limit is 0.5‰, then water is considered **brackish** for the 0.5‰–30‰ range, above that we have **saline** water in the 30‰–50‰, and **brine** with more than 50‰.

Examine **lab1/salinity.txt** file. It consists of four lines of 10 numbers each. It corresponds to salinity of water from 40 measurements at Bayou Chico, a small estuary in Pensacola Bay, FL. The values are from the same location and taken every 15 min. The salinity.txt file is also a single variable but given in 10 values per line with blank separations (Figure 1.14). We will practice how to scan a file and plot histograms using this dataset. Create an object containing these data. What is the length of the object? Obtain a stem-and-leaf plot. Obtain a histogram. Save the graph as a Jpeg file.

```
> x <- scan("lab1/salinity.txt")
Read 40 items
> x
 [1] 24.2 23.9 24.0 24.0 24.2 24.1 24.2 24.0 24.0 23.8 23.9 23.8 23.8
     23.8 23.8
[16] 23.7 23.6 23.5 23.3 23.2 23.3 23.2 23.1 23.1 23.1 23.2 23.0
     22.8 22.8 22.8
[31] 22.8 22.7 22.7 22.7 22.7 22.7 22.7 22.7 22.7 22.8
> length(x)
[1] 40
> stem(x)

The decimal point is 1 digit(s) to the left of the |

226 | 00000000
228 | 00000
230 | 0000
232 | 00000
234 | 0
236 | 00
238 | 0000000
240 | 00000
242 | 000

> hist(x)
> round(mean(x),1)
[1] 23.4
> round(var(x),1)
[1] 0.3
> round(sd(x),1)
[1] 0.5
>
>
```

To import the data we are using the scan command because, even though there are rows and columns in the data file, all of the numbers will be read as only **one** data stream; that is to say, a one-dimensional array. See the histogram in Figure 1.15. This water is brackish and does not get to be saline because it is under 30‰.

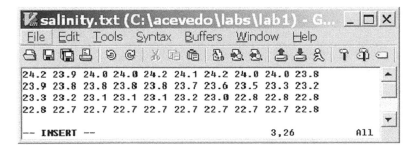

FIGURE 1.14   File with single variable, but several fields per line.

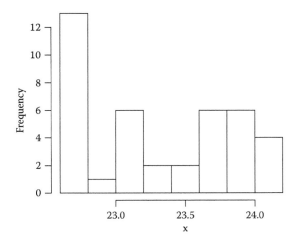

FIGURE 1.15   Histogram of salinity.

1.16.16   CSV Text Files

A CSV file is a text file with fields separated by commas. CSV stands for the format of **comma separated values**. In Windows, a default program to open the **CSV** files is Excel. Double click on the **salinity.csv** file to obtain Figure 1.16. To see more numbers you would have to scroll to the right. A CSV file is just a **text** file, and therefore it also opens with the notepad and Vim. Right click on the file name, select open with, and then select Vim. As you can see, commas separate the numbers and the lines are "word wrapped" (Figure 1.17). If using the notepad, you can choose the option **Format|Word wrap** to show all the numbers.

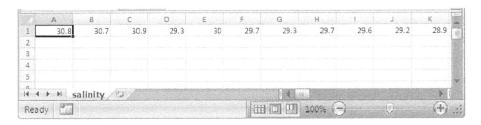

FIGURE 1.16   Example of a CSV file opened in MS Excel.

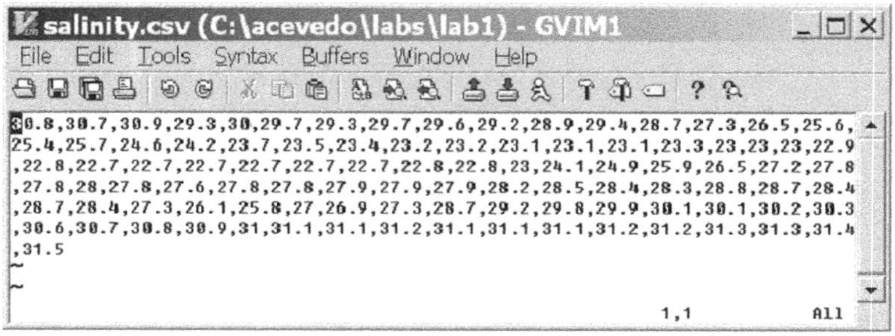

```
salinity.csv (C:\acevedo\labs\lab1) - GVIM1              _□ X
File  Edit  Tools  Syntax  Buffers  Window  Help

30.8,30.7,30.9,29.3,30,29.7,29.3,29.7,29.6,29.2,28.9,29.4,28.7,27.3,26.5,25.6,
25.4,25.7,24.6,24.2,23.7,23.5,23.4,23.2,23.2,23.1,23.1,23.1,23.3,23,23,23,22.9
,22.8,22.7,22.7,22.7,22.7,22.7,22.7,22.8,22.8,23,24.1,24.9,25.9,26.5,27.2,27.8
,27.8,28,27.8,27.6,27.8,27.8,27.9,27.9,27.9,28.2,28.5,28.4,28.3,28.8,28.7,28.4
,28.7,28.4,27.3,26.1,25.8,27,26.9,27.3,28.7,29.2,29.8,29.9,30.1,30.1,30.2,30.3
,30.6,30.7,30.8,30.9,31,31.1,31.1,31.2,31.1,31.1,31.1,31.2,31.2,31.3,31.3,31.4
,31.5

                                                    1,1                All
```

FIGURE 1.17   A CSV file opened in the notepad.

To read a CSV file into R, all you have to do is use scan with `sep= ","` argument.

```
> x<- scan("lab1/salinity.csv", sep=",")
Read 98 items
```

Then operate on object x as before

```
> length(x)
[1] 98
> stem(x)

The decimal point is at the |

22 | 7777778889
23 | 0000111223457
24 | 1269
25 | 46789
26 | 1559
27 | 02333688888999
28 | 0234445777789
29 | 2233467789
30 | 011236778899
31 | 011111223345

> hist(x)
> round(mean(x),1)
[1] 27.4
> round(var(x),1)
[1] 9
> round(sd(x),1)
[1] 3
>
```

I am not showing the resulting histogram for the sake of saving space. These values indicate that this water is brackish on the average but close to saline. In fact 25 observations out of 98 were 30 or above.

1.16.17   Store Your Data Files and Objects

The workspace with "objects", functions, and results from object-making commands can be stored in a file **.Rdata**. Use **File|Save workspace** menu to store an image of your objects. File **.Rdata** is

created in the launch folder specified in the **Start In** field of the R shortcut or you can browse to find
the desired working folder to store. For example, **c:/labs**, the console will inform of the save operation

```
> save.image("C:/labs/.RData")
```

To follow this book, it is convenient to store the workspace in your working folder. Once you save
the workspace, right after opening the commands window, you will see the following message:

```
[Previously saved workspace restored]
```

When done this way, **.Rdata** resides in your working drive and you can use this **.Rdata** file for
all computer sessions in order to facilitate access to functions created by various exercises. After
launching the program, you could load the workspace using **File|Load Workspace** and browse to
find the desired **.Rdata** file. Alternatively, you can also double click on the **.Rdata** file to launch the
program and load the workspace.

You may want to have control of where the **.Rdata** file is stored and to store objects in different
.Rdata files. You can use different files for storing objects. For example, we could use file name
other.Rdata to save the workspace related to a different project. After launching the console win-
dow, you could load the workspace using **File|Load Workspace** and browse to find the desired
.Rdata file. Again, for now I recommend that you use the same **.Rdata** folder in the launch folder
for all lab sessions in order to facilitate access to functions you will create using the various computer exercises of this book.

To list objects in your workspace type ls() or using **Misc|List objects** of the RGUI menu.
Check your objects with ls() or with objects(); if this is your first run, then you will not have
existing objects and you will get character(0).

```
> ls()
character(0)
> objects()
character(0)
>
```

1.16.18   COMMAND HISTORY AND LONG SEQUENCES OF COMMANDS

When editing through a long sequence of commands by using the arrow keys, one could use **History**
to visualize all the previous commands at once. However, it is typically more convenient (especially
if writing functions) to type the commands first in a script using the script editor or a text editor (say
Vim). Then, Copy and Paste from the text editor to the R console to execute.

1.16.19   EDITING DATA IN OBJECTS

Example: want to edit object x100. Use

```
>edit(x100)
```

to invoke data editor or from **Edit** menu item GUI use **data editor**.

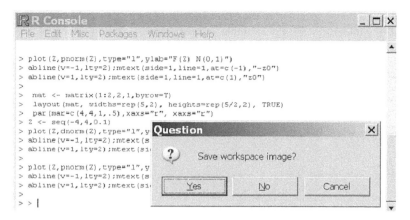

FIGURE 1.18   Closing the R session: reply yes.

1.16.20 CLEANUP AND CLOSE R SESSION

Many times we generate objects that may not be needed after we use them. In this case, it is good practice to clean up after a session by removing objects using the rm function. A convenient way of doing this is to get a list of objects with ls() and then see what we need to remove.

For example, suppose at this point we may want to keep object x100 because they contain data we may need later but remove object x.

```
> ls()
[1] "x100" .."x"
> rm(x)
```

You can also confirm that objects were indeed removed and get an update of the list to see if there is some more cleanup required.

```
> ls()
[1] "x100"
```

The objects can also be listed using **Misc|List Objects** in the Rgui menu.

Some of the clutter can be avoided by being careful about not generating objects unnecessarily. We will discuss how to do this later.

You can use q() or **File|Exit** to finalize R and close the session. When you do this, you will be prompted to save the workspace (Figure 1.18). Reply yes. This way you will have the objects created available for the next time you use R.

1.16.21 COMPUTER EXERCISES

Exercise 1.4
To make sure you understand the workspace. Save your workspace .**Rdata** file. Then close R and start a new R session, Load the workspace, make sure you have the objects created before.

Exercise 1.5
Use the notepad or Vim to create a simple text file **myfile.txt**. Type 10 numbers in a row separated by a blank space, trying to type numbers around a value of 10. Save in folder **lab1**. Now read the

file using scan, calculate sample mean, variance, and standard deviation, plot a stem-and-leaf diagram and a histogram and discuss.

Exercise 1.6
Use file **lab1\exercise.csv**. Examine the file contents using the notepad or Vim. Read the file, list numbers on the R Console rounding to 2 decimals. Calculate sample mean, variance, and standard deviation, plot a stem-and-leaf diagram and a histogram and discuss.

Exercise 1.7
Separate the first 20 and last 20 elements of salinity × array into two objects. Plot a stem-and-leaf plot and a histogram for each.

SUPPLEMENTARY READING

Several textbooks cover similar and related topics and can be used for a tutorial and supplementary reading for geography and geosciences (Burt et al., 2009; Carr, 1995, 2002; Davis, 2002; Fotheringham et al., 2000; Jensen et al., 1997; Rogerson, 2001), ecology and environmental science (Gotelli and Ellison, 2004; Manly, 2009; Qian, 2010; Quinn and Keogh, 2002; Reimann et al., 2008; Zuur et al., 2009), and engineering (DeCoursey, 2003; Ledolter and Hogg, 2010; Petruccelli et al., 1999; Schiff and D'Agostino, 1996; Wadsworth, 1998).

In recent years, there has been an increase in books using R including empirical and mechanistic models (Bolker, 2008; Clark, 2007; Crawley, 2005; Dalgaard, 2008; Everitt and Hothorn, 2010; Jones et al., 2009; Manly, 2009; Qian, 2010; Reimann et al., 2008; Soetaert and Herman, 2009; Stevens, 2009; Zuur et al., 2009).

Software papers, manuals, and books are very useful to supplement the computer sessions (Chambers and Hastie, 1993; Clark, 2007; Venables et al. 2012; Crawley, 2002; Deutsch and Journel, 1992; Fox, 2005; Kaluzny et al., 1996; MathSoft, 1999; Middleton, 2000; Oksanen, 2011).

Several introductory texts serve to review basic concepts (Drake, 1967; Gonick and Smith, 1993; Griffith and Amrhein, 1991; Mann Prem, 1998; Sprinthall, 1990; Sullivan, 2004).

2 Probability Theory

2.1 EVENTS AND PROBABILITIES

Probability theory is the basis for the analysis of uncertainty in science and engineering. The concept of probability ties to a numerical measure of the likelihood of an **event**, which is one outcome of an experiment or measurement. The **sample space** is the set of all possible outcomes, and therefore an event is a subset of the sample space. Probability can thus be defined as a real number between zero and one (0 and 1 included) assigned to the likelihood of an event. As a shorthand for the probability of an event, we can write Pr[event] or P[event]. For example, for event A, $Pr[A]$ or $P[A]$, is a real number between 0 and 1 (0 and 1 included).

It is common to give examples of games of chance when illustrating probability. Consider rolling a six-sided die. The sample space has six possible outcomes, $U = \{$side facing up is 1, side facing up is 2, …, side facing up is 6$\}$. Note the use of curly brackets to define set of events. Define event $A = \{$side facing up is number 3$\}$, then $P[A] = 1/6$ or 1 out of 6 possible and equally likely outcomes.

2.2 ALGEBRA OF EVENTS

For didactic purposes, events are usually illustrated using Venn diagrams and set theory. Events are represented by shapes or areas located in a box or domain. The **universal** event is the sample space U (includes all possible events), and therefore occurs with absolute certainty $P[U] = 1$. For example, $U = \{$any number 1 to 6 faces up after rolling a die$\}$. See Figure 2.1.

The **null** event is an impossible event or one that includes none of the possible events. Therefore, its probability is zero, $P[\phi] = 0$. For example, $\phi = \{$the side with number 0 will face up$\}$. This is not possible because the die does not have a side with number 0.

An oval shape represents an event A within U as shown in Figure 2.1. We also refer to B as the **complement** of A, i.e., the only other event that could occur. Therefore, the only outcomes are that A happens or B happens. Also, A and B are **mutually exclusive** and collectively exhaustive. The complement is an important concept often used to simplify solving problems. It is the same as B is NOT A, which in shorthand is $B = \bar{A}$ where the bar on top of the event means complement or logical operation NOT.

In the Venn diagram of Figure 2.1, B is shaded. The box represents U and the clear oval represents A. The key numeric relation is

$$P[B] = 1 - P[A] \qquad (2.1)$$

Also, note that the complement of U is the null event.

Example from rolling a six-sided die, define $B = \{$any side up except a six$\}$, $A = \{$side six faces up$\}$, determine $P[B]$. Solution: first note that $B = \bar{A}$ and therefore we can use $P[B] = 1 - 1/6 = 5/6$. We did not have to enumerate B with detail, just subtracted from 1.

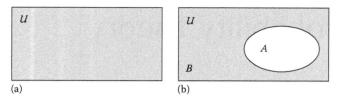

FIGURE 2.1   Universal event U or sample space (a), event A in U (b) showing B as the complement of A.

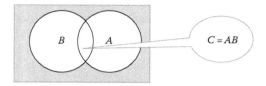

FIGURE 2.2   Intersection of two events; C is the sliver shared by events A and B.

When two events share common outcomes, we define the **intersection** of two events as the common or shared events. In other words, the intersection of A and B is the event C that is in both A and B. Denote the intersection by $C = AB$, then the probability of the intersection is $P[C] = P[AB]$. In the popular diagram illustrated in Figure 2.2, AB is contained in A and in B. It corresponds to the AND logical operation.

Back to the rolling die example. Define $A =$ {side 1 faces up, side 2 faces up}, $B =$ {side 2 faces up, side 3 faces up}. Obviously event {side 2 faces up} is common to A and B, therefore $C = AB =$ {side 2 faces up} and we know that this event has probability 1/6, thus $P[C] = 1/6$.

When A and B do not intersect, then AB is the null event $AB = \phi$ and therefore $P[AB] = 0$.

Example: $A =$ {side 1 faces up, side 2 faces up}, $B =$ {side 3 faces up, side 4 faces up}, $C =$ {null} and this event has probability 0, thus $P[C] = 0$.

The **union** of A and B is the event C defined as A happens or B happens. It is the OR logical operation and is denoted by $A + B$. In reference to Figure 2.2, it would be the addition of the two circles but we have to avoid double counting the sliver of the intersection. Therefore, we discount the intersection AB once.

$$P[C] = P[A + B] = P[A] + P[B] - P[AB] \tag{2.2}$$

Example, $A =$ {side 1 faces up, side 2 faces up}, $B =$ {side 2 faces up, side 3 faces up}, then $C =$ {side 1 faces up, side 2 faces up, side 3 faces up} and $AB =$ {side 2 faces up}. Assigning probabilities, $P[A] = 2/6$, $P[B] = 2/6$, $P[AB] = 1/6$, $P[A + B] = 2/6 + 2/6 - 1/6 = 3/6$.

An event B is included in A when event B is a subset of A, in set notation $B \subset A$ and therefore $P[B] < P[A]$. See Figure 2.3.

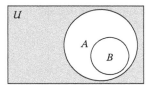

FIGURE 2.3   Event *B* is included in event *A*.

2.3   COMBINATIONS

When we complicate the experiment, for example, tossing a coin three times in a sequence, rolling a die five times in a sequence, we can combine the probabilities from the simpler components of the experiments to obtain probabilities of the more complex outcomes. The number n of independent repetitions (trials) and the number k of outcomes of each repetition determine the total number of possible outcomes. In general,

$$N = k^n \tag{2.3}$$

For example, consider tossing a coin twice and denote H for head and T for tail. We have two outcomes $k = 2$ for each trial and $n = 2$ trials. The outcome of a toss is independent of the other. Possible combinations lead to four events $N = 2^2$ and this constitutes the sample space $U = \{HH, HT, TH, TT\}$. Each outcome is equally likely with probability ¼. To see this we reason that the probability of getting H in first toss, $P[H] = 1/2$, and to get H in the second toss is the same because of independence, therefore $P[HH] = 1/4$.

The combinations of n items taken r at a time are of great interest

$$\binom{n}{r} = \frac{n!}{r!(n-r)!} \tag{2.4}$$

The exclamation "!" symbol is a factorial operation defined as

$$n! = n \times (n-1) \times (n-2) \times \ldots \times 2 \times 1 \tag{2.5}$$

For example, how many events have exactly one tail in two tosses of a coin? What is the probability of obtaining event $A = \{$exactly one tail in two coin tosses$\}$? Using Equation 2.4 yields

$$\binom{2}{1} = \frac{2!}{1!(2-1)!} = \frac{1 \times 2}{1 \times 1} = 2.$$

Thus, there are two possible combinations of one head in two tosses. This makes sense because from the previous example we know that we have four possible events. Only a set of these would have one tail in two trials and we can count them HT, TH. We can calculate the probability as $P[A] = P[HT] + P[TH] = 0.25 + 0.25 = 0.5 = 1/2$.

Just to practice, consider one more example of this situation but a more complicated experiment of three tosses (Exercise 2.5 at the end of the chapter).

2.4   PROBABILITY TREES

Another great visual aid in probability theory is a **tree**. The basic unit is a node from which we branch in arcs denoting events. Next to the arc we write the probability of the event and the name or code of the event at the tip of the arc. For example in the coin toss, we have the basic branch shown in Figure 2.4.

This basic unit can be iterated and combined to visualize situations that are more complex. For example, the two-toss coin experiment shown in Figure 2.5. Here, the end branches of the tree correspond to the four outcomes. Multiplication of the probabilities of all the arcs traversed yields the probability of each path. Thus,

$$P[HH] = 0.5 \times 0.5 = 0.25$$
$$P[HT] = 0.5 \times 0.5 = 0.25$$
$$P[TH] = 0.5 \times 0.5 = 0.25$$
$$P[TT] = 0.5 \times 0.5 = 0.25$$

(2.6)

As we can see, the probability of each path is the same and the sum of all probabilities is equal to one.

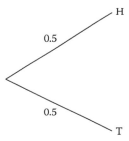

FIGURE 2.4   Basic element of a probability tree: node and set of arcs.

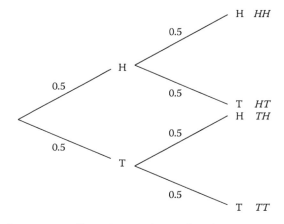

FIGURE 2.5   Example of a sequence of branches: tossing a coin twice.

2.5 CONDITIONAL PROBABILITY

Consider an experiment performed in two steps. At the first step, an event can occur with probability $P[A]$ and then at the second step the occurrence of an event C is **dependent** on whether A occurred. We use a vertical bar to denote the occurrence of one event conditioned on another

$$\Pr[C \text{ occurs given } A] = \Pr[C|A] = P[C|A]$$

The key relation here is that A and C must have non-null intersection $P[AC] \neq 0$ and that the ratio of the probability of this intersection to the probability of A will determine P[C|A]; this is to say,

$$P[C|A] = \frac{P[AC]}{P[A]} \tag{2.7}$$

A Venn diagram (Figure 2.6) and a tree (Figure 2.7) help illustrate this relation. The events at the end of the second set of arcs of Figure 2.7 are intersections and their probabilities can be found using Equation 2.7, rewritten as

$$P[AC] = P[C|A]\,P[A] \tag{2.8}$$

Example: Let us consider a sequence of two days. In the first day define event: $A = \{$rains the first day$\}$, with $P[A] = 0.5$ and event $C = \{$rains the second day$\}$. Assume that the probability of raining the second day given that it rains the first is 0.7. What is the probability that it rains both days? In this case, the intersection event is $AC = \{$rains the first day AND rains the second day$\}$. Let us calculate $P[AC]$ using conditional probability. Applying Equation 2.8 $P[AC] = P[C|A]\,P[A] = 0.7 \times 0.5 = 0.35$.

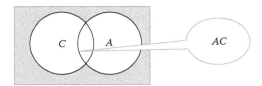

FIGURE 2.6   Intersection of events C and A: shows that C is conditioned on A.

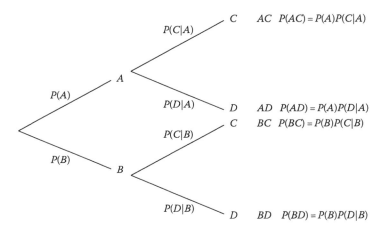

FIGURE 2.7   Second set of arcs show conditional probabilities.

If events A and C are independent, then $P[C|A]$ is just $P[C]$ and therefore

$$P[AC] = P[C|A]P[C] = P[C]P[A] \tag{2.9}$$

Example: Consider a sequence of two tosses of a fair coin. In the first toss define event A = {heads in first toss}, with $P[A]$ = 0.5 and event C = {heads in second toss}. What is the probability of getting a head in both tosses? The probability of getting a head in the second toss is independent of whatever we got in the first toss, thus $P[C|A]$ = $P[C]$ = 0.5. In this case, the event AC = {head the first toss AND head the second toss}. Then, calculate $P[AC]$ using conditional probability using Equation 2.9 $P[AC]$ = $P[C]P[A]$ = $0.5 \times 0.5 = 0.25$.

2.6   TESTING WATER QUALITY: FALSE NEGATIVE AND FALSE POSITIVE

Let us consider an example from water quality.

Example (Carr, 1995), a water quality test is conducted to decide whether water of a site is contaminated (event A) or not (event B). Assume that 20% of sampling sites are contaminated, then $P(A)$ = 0.2. Define C = the test result is negative, D = the test result is positive. Suppose that the test yields a **false negative** (i.e., fails to determine contaminated water) 3% of the time and that it yields a **false positive** 7% of the time. We can think of testing as a two-step experiment: the outcomes of the first step correspond to whether the water is contaminated or not, the outcomes of the second step correspond to the **results of the test**. Note that 0.03 and 0.07 are probabilities for test errors. We build the tree shown in Figure 2.8 using this information. In the tree the probability that the test is negative if the water is contaminated is $P[C|A]$ = 0.03, and that the water is indeed contaminated $P[A]$ = 0.2; then the probability that the test yielded negative results AND that the water was contaminated is $P[AC]$ = $0.03 \times 0.2 = 0.006$. Note that events C and D are dependent on both A and B.

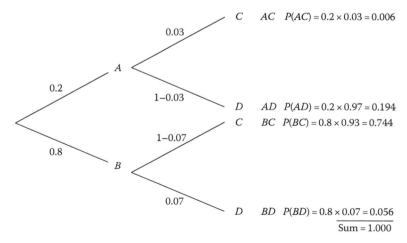

FIGURE 2.8   Water quality test: positive and negative results.

2.7 BAYES' THEOREM

Bayes' rule connects the conditional probability of an event A given C to its "inverse," or the conditional probability of C given A; in other words provides a link between $P[A|C]$ and $P[C|A]$. It is actually quite simple to derive it. From the conditional probability equation, we have

$$P[AC] = P[C|A]P[A] \qquad (2.10)$$

In addition, A depends on C, and thus we can write

$$P[A|C] = \frac{P[AC]}{P[C]} \qquad (2.11)$$

Or in the same form as Equation 2.10

$$P[AC] = P[A|C]P[C] \qquad (2.12)$$

Equating the two relations (Equations 2.10 and 2.12) for $P[AC]$

$$P[A|C]P[C] = P[C|A]P[A] \qquad (2.13)$$

And solving for $P[C|A]$

$$P[C|A] = \frac{P[A|C]P[C]}{P[A]} \qquad (2.14)$$

This relation is Bayes' theorem or rule for two events and is of great use, as we will see in the next examples.

In the tree of Figure 2.8 consider known probabilities for A and B, and for $P[C|A]$ and $P[C|B]$. Can we back-calculate? For example, what is the probability that we have contaminated water given a negative test result? In other words, what is the probability that A happens if we know that C happened? That is, what is $P[A|C]$? Use Bayes' theorem to calculate

$$P[A|C] = \frac{P[AC]}{P[C]} = \frac{P[C|A]P[A]}{P[C]} \qquad (2.15)$$

but we dont know P[C] yet. This probability of C should be obtained from either A or B happening

$$P[C] = P[C|A]P[A] + P[C|B]P[B] \qquad (2.16)$$

That is, adding the two paths leading to a C at the end. Now substitute Equation 2.16 in the denominator of Equation 2.15 to obtain

$$P[A|C] = \frac{P[AC]}{P[C]} = \frac{P[C|A]P[A]}{P[C|A]P[A] + P[C|B]P[B]} \qquad (2.17)$$

Use complementary events

$$P[C|B] = 1 - P[D|B] \quad \text{and} \quad P[B] = 1 - P[A] \qquad (2.18)$$

And substitute to obtain the final answer

$$P[A|C] = \frac{P[AC]}{P[C]} = \frac{P[C|A]P[A]}{P[C|A]P[A] + (1 - P[D|B])(1 - P[A])} \tag{2.19}$$

Example, in the tree of Figure 2.8, $P[AC] = 0.2 \times 0.03 = 0.006$, $P[BC] = (1 - 0.2) \times (1 - 0.07) = 0.8 \times 0.93 = 0.744$, and $P[C] = 0.2 \times 0.03 + 0.8 \times 0.93 = 0.75$. Then, using Equation 2.17 we get $P[A|C] = 0.006/0.75 = 0.008$. That is, if the test result is negative, we still have a 0.8% probability that the water is contaminated.

2.8   GENERALIZATION OF BAYES' RULE TO MANY EVENTS

Suppose there are several events B_1, B_2, ... that condition event C. For each event B_i, where $i = 1, n$, we can write an equation like Equation 2.10

$$P[CB_i] = P[C|B_i]P[B_i] \tag{2.20}$$

and using Bayes' rule for each event B_i we have

$$P[B_i|C] = \frac{P[C|B_i]P[B_i]}{P[C]} \tag{2.21}$$

Note that if events B_i account for all ways in which we can get event C, then by adding Equation 2.20 for all events B_i we get

$$P[C] = \sum_{i=1}^{n} P[CB_i] = \sum_{i=1}^{n} P[C|B_i]P[B_i] \tag{2.22}$$

And by substituting this last expression in the denominator of Bayes' rule (Equation 2.21), we can derive an extension to Bayes' rule like Equation 2.17

$$P[B_i|C] = \frac{P[C|B_i]P[B_i]}{P[C]} = \frac{P[C|B_i]P[B_i]}{\sum_{i=1}^{n} P[C|B_i]P[B_i]} \tag{2.23}$$

2.9   BIO-SENSING

Let us consider an application to bio-sensing water quality. In the mid-1990s, we started developing a method to detect contamination or other water quality issues by electronically monitoring the gape of clams (Allen et al., 1996). The working concept is that organisms integrate many signals from the environment and therefore are excellent environmental sentinels. Because of variability in behavioral response to stress, it is necessary to setup the monitor using more than one individual; we used 10 or more clams for water quality.

For the sake of a simple example, suppose we use two clams to sense water quality and that we are only considering valves completely shut or open. At any one measurement time, there are three events from measuring clam gape, B_1 = {two animals have valves shut}, B_2 = {one animal has shut and the other is open}, and B_3 = {both animals have open valves}. Define C as the event of contaminated water. Take an interval of 100 measurements and assume that 70% result in event B_1, 20% in B_2, and 10% in B_3. The probabilities of false positive errors are 0.1, 0.2, and 0.9 for B_1, B_2, and B_3 respectively. What is the probability that the water is contaminated? What is the probability that two animals have valves shut if the water is contaminated? The probability of contaminated water is $P[C]$, thus first apply Equation 2.22 and use the complement for all error probabilities because we are giving the false positive.

$$P[C] = \sum_{i=1}^{n} P[C|B_i]P[B_i] = \sum_{i} \left(1 - P[\bar{C}|B_i]\right)P[B_i] =$$

$$= (1-0.1)\times 0.7 + (1-0.2)\times 0.2 + (1-0.9)\times 0.1 =$$

$$= 0.8$$

The probability that two animals are shut if the water is contaminated is $P[B_1|C]$, and then apply Equation 2.23

$$P[B_1|C] = \frac{P[C|B_1]P[B_1]}{P[C]} = \frac{(1-0.1)\times 0.7}{0.8} = 0.79$$

2.10   DECISION MAKING

Probability theory is one of the bases of decision theory. We can frame a decision problem as selecting the most promising alternative **action** or **option** given the uncertainties. The major concept is that events occurring with given probabilities follow the options. Say, we choose option A_1 and event E_1 occurs with probability p, whereas event E_2 occurs with probability $1 - p$. Therefore, if we associate a cost or a loss to a combination of action and event (say A_1E_1), we can weigh the cost by its probability to calculate an **expected** cost or loss for this branch of the decision. We do this for all options, and then select the option with minimum expected loss or cost. Depending on the situation, we would rather formulate the decision in terms of benefits. For example, assign a profit or gain to the outcomes and tackle the decision by selecting the alternative with the maximum profit or gain.

A very simple example is one with two alternative actions A_1 and A_2 and two events E_1, E_2 as shown in the tree of Figure 2.9. This is a **decision tree** and is similar to a probability tree, except that some nodes (marked with squares) are decision nodes and others are event nodes (marked with circles). Suppose the alternative actions are to invest in an environmental protection at an investment cost of I_1 or at a lower level I_2 to prevent the detrimental ecosystem effect of a potential spill. Suppose the events are E_1 (contaminant spilled) with probability $p = P[E_1]$ and E_2 (no spill). If we implement prevention measures, the mitigation or restoration costs would be M_1, otherwise it would be M_2. The combination of decisions and outcomes are four A_1E_1 = invest more and spill occurs,

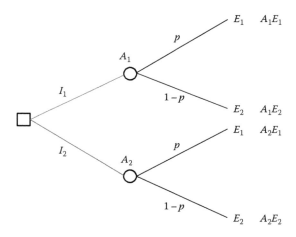

FIGURE 2.9   Decision tree: a simple example from environmental protection.

A_1E_2 = invest more and spill does not occur, A_2E_1 = invest less and spill occurs, and finally A_2E_2 = invest less and spill does not occur. The costs for each are

$$C(A_1E_1) = (I_1 + M_1) \times p$$
$$C(A_1E_2) = I_1 \times (1 - p)$$
$$C(A_2E_1) = (I_2 + M_2) \times p$$
$$C(A_2E_2) = I_2 \times (1 - p)$$

(2.24)

Adding up the costs for each alternative, we have

$$C(A_1) = C(A_1E_1) + C(A_1E_2) = (I_1 + M_1) \times p + I_1 \times (1 - p)$$
$$C(A_2) = C(A_2E_1) + C(A_2E_2) = (I_2 + M_2) \times p + I_2 \times (1 - p)$$

(2.25)

We would decide for A_1 or A_2 depending on which one is lower $C(A_1)$ or $C(A_2)$. A useful compact way of combining these costs is to take the cost difference $\Delta C = C(A_1) - C(A_2)$. Therefore, when ΔC is positive, we decide for A_2 but when ΔC is negative, we decide for A_1.

$$\Delta C = C(A_1) - C(A_2) = (I_1 + M_1 - I_2 - M_2) \times p + (I_1 - I_2) \times (1 - p) =$$
$$= (I_1 - I_2)p + (M_1 - M_2) \times p + (I_1 - I_2) \times (1 - p) =$$
$$= (I_1 - I_2) + (M_1 - M_2) \times p$$

(2.26)

An interesting situation is when your investment is higher for more protective measures, $I_1 > I_2$, and the mitigation costs are higher for the less protective measure, i.e., $M_2 > M_1$. In this case, it is convenient to express the differences $\Delta I = I_1 - I_2$ and $\Delta M = M_2 - M_1$ as positive numbers

$$\Delta C = (I_1 - I_2) - (M_2 - M_1) \times p = \Delta I - \Delta M \times p$$

(2.27)

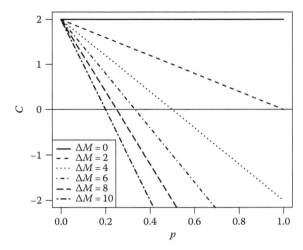

FIGURE 2.10   Cost vs. probability for several values of ΔM.

To further exemplify, suppose $\Delta I = \$2$ million, $\Delta M = \$10$ million. Moreover, assume the spill is relatively unlikely $p = 0.2$. Calculating the cost using Equation 2.27 we get $\Delta C = \Delta I - \Delta M \times p = 2 - 20 \times 0.2 = -2$ and because it is negative we decide to invest in the protective measure. We can investigate what would happen if the probability p were higher or lower by plotting ΔC vs. p (Figure 2.10). We see that when $0 \leq \Delta M < \$2$ million, the cost difference ΔC is never negative and we always choose A_2. Then when $\Delta M = \$2$ million, the spill would have to occur with certainty ($p = 1$) for the options to be equal. As we increase ΔM the value of p at which the line crosses zero, decreases; for example, we would choose option A_1 for $p > 0.2$.

This example is just a cartoon of a realistic situation, but illustrates the major points. This is similar to problems of insurance, risk, and many other practical issues. Note that we are only addressing the monetary value of the cost (what many would call the "bottom-line") and ignoring other aspects such as the ecological value of the resource affected.

2.11   EXERCISES

Exercise 2.1
Suppose we flip a fair coin to obtain heads or tails. Define the sample space and the possible outcomes. Define events and the probabilities of each.

Exercise 2.2
Define event $A = \{\text{rain today}\}$ with probability 0.2. Define the complement of event A. What is the probability of the complement?

Exercise 2.3
Define $A = \{\text{rains less than 2 cm}\}$ $B = \{\text{rains more than 1 cm}\}$. What is the intersection event C?

Exercise 2.4
A pixel of a remote sensing image can be classified as grassland, forest, or residential. Define $A = \{\text{land cover is grassland}\}$ $B = \{\text{land cover is forest}\}$. What is the union event C? What is $D = $ the complement of C?

Exercise 2.5

Assume we flip a coin three times in sequence. The outcome of a toss is independent of the others. Calculate and enumerate the possible combinations and their probabilities.

Exercise 2.6

Assume we take water samples from four water wells to determine if the water is contaminated and that they are independent. Calculate the number and enumerate the possible events of contamination results. Calculate the number and enumerate those that would have exactly two contaminated wells in the four trials.

Exercise 2.7

Using the tree of Figure 2.8 calculate the probability that test is positive **and** that water was not contaminated $P[BD]$? What is the total probability of the test is in error? Hint: BD **or** AC. What is the probability that the test is correct?

Exercise 2.8

Using Figure 2.8 and Bayes' theorem: What is the probability that the water is contaminated given a positive test result? Hint: calculate $P[A|D]$.

Exercise 2.9

Assume 20% of an area is grassland. We have a remote sensing image of the area. An image classification method yields correct grass class with probability = 0.9 and correct non-grass class with probability = 0.9. What is the probability that the true vegetation of a pixel classified as grass is grass? Repeat assuming that grasslands is 50% of the area? Which one is higher and why?

2.12   COMPUTER SESSION: INTRODUCTION TO Rcmdr, PROGRAMMING, AND MULTIPLE PLOTS

The objective of this session is to learn the Rcmdr package and to perform more complicated processing tasks in R, namely, how to program a loop to calculate an expression repetitively and how to plot multiple variables in a single plot.

2.12.1   R COMMANDER

As mentioned in the preface, R has an extensive collection of packages for more specialized functions; nearly 3600 at the time of this writing. An example of a useful package is the **R commander** or **Rcmdr** (Fox, 2005). This package provides a Graphical User Interface (GUI) that facilitates entering command, data input/output, and other tasks. It uses TclTk, which is also an open-source software. Once loaded, you can execute the functions in the GUI (Figure 2.11). First, let us discuss how to install a package from the CRAN website.

2.12.2   PACKAGE INSTALLATION AND LOADING

Packages are set up to use by the following two steps: (1) **Installation** from the Internet and (2) **Loading** it for use. First, to install from the Rgui, go to **Packages|Install package(s)**, select a mirror site depending on your location, and select Install package (Figure 2.12). For example, select the Rcmdr package (Figure 2.13). The download process starts and will give you messages about progress and success or not of the package installation. You do not need to repeat the installation as long as you do not re-install the R software or want to update the package.

Second, to load manually (on demand) from the Rgui, go to **Packages|Load package** and then select the package. For example, browse and select **Rcmdr** (Figure 2.14). Alternatively, you can run the function **library** from the R console. One way to load the R Commander automatically every

FIGURE 2.11   R Commander (Rcmdr) GUI.

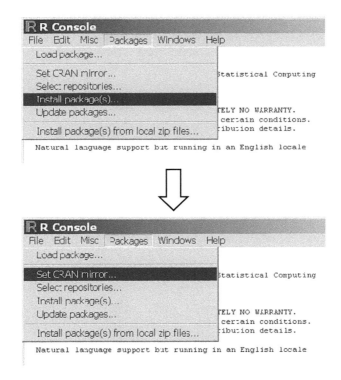

FIGURE 2.12   Selecting the mirror site and Install Package menu item.

time you start the Rgui is to insert **library(Rcmdr)** in a .**First** object which you save in the work-space, the commander will restart automatically next time you start R. It is also possible to load it automatically every time you start R by adding the following segment of code to the **Rprofile.site** file in R's **etc** directory:

```
local({
  old <- getOption("defaultPackages")
  options(defaultPackages = c(old, "Rcmdr"))
})
```

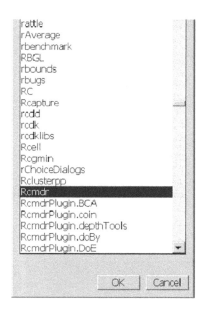

FIGURE 2.13   Selecting a package to install.

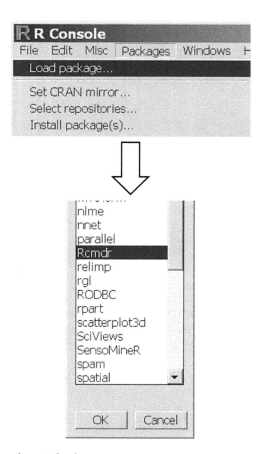

FIGURE 2.14   Selecting a package to load.

2.12.3 R GUI SDI Option: Best for R Commander

For Windows, the Rcmdr package works better with the **single-document** Rgui interface (**SDI**). Under the default multiple-document interface (MDI), Tk dialog boxes do not stay on top of the main R window. To enable the SDI, modify the R shortcut. As is often the case, it is better to make a copy of the shortcut, modify it, and keep the original as backup. To copy the shortcut, drag it to a different location or to the desktop. To modify the copy of the shortcut, Right-click and select **Properties**. Add—**sdi** (preceded by a space) to the **Target** field on the **Shortcut** tab of the dialog box. The field should read something like

```
"C:\Program Files\R\R-2.14.0\bin\i386\Rgui.exe" --sdi
```

The details vary according to what version you are using. If you wish, change the name of the icon on the **General** tab for you to remember and others to know. Click OK, and now you are ready to use the Shortcut.

2.12.4 How to Import a Text Data File Using Rcmdr

Once Rcmdr is loaded, you can execute the functions in the package (Figure 2.11). You can learn more using Help and then Introduction to the Rcmdr. In this section, we will study some basic operations such as reading data files, manipulate datasets and arrays, and make graphs using the R commander.

Take, for example, the dataset of 100 numbers that we used in Chapter 1. Make sure your working folder is **labs** so that the path to the file is relative to this folder. Recall that to do this you use **File|Change Dir** to the working folder **labs**. If you have personalized the R shortcut to start in folder **labs,** then the path to the file is relative to this folder, for example, in this case, **lab1\ test100.txt**.

Go to Rcmdr menu **Data|Import data|from text file** (Figure 2.15), enter a name in the dialog box, i.e., test100, unmark the names in file, select white space, and click ok. Then browse to lab1 and select **test100.txt**. Rcmdr creates object test100 as a **data.frame**. A nice feature of Rcmdr is that in the Rcmdr's script window you see the commands just applied. This helps you learn more about R language. Some of these commands are part of the Tcl and Tk languages.

You can use button **View data set** in the toolbar. This will open a new window with the data. Note that all numbers have been put in one column under the variable name V1 (Figure 2.16). To address a **variable** within a **data frame**, you use the dollar sign. For example, in this case, test100 is the data frame and V1 is the variable; therefore, we use test100$V1. Note that from the R console we can access objects created by the Rcmdr; for example,

```
> test100$V1
  [1] 48 38 44 41 56 45 39 43 38 57 42 31 40 56 42 56 42 46 35 40 30
49 36 28 55
 [26] 29 40 53 49 45 32 35 38 38 26 38 26 49 45 30 40 38 38 36 45 41
42 35 35 25
 [51] 44 39 42 23 44 42 52 55 46 44 36 26 42 31 44 49 32 39 42 41 45
50 39 55 48
 [76] 49 26 50 46 56 31 54 26 29 32 34 40 53 37 27 45 37 34 32 33 35
50 37 74 44
```

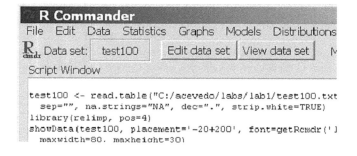

FIGURE 2.15   Importing data and dialog window to create dataset.

FIGURE 2.16   View dataset just created.

As we can see, this is simply an array. Of course, we can rename `test100$V1` for simplicity of reference, say

```
>x <-test100$V1
```

2.12.5 SIMPLE GRAPHS ON A TEXT WINDOW

The Rcmdr includes stem-and-leaf plots. Use **Graphs|Stem** and **Leaf display**. Then pick variable `V1`. For options, one could unmark trim outliers, show depths, and reverse negative values (Figure 2.17). On the Rcmdr output windows we obtain

```
> stem.leaf(test100$V1, style="bare", trim.outliers=FALSE,
depths=FALSE, reverse.negative.leaves=FALSE)
1 | 2: represents 12
 leaf unit: 1
        n: 100

   2 | 3
   2 | 5666667899
   3 | 001112222344
   3 | 5555566677788888889999
   4 | 000001111222222223444444
   4 | 5555556668899999
   5 | 0002334
   5 | 55566667
   6 |
   6 |
   7 | 4
```

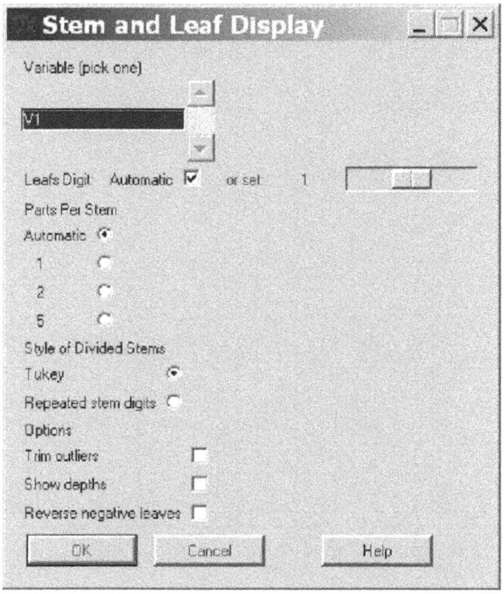

FIGURE 2.17   Using Rcmdr to plot stem and leaf.

2.12.6   SIMPLE GRAPHS ON A GRAPHICS WINDOW: HISTOGRAMS

Using the object created using Rcmdr `test100$V1`, which we renamed as x we can from the R console ask for a histogram

```
>hist(x)
```

On the window corresponding to graphics device, you will see a histogram like Figure 2.18. You could use a probability density scale (the bar height times the width would be a probability, i.e., from 0 to 1) on the vertical axis by using **hist** with an option

```
>hist(x, prob=T)
```

R handles many options in a similar manner: gives a logical variable the value **T** (for **T**rue); it is **F** (for **F**alse by default). Here, the default is **F** and corresponds to bar height equal to the count (Figure 2.19). The argument probability can be abbreviated here as prob.

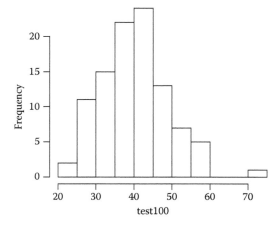

FIGURE 2.18   Histogram in frequency units.

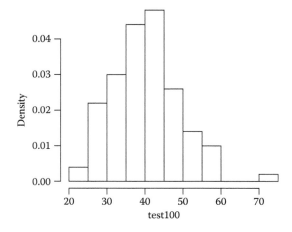

FIGURE 2.19   Histogram using probability units.

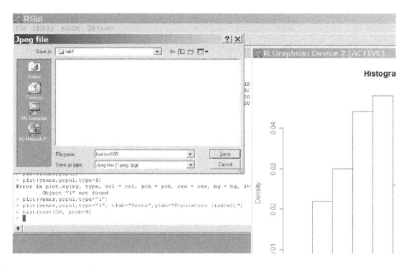

FIGURE 2.20   Saving a graph as a Jpeg file.

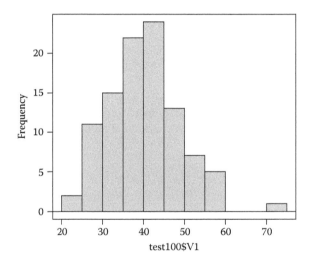

FIGURE 2.21   Histogram using Rcmdr.

Save the contents of this graphics window, say in Jpeg format. Save in **labs\lab1**. Use **File|Save as**, then select Jpeg and one of three quality values (Figure 2.20). This graphics file can be used in an application, to produce a paper or a report. In word and power point, use **Insert|Picture** and select filename. You could simply use Copy and Paste from the graphics window. Alternatively, you can do **File|Copy to clipboard** and then paste to the target file.

Using the Rcmdr, go to **Graphs|Histograms** | pick variable (in this case there is only one, V1), use **Auto** for number of bins, and select frequency. The output goes to the graphics window (Figure 2.21). Repeat with density as we did with the R console.

2.12.7   More than One Variable: Reading Files and Plot Variables

Let us use an example when we have more than one variable to read from file, typically organized in more than one column. Use downloaded file **lab2\dosonde.csv**. Verify contents using Vim. The few first lines look like this

```
Time,Temp,pH,Salinity,DO,Depth
143000,27.45,7.52,24.2,2.86,3.3
144500,27.58,7.54,23.9,3.03,3.3
150000,27.64,7.57,24.0,3.26,3.3
151500,27.68,7.58,24.0,3.41,3.3
153000,27.58,7.54,24.2,3.20,3.3
. ...
```

We see that is a CSV file of six columns corresponding to Time, Temp, pH, Salinity, DO, and Depth as declared in the header (first row). Vim tells us that we have 107 lines; therefore, discounting the header we have 106 lines of six data values each. We will see how to read this file by doing this in two different ways: using the R console and using the Rcmdr.

2.12.7.1   Using the R Console

We want to make an object named x from the data in the file named **datasonde.csv**. First, create an object using `read.table` function with argument `header` = `T` so that we use the header to give names to the columns, and the separator is the comma character.

```
> x <- read.table("lab2/datasonde.csv",header=T,sep=",")
```

Here, x is a data.frame configured from the data by rows. The components of x have names given in header. Now you can use `dim` command to check the dimensions of the matrix

```
> dim(x)
[1] 106 6
```

We can use the names of the components once we `attach` the data frame.

```
> attach(x)
```

Once we attach x we can use the names of components of x. Say, to plot variable DO as a function of Time (Figure 2.22)

```
>plot(Time, DO)
```

You can use a line graph by adding an argument `type` `="l"` to the plot function. Be careful that this is a letter "l" for line not number one "1" (Figure 2.23).

```
> plot(Time, DO, type="l")
```

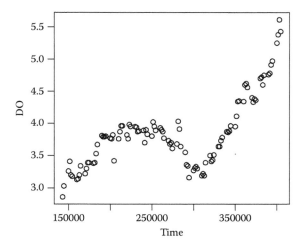

FIGURE 2.22 Plot as x–y using points.

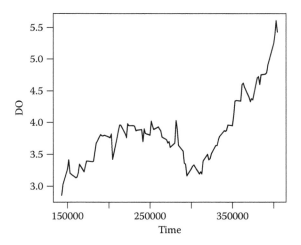

FIGURE 2.23 Plot as x–y using lines.

By default, the x- and y-axes are labeled with the variable name. These can be changed using `xlab` and `ylab` arguments. For example, we obtain Figure 2.24 applying

```
> plot(Time, DO, type="l",xlab="Time [hhmmss]", ylab="Dissolved
Oxygen [mg/l]")
```

Then you can save, for example, as Jpeg.

The limits of x- and y-axes can be changed using `xlim = c()` and `ylim = c()`, where `c()` denotes a one-dimensional array. For example, `xlim = c(150000,240000)` establishes the range from 150,000 to 240,000 and `ylim = c(0,5)` has two elements, a minimum of 0 and a maximum of 5.

```
> plot(Time, DO, type="l",xlab="Time [hhmmss]", ylab="Dissolved
Oxygen [mg/l]", xlim=c(150000,240000),ylim=c(0,5))
```

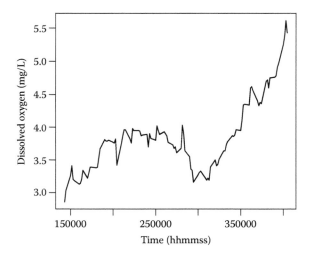

FIGURE 2.24   Plot with customized labels.

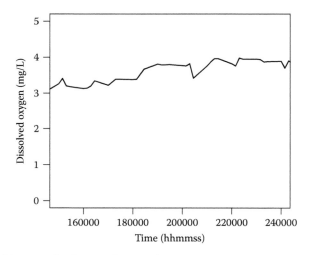

FIGURE 2.25   Plot with customized ranges for x- and y-axes.

See Figure 2.25. One more detail to mention now is that the axis can intersect at their minimum values by using argument xaxs = "i" and yaxs = "i"

```
plot(Time, DO, type="l",xlab="Time [hhmmss]", ylab="Dissolved Oxygen
[mg/l]", xlim=c(150000,240000),ylim=c(0,5), xaxs="i",yaxs="i")
```

as seen in Figure 2.26.

To visualize several variables, use function matplot and add a legend to identify the curves. The col argument is the color for the lines. In this case, col = 1 will be black color for all lines.

```
> matplot(Time,x[,-1], type="l", col=1, ylab="Water Quality")
```

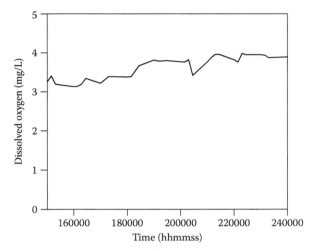

FIGURE 2.26 Plot with axis intersecting at the origin.

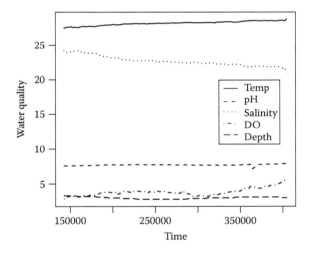

FIGURE 2.27 Plot with a family of lines.

We can place a legend using legend function.

```
> legend(350000,20,leg=names(x)[-1],lty=1:length(names(x)[-1]))
```

The first two arguments of legend are the x,y coordinates to place the legend, then leg argument is an array with the labels; here we have used the names of the dataset except the first, which is Time (that is why we write −1). Lastly, lty declares the number of line types which here we extract from the length of the names of the dataset except the first name (Figure 2.27).

2.12.7.2   Using the R Commander

Go to menu **Data|Import data|from text file..|** enter a name in the dialog box, i.e., x, keep the checkmark for the names in file, select commas for field separator, click ok, browse to the file, and open. The object x has been created. Use button **View data set** to inspect the data. As we have just generated this dataset x, it shows as the active Dataset in the upper left-hand-side corner of their

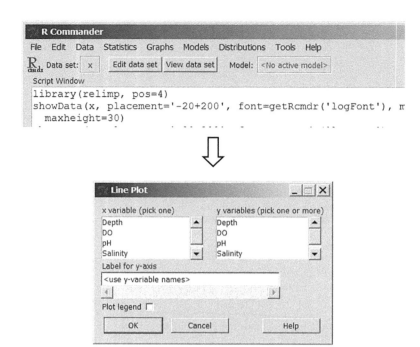

FIGURE 2.28   Line graph setup in Rcmdr.

Rcmdr GUI (Figure 2.28). To plot go to menu **Graphs|Line Graph** Select Time for x variable and
DO for y variable (Figure 2.28). This will produce the plot in an R graphics window (Figure 2.29).
Note that it uses both lines and markers.

We can do multiple plots using Rcmdr but the graphics are not as satisfactory as with the R con-
sole. Use **Graphs|Line graph** as before, select Time for x variable and the choose several for y vari-
able, type a name for Y axis, say "Water Quality", mark that you want a legend, and OK. Compare
the graphics to the plots we obtained using the console (Figure 2.27). We have more control to produce graphics using the R console.

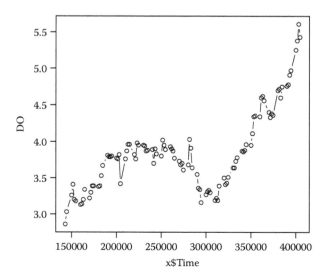

FIGURE 2.29   Line graph produced by Rcmdr.

2.12.8   PROGRAMMING LOOPS

We will use the simple example of evaluating a function of one variable for different values of a parameter. Say, evaluate the exponential function $x = \exp(rt)$ for several values of the coefficient r. First, declare the sequence of values of t

```
> t<-seq(0,10,0.1)
```

Then store values of r in an array

```
>r <- c(-0.1,0,0.1)
```

Now declare a two-dimensional array (matrix) to store the results, each column corresponds to the values of x for a given value of r

```
> x <- matrix(nrow=length(t), ncol=length(r))
```

Here, nrow and ncol denote the number of rows and columns, respectively.
   Now use function for to perform a loop

```
> for(i in 1:3) x[,i]<-exp(r[i]*t)
```

We can ask to see the results; the first few lines are

```
> x
          [,1] [,2]    [,3]
[1,] 1.0000000    1 1.000000
[2,] 0.9900498    1 1.010050
[3,] 0.9801987    1 1.020201
[4,] 0.9704455    1 1.030455
[5,] 0.9607894    1 1.040811
[6,] 0.9512294    1 1.051271
```

We can round to two decimal places

```
> round(x,2)
        [,1]  [,2] [,3]
 [1,] 1.00     1 1.00
 [2,] 0.99     1 1.01
 [3,] 0.98     1 1.02
 [4,] 0.97     1 1.03
 [5,] 0.96     1 1.04
 [6,] 0.95     1 1.05
 [7,] 0.94     1 1.06
 [8,] 0.93     1 1.07
 [9,] 0.92     1 1.08
[10,] 0.91     1 1.09
```

2.12.9 APPLICATION: BAYES' THEOREM

As an example of integrating several of the tools learned, consider Bayes' rule for two events A (contaminated water) and B (uncontaminated water), and a water quality test with C (test negative) and D (test positive).

$$P[A|C] = \frac{P[AC]}{P[C]} = \frac{P[C|A]P[A]}{P[C|A]P[A] + P[C|B]P[B]}$$

Suppose we know $P[A] = 0.2$ and want to explore how $P[A|C]$ varies as we change the probabilities of false negative $P[C|A]$ and false positive $P[D|B]$. The following script automates the calculation and produces graphical output (Figure 2.30).

```
# pA =contamination p[A]
# Fneg = false negative p[C|A]
# Fpos = false positive p[D|B]
# fix pA and explore changes of p[A|C]
# as we vary Fpos and Fneg

# fix pA
pA=0.2
# sequence of values
Fpos <- seq(0,1,0.05); Fneg <- seq(0,1,0.2)
# array to store results
Cont.neg <- matrix(nrow=length(Fpos),ncol=length(Fneg))
# Bayes theorem
for(i in 1:length(Fneg))
  Cont.neg[,i] <- Fneg[i]*pA/(Fneg[i]*pA + (1-Fpos)*(1-pA))
# plot
  matplot(Fpos,Cont.neg, type="l",lty=1:length(Fneg), col=1,
    xlab="False Positive Error", ylab="Prob(Contaminated | test
    negative)")
  legend(0,1, paste("Fneg=",as.character(Fneg)), lty=1:length(Fneg),
    col=1)
```

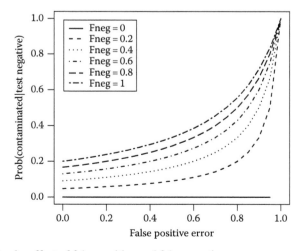

FIGURE 2.30 Bayes' rule: effect of false positive and false negative errors.

There are a couple of things to say about this brief script. First, same as we did in the previous example, we define the array `Cont.neg` to store results (before actually calculating them) using the `matrix` function. Second, inside the **for** loop we use `i` for variable `Cont.neg` (in the column position), and `Fneg`, but not for `Fpos`. Thus the calculation for each pass through the loop is done for the entire array `Fpos`.

2.12.10 APPLICATION: DECISION MAKING

Consider Equation 2.27 and let us build graphs of how C varies with the deltas and the probability p. In the following script, I have fixed ΔI and varied ΔM and p. This script produces the graph already shown in Figure 2.10.

```
# fix delta I
dI <- 2
# sequences for delta M and p
dM <- seq(0,10,2); nM <- length(dM)
p <- seq(0,1,0.01); np <- length(p)
# prepare a 2D array to store results
C <- matrix(nrow=np, ncol=nM)
# loop to calculate C for various dM
for(i in 1:nM) C[,i] <- dI-dM[i]*p
# plot the family of lines
matplot(p,C,type="l",lty=1:nM,col=1,ylim=c(-dI,dI))
# draw horizontal line at 0 to visualize crossover
abline(h=0)
# legend to identify the lines, use a keyword to position it
legend("bottomleft",leg=paste("dM=",dM),lty=1:nM,col=1)
```

There is not much more to say about this script beyond the explanatory remarks given for each line.

2.12.11 MORE ON GRAPHICS WINDOWS

We can open more graphics device for more plots. This can be done typing commands `win.graph()` or `windows`. Width and height are arguments to control the size of the graphics windows. For example, `windows(4,4)` opens up a new graphics window of width 4 and height 4 in. The current graph window can be closed with `dev.off()`; all graphics windows can be closed with `graphics.off()`.

You could also direct graphics to specific formats. For example, the water quality x-y plot can be directed to a PDF using

```
> pdf(file="lab2/datasonde.pdf")
> matplot(Time,x[,-1], type="l", col=1, ylab="Water Quality")
legend(350000,20,leg=names(x)[-1],lty=1:length(names(x)[-1]))
> dev.off()
```

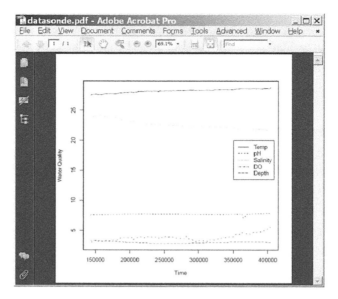

FIGURE 2.31   Graph as a pdf file.

When the PDF file is opened using Adobe Acrobat, we get Figure 2.31. Several graphs commands between the pdf() and the dev.off() commands are converted to multiple pages of the PDF file. For example,

```
> pdf(file="lab2/test.pdf")
plot(Time, DO, type="l",xlab="Time [hhmmss]", ylab="Dissolved Oxygen
[mg/l]",
    xlim=c(150000,240000),ylim=c(0,5), xaxs="i",yaxs="i")
matplot(Time,x[,-1], type="l", col=1, ylab="Water Quality")
legend(350000,20,leg=names(x)[-1],lty=1:length(names(x)[-1]))
dev.off()
```

produce Figure 2.32.

2.12.12   EDITING DATA IN OBJECTS

Example: Suppose we want to edit object x. Use

```
>edit(x)
```

to invoke the data editor or alternatively from the Rgui **Edit|data editor** and the **Edit Data set** in the Rcmdr.

2.12.13   CLEAN UP AND EXIT

As explained in Chapter 1, many times we generate objects that may not be needed after we use them. In this case, it is good practice to clean up after a session by removing objects with rm command. A convenient way of doing this is to get a list of objects with ls() and then see what we need to remove. The objects can also be listed using **Misc|List Objects** in the Rgui menu. Some of the

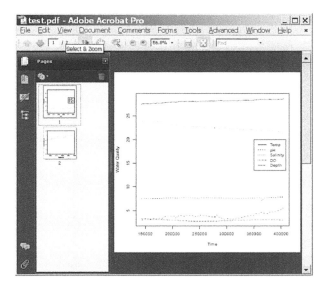

FIGURE 2.32   Multiple graphs in a pdf file with several pages.

clutter can be avoided by being careful about not generating objects unnecessarily. You can use q()
or **File|Exit** to finalize R and close the session. When you do this, you will be prompted for saving the
workspace. Reply yes. This way you will have the objects created available for the next time you use R.

2.12.14   ADDITIONAL GUIs TO USE R

There are two other GUIs available to simplify or extend R. A **web interface** available at http://
rss.acs.unt.edu/cgi-bin/R/Rprog. This interface can be used to enter commands over a web browser
(Herrington, 2002, 2003). A GUI called **SciViews** available at http://www.sciviews.org/SciViews-R/,
written as a GUI on top of the Rconsole.

2.12.15   MODIFYING THE R COMMANDER

One nice feature of the Rcmdr is that one can customize the menu and create a plug-in. For the
purpose of this book, I have extended the Rcmdr to perform more functions. This plug-in is
RcmdrPlugin.Seeg, which you can download, install, and load. You can find the latest updates of
this Plug-in at the URLs given in the preface. Further instructions are provided there.

2.12.16   OTHER PACKAGES TO BE USED IN THE BOOK

You can install packages as needed or alternatively install all packages required for this book so that
you do not have to repeat this process every time you start the R GUI. The packages are

```
rgl
Rcmdr
spatstat
sgeostat
maptools
tripack
spdep
vegan
MASS
```

You can modify the Rprofile.site so that packages will be loaded automatically every time you start the Rgui. Add the following segment of code to the **Rprofile.site** file in R's etc directory. Note that this replaces the previously used one for the Rcmdr.

```
# loading packages automatically
local({
old <- getOption("defaultPackages")
options(defaultPackages = c(old, "spdep",
"MASS", "spatstat", "sgeostat", "vegan","Rcmdr"))
})
```

This code loads all packages needed.

2.12.17   COMPUTER EXERCISES

Exercise 2.10
Plot a histogram in probability density scale for DO variable of the x object from **datasonde.csv**. Save the graph as a Jpeg file. Insert to an application.

Exercise 2.11
Read file **lab2/lake-lewisville.csv** to a data frame. Use both Rcmdr and Rconsole.

Exercise 2.12
Plot variables of data frame created in Exercise 2.11.

Exercise 2.13
Generate a linear function $y = ax + b$. Using $a = 0.1$, $b = 0.1$. Plot y for values of x in 0 to 1. Limit y-axis to go from 0 to the maximum of y.

Exercise 2.14
Generate a linear function $y = ax + b$ using $b = 0.1$ and two values of a, $a = 0.1$ and $a = -0.1$.
    Plot y for values of x in the interval [0,1]. Limit the y-axis to the interval [minimum of y, maximum of y]. Place a legend.

Exercise 2.15
This exercise refers to the Bayes' rule script in Section 2.12.9. Change probability of contamination $P[A]$ to 0.3. Plot the probability of contamination given that a test is negative $P[A|C]$ vs. false negative error with false positive error as a parameter. Hint: modify the script given in Section 2.12.9 for Bayes' rule to reverse the roles of `Fneg` and `Fpos`.

Exercise 2.16
On the decision-making script in Section 2.12.10. Change ΔI to 4 and plot again. Discuss the changes obtained for the values of p at which we would decide for alternative A_1.

SUPPLEMENTARY READING

For supplemental reading on the material covered in this chapter, you can see Davis, 2002, Chapter 2; Rogerson, 2001, Chapters 1 and 2; Carr, 1995, Chapter 4; Griffith and Amrhein, 1991, Chapter 6 and 7, pp. 147–196; Drake, 1967, Chapter 1, pp. 1–39. For reading on specific topics, see on events and probability: Davis, 2002, pp. 11–17; Rogerson, 2001, pp. 23–25; Carr, 1995, p. 67; on algebra of events: Drake, 1967, Chapter 1, on trees: Griffith and Amrhein, 1991, p. 151; on conditional probability and Bayes' rule: Davis, 2002, pp. 22–23, Carr, 1995, pp. 67–68.

3 Random Variables, Distributions, Moments, and Statistics

3.1 RANDOM VARIABLES

We can define a random variable (RV) once we have defined the sample space, based on the possible events and their probabilities. An RV is a rule, or a function, or a map associating a number to each event in the sample space (see Figure 3.1).

> Example: In the roll of six-sided die, the events are A_i = side facing up is i where $i = 1, 2, ...,$ 6 with $P[A_i] = 1/6$. We can make a "discrete" RV, denoted by X, associating each event with the value of the RV; for example, X taking values X = {1, 2, 3, 4, 5, 6} each with $P[x_i] = 1/6$.

We refer to the type of RV described in the example as **discrete** because its values are discrete, that is, a set of numbers, in this case integers 1, 2, ..., 6. The events can be defined from intervals contained in a range of **real** values from a to b. In this case, the values of RV X are continuous in this range. We call this type of RV **continuous**.

> Example: We measure concentration of a mineral (in ppm) at a given location and it can take values between 0 and 10,000 ppm. Continuous random variable X is concentration. An event could be defined as A = measured concentration is in the interval 10–15 ppm.

3.2 DISTRIBUTIONS

RV distributions are defined in the following manner. Here we assume that X is an RV.

3.2.1 PROBABILITY MASS AND DENSITY FUNCTIONS (pmf and pdf)

A *discrete* distribution or probability "mass" function (pmf) $p(X)$ is a set of probabilities, one for each value of X. More precisely, denoting x_i as the values of X

$$p(x_i) = P[X = x_i] \tag{3.1}$$

for all values x_i of X

$$0 \le p(x_i) \le 1 \quad \text{for all } i \tag{3.2}$$

$$\sum_i p(x_i) = 1 \tag{3.3}$$

59

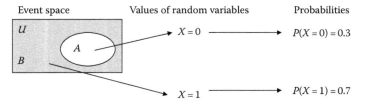

FIGURE 3.1 Constructing a random variable.

The last equation says that the total probability (sample space) must be equal to 1 when all probabilities are summed over all i.

> Example: Toss a coin. Assign 0 to T and 1 to H. $p(0) = 0.5$, $p(1) = 0.5$. This is an example of a uniform discrete RV: probabilities of each event are the same.

We can represent the probabilities as a graph as illustrated in Figure 3.2 for the example shown earlier. Vertical thick arrows or bars represent a spike or impulse with intensity given by the height of the spike and equals the probability of that particular value. Alternatively, it can be represented as a bar graph where the height of each bar represents the probability (see Figure 3.2).

> Example: Roll a six-side die. Then the pmf is $p(x_i) = 1/6$ where $x_i = \{1, 2, \ldots, 6\}$. This is also a uniform discrete RV (see Figure 3.3).

For more examples of illustrations of pmfs, see Davis, 2002, Chapter 2.

A **continuous** distribution or probability "density" function (pdf) $p(X)$ is defined based on intervals; the probability of the value being in an "infinitesimal" (this is a calculus concept, basically means "very, very small") interval of X between x and $x + dx$, this to say

$$p(x)dx = P[x < X \leq x + dx] \tag{3.4}$$

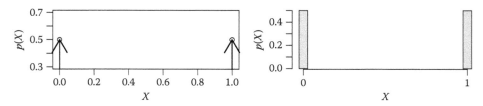

FIGURE 3.2 pmf of a discrete RV represented as a spike graph and as bar graph.

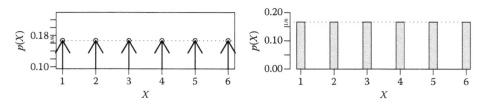

FIGURE 3.3 pmf of a discrete RV for a six-side die. Spike and bar graphs.

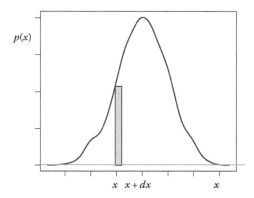

FIGURE 3.4 An illustration of a pdf for a continuous RV. Probability is area under the curve in between two values separated by a very small difference.

where $p(x)$ is always positive or zero, that is $p(x) \geq 0$. Probability is the area under the curve in between the two values x and $x + dx$ (see Figure 3.4).

The probability of a value being in an interval of X between a and b can be found using the integral (this is a calculus concept, the continuous analog of a sum)

$$P[a < X \leq b] = \int_a^b p(x)dx \qquad (3.5)$$

The area under the curve is the integral that can be approximated by the sum of small rectangles of width dx and the height given by the value of density function (Figure 3.5). At this point, a review of integration would be useful. For example, Pages 1–9 of Carr (1995), Chapter 1: "A review of calculus". The integral represents the area under the curve $p(x)$ in a given interval from a to b. The area is calculated as a summation of many small rectangles of width dx and height $p(x_i)$ when dx is infinitesimal.

When the interval is the whole range of values of X, then the value of the integral should be 1

$$\int_{-\infty}^{+\infty} p(x)dx = 1 \qquad (3.6)$$

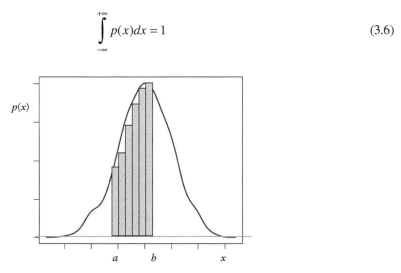

FIGURE 3.5 Probability of X having a value in between a and b is the area under the curve between these two values.

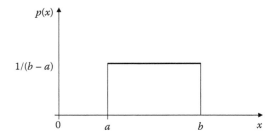

FIGURE 3.6   pdf of a uniform RV.

We have indicated the entire range of real values by selecting the limits from minus infinity (−∞) to plus infinity (+∞), or from a very large negative value to a very large positive value.

As an example, let us examine the **uniform** continuous distribution. It is a very important one. The density is the same over the range a, b and 0 outside this range

$$U_{a,b}(x) = \frac{1}{b-a} \quad \text{when } a \leq x \leq b$$

$$= 0 \qquad \text{otherwise} \tag{3.7}$$

as illustrated in Figure 3.6.

3.2.2   CUMULATIVE FUNCTIONS (cmf and cdf)

The "cumulative" density (cdf) and mass (cmf) functions at a given value are defined by "accumulating" all probabilities up to that value. Accumulation is simply a summation in the case of discrete RV or integration in the case of continuous RV.

$$F(x) = P[X \leq x] = \sum_{i=1}^{x} p(x_i) \tag{3.8}$$

$$F(x) = P[X \leq x] = \int_{-\infty}^{x} p(s)ds \tag{3.9}$$

Please note that the value at which we evaluate the cumulative is the upper limit of the accumulation (regardless of whether it is summation or integration). The variable s is a "dummy" variable to avoid confusion with x. The cdf $F(x)$ at a value x is the area under the density curve up to that value. The value of the cumulative for the largest value of X is the largest value of the cumulative and should be equal to 1.

Examples: Toss of a coin and roll of a die. The cmf is a stepwise function (see Figure 3.7). Uniform continuous $U_{a,b}(x)$ is a ramp with slope $1/(b-a)$ (see Figure 3.8).

3.2.3   HISTOGRAMS

As discussed earlier, a histogram is a graphical display of the distribution of a sample; it is a natural way of displaying samples of pmf and discrete approximation of continuous pdf. It graphs the frequency with which you obtain a value in a sample (if discrete) or values falling in intervals ("bins")

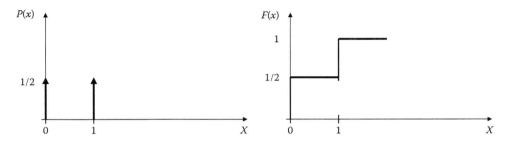

FIGURE 3.7 Integration of pmf to obtain cmf. At each value, we add the intensity of a spike to obtain a staircase for the cmf.

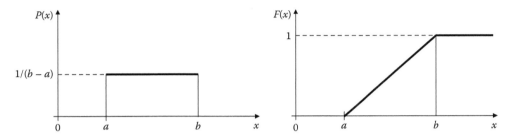

FIGURE 3.8 pdf and cdf of a uniform RV. Integration of a constant yields a linear increase (ramp function).

of the range of the variable (if continuous). Given a large enough sample, a histogram can help to characterize the pdf of a variable. A cumulative relative frequency can also be displayed and represents an approximation to the cmf or cdf.

3.3 MOMENTS

3.3.1 First Moment or Mean

The **first moment** of X is the **"expected value"** of X (Carr, 1995, p. 69) denoted by the operator $E[\ ]$ applied to X, this is $E[X]$. The first moment is the same as the **mean** of X. When X is discrete the **mean** is

$$\mu_X = E[X] = \sum_i x_i p(x_i) \tag{3.10}$$

where the sum is over all values of X. It is equivalent to the location of the center of mass of the pmf.

Examples: Toss a coin

$$E(X) = 0 \times 0.5 + 1 \times 0.5 = 0.5$$

this is the value of X in between the two spikes or bars in Figure 3.2.

Example: Roll of a die

$$E(X) = 1 \times 1/6 + 2 \times 1/6 + 3 \times 1/6 + 4 \times 1/6 + 5 \times 1/6 + 6 \times 1/6$$

$$E(X) = (1 + 2 + 3 + 4 + 5 + 6) \times 1/6 = 21/6 = 3.5$$

This is the value of X in between the two spikes or bars in Figure 3.3.

When X is continuous

$$\mu_X = E[X] = \int\limits_{-\infty}^{+\infty} xp(x)dx \tag{3.11}$$

Note that the integration is over all values of X.

Example: Uniform in [0, 1]. In this case $b = 1$, $a = 0$. We know that $p(x) = 1/(b - a) = 1$.

$$\mu_X = E[X] = \int\limits_0^1 x\,dx = \frac{1}{2}x^2\bigg|_0^1 = \frac{1}{2} \tag{3.12}$$

You need to recall integration from calculus.

The expected value or mean is a **population** concept. To calculate it we need its density or mass function. The mean is not the same as the **statistic** known as the **"sample mean"** (Davis, 2002, pp. 28–29; p. 65; Carr, 1995). As explained in Chapter 1, the **sample mean** is the arithmetic average of n data values x_i comprising a sample

$$\bar{X} = \frac{1}{n}\sum_{i=1}^{n} x_i \tag{3.13}$$

Note that the mean (first moment) is denoted with the Greek letter μ, whereas the sample mean or average is denoted with a bar on top of X, that is $\bar{X}$.

Example: Suppose we perform a coin toss 10 times and that we assign H = 1, T = 0 as in example one. Suppose we get six heads and four tails. Then the average is

$$\bar{X} = \frac{1}{10}\times 6 = 0.6$$

This sample mean is different from the population mean that we calculated to be $\mu_X = 0.5$.

See also example and table 2-1 in Davis, 2002, p. 35.

3.3.2 SECOND CENTRAL MOMENT OR VARIANCE

The second **central** (i.e., with respect to the mean) moment is the **variance** or the expected value of the square of the difference with respect to the mean (Carr, 1995, p. 70)

$$\sigma_X^2 = E[(X - \mu_X)^2] \tag{3.14}$$

If X is discrete,

$$\sigma_X^2 = E[(X - \mu_X)^2] = \sum_i (x_i - \mu_X)^2 p(x_i) \tag{3.15}$$

If X is continuous,

$$\sigma_X^2 = E[(X - \mu_X)^2] = \int_{-\infty}^{+\infty} (x - \mu_X)^2 p(x) dx \tag{3.16}$$

The expectation $E[.]$ is calculated by a sum when the RV is discrete and by an integral when the RV is continuous. The variance is a population concept. To calculate it we need its density or mass function. The **standard deviation** is the square root of the variance.

From the definition of variance in Equation 3.14, we can derive a more practical expression by substituting $\mu_X = E[X]$ and expanding the square of a sum to obtain

$$\sigma_X^2 = E[(X - E(X))^2] = E[X^2 - 2XE(X) + E(X)^2]$$

Then take the expected value of each term and use the fact that the expected value of a constant is the same constant

$$\sigma_X^2 = E[X^2] - 2E(X)E(X) + E(X)^2 = E[X^2] - 2E(X)^2 + E(X)^2$$

and finally

$$\sigma_X^2 = E[X^2] - E(X)^2 = E[X^2] - \mu_X^2 \tag{3.17}$$

Example: The variance of RV from toss of a coin $x_i = \{0,1\}$ and $p(x_i) = 0.5$ for all i. Using the definition in Equation 3.14

$$\sigma_X^2 = E[(X - \mu_X)^2] = (0 - 0.5)^2 \times 0.5 + (1 - 0.5)^2 \times 0.5 = 0.25 \times 0.5 + 0.25 \times 0.5 = 0.25$$

using the simplified expression in Equation 3.17

$$\sigma_X^2 = E[X^2] - E[X]^2 = \{(0)^2 \times 0.5 + (1)^2 \times 0.5\} - 0.5^2 = 0.5 - 0.25 = 0.25$$

The standard deviation is the square root

$$\sigma_X = \sqrt{\sigma_X^2} = \sqrt{0.25} = 0.5$$

The variance or second central moment is not the same as the **statistic** known as the **"sample variance"** (Davis, 2002, pp. 35–37; p. 66, Carr, 1995). As explained in Chapter 1, the sample variance is the variability measured relative to the arithmetic average of n data values x_i comprising a sample

$$\text{var}(X) = s_X^2 = \frac{1}{n} \sum_{i=1}^{n} (x_i - \bar{X})^2 \tag{3.18}$$

This is the average of the square of the deviations from the sample mean. Alternatively,

$$\text{var}(X) = s_X^2 = \frac{1}{n-1} \sum_{i=1}^{n} (x_i - \bar{X})^2 \tag{3.19}$$

where $n - 1$ is used to account for the fact that the sample mean was already estimated from the n values. We write s_X^2 to denote the sample variance to distinguish from the variance σ_X^2.

This equation can be converted in a more practical one by using Equation 3.13 and doing algebra to obtain

$$\text{var}(X) = s_X^2 = \frac{1}{n-1} \left[\sum_{i=1}^{n} x_i^2 - \frac{1}{n} \left(\sum_{i=1}^{n} x_i \right)^2 \right] \tag{3.20}$$

This is easier to calculate because we can sum the squares of x_i and subtract the square of the sum of the x_i. Recall that standard deviation is the square root of the variance. Again, you could have the standard deviation of a sample, which is different from the population concept.

Example: Suppose we get 6 heads in 10 tosses of a coin. The sum of squares is 6 and the sum of the x_i is also 6, then the sample variance is

$$s_X^2 = \frac{1}{9} \times \left[6 - \frac{1}{10} \times 6^2 \right] = \frac{1}{9} \times [6 - 3.6] = 0.26$$

The sample standard deviation is $s_X = \sqrt{.26} = 0.509$.

Note that the population variance is 0.25 and standard deviation 0.5 according to calculation in previous exercise. Therefore, the statistic sample variance has overestimated the variance.

For another example of calculation of sample variance, see Table 2-2 in Davis, 2002, p. 38.

3.3.3 POPULATION AND SAMPLE

The first and second central moments (i.e., mean and variance) are also referred to as **parameters** of the RV (Davis, 2002, p. 29). You have to be careful to avoid confusion with the term parameter as applied to an equation or model. If you employ the term **parameters** for the mean and variance, then you associate the term to a **population** and they are different from the **statistics**, which are associated with the **sample** (Davis, 2002, p. 29).

Another way of looking at this is to think of the pmf or pdf as theoretical models expressing the underlying probability structure of the RV. These functions allow calculation of the moments. However, the statistics are calculated from observed data and are used to estimate the moments.

3.3.4 Other Statistics and Ways of Characterizing a Sample

To characterize a sample we can use the sample mean and sample variance. Other statistics are the **median**, which is the middle data value, or the value that divides the area under the histogram in two equal parts. In terms of the population, this is the value midway in the pmf or pdf or the value x at which the cmf or cdf attain $F(x) = 0.5$.

There are other ways of dividing the area of the histogram into equal parts. For example, if we divide in four equal parts we obtain **quartiles** and in this case $F(x_1) = 0.25$, $F(x_2) = 0.50$, $F(x_3) = 0.75$ where x_1, x_2, x_3 are the quartiles (see Figure 3.9). Note that the inter-quartile interval $x_3 - x_2$, or difference between the third quartile and the first quartile has probability $= 0.75 - 0.25 = 0.5$.

If we use 100 equal parts, we obtain **percentiles** $x_1, x_2, \ldots, x_{99}$ where $F(x_1) = 0.01$, $F(x_2) = 0.02$, $\ldots$ and so on until $F(x_{99}) = 0.99$. The general term is **quantile**. See Davis, 2002, pp. 32–33.

The ith-order statistic of a sample is equal to the kth-smallest value and is denoted by $x^{(i)}$. The notation is obtained by using parenthesis around the subscript. Note that the minimum is the first-order statistic $x_{(1)}$ and the maximum is the nth-order statistic $x^{(n)}$ in a sample of size n. The quantiles are equal to the order statistics when we calculate n quantiles in a sample of size n. In addition, when $n = 2m + 1$, then n is odd and the median is the mth-order statistic $x^{(m+1)}$. However, when n is even, say $n = 2m$, then the median is the average of $x^{(m)}$ and $x^{(m+1)}$ but is not itself an order statistic.

Another important way to characterize the sample is the **mode**, which is the most common value or the maximum of the histogram or the frequency distribution (Davis, 2002, p. 34; Carr, 1995, pp. 65–66).

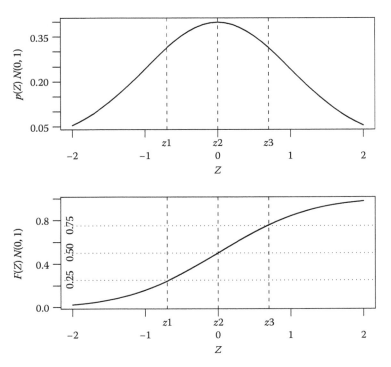

FIGURE 3.9 Quartiles of a normal distribution.

Variability can be expressed as a **coefficient of variation**, which is defined as the ratio of the sample standard deviation to the sample mean (Davis, 2002, p. 39).

$$Cv = \frac{s_X}{\overline{X}} \tag{3.21}$$

3.4   SOME IMPORTANT RV AND DISTRIBUTIONS

Uniform
Continuous pdf

$$U_{a,b}(x) = \frac{1}{b-a} \quad \text{when } a \leq x \leq b$$

$$= 0 \qquad \text{otherwise} \tag{3.22}$$

The mean is

$$\mu_X = \frac{b+a}{2} \tag{3.23}$$

and the variance is

$$\sigma_X^2 = \frac{(b-a)^2}{12} \tag{3.24}$$

Example: Uniform between $a = 0$, $b = 1$. The mean is 1/2 and variance 1/12.

Normal or Gaussian
Continuous, pdf of mean μ and variance σ^2

$$N_{\mu,\sigma}(x) = \frac{1}{\sqrt{2\pi}\sigma} \exp\left[-\frac{(x-\mu)^2}{2\sigma^2}\right] \quad -\infty < X < +\infty \tag{3.25}$$

This is a symmetrical pdf, that is, the mean is equal to the median. The area under the curve left of the mean is the same as area under the curve right of the mean. At one standard deviation from the mean on both sides $\mu \pm \sigma$ the area is 0.68, at two standard deviations $\mu \pm 2\sigma$ is 0.95, and at three standard deviations $\mu \pm 3\sigma$ is 0.99. We will explain these numbers in the next section.

Example: A normal variable with mean = 1 and variance = 0.25. What is the probability of obtaining a value in between 0.5 and 1.5? The standard deviation is 0.5. This interval is one standard deviation away from the mean on each side. Therefore, the probability is 0.68.

A normally distributed sample is symmetric. That is to say, the mean is equal to the median. Very commonly, the data are not symmetrical; that is, there is higher frequency left or right of the mean. Two important cases: positive (mean < median) or values biased toward the right and negative (median < mean) or biased toward the left.

Standard Normal

The standard normal RV, denoted by Z, has zero mean, that is to say $\mu_z = E(Z) = 0$, and standard deviation and variance equal to unity, that is to say $\sigma_Z = 1$. We write the pdf as $p(Z) = N_{0,1}(Z)$ and simplify the normal or Gaussian to be

$$N_{0,1}(z) = \frac{1}{\sqrt{2\pi}} \exp\left(\frac{-z^2}{2}\right) \tag{3.26}$$

A normal $N_{\mu,\sigma}(x)$ is standardized to $N_{0,1}(Z)$ by **subtracting the mean and dividing by the standard deviation**. That is to say,

$$z = \frac{x - \mu_X}{\sigma_X} \tag{3.27}$$

All values to the left of the mean are negative ($z < 0$) and all values to the right of the mean are positive ($z > 0$). Note that z is scaled in units of number of standard deviations.

Because the normal is symmetric, calculating the area under the standard pdf curve from $-\infty$ up to a value $-z_0$ (left of the mean) is the same as calculating the area under the curve from that value made positive $+z_0$ to $+\infty$ (right of the mean) (see Figure 3.10).

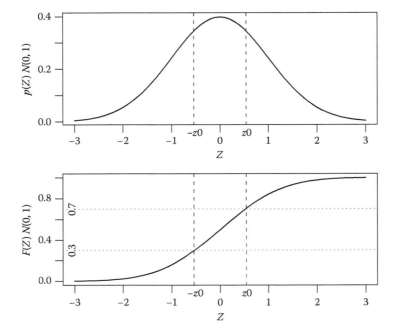

FIGURE 3.10 Symmetry of standard normal distribution around the mean.

The cdf is tabulated as the area under the curve up to z_0 (in units of σ) from zero (the mean). The area under the curve right of the point z_0 is the same as $1 - F(z_0)$, where $F(z)$ is the cdf.

Example: $F(0) = 0.5000$, $F(1) = 0.8413$, $F(2) = 0.9772$, $F(3) = 0.9987$ (see Figure 3.11). What is the probability that the variable is within the interval $\pm k\sigma$ around the mean? Where $k = 1$, 2, 3. When $k = 2$: what is the probability that the variable is within the interval $\pm 2\sigma$ around the mean? That is to say, what is

$$P(|z - \mu| \le k\sigma) \quad \text{for } k = 1, 2, 3?$$

For $k = 1$, to obtain the probability of the variable taking values right of the mean and up to $+\sigma$ is $F(1) - F(0) = 0.8413 - 0.5 = 0.3413$ (see Figure 3.11). Due to symmetry, this should be the same as left of the mean 0.3413, then

$$P(|z - \mu| \le \sigma) = 2[F(1) - F(0)] = 2 \times (0.8413 - 0.5000) = 0.682$$

For $k = 2$

$$P(|z - \mu| \le 2\sigma) = 2[F(2) - F(0)] = 2 \times (0.9772 - 0.500) = 0.954$$

and for $k = 3$,

$$P(|z - \mu| \le 3\sigma) = 2[F(3) - F(0)] = 2 \times (0.9987 - 0.500) = 0.997$$

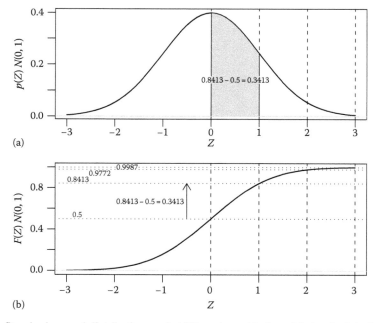

FIGURE 3.11 Standard normal distribution: probability values at 1, 2, and 3 standard deviation units (b). It also illustrates the equivalence between the area under the density curve between two points (a) and the difference in height under the cumulative curve (b).

These values are well known and worth remembering. For a normal RV, there is 68% probability for the variable to be in the interval $\pm\sigma$ around the mean, 95% probability of being in the interval $\pm2\sigma$ around the mean, and 99% of being in the interval $\pm3\sigma$ around the mean. (Davis, 2002, p. 37; Carr, 1995, pp. 70–71). The values at the 0.1, 0.2, 0.3, ..., 0.9 quantiles or 10, 20, 30, 40, ..., 90 percentiles are also notable, approximately −1.28, −0.84, −0.53, −0.25, 0.00, 0.25, 0.53, 0.84, 1.28.

Binomial

The binomial is a discrete RV. The possible values of r are 0, 1, 2, ..., n (values represent number of successes in n trials). The probability of r "successes" in n independent trials given that probability of a success is p. In the following $P[X = r] = p(r)$

$$p(r) = \binom{n}{r} p^r (1-p)^{n-r} \tag{3.28}$$

The mean is

$$\mu_X = np \tag{3.29}$$

and the variance is

$$\sigma_X^2 = np(1-p) \tag{3.30}$$

Example: Assume a binomial RV (Davis, 2002, pp. 13–16) with probability of success $p = 0.2$ and $n = 3$. Use a calculator and evaluate the binomial at $r = 0, 1, 2, 3$.

$$p(0) = \binom{3}{0} \times 0.2^0 \times 0.8^3 = 1 \times 1 \times 0.512 = 0.512$$

$$p(1) = \binom{3}{1} \times 0.2^1 \times 0.8^2 = 3 \times 0.2 \times 0.64 = 0.384$$

$$p(2) = \binom{3}{2} \times 0.2^2 \times 0.8^1 = 3 \times 0.04 \times 0.8 = 0.096$$

$$p(3) = \binom{3}{3} \times 0.2^3 \times 0.8^0 = 1 \times 0.008 \times 1 = 0.008$$

Note that these probabilities add up to 1.000. The mean is $3 \times 0.2 = 0.6$ and the variance is $0.6(1 - 0.2) = 0.6 \times 0.8 = 0.48$.

Poisson

This is a discrete RV to model the number of events that occur in a given time or space interval. A Poisson is a binomial with a very small probability p of success in one of many (n) small intervals. The pdf is

$$p(x) = \frac{a^x e^{-a}}{x!} \tag{3.31}$$

where $a = np$ is a positive rate of occurrence and $x = 0, 1, 2, ...$ The mean and the variance are equal to the rate a. We can deduct this from Equation 3.30 when p is very small.

Example: Number of earthquakes that occur in 1 year. Suppose the rate is $a = 1$. The mean is 1. We expect one earthquake per year. Calculate the probability of two earthquakes the same year

$$p(x = 2) = \frac{e^{-1}1^2}{2!} = \frac{0.37}{2} = 0.18$$

and of four earthquakes

$$p(x = 2) = \frac{e^{-1}1^4}{4!} = \frac{0.37}{4} = 0.09$$

3.5   APPLICATION EXAMPLES: SPECIES DIVERSITY

Species diversity is measured by species richness n and by some index of the relative abundances of the species. Richness is simply the number of coexisting species. In the case of three-species communities, richness can be 1, 2, or 3. For an index of relative abundances, we need the species composition, given by the relative abundances of the species. That is, the proportions p_i of each species with respect to the total.

Diversity of species composition is calculated by indices based on the functions of the species distribution p_i. Several indices are common, among them the Simpson diversity index and the Shannon diversity index. The latter is popular and used frequently and termed evenness. It is derived from the concept of information

$$E = -\sum_{i=1}^{n} p_i \ln(p_i) \tag{3.32}$$

Here, we have used natural logarithms, but actually the base of the logarithms determines the units. For example, if we select base 2, we obtain binary units. As an example, let us calculate evenness for three species, uniformly distributed. In this case $p_i = 1/3$. Equation 3.32 yields $E = -1.098$.

3.6   CENTRAL LIMIT THEOREM

This is an amazing result. Add many RV and get a normally distributed RV regardless of the original distributions! Take a set of samples from a population. Calculate the sample mean for each sample to obtain a set of sample means. The sample means are normally distributed. The mean of the sample means is the population mean. The variance of the sample mean is the population variance divided by the size (n) of the sample. The square root of the variance of the sample means is the **standard error** of the estimate of the mean.

Define a sample of size n as a set of n RVs ($X_1, X_2, ..., X_n$), where X_i is the RV for the ith performance of the experiment modeled by RV X. All X_i have the same mean $E[X]$. A compound pdf can be specified for the sample and therefore the expected value for the mean and variance of a statistic can be calculated. For example, the statistic known as the sample mean

$$\bar{X} = \frac{1}{n} \sum_{i=1}^{n} x_i \tag{3.33}$$

has an expected value or mean of

$$E(\bar{X}) = E\left(\frac{1}{n}\sum_{i=1}^{n} X_i\right) = \frac{1}{n}\sum_{i=1}^{n} E[X_i] = \frac{1}{n} \times n \times E(X) = E(X) = \mu_X \quad (3.34)$$

That is, the expected value of the sample mean $\bar{X}$ of X is equal to the mean μ_X of X. The variance of the sample mean of X is the variance of X divided by n.

$$\sigma_{\bar{X}}^2 = E[(\bar{X} - \mu_{\bar{X}})^2] = \frac{\sigma_X^2}{n} \quad (3.35)$$

The square root of this quantity is the standard error of the estimate of the mean, commonly referred to as the standard error.

$$\sigma_e = \sqrt{\sigma_{\bar{X}}^2} = \sqrt{\frac{\sigma_X^2}{n}} = \frac{\sigma_X}{\sqrt{n}} \quad (3.36)$$

The expected value of the sample variance is

$$E\left(s_{\bar{X}}^2\right) = \frac{n-1}{n}\sigma_X^2 \quad (3.37)$$

For a large value of n, the sample mean approaches the mean and the mean of the sample variance is approximately the same as the variance.

3.7 RANDOM NUMBER GENERATION

These are **pseudo-random** numbers or long sequence of integers generated by digital computer. The sequence is so long that "behaves" as it was random. It needs to be "seeded" (started) from different numbers. A sequence of values drawn from a distribution behaves as one **realization** of a **stochastic** process. Choosing numbers at random consists of generating a pseudo-random number sequence for the values of the RV. **Monte Carlo simulation** consists of drawing many random samples from a theoretical distribution and looking at the empirical distribution and its moments.

There are many methods to generate random numbers. Most programming languages provide a basic function to generate numbers that would seem to be drawn from an RV with uniform pdf between 0 and 1, that is $U(0,1)$. For example in R, this is done with **runif(1,0,1)**. Multiple values, say n, can be obtained by **runif(n,0,1)**.

We can transform the RV U with $U(0,1)$ to obtain other distributions, for example, an RV X with $U(a,b)$ can be generated from an RV U with $U(0,1)$ by first drawing random value u and then applying the expression

$$x = (b-a)u + a \quad (3.38)$$

When the cdf of the desired RV has an explicit formula, a useful method to draw numbers from this RV distribution is to transform via the inverse of the cdf. First, generate u from $U(0,1)$ then apply $x = F^{-1}(u)$. For example, for the exponential pdf,

$$
\begin{aligned}
u &= F(x) = 1 - \exp(-ax) \\
1 - u &= \exp(-ax) \\
\ln(1-u) &= -ax \\
x &= -\left(\frac{1}{a}\right)\ln(1-u) = F^{-1}(u)
\end{aligned}
\quad (3.39)
$$

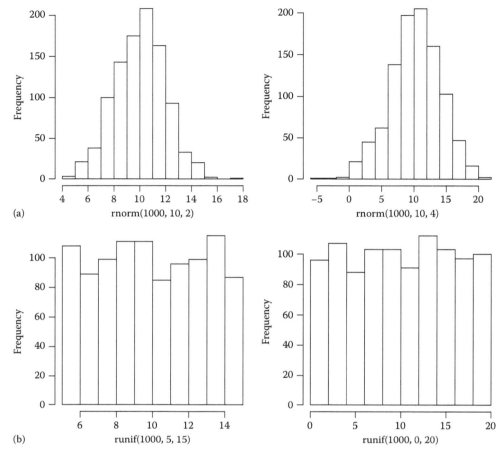

FIGURE 3.12 Normal and uniform distributions obtained by $n = 1,000$ pseudo-random number generation using R. (a) Normal with mean 10, standard deviation 2 (left) and 4 (right). (b) Uniform from 5 to 15 (left) and 0 to 20 (right).

For practical purposes, it is the same to use u instead of $1 - u$. So, first generate u and then apply the inverse $x = -\ln(u)/a$.

A value z from a standard normal Z can be generated from two u_1, u_2 numbers generated from $U(0,1)$ using

$$z = \cos(2\pi u_1)\sqrt{-2\ln(u_2)} \tag{3.40}$$

In R, a uniform RV can be in any range [a, b] directly by using **runif(1,a,b)**. In addition, there are a variety of functions for other distributions, such as the normal and the exponential. For example, the normal of mean u and standard deviation s can be generated with **rnorm(n,u,s)** where n is the number of draws. Figure 3.12 provides an example and we will explain how to use these functions in the computer session part of this chapter.

3.8   EXERCISES

Exercise 3.1
Define an RV based on outcomes of classification of a pixel of a remote sensing image as grassland (prob = 0.2), forest (prob = 0.4), or residential (prob = ?). Is this RV discrete or continuous? Plot the distributions (density or mass) and cumulative. Calculate the mean and variance.

Exercise 3.2

Define an RV from the outcome of soil moisture measurements in the range of 20%–40% in volume. Give an example of an event. Is this RV discrete or continuous? Assuming that it can take values in [20,40] uniformly, plot appropriate distributions (density or mass) and cumulative. Calculate the mean and variance.

Exercise 3.3

Consider an RV uniformly distributed between $a = 5$ and $b = 10$. Calculate the mean and variance using Equations 3.23 and 3.24. Plot the pdf and cdf.

Exercise 3.4

Plot the pmf and cmf of a binomial with $n = 3$ trials for three values of p, 0.2, 0.5, 0.8. Discuss the differences.

Exercise 3.5

Calculate the sample mean, variance, and standard deviation of hypothetical data drawn from the RV of Exercise 3.1. The data obtained were 300 grass pixels, 500 forest, and 200 residential out of 1000 pixels.

Exercise 3.6

At a site, monthly air temperature is normally distributed. It averages to 20°C with standard deviation 4°C. What is the probability that a value of air temperature in a given month exceeds 24°C? What is the probability if it is below 16°C or above 24°C?

Exercise 3.7

Assume 60% of a landfill is contaminated. Suppose that we randomly take three soil samples to test for contamination. We define event C = soil sample contaminated. We define X to be an RV where x = number of contaminated soil samples. Determine all possible values of X. What distribution do we get for X? Calculate the values of pmf and cmf for all values of X. Graph the pmf and cmf. Calculate the mean and the variance.

3.9 COMPUTER SESSION: PROBABILITY AND DESCRIPTIVE STATISTICS

3.9.1 DESCRIPTIVE STATISTICS: CATEGORICAL DATA, TABLE, AND PIE CHART

In the previous computer session (Chapter 2), we started some basic descriptive statistics by using stem-and-leaf plot and histogram. Now we look at another simple tool: how to table data and make a pie chart.

Suppose you want to start investigating the distribution of availability of cars in a neighborhood to plan transportation. For 100 households, you found the number of cars per household. Data in file **lab3/num-cars.csv** are hypothetical and for illustration only. For a real example, see Sullivan (2004) and United States Bureau of the Census. We will compute descriptive statistics and perform exploratory analysis.

As always, start by examining the data file. We note that there is no header and that data are given in 100 rows of 1 columns. Using the R console we can just write

```
> n.cars <- scan("lab3/num-cars.csv")
Read 100 items
> n.cars
  [1] 0 0 2 3 1 3 4 2 3 0 2 4 2 3 1 1 4 etc
>
```

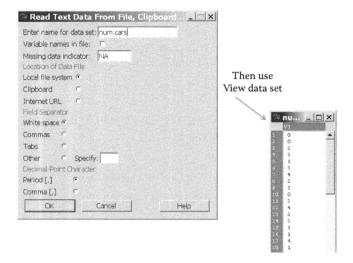

FIGURE 3.13 Using Rcmdr to read file and then to view the dataset created.

using the Rcmdr we can go to **Data|Import Data|from text file, ...** name the dataset num.cars (Figure 3.13). Browse to **lab3** and **num-cars.txt** file and select it. Note that num.cars is a data frame and thus its variable V1 is an array

```
> num.cars$V1
  [1] 0 0 2 3 1 3 4 2 3 0 2 4 2 3 1 1 4 1 3 3 1 1 2 3 1 1 2 3 2 3 0 2
  2 3 2 2 1
[38] 4 1 0 3 3 2 5 2 2 3 2 3 0 3 4 3 4 1 0 1 3 2 1 2 4 2 3 2 3 3 4 3
  3 4 1 1 1
[75] 3 2 3 2 0 1 2 3 2 2 3 4 1 2 1 0 1 2 2 2 1 2 3 1 2 4
>
```

This is equal to n.cars as an array read from the Rconsole. The variable is categorical. The simplest way to summarize is to give a table of number of observations of each class. Using the R Console

```
> table(n.cars)
n.cars
 0  1  2  3  4  5
 9 22 30 27 11  1
>
```

Note that the most common value is 2 with 30 observations. This is the **mode**, which can be calculated by

```
> max(table(n.cars))
[1] 30
```

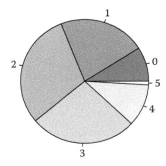

FIGURE 3.14 Pie chart.

Now we can do a pie chart (Figure 3.14) using a gray scale of colors

```
> pie(table(n.cars),col=gray(seq(0.4,1,length=6)))
```

and can also do a bar graph (Figure 3.15)

```
> barplot(table(n.cars))
```

Sometimes we may want to have two graphs on the same page. For example, we can use the graphics function par(mfrow = c(1,2)) to establish a two-panel graph of one row and two columns (Figure 3.16).

```
> par(mfrow=c(1,2))
> pie(table(n.cars))
> barplot(table(n.cars))
```

Also, we could use the graphics function par (mfrow = c(2,1)) to establish a two-panel graph but two rows and one column.

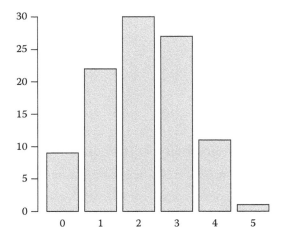

FIGURE 3.15 Barplot.

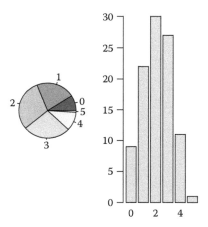

FIGURE 3.16 Pie chart and barplot on the same page. Two panels on the same page arranged as one row and two columns.

3.9.2 USING A PREVIOUSLY GENERATED OBJECT OR A DATASET

Using the Rcmdr select the dataset `test100` from last session. You do this by using **Data|Active dataset|Select active dataset** and then browse to select the object as shown in Figure 3.17. Then you should see `test100` as active dataset and can use **View dataset** button to view. Alternatively, using the R Console you can just type

```
> test100
    V1
1   48
2   38
3   44
4   41
5   56
etc
```

to assure yourself the dataset is there. If not you may have not restored your workspace. Try Load workspace or remake the object. Then we can rename for simplicity

```
> x <- test100$V1
> x
[1] 48 38 44 41 56 45 39 43 38 57 42 31 40 56 42 56 42 46 35 40 30
   49 36 28 55
etc
```

3.9.3 SUMMARY OF DESCRIPTIVE STATISTICS AND HISTOGRAM

Using the Rconsole, another option of the `hist` function is to explicitly declare or define the number of intervals, for example,

```
>hist(x, nclass=10, prob=T, main="")
```

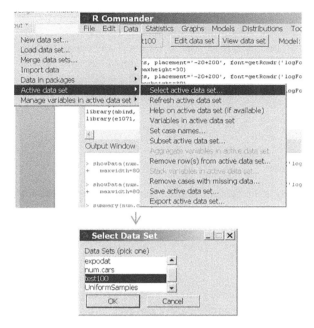

FIGURE 3.17 Selecting a previously generated dataset.

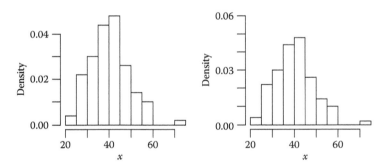

FIGURE 3.18 Histogram in probability units using $n = 10$ intervals and upper limit in y-axis.

to obtain Figure 3.18a. Here, we have also used main = "" to print blank as the main title.

You will often have to make a decision on the best number of intervals to use according to the trade-off between smoothness and detail. Also, you can establish the maximum value to scale the y-axis with ylim = c(y1,y2) in the following manner:

```
>hist(x, nclass=10, prob=T, ylim=c(0,0.06), main="")
```

The maximum value on the y-axis will be set to 0.06, the minimum at 0.00 (Figure 3.18b). The subsequent command to plot will use this maximum value and therefore you may have to reset it.

To get a summary of the descriptive stats of the variable, use the summary function

```
>summary(x)
   Min. 1st Qu.  Median    Mean 3rd Qu.    Max.
  23.00   35.00   40.50   40.86   46.00   74.00
>
```

that yields the minimum and maximum, mean and median, first quartile and third quartile. In this case, the median is close to the mean indicating symmetry.

Recall that we can also get the sample mean, variance, and standard deviation using `mean var` and `sd` functions. We can concatenate commands in a single line by using the semicolon ";" character

```
> mean(x); round(var(x),2); round(sd(x),2)
[1]  40.86
[1]  81.62
[1]  9.03
```

The coefficient of variation is the ratio of sample standard deviation to sample mean

```
> sd(x)/mean(x)
[1]  0.2210
```

Let us look at how to perform the above functions using the Rcmdr. Assume `test100` is the active dataset. Then go to **Statistics|Summaries|Active dataset**.

In the output window of the Rcmdr, you will see

```
summary(test100)
       V1
 Min.    :23.00
 1st Qu.:35.00
 Median :40.50
 Mean    :40.86
 3rd Qu.:46.00
 Max.    :74.00
```

Which are the same values we obtained before using the R console. We can do **Statistics|Summaries|Numerical summaries ....** and mark coefficient of variation, skewness, and kurtosis as shown in Figure 3.19.

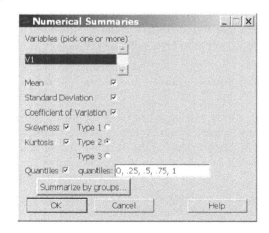

FIGURE 3.19   Obtaining numerical summaries using Rcmdr.

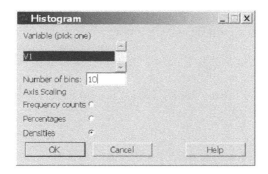

FIGURE 3.20 Selecting number of bins and density using Rcmdr.

```
> numSummary(test100[,"V1"], statistics=c("mean", "sd", "quantiles",
"cv",
+    "skewness", "kurtosis"), quantiles=c(0,.25,.5,.75,1), type="2")
   mean       sd        cv   skewness kurtosis 0%  25%  50%  75% 100%   n
  40.86 9.034189 0.221101 0.4426246 0.702436 23   35 40.5   46   74 100
```

To plot a histogram go to **Graphs|Histogram...** and then customize the histogram to 10 bins and using densities as shown in Figure 3.20.

3.9.4 DENSITY APPROXIMATION

To convert the histogram to an estimate of the density, we can use commands `density` and `plot` from the R Console

```
> plot(density(x,bw='SJ'))
```

will generate Figure 3.21a. This estimate of the density was obtained by using option `bw`. This option `bw` can be given numeric values directly or by a character string denoting a method to compute it. An alternative method is derived from S. Select a value of argument `width` equal to twice

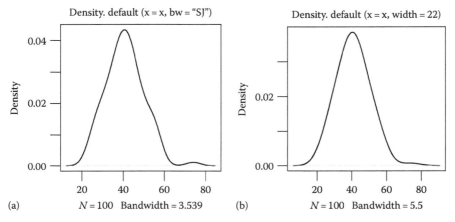

FIGURE 3.21 Density of `test100$V1` smoothed using (a) `bw = "SJ"` and (b) `width = iqd`.

the inter-quartile distance (width = 2*iqd). To get the iqd, get a summary of the descriptive statistics of the variable; you do this by using summary function

```
> summary(x)
  Min. 1st Qu. Median  Mean 3rd Qu.  Max.
    23      35   40.5 40.86      46    74
```

and subtracting first quartile from the third quartile, iqd = 3rd Qu − 1st Qu = 46 − 35 = 11.

This iqd value in the last command can be implemented directly as we will see later. Therefore, using width = 2 × 11 = 22 should smooth the density plot as shown in Figure 3.21b by means of argument width

```
>plot(density(x, width=22))
```

3.9.5 Theoretical Distribution: Example Binomial Distribution

Assume a binomial RV with probability of success p and n trials

$$p(r) = \binom{n}{r} p^r (1-p)^{n-r} \tag{3.41}$$

The possible values of r are 0, 1, 2, ..., n (values represent number of successes in n trials).

For example, let us use $n = 3$, $p = 0.2$ plot a "probability mass function" (or a discrete distribution or a histogram) and cumulative probability distribution for this RV. Create a vector object containing the values (quantiles or intervals) 0, 1, 2, 3

```
>xbinom <- c(0:3)
```

In the following, functions dname, pname, qname, rname apply to RV name, for example, name could be binom for binomial, norm for normal, etc.

Create a vector object containing the probabilities for the pmf by applying density function dbinom(x, size, prob), where x are quantiles, size is number of trials, and prob is probability of success p

```
>pmfbinom <- dbinom(xbinom, 3, 0.2)
```

Then the entries of the pmf are

```
> pmfbinom
[1] 0.512 0.384 0.096 0.008
```

Note that the expected value (mean) would be the sum of entries of `pmfbinom*xbinom = 0.6`

```
> pmfbinom*xbinom
[1] 0.000 0.384 0.192 0.024
> sum(pmfbinom*xbinom)
[1] 0.6
```

Note that indeed $0.384 + 0.192 + 0.024 = 0.6$.

Use `barplot` function, with categories `xbinom` having heights `pmfbinom`,

```
> barplot(pmfbinom, names.arg=xbinom, axis.lty=1,xlab="X",
  ylab="P(X)")
```

to get the pmf plot (Figure 3.22a).

Create a vector with the cumulative probabilities for the cmf by applying function `pbinom(x, n, prob)`

```
>cpmfbinom <- pbinom(xbinom, 3, 0.2)
[1] 0.512 0.896 0.992 1.000
```

Now simply plot the cmf

```
> barplot(cpmfbinom, names.arg=xbinom, axis.lty=1,xlab="X",
  ylab="F(X)")
```

to obtain (Figure 3.22b).

The Rcmdr has a menu item for **Distributions**, which includes both continuous and **discrete** variables. For continuous, it includes the normal, the t, the chi-square, the F, and several others (see Figure 3.23). For discrete it has the binomial, Poisson, and others. In this chapter, we will concentrate on the uniform, normal, and binomial. We will discuss t, chi-square, F, and Poisson in later chapters.

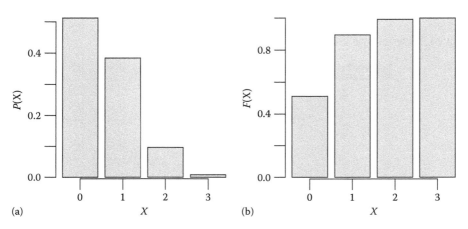

FIGURE 3.22 (a) pmf and (b) cmf of a binomial RV.

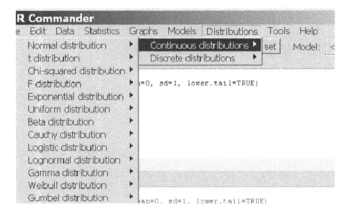

FIGURE 3.23   Rcmdr distribution menu items for continuous distributions.

For example, go to **Continuous distributions|Normal distribution** and use the quantiles option (Figure 3.24) to obtain quantiles given probabilities. We can determine the quartiles or quantiles for probabilities 0.25, 0.50, 0.75 (see Figure 3.25). We get the result in the output window

```
> qnorm(c(.25,.50,.75), mean=0, sd=1, lower.tail=TRUE)
[1] -0.6744898   0.0000000   0.6744898
```

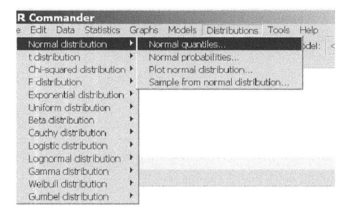

FIGURE 3.24   Selecting quantiles option for normal distribution.

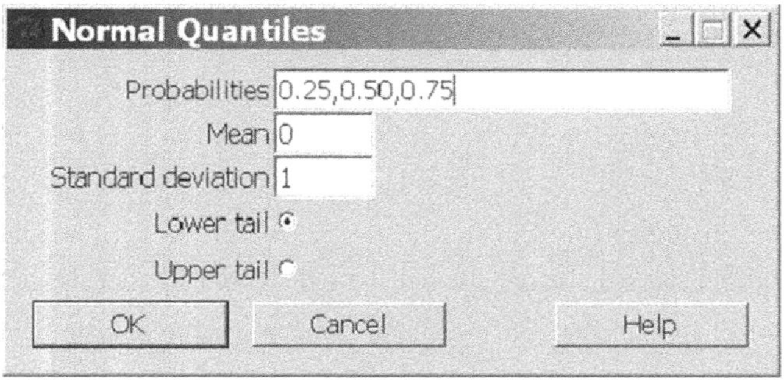

FIGURE 3.25   Calculating quartiles for the normal distribution.

Conversely, we can obtain probabilities from quantiles using the probabilities option. In addition, using the plot option (Figure 3.26) we get a graph of the distribution (Figure 3.27).

Using the **Discrete distributions|Binomial** menu item, we can calculate probabilities for $n = 3$ trials and $p = 0.2$ as probability of success (Figure 3.28). The response in the output window is

```
> .Table
      Pr
0 0.512
1 0.384
2 0.096
3 0.008
```

which are the same values we obtained using the Rconsole.

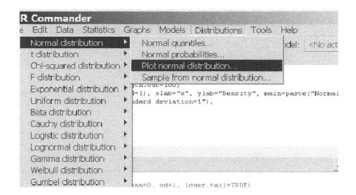

FIGURE 3.26 Option to plot.

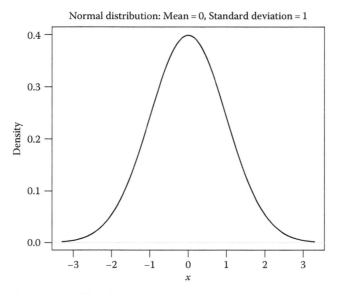

FIGURE 3.27 Plot of normal from Rcmdr.

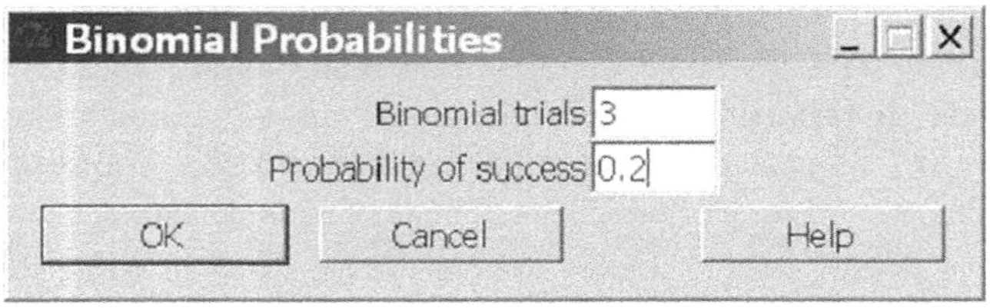

FIGURE 3.28 Calculating a binomial.

3.9.6 APPLICATION EXAMPLE: SPECIES DIVERSITY

Let us compare evenness for a binomial distribution of size 10 and $p = 0.5$ and one of lower probability $p = 0.2$ and higher probability $p = 0.8$.

```
> n=10
> p <- dbinom(x=seq(0,n),size=n, prob=0.2)
> print(E <- - sum(p*log(p)))
[1] 1.621929
> p <- dbinom(x=seq(0,n),size=n, prob=0.5)
> print(E <- - sum(p*log(p)))
[1] 1.875954
> p <- dbinom(x=seq(0,n),size=n, prob=0.8)
> print(E <- - sum(p*log(p)))
[1] 1.621929
>
```

We can see how the binomial distribution has higher Shannon evenness at $p = 0.5$ than at both lower and higher probabilities. In addition, those probabilities have the same evenness because for $p = 0.8$, the complement $1 - 0.8$ is the same as $p = 0.2$. Thus, a binomial has maximum evenness at $p = 0.5$. You can verify this for several combinations of p.

Let us compare evenness for a binomial distribution of size $n = 10$ and $p = 0.5$ and one of the same size but uniform.

```
> n = 10
> p <- dbinom(x=seq(0,n),size=n, prob=0.5)
> print(E <- - sum(p*log(p)))
[1] 1.875954
> p <- rep(1/(n+1),(n+1))
> print(E <- - sum(p*log(p)))
[1] 2.397895
>
```

We can see how the uniform distribution has more Shannon evenness than a binomial with maximum evenness ($p = 0.5$). In fact, the uniform would always yield maximum evenness for the same richness (n).

3.9.7 RANDOM NUMBER GENERATION

Random numbers from a variety of distributions are generated with functions **r**_name_, where _name_ applies to an RV name. For example, name could be binom for binomial, to obtain function rbinom. It could be norm for normal to obtain rnorm, also unif for uniform to get runif. All use argument n for how many values to generate. Other arguments vary according to the function as follows:

- Uniform, runif(n, min = 0, max = 1), min and max are lower and upper bounds of interval, with default 0, 1.
- Normal, rnorm(n, mean = 0, sd = 1), mean and sd are mean and standard deviation, with default 0, 1.
- Binomial, rbinom(n, size, prob), size and prob are number of trials and p respectively, with no default values.
- Exponential, rexp(n, rate = 1), rate parameter with default 1.

For example, runif(100) will be 100 numbers uniformly distributed between 0 and 1 (by default). That is, a sample of size 100.

```
> x <- runif(100)
> hist(x,prob=T,xlab="x, n=100, U(0,1)",main="")
```

We are following it with command hist to obtain the histogram of this sample x (Figure 3.29a). Increasing the sample size to 1000 makes the sample approach a uniform (Figure 3.29b)

```
> x <- runif(1000,0,1)
> hist(x,prob=T,xlab="x, n=1000, U(0,1)", main="")
```

A summary of these data is obtained with the following command

```
> summary(x)
    Min. 1st Qu.  Median    Mean 3rd Qu.    Max.
 0.00104 0.21840 0.44700 0.47230 0.71930 0.99950
>
```

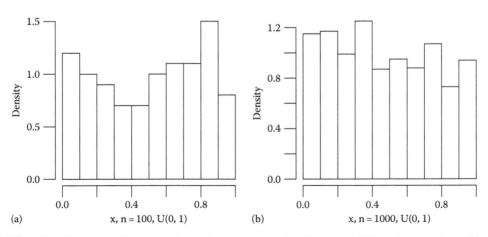

FIGURE 3.29 Histogram of a sample drawn from a uniform distribution. (a) 100 random numbers. (b) 1000 random numbers.

This summary indicates that the mean is close to the median. The min and max are close to 0 and 1 and the first and third quartiles are close to 0.25 and 0.75.

Let us try a normal with mean = 0 and sd = 1, plot a histogram and a density estimate

```
> xnorm <- rnorm(1000,0,1)
> hist(xnorm,prob=T,xlab="x, n=1000, N(0,1)", main="",cex.lab=0.7)
> plot(density(xnorm), cex.main=0.7,cex.lab=0.7)
```

as shown in Figure 3.30.

The Rcmdr includes random number generation under the Distribution Menu. For each distribution, there is an item to draw a **sample** from that distribution. For example, for normal, it is sample from normal distribution. For example, for the normal $N(0, 1)$, see Figure 3.31a to generate one sample of 100 numbers and the result of **view dataset** (Figure 3.31b).

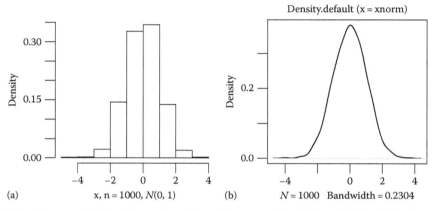

FIGURE 3.30   Histogram (a) of random numbers drawn from normal distribution $N(0, 1)$ and density estimate (b).

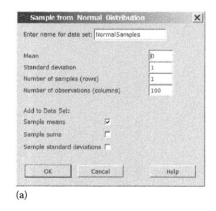

FIGURE 3.31   Random number generation using Rcmdr. (a) Dialog window (b) Data.

3.9.8 COMPARING SAMPLE AND THEORETICAL DISTRIBUTIONS: EXAMPLE BINOMIAL

Generate a random sample from a binomial with $p = 0.2$. Verify that sample mean is approximately equal to expected value. Run hist on the object you generated to demonstrate that the histogram indeed looks like a binomial pmf.

Generate random sample of size 100 and store it in a vector object

```
>sambinom100 <- rbinom(100,3,0.2)
```

The arithmetic average (sample mean) of these 100 numbers is about 0.6 but not exactly. For example,

```
> mean(sambinom100)
[1] 0.57
```

Of course, it depends on the sample. Now plot histogram of the random sample

```
>hist(sambinom100, probability=T, breaks=-1:3,main="")
```

this is the histogram for sample size of 100 which looks binomial but does not match the theoretical that well because the sample size (100) is not too large (Figure 3.32a). Now increase the sample size. Try

```
> sambinom1000 <- rbinom(1000,3,0.2)
> mean(sambinom1000)
[1] 0.576
```

we see that the mean approaches 0.6; we can plot the histogram again

```
> hist(sambinom1000, probability=T, breaks=-1:3,main="")
```

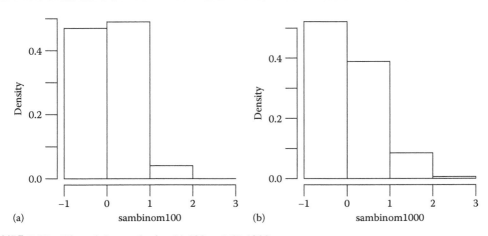

FIGURE 3.32 Binomial: sample size (a) 100 and (b) 1000.

for 1000 sample size (Figure 3.32b). Increasing sample size leads to a histogram that approximates the binomial pmf. This example should have made evident the difference between **histogram** (sample) and theoretical **pmf** (population).

3.9.9 Programming Application: Central Limit Theorem

In this section, we will work with the Rconsole. We will write a short script to calculate **sample means** for samples of various sizes. We will work with four values of sample size n: 1, 10, 100, and 1000. The sample size is the number of variables summed to obtain the mean. First, we prepare the graphics device to have four panels in order to plot a histogram in each and define the number of samples (m) and sample size array. Then, we create an array x to store the sample means (or means of the sum of the variables), and a matrix z to store the results. Then, we write a loop to vary the **sample size** n and an inner loop to vary the **number of samples**. We use m for the number of samples in the inner loop.

```
par(mfrow=c(2,2))
# num of samples and various sample sizes
m=100;n=c(1,10,100,1000)
# matrix to store result
x <- array(); z <- matrix(nrow=length(n),ncol=4)
# loop for various sample size
for(i in 1:length(n)){
# sample means
for(j in 1:m) x[j] <- mean(runif(n[i]))
# histograms
hist(x, prob=T,main=paste("n=",n[i]),cex.main=0.8)
# store results
z[i,1] <- n[i]; z[i,2:4] <- c(mean(x),var(x),sd(x))
}
```

This inner loop repeats 100 times the calculation of the mean of a sample of size n of an RV, with RV uniform between 0 and 1 (default). The `hist` function is applied for each value of n once the inner loop is completed (Figure 3.33). Then, for each value of n, we store its value together with the sample mean, variance, and standard deviation of the sample means.

The standard deviation of the sample means is the **standard error**. Recall that the variance of the sample means is σ^2/n, where n is sample size. In addition, the standard error is the square root of the variance of the sample means.

Let us look at the results stored in z

```
> z
      [,1]      [,2]         [,3]          [,4]
[1,]     1 0.4913332 8.455316e-02 0.290780253
[2,]    10 0.4951969 7.953786e-03 0.089184001
[3,]   100 0.5054084 7.397314e-04 0.027198004
[4,]  1000 0.4994328 8.198561e-05 0.009054591
>
```

Recall that for $U(0, 1)$ the mean is $\mu = 0.5$ and variance is $\sigma^2 = 1/12 = 0.083$ and standard deviation is $\sigma = 0.288$. Therefore, in this case with $n = 1$, the sample variance 0.0845 of the sample means is

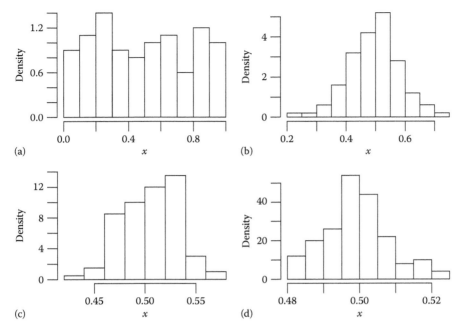

FIGURE 3.33 Sample means for sample size (a) 1, (b) 10, (c) 100, and (d) 1000.

close to the variance 0.083. The standard error 0.2907 is close to the 0.288 value. For $n = 10$, the sample variance of the sample means 0.0079 is close to $\sigma^2/10 = 0.083/10 = 0.0083$ and standard error 0.089 is close to $\sqrt{0.0083} = 0.0911$. You can continue to examine and compare the results for $n = 100$ and $n = 1000$. As we can see in Figure 3.33, we are getting closer to a normal distribution as we increase the sample size.

An alternative method to do the same program and avoid the inner loop

```
# sample means
  for(j in 1:m) x[j] <- mean(runif(n[i]))
```

is to use function `apply` on a matrix of samples. Function `apply(X, margin, fun,…)` applies a function `fun` to entries of a matrix defined by `margin` row (1), column (2) or row and column (c(1, 2)). Therefore, we build a matrix `y` with the samples and use `apply` over `margin` 2 (columns) invoking `fun = mean`.

```
# samples from uniform
y <- matrix(runif(n[i]*m), ncol=m)
# sample means
x <- apply(y[,1:m,drop=F],2,mean)
```

The equivalent process of obtaining samples and sample means can be performed using the Rcmdr. For example, go to **Distributions|Continuous|Uniform** and select option **Sample from a uniform distribution** and select for instance number of samples (rows) 10 and number of

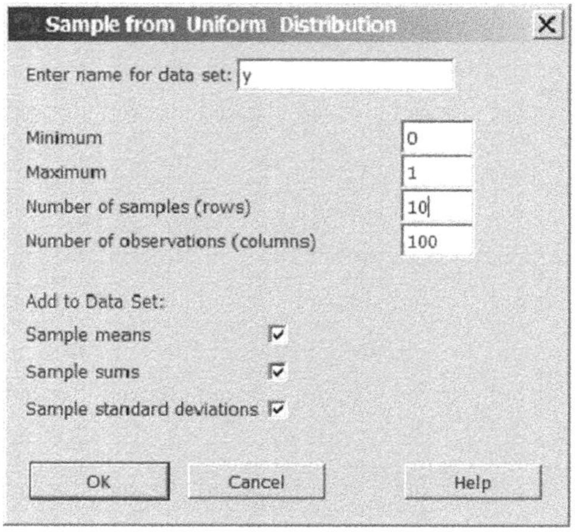

FIGURE 3.34   Sample from uniform using Rcmdr.

observations (columns) 100. You can check calculation options to obtain sample means, sample sums, and sample standard deviations (Figure 3.34).

3.9.10 SAMPLING: FUNCTION SAMPLE

Suppose we want to sample from a number N of elements (e.g., $N = 1000$) and that you have the label for these elements in a sequence from 1 to 1000. You want to obtain a sample of size n given of some percent of N (say, 10%, i.e., $n = 100$). To select a random sample without replacement, we can use function `sample` to determine the labels of the sample.

First generate the label with `seq(1,1000,1)` and then since 10% of 1000 is 100, we need to select 100 elements. Using function `sample`

```
> sample(seq(1,1000,1), 100, replace=F)
```

then this will give you a list of 100 elements.

3.9.11 CLEANUP AND CLOSE R SESSION

It is a good practice to clean up after a session. From this lab session, you could remove the following

```
> rm(sambinom, xbinom, pmfbinom, cpmfbinom, x, numbers, relfreq,
  probs)
```

As before, we could also have avoided cluttering by using nested commands and referring to components of objects instead of generating new ones.

Recall, you can use `q()` or **File|Exit** to finalize R and close the session. When you do this, you will be prompted for saving the workspace. Reply yes. This way you will have the objects created available for the next time you use R.

3.9.12 COMPUTER EXERCISES

Exercise 3.8

Read data in file **lab3/hintense.txt,** which contains the number of intense hurricanes by year in the period 1945–1999 (Landsea, 2012). Table the data, plot a pie chart and a barplot. What is the most common number of intense hurricanes per season in this time period? Determine the mean, standard deviation, and coefficient of variation.

Exercise 3.9

Generate a sample of 100 random numbers from a $N(0, 1)$. Obtain an estimate of the density function. Calculate the interquartile distance (**iqd**). Smooth the density using **iqd**. Calculate sample mean, variance, and standard deviation. Compare the sample statistics to the theoretical moments of $N(0, 1)$ and the density estimate to the theoretical density.

Exercise 3.10

Suppose that on the average we get 1/3 intense hurricanes out of the total Atlantic hurricanes in a season. Suppose a binomial distribution for the number of intense hurricanes in a season. Calculate the probabilities of the possible values of the number of intense hurricanes in a season of 10 hurricanes. Calculate the standard deviation.

Exercise 3.11

Calculate mean and variance of sample means of 100 samples of size 10 from a normal distribution. Calculate the standard error.

Exercise 3.12

Write a script to calculate the species evenness of a discrete uniform distribution from 1 to n, for $n = 2, 3, \ldots, 10$. Here n is species richness. Plot evenness vs. richness and discuss what happens to evenness as you change species richness.

Exercise 3.13

There are 130 residents in an area where you are researching their attitudes toward the environment and natural resources. Suppose you assign a code number in a sequence from 1 to 130. You want to conduct a survey that covers 10% of these residents. Select a random sample without replacement. List the code numbers of the residents to interview.

SUPPLEMENTARY READING

General topics of this chapter are covered by Davis, 2002, Chapter 2, pp. 25–50; Carr, 1995, Chapter 1, pp. 1–8, and p. 14, Chapter 4, pp. 65–66, 68–90, except Sections 4.5, 4.6, 4.7, 4.8.3, 4.8.4; Griffith and Amrhein, 1991, Chapter 6 and 7, pp. 147–196; and Rogerson, 2001, Chapter 2. The following readings would be useful on specific topics. On random variables: Davis, 2002, pp. 25–29, Carr, 1995, pp. 68–69. On cumulative functions: Davis, 2002, pp. 18–19, pp. 31–32. On histograms Davis, 2002, pp. 29–30, Carr, 1995, p. 71. On Poisson: Davis, 2002, p. 19, pp. 184–185, Carr, 1995, pp. 70–71. On the central limit theorem: Davis, 2002, pp. 58–60, Drake, 1967, pp. 232–234.

4 Exploratory Analysis and Introduction to Inferential Statistics

4.1 EXPLORATORY DATA ANALYSIS (EDA)

Classical parametric statistical inference depends on outlier-free and nearly Gaussian data. Before applying these inferential methods, it is a good idea to explore the data and see if they conform to the assumptions (MathSoft, 1999, pp. 3-6–3-8 and 3-14–3-15). One way of accomplishing this is by visual inspection of several plots. Because it is a first look at the data, we can refer to this process as **exploratory**, thus the term exploratory data analysis (EDA); in other words, use graphs to check the assumptions before proceeding to formal analysis. These plots are the following:

- **Index** plot, where the observations are arranged serially. From this we can visualize variability and potential outliers
- **Histogram**, to get a sense of the shape of the distribution
- **Density**, for the same purpose as the histogram
- **Boxplot**, or **box and whiskers** plot to indicate where the median lies with respect to the mean (symmetry), to visualize outliers and where the bulk of the data is located
- **Cumulative** plot or empirical cdf
- **Quantile–quantile plot** (**qq plot** for short), to qualitatively visualize if data are normal

We have already studied the histogram and the density. Now we will study the other plots just listed.

4.1.1 INDEX PLOT

This is a simple plot where observations are arranged serially with an index given to the number of the observation. The observations are depicted either as simple points (Figure 4.1a) or as spikes (Figure 4.1b). From this type of graph, we can see the variability of the data and identify potential outliers. For example, we can tell that there are three observations (10, 53, and 74) that have very low values (around and less than 10) and one observation (38) that has a very high value (near 100). In the computer session, we will see how to identify the observations on a plot.

4.1.2 BOXPLOT

The **boxplot** or **box and whiskers** plot (Figure 4.2) is a display of the main features of the descriptive summary: the median (a line inside the box), the first and third quartiles or lower and upper **hinges** (edges of the box), and the minimum and maximum nonoutlier values (the whiskers). These last two values are determined from the extremes of the **range** (or fence), which are the hinge (lower and upper respectively) minus or plus a factor (e.g., 1.5) of the inter-quartile distance (iqd, for short). The upper whisker is at the largest value within the range and the lower whisker is

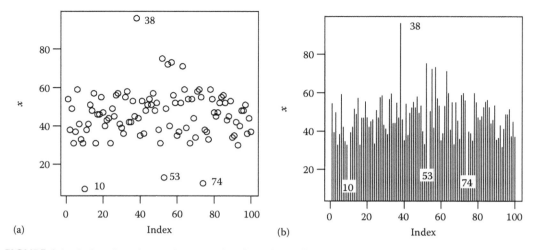

(a)   (b)

FIGURE 4.1   Index plot: observations as points (a) and as spikes (b).

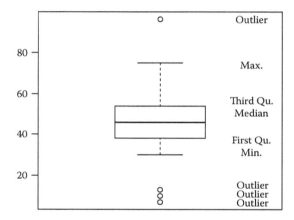

FIGURE 4.2   Boxplot or box and whiskers plot.

the smallest value within the range. Values above or below the extremes of the range are outliers and identified as circles on the plot.

For example, for the 100 observations used for the boxplot of Figure 4.2, the following values are used: lower hinge (first quartile) = 38, upper hinge (third quartile) = 54, and median = 46. In this case, the iqd is $54 - 38 = 16$, and therefore using $1.5 \times 16 = 24$ for the range, we obtain $38 - 24 = 14$ and $54 + 24 = 78$ for the extremes of the range. The lowest value contained within the range is 30 (this sets the lower whisker) and the largest value is 75 (upper whisker). In this case, below 14 we have three values (7, 10, 13) and above 78 we have one value (96). All these four values are outliers and displayed as small circles (Figure 4.2). It is helpful to label the outliers with the observation number (Figure 4.3).

4.1.3   EMPIRICAL CUMULATIVE DISTRIBUTION FUNCTION (ecdf)

The empirical cdf or ecdf is a visual aid to explore a sample. We first sort observations from smallest to largest to decide their position on the horizontal axes. Recall from Chapter 3 that these are the ith-order statistics. Once sorted, we rank the observations. Then, we divide these ranks by the number of observations to obtain fractions of 1. Finally, these fractions go in the vertical axes (Figure 4.4a). Naturally, 100 could multiply these fractions if we want the information in percentiles.

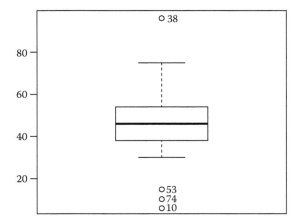

FIGURE 4.3 Boxplot with outlier identification.

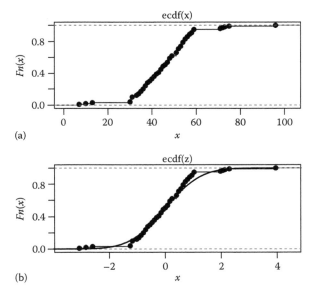

FIGURE 4.4 Empirical cdf (a) and comparing empirical and theoretical cdf (b).

We can explore normality by converting to Z scores, calculating the ecdf of the standardized sample, and comparing to the theoretical cdf of a standard normal. To obtain the z_i score for the data value x_i, we use

$$z_i = \frac{x_i - \overline{X}}{s_X} \tag{4.1}$$

where s_X is the sample standard deviation (Davis, 2002, p. 57). In essence, we center the data at 0 by subtracting the sample mean (average), thus shifting the distribution, and then scaling up or down, according to s_X.

For example, the ecdf of the standardized dataset of 100 numbers used in the boxplot is shown in Figure 4.4b, together with the theoretical cdf of the standard normal. In this graph, we can see that the ecdf follows relatively well the expected or theoretical cdf for the normal distribution in the interval $(-1, 1)$ but departs significantly outside this interval.

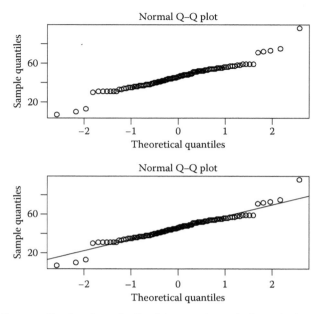

FIGURE 4.5   Quantile–quantile plot of standardized data vs. theoretical standard normal.

4.1.4   Quantile–Quantile (q–q) Plots

Another tool to visualize whether the data follow a theoretical distribution is the **probability plot** or **quantile–quantile plot**, which many call the **q–q plot** for short. For example, we can compare the quantiles of the data to the theoretical quantiles of a normal distribution. This is done using the standard normal Z: the z scores are ranked and the ranks converted to fractions (values in between 0 and 1 or percentiles values in between 0 and 100) in order to obtain quantiles. Then these values are plotted in the vertical axis corresponding to the values of a standard normal at the theoretical quantiles used in the horizontal axis (Figure 4.5, top panel). If the data are normal, then the plot should follow a straight line (Figure 4.5, bottom panel). The data used in Figure 4.5 are the same we show before for the boxplot and the ecdf. It is clear that the outliers we saw in the boxplot are those observations that depart of the line in Figure 4.5.

4.1.5   Combining Plots for Exploratory Data Analysis (EDA)

We now combine the plots explained so far for exploratory data analysis as illustrated in Figure 4.6 using as an example the dataset we have discussed earlier. We conclude that the sample has some outliers making it depart from a normal distribution.

For comparison, refer to the example in Figure 4.7 where we see a better fit of the data to a normal distribution. There are no outliers and the boxplot shows almost similar spread in range. The histogram and density show a slight asymmetry. The latter is noticeable also in the qq plot and the ecdf.

4.2   RELATIONSHIPS: COVARIANCE AND CORRELATION

In many cases, we are interested in how variables relate to each other. That is to say, inter-variable relationships (Davis, 2002, pp. 40–54, Carr, 1995, pp. 77–78 and 80–81). A simple case is that of **two** variables, called **bivariate**, which we will discuss next in this section. When we have more than two variables, then we conduct **multivariable** analysis, which we will cover after Chapter 9.

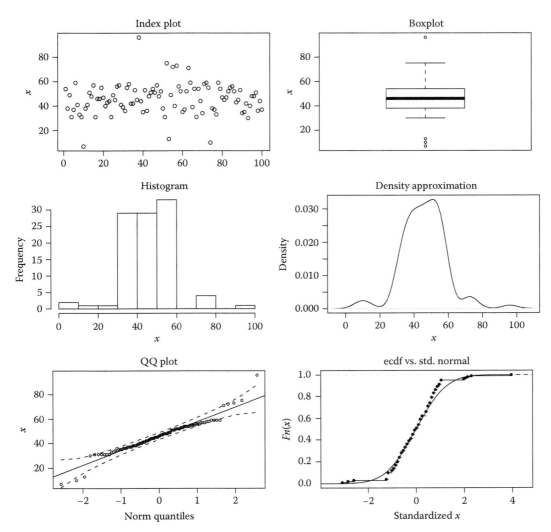

FIGURE 4.6 Index plot, boxplot, index histogram, density approximation, qq-plots, and ecdf vs. standard normal plots to explore data.

Suppose we have two RV's X and Y. Each one of these variables has first and second moments, as defined in chapter 3. An important concept is the **joint variation** or the expected value of the product of the two variables, where each one is centered at the mean. This is the **covariance** and can be written as

$$\text{cov}(X,Y) = E\left[(X - \mu_X)(Y - \mu_Y)\right] \tag{4.2}$$

Please note this is a population concept, since the expectation operator implies using the distribution of the product. Therefore, we require the joint pdf or pmf of X and Y. A derived concept is the **correlation coefficient** obtained by scaling the covariance to values less or equal than 1. The scaling factor is the product of the two standard deviations

$$\rho = \frac{\text{cov}(X,Y)}{\sigma_X \sigma_Y} \tag{4.3}$$

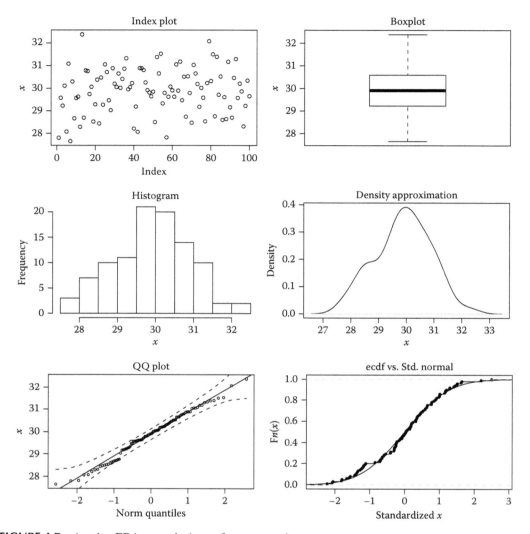

FIGURE 4.7   Another EDA example: better fit to a normal.

Because this product is always larger than the expected value of the product, then the ratio is always less than 1. We can note this also by calculating the correlation coefficient for maximum covariance, which occurs when X and Y are identical

$$\rho = \frac{\text{cov}(X,Y)}{\sigma_X \sigma_Y} = \frac{\text{cov}(X,X)}{\sigma_X \sigma_X} = \frac{\sigma_X^2}{\sigma_X^2} = 1 \tag{4.4}$$

Expanding Equation 4.2, and since the expectation of a constant yields the same constant, we obtain

$$\text{cov}(X,Y) = E\left[(XY - Y\mu_X - X\mu_Y + \mu_X\mu_Y)\right]$$

$$= E[XY] - E[Y]\mu_X - E[X]\mu_Y + \mu_X\mu_Y$$

$$= E[XY] - \mu_Y\mu_X - \mu_X\mu_Y + \mu_X\mu_Y$$

$$= E[XY] - \mu_X\mu_Y \tag{4.5}$$

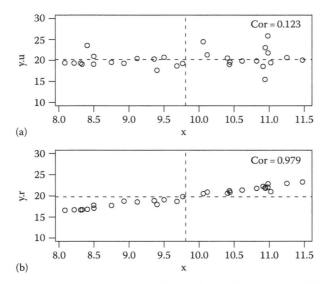

FIGURE 4.8 Scatter plots for two pairs of samples. Uncorrelated (a) and correlated (b).

The same idea applied to a sample leads to the **sample correlation coefficient** r where the covariance is a **sample covariance** and the standard deviations in the denominator correspond to the sample.

For illustration, we see two contrasting situations in Figure 4.8. These plots are called **scatter plots**; each pair of values x_i, y_i is marked with a symbol on the x–y plane. In Figure 4.8a, we have y.u vs. x, where x.u is a sample uncorrelated to x; the sample correlation coefficient is r = 0.123. In Figure 4.8b, we have y.r vs. x, where y.r is correlated to x; the sample correlation coefficient is r = 0.979, much closer to 1. In addition, we have drawn dotted lines at the mean of each sample.

4.2.1 SERIAL DATA: TIME SERIES AND AUTOCORRELATION

An important application of covariance is to serial data. This is to say, to data collected in a time (or space) sequence or series. For example, air temperature taken every hour, vegetation cover taken along a transect every 10 m. As a specific example, consider the number of Atlantic hurricanes per season from 1944 to 1999 shown in Figure 4.9 (top panel). This is a **time series** plot; it allows seeing patterns in successive data values.

A special application of the covariance and correlation concepts is to calculate the **co-variation between pair of values separated by a lag** L (either time or space) by means of the **autocovariance** function

$$C_X(L) = \text{cov}(x_{i+L}, x_i) \tag{4.6}$$

Note that we use the term function for autocovariance because it takes a different value for each value of lag L. In the example given earlier, the sequence is given in years and the lag can take values 0, 1, 2, … years. The autocovariance may tell us something about periodic variations in hurricane frequency. We often work directly with the **autocorrelation** function, which is derived in the following paragraph.

Note that when the lag is zero, $L = 0$, the two values of X should correspond to the same time instant or distance, and we obtain

$$C_X(0) = \text{cov}(x_{i+0}, x_i) = \text{cov}(x_i, x_i) = \sigma_X^2 \tag{4.7}$$

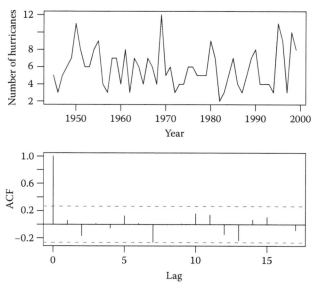

FIGURE 4.9　Number of Atlantic hurricanes. Time series plot and autocorrelation plot. (Source: Landsea, 2012.)

that is to say, the **autocovariance at zero lag is the variance** of X. In addition, this is the **maximum** value of the autocovariance function. Now, scaling the covariance function by this maximum, we define the **autocorrelation** or the correlation of lagged values for one variable. This to say,

$$\rho_X(L) = \frac{\text{cov}(x_{i+L}, x_i)}{\sigma_X^2} \tag{4.8}$$

Note that for $L = 0$ the autocorrelation will attain the maximum value of 1

$$\rho_X(0) = \frac{C_X(0)}{\sigma_X^2} = \frac{\text{cov}(x_i, x_i)}{\sigma_X^2} = 1 \tag{4.9}$$

It is very important to realize that X is a function of time or distance and that both autocovariance and autocorrelation are functions of the lag L. A time series plot is a plot of the variable in the vertical axis and time in the horizontal axis. A **correlogram** or **autocorrelation plot** is a plot of the sample autocorrelation in the vertical axis and lag in the horizontal axis. The autocorrelation plot helps visualize the relation between successive data values (see Figure 4.9, bottom panel). Note how the maximum value of autocorrelation is 1 and that it occurs for lag = 0. Again, the lag does not have to be time. It could also be a distance. For example, values of soil moisture taken at every depth, and values of vegetation cover taken every 10 m.

In later chapters, we will be studying how to analyze serial data with more detail. In this chapter, we will only be concerned with checking that the data are not serially correlated in order to make sure we satisfy assumptions of parametric inference. For example, the results displayed in Figure 4.9 would suggest that there is no serial structure in the data.

4.3　STATISTICAL INFERENCE

The goal of inference is to obtain probability statements about population parameters (mean, variance) from data (Davis, 2002, chapter 2; Qian, 2010). It involves two important concepts: hypothesis testing and confidence intervals. We seek to answer questions like these: Is the sample drawn for a normal distribution? What is the uncertainty of an estimate of the population mean? For practicality, in the following we will explain some simple methods of inference. However, the reader must be aware that the basis of inference is a complicated subject with roots in the debate between the theories of Fisher and of Neyman–Pearson (Qian, 2010; Lehmann, 1993).

4.3.1 Hypothesis Testing

In the simplest approach, α is the probability of an unlikely event that we set up from the hypothetical distribution pdf. It is a conditional probability of an unlikely event (interval called the **critical region**) given that the hypothesis is correct (Figures 4.10 and 4.11). Once we conduct the experiment, if the result is within what we considered an unlikely event, given that the hypothesis was true, then we believe that it is unlikely that the hypothesis is correct and we reject it. On the contrary, if the experimental result is different from the improbable event, then there is no reason to reject the hypothesis.

Let us explain these ideas using an example (following Drake, 1967, p. 235): toss a coin $m = 10,000$ times; use tail = 0, head = 1; we assume that the coin is fair and therefore the probability of "success" (i.e., obtaining a head) is $p_h = 0.5$. The mean of each toss is p_h and the variance is $p_h(1 - p_h)$. Select as statistic N_h = number of heads. Then from the central limit theorem we know that N_h is normal with mean $mp_h = 5000$ and variance $mp_h(1 - p_h) = 2500$. The standard deviation is the square root of the variance and therefore is 50.

Select $\alpha = 0.05$, and set up an unlikely event with this probability. We divide 0.05 in two equal parts of 0.025 because the event is defined such that the statistic N_h is either above a critical value or

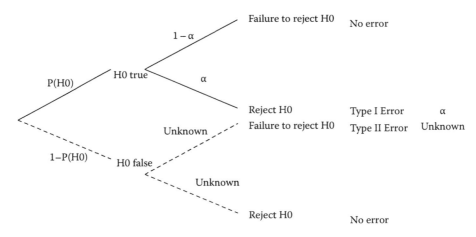

FIGURE 4.10   Decision tree for hypothesis testing.

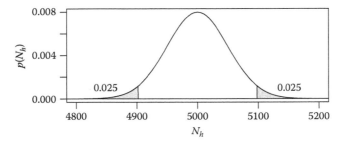

FIGURE 4.11   Two-tail critical region for hypothesis testing.

below another critical value. We call this type of situation **two-tail**, one on the left and one on the right (Figure 4.11). These are distribution tails and should not be confused with a tail as an outcome of a coin toss. The two distribution tails comprise the critical region and we need the knowledge of the normal distribution to find the interval. Therefore, we find the values for a normal distribution with these quantiles. We can obtain the interval using R,

```
> qnorm(c(0.025, 0.975), 5000, 50)
[1] 4902.002 5097.998
```

Therefore, the unlikely event is that number of heads exceeds 5098 (upper tail) or is less than 4902 (lower tail). Note that to establish the critical region we needed the hypothetical pdf for the fair coin (Figure 4.11).

Now, when we perform the experiment and obtain more than 5098 heads or less than 4902 heads, we conclude that the hypothesis of a fair coin ($p_h = 0.5$) is unlikely to be correct. We have at most a 5% probability of error ($\alpha = 0.05$) when rejecting that the coin is fair ($p_h \neq 0.5$) given that it is. This is a type I error. If the number obtained is more than 4092 and less than 5098, then we cannot reject the hypothesis. In this case, there is no way of telling what the type II error of falsely failing to reject that the coin is fair (given that it is not) because we did not set up any specific alternative hypothesis.

We can simulate the experiment using R

```
> x <- runif(10000)
> y <- x
> for(i in 1:10000) if(x[i]<=0.5) y[i] <- 1 else y[i] <- 0
> Nh <- sum(y)
> Nh
[1] 4939
```

Therefore, we cannot reject the H0 of a fair coin because the value is not in the improbable interval. Please realize that we may be in error by not rejecting it; this is the probability of type II error but we do not know what it is.

Note that if we bias the coin using $p = 0.4$

```
> for(i in 1:10000) if(x[i]<=0.4) y[i] <- 1 else y[i] <- 0
> Nh <- sum(y)
> Nh
[1] 3932
```

This value is in the left tail of the improbable interval; that is, less than 4902. Therefore, we can reject the H0 of a fair coin with type I error of at most $\alpha = 0.05$. The value α is a **significance level** or the maximum error we are willing to accept. You should note that 5% is an arbitrary number that has remained popular in the literature as a convenient level of significance. The advantage of rejecting it is that we limit the probability of error.

4.3.2   *p*-Value

Furthermore, we can calculate the **p-value** or probability that the measured outcome would occur. Thus, the *p*-value is the value at which the H0 can be rejected given the result of the experiment. See Davis (2002) and Qian (2010). Using the *p*-value there is no need to establish a significance level a priori but just judge the magnitude of the *p*-value. If we believe the *p*-value is small enough, then we can reject the null. Of course, the issue is how small the *p*-value should be for us to reject the null.

It has become popular to mix the *p*-value and significance level approaches and judge the *p*-value small when it is lower than the significance level. Using this combined approach, when the measured value falls in the improbable interval, its probability of occurrence (*p*-value) should be lower than α. Otherwise, it should be higher than α. However, we should not confuse the *p*-value with the type I error rate (Qian, 2010).

> For example, in the 10,000 coin tosses, suppose we get 4850 heads. We can calculate the probability that we get a value as low as 4850 given that the coin is fair.

```
> pnorm(c(4850),5000,50)
[1] 0.001349898
```

> For a two-tail we double this probability to account for the right tail

```
> 0.00135*2
[1] 0.0027
```

> Therefore, the *p*-value is 0.0027, meaning we can reject the H0 of a fair coin with probability of error 0.27% given the result of the experiment. Note that this is lower than the type I error rate $\alpha = 5\%$.

4.3.3   Power

To address the type II error of falsely failing to reject the H0, we set up two mutually exclusive events: the null hypothesis H0 and the alternative hypothesis H1. The probability of error in rejecting H0 given that it is actually true is the **type I error (α)**. The probability of error in rejecting H1 when it is actually true is the **type II error (β)**. The power is **1-β** or probability of accepting H1 given that it is true. See the tree diagram (Figure 4.12).

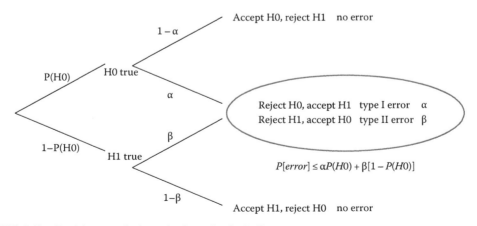

FIGURE 4.12   Decision tree for hypothesis testing including power.

Then the total probability of error is the sum of probability of false rejection of the null (false positive) and the probability of false rejection of the alternative (same as falsely accepting the null or false negative). These two events are in the middle of the ending branches of the tree.

$$P[error] = \alpha P(H0) + \beta[1 - P(H0)] \tag{4.10}$$

For a given α we select the critical region in such a way that we minimize β or maximize power $1 - \beta$. In other words, we fix error type I a priori, and then design the test to minimize type II error β given this error type I.

Expanding the example of the previous two sections about testing a fair coin, let us continue to assume that $m = 10,000$. Now suppose H0 is $p_h \leq 0.5$ and select a specific H1 to be $p_h = 0.51$, which is larger than 0.5. For this H1 we have a normal with mean $m \times 0.51 = 5100$ and $\sigma = 49.99$. If H0 is true, we can select $\alpha = 0.05$ as the type I error rate and consider the upper tail because we are interested only in p_h being less than 0.5. We call this type of situation **one-tail** (Figure 4.13). The tail in the right defines the critical region to reject H0 and we need the knowledge of the normal distribution to find the interval. We can obtain the interval using R,

```
> > qnorm(0.950, 5000, 50)
[1] 5082.243
```

Therefore, the unlikely event is that number of heads exceeds 5082. Its area is indicated as solid shade in Figure 4.13. Now, we determine the area under the curve for the normal for H1 (Figure 4.13a) left of the critical value 5082. This area determines the probability $\beta = 0.361$ that we accept H0 (reject H1) given that H1 is true. In this case, the power is $1 - \beta = 1 - 0.361 = 0.639$. The difference between the two means 5000 and 5100 is the **effect size**.

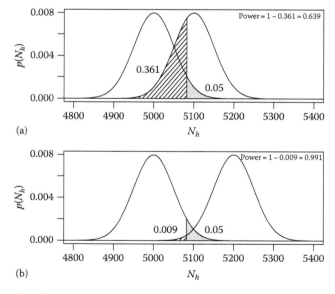

FIGURE 4.13 One-tail critical region for hypothesis testing and power. (a) Small effect size and (b) large effect size.

As a contrasting case, consider a H1 of $p_h = 0.52$. In this case, the mean is 5200 and $\sigma = 49.96$. Again, we determine the area under the curve for the normal for H1 (Figure 4.13b) left of the critical value 5082. This area determines the probability $\beta = 0.009$ that we accept H0 (reject H1) given that H1 is true. In this case, the power is $1 - \beta = 1 - 0.009 = 0.991$. The effect size (difference between the two means) is now larger 5000 and 5200.

We conclude that for H1: $p_h = 0.51$ we have only 63.9% probability of selecting H1 when it is true, but that we have 99.1% probability of correctly selecting H1: $p_h = 0.52$. By using additional examples, we can determine that power increases with effect size. Intuitively, for the same α, the area under the H1 density curve left of the critical point decreases as we shift the H1 mean to the right.

Consider H1: $p_h = 0.51$, when we perform the experiment and obtain more than 5082 heads we conclude that the null hypothesis H0: $p_h \leq 0.5$ is unlikely to be correct and accept H1. We have at most a 5% probability of error ($\alpha = 0.05$) when rejecting H0 given that it is true. If the number obtained is less than 5082, then we cannot reject the null but reject H1. In this case, the type II error of falsely rejecting H1 is 63.9%.

We can simulate the experiment using R

```
> x <- runif(10000)
> y <- x
> for(i in 1:10000) if(x[i]<=0.5) y[i] <- 1 else y[i] <- 0
> Nh <- sum(y)
> Nh
[1] 4939
```

Therefore, we cannot reject the H0 because the value is less than 5082 and the type II error rate is 63.9%. Note that if we bias the coin using $p_h = 0.51$

```
> for(i in 1:10000) if(x[i]<=0.51) y[i] <- 1 else y[i] <- 0
> Nh <- sum(y)
> Nh
[1] 5122
```

This value is more than 5082. Therefore, we can reject the H0 and accept H1 with type I error rate of $\alpha = 0.05$.

A detailed explanation of hypothesis testing and the relationship between power, significance level, and p-value is given in Qian (2010, pp. 80–87; pp. 61–69). Type I and type II errors are very relevant to environmental protection and the criminal justice system in the United States (Roger, 2011).

4.3.4 Confidence Intervals

Recall from Chapter 3 that the sample mean distribution is approximately normal for large sample size n. In addition, the population mean is the mean of the sample mean distribution. The variance of the sample mean distribution is the population variance divided by the sample size n. The square root of the variance of the sample means is the **standard error** of the sample mean. Therefore, the sample means have a distribution $\bar{X} \sim N(\mu_X, \sigma_X/n)$.

Convert this distribution to a standard normal $N(0, 1)$

$$Z = \frac{\bar{X} - \mu_X}{\sigma_X / \sqrt{n}} \tag{4.11}$$

Now we can say that 95% of the values of Z fall in the interval $[-2, 2]$. Thus, the probability that this interval contains the mean is 0.95. We define this as the 95% **confidence interval**. In general, we can use other quantiles of the standard normal. For quantile $z_{\alpha/2}$ we have the $100 \times (1 - \alpha)\%$ confidence interval.

Note that this is defined as an error rate α, i.e., the probability that an interval [a,b] fails to include the mean true value; common values for α are 0.01, 0.05, 0.1, or 1%, 5%, and 10%. The **confidence level** $(1 - \alpha)$ is the probability that the interval [a, b] includes the parameter true value. In other words, as we repeat an experiment and calculate the 95% confidence interval each time, we expect that 5% of the time the interval will fail to include the mean or that 95% of the time the interval will include the mean.

As an example consider the experiment of tossing a coin which we used already to discuss hypothesis testing. The number of heads is normally distributed with mean 5000 and standard deviation 50. Using Monte Carlo simulation, we perform the experiment of 10,000 tosses $n = 50$ times and obtain $\bar{X} = 4985$ then using Equation 4.11 and $\alpha = 0.05$ we obtain interval [4971, 4999] which does not contain the true mean. Another repetition of the experiment yields $\bar{X} = 5002$, then using Equation 4.11 and $\alpha = 0.05$ we obtain interval [4988, 5015] which contains the true mean. If we repeat drawing sample means and confidence interval many times, the interval would contain the true mean 95% of the time. Figure 4.14 shows the results of 100 repetitions of $n = 50$ sample size. In this case, the confidence interval was in error 4 times out of 100.

When we have a line or a curve, we can plot confidence intervals for each point of the line or curve resulting in an envelope around the line or curve. For example, we can plot envelope of the q–q plot to determine which points lay within a reasonable spread away from the line. In Figure 4.15, we see the envelopes corresponding to the confidence interval. These plots are the same as Figure 4.5 except for the addition of the confidence intervals. Note that for the top panel we clearly see those outliers that wander off the envelope. However, for the bottom panel, all points lay within the envelope. For further reading on confidence intervals, see Davis (2002, pp. 66–68), and Qian (2010, pp. 50–61).

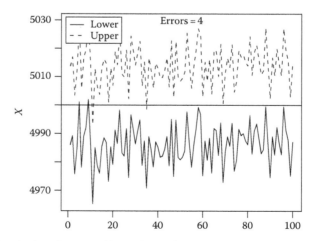

FIGURE 4.14 One hundred random repetitions of the 95% confidence interval calculation. In this case, the interval includes the true mean (5000) 96 times out of 100.

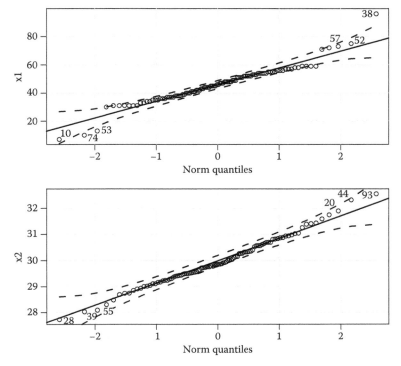

FIGURE 4.15 Quantile–quantile plot of standardized data versus theoretical standard normal and envelope showing confidence interval for each point.

4.4 STATISTICAL METHODS

As a broad classification, we can think of available methods by grouping them according to the following types (MathSoft, 1999, pp. 3-10–3-11):

- **Classical** inferential statistics proceeds as outlined so far in this chapter. It can be of two major types: parametric and nonparametric. **Parametric**: the pdf for the statistic is based on a parameter and assumes normal, outlier-free, nonserially correlated data. **Nonparametric**: the method independent of the distribution (normality not needed) and usually based on ranks of observations.
- **Robust** methods are such that conclusions are not affected by deviance from assumptions, i.e., outliers and non-Gaussian (serial correlation not included). The emphasis is on reducing sensitivity to outliers. One method is to trim outliers by a given fraction a of the lowest and a fraction a of the largest observations and to take the mean of remaining ones: the mid $1–2a$ fraction of observations.
- **Resampling**: Distribution free, based on random-generation of many new samples taken from the sample. This is computer intensive. Examples are bootstrap and jackknife methods.
- **Bayesian analysis**: Based on applying a priori knowledge and updating it with data from outcomes of the experiment.

In this book, we cover mostly classical tests. Note that normality is an important assumption for parametric methods. In this chapter, we will decide to apply parametric methods based on a qualitative EDA. However, in the following chapter when we study goodness of fit methods, we will see how to perform statistical tests to make a quantitative decision on normality.

4.5 PARAMETRIC METHODS

4.5.1 Z TEST OR STANDARD NORMAL

We use this parametric method to test whether the sample mean is different from the population mean (parameter). It assumes a known distribution (normal) with known parameters mean μ_X and variance $= \sigma_X^2$. We build the statistic z by the ratio of deviations of the sample mean from hypothetical or population mean to the standard error σ_e.

$$Z = \frac{\bar{X} - \mu_X}{\sigma_e} = \frac{\bar{X} - \mu_X}{\sigma_X / \sqrt{n}} \tag{4.12}$$

The null hypothesis is H0: $\bar{X} = \mu_X$. If z is large for given α, then reject H0; if the z value indicates that H0 cannot be rejected, then there is no evidence that the sample mean is different from the population mean (Davis, 2002, pp. 60–64).

An example is the result of one of the repetitions of $n = 50$ of the 10,000 coin toss experiment discussed in the confidence interval section. The number of heads is normally distributed with mean 5000 and standard deviation 50. When we obtain $\bar{X} = 4985$, then using Equation 4.12 we obtain $z = -2.12$. This value of z is outside the improbable region $[-2, 2]$ for $\alpha = 0.05$ and we would reject the null in error. Using the p-value approach, this value of z corresponds to a probability pnorm(z,0,1) of 0.017. Thus, for two tails the p-value is 0.034 from this sample. We would reject the null in error. As another example when we obtain $\bar{X} = 5001$ using Equation 4.12 we obtain $z = 0.14$, which is well inside the $[-2, 2]$ interval for $\alpha = 0.05$ and we would not reject H0. The probability is 1-pnorm(z,0,1) or 0.44 and p-value is $2 \times 0.44 = 0.88$ also suggesting not to reject H0.

4.5.2 THE t-TEST

This is a parametric test that uses the sample to estimate unknown parameters and therefore it has reduced **degrees of freedom (df)** to compensate for use of data in estimation of the parameters. In this case, we estimate the standard error from the sample standard deviation s_X. The number of degrees of freedom is $df =$ number of observations in sample − number of parameters estimated from the sample.

The t distribution is similar to the normal but with heavier tails, which depend on size (n) of the sample or degrees of freedom $df = n - 1$; the t statistic approaches the normal if the sample size is large (Figure 4.16)

$$t = \frac{\bar{X} - \mu_X}{s_e} = \frac{\bar{X} - \mu_X}{s_X / \sqrt{n}} \tag{4.13}$$

This distribution can be used to test that the sample mean is equal to the true mean. The null H0: $\bar{X} = \mu_X$. Reject H0 if t large for a given α. It assumes normality of the data. The number of degrees of freedom $df = n - 1$, since s_X is estimated from the sample.

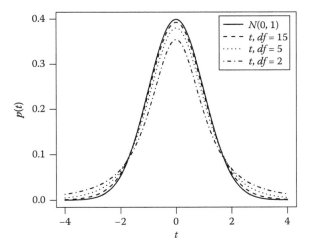

FIGURE 4.16 Comparing the t distribution for several values of df and to the standard normal.

As an example suppose we have $n = 20$ and the sample is 1.42, 2.39, 0.69, 0.33, 1.51, −0.02, 1.35, 0.79, 0.60, 0.32, −0.34, 0.32, 0.51, 1.24, 0.39, 0.13, 1.02, 0.11, −0.20, 1.09. The H0 is that the mean $\mu_X =$ is 0. The critical values for $\alpha = 0.05$ are −2.09 and 2.09 for a t density with $df = 19$. This can be determined with `qt(c(0.025,0.975),df = 19)`. Calculating t from Equation 4.13, we get $t = (0.68 − 0)/(0.67/20) = 4.52$; therefore, we can reject H0 at $\alpha = 0.05$. The probability of this t value is `1- pt(4.52,df = n-1)` yielding 0.000116 and a p-value of 0.00023 for two-tails. With this low p-value we would reject H0 given this outcome.

The test can also be used for one-sided situations; that is to say, to see if the sample mean is less than the true mean or to see if the sample mean is greater than the true mean. In addition, the t distribution can be used to test the equivalency of two samples (t statistic is redefined, s_e depends on n_1 and n_2). In this case, the null H0: $\mu1 = \mu2$. The degrees of freedom $df = n_1 + n_2 − 2$, since two parameters are estimated from the sample (Davis, 2002, pp. 68–75).

4.5.3　THE *F* TEST

The F test is a parametric test for the equality of variances of two samples. Based on the F statistic given by ratio of two sample variances

$$F = \frac{s_1^2}{s_2^2} \tag{4.14}$$

The degrees of freedom are $df_1 = n_1 − 1$ and $df_2 = n_2 − 1$. The null H0: $\sigma_1^2 = \sigma_2^2$, the alternative is Reject H0 if F is large for a given α. If we cannot reject H0, then there is no evidence to say that the variances are different. This test is performed prior to the t-test for two samples since the t-test assumes similar variances (Davis, 2002, pp. 75–78). We will see examples later in the computer session of this chapter.

4.5.4 Correlation

This test allows establishing the statistical significance of the correlation coefficient. The parametric version is the Pearson's classical product-moment measure of correlation. The sample correlation r is an estimate of the correlation coefficient ρ. We test the significance of r. Assume normality for both variables. The null hypothesis is that the samples are uncorrelated; this is to say H0: $\rho = 0$. A t statistic is

$$t = \frac{r\sqrt{n-2}}{1-r^2} \tag{4.15}$$

with $df = n - 2$. We reject H0 if t is large for a given α; concluding that samples are correlated. However, if we cannot reject H0, then there is no evidence to say that the samples are correlated.

4.6 NONPARAMETRIC METHODS

Many of these methods are based on a **rank** transformation of the data; values are sorted from smallest to largest. Thus, the actual values are no longer relevant eliminating the impact of outliers. Let us denote $R(x_i)$ as the rank of observation x_i of the sample. There are several ways of resolving ties. For example, the ranks can be assigned based on the average of tied observations (Davis, 2002, pp. 102–105; Qian, 2010, p. 73–79).

4.6.1 Mann–Whitney or Wilcoxon Rank Sum Test

This test is a nonparametric alternative to the t-test for equality of means of two samples of size n and m. In this case, we test for the equality of medians. The ranks should spread out uniformly if it is true that the samples come from the same distribution. The test assumes that observations are independent. First we assign ranks $R(x_i)$ to the combined samples (observations from both samples). The test statistic W is the sum of ranks of one sample and it can be compared to the value that would be obtained for an area with probability α. The test statistic is approximated by a standard normal for large samples sizes.

4.6.2 Wilcoxon Signed Rank Test

This test is for one sample and is used when we want to see if the median is equal to a given value. We take the difference between the observations and the hypothetical median. Denote these differences by x_i. Then assign ranks $R(x_i)$ based on the absolute values $|x_i|$ of the differences x_i. However, to compute the test statistic V we only sum those ranks corresponding to positive differences $x_i > 0$. The test statistic value is compared to the critical value for given α. The test statistic is approximated by a standard normal for large sample size.

4.6.3 Spearman Correlation

A popular method is the rank-based Spearman's measure of correlation. It is used to test for similarity between two sets of ranks. Spearman's rank correlation coefficient is based on summing the square of differences in rank

$$r' = 1 - \frac{6\sum_{i=1}^{n}[R(x_i) - R(y_i)]^2}{n(n^2-1)} \tag{4.16}$$

H0: $\rho' = 0$ Reject if r' large for given α (Davis, 2002, pp. 74–75 and pp. 105–107).

4.7   EXERCISES

Exercise 4.1

At a site, monthly air temperature is normally distributed. It averages to 20°C with standard deviation 2°C.

1. What is the probability that a value of air temperature in a given month exceeds 24°C?
2. What is the probability that it is below 16°C or above 24°C?
3. What is the probability that it is below 18°C or above 26°C?

Exercise 4.2

Is it true that covariance of X and Y is equal to variance of X when $X = Y$? Explain why. Demonstrate. Explain why the autocovariance of X for any lag cannot exceed the value at lag 0.

Exercise 4.3

Suppose we have collected 50 values for a sample of ozone and the average is 2.00, with standard deviation of 0.5. What would be the value of t when testing that the mean is equal to 2.5?

Exercise 4.4

Monthly rainfall at a site is classified in two groups: one group for El Niño months and the other for La Niña months (defined according to sea surface temperature in the Pacific Ocean). We have 100 months for each group. The variance of each group is the same. Is it true that rainfall during El Niño is different to that during La Niña? What type of test would you run? What is H0? Suppose you get a p-value $= 0.045$. What is the conclusion of the study?

4.8   COMPUTER SESSION: EXPLORATORY ANALYSIS AND INFERENTIAL STATISTICS

4.8.1   Create an Example Dataset

We will input an example dataset contained in file **lab4/example1.txt**. It has only 1 column and 100 rows. To read these data, we can simply use scan in the R console

```
>x<- scan("lab4/example1.txt")
```

or, if you prefer, from the Rcmdr, go to **Data|Import data|from text file..|** enter the name xd (to distinguish from x created in Console) in the dialog box, clear the checkmark for the names in file, select white space for field separator, click ok, browse to the file **lab4/example1.txt**, and open. The object xd has been created. Use button **View dataset** to inspect the data. As we have just generated this dataset xd, it shows as the active Dataset in the upper left-hand-side corner of the Rcmdr GUI.

We will learn how to produce graphs to check the assumptions of classical tests before proceeding to formal inference. Let us use the array x read from the console or xd$V1 read from Rcmdr.

4.8.2   Index Plot

From the Rcmdr, with xd selected as active dataset, use the **Graphs|Index plot** menu. At the dialog window, select V1, points, and option **Identify outliers with mouse**. It will be useful, as you will see later. A graph similar to the one in Figure 4.1a is produced in the graphics window. As we can see, the observations are arranged serially as already explained. A useful feature is the tool to identify the observations: use the cross hair to click on the potential outlier observations, which displays the observation numbers (38, 10, 53, 74). Press stop to finish with the identification process.

Similarly, to get a spike Index plot from the Rcmdr, use the **Graphs|Index plot** menu, at the dialog window, select V1, spikes, and option **Identify outliers with mouse**. A graph similar to Figure 4.1b is produced in the graphics window. Proceed to identify as before and stop when finished.

It is also simple to produce the index plot using the R Console using the function `plot`. For points,

```
>plot(x)
>identify(x)
```

proceed to identify points and press stop once you are done. For spikes,

```
>plot(x,type="h")
>identify(x)
```

and conduct the identification. The results are the same as before and as shown in Figure 4.1.

4.8.3   BOXPLOT

From the Rcmdr use the **Graphs|Boxplot** menu. A dialog window pops up, select V1, click on the option identify outliers with mouse. The graph is produced on the graphics window, use the cross hair to click on the outlier observations, and press stop to finish with identification (see Figure 4.3).

To produce the boxplot using the R Console use function `boxplot`, and to identify outliers you can use 1 for x-coordinates and the values of x for the y-coordinate. This is easy to do using the `rep` function

```
>boxplot(x)
>identify(rep(1, length(x)), x)
```

which can be stopped using stop button. The result is the same as before (Figure 4.3).

4.8.4   EMPIRICAL CUMULATIVE PLOT

Function `ecdf` and it is part of package `stats`. It should already be in your installation. From the Rconsole, you can also execute this function as

```
>plot(ecdf(x))
```

to obtain the cumulative plot of Figure 4.4a that shows how some observations are outliers as we already learned from the index and boxplot.

Now, to standardize a variable we subtract the mean and divide by the standard deviation. After plotting it, we can add the theoretical $N(0, 1)$ using lines

```
# standardize observations before plot
z <- (x-mean(x))/sd(x); plot(ecdf(z))
# plot theoretical standard normal N(0,1)
Z <- seq(-4,+4,0.1); lines(Z, pnorm(Z,0,1),lwd=2)
```

See Figure 4.4b. In the graph, we can see how well or not the cdf of the data (the empirical cdf) follows the expected or theoretical cdf for the normal distribution.

4.8.5 Functions

We can enhance existing R functions by creating additional functions. When we develop a program into a function with proper arguments, it will allow using the program without rewriting all the statements. One can edit a function in a **script window** and save it as an ***.R** file. The set of statements follows this syntax

```
fnName <- function(x, other)
{
        Body of the function: a set of commands
        These commands operate on x and other arguments
        Arguments are data objects
}
```

All we have to do is declare object `fnName` as a function and then collect the statements within curly brackets "{" to mark the start point and "}" to mark the end point. Indenting statements within the brackets makes the program more readable. Here x and other are arguments to the function.

Once the file ***.R** is saved, we can use **File|Source R Code** from the R GUI menu to source it. Alternatively, we can apply the `source` function to this file name. For example, if we save the file as **fnName.R**, then

```
> source("fnName.R")
```

You can use any text editor to write the script, for example, the notepad, or Vim. However, the R script editor is convenient because you can run lines or set of lines directly.

Once your function is sourced, to use it just type its name with its arguments. For example,

```
> fnName(data object for x, data object for other)
```

4.8.6 Building a Function: Example

Next, we will illustrate how to build a function to plot the ecdf from scratch for two reasons: (1) to make sure you understand the ecdf, and (2) to learn how to create a function using R. You can type it in a new script window

```
cdf.plot <- function(x){
        nx <- length(x)                        # number of observations
        vx <- sort(x)                          # sort the observations
          for horizontal axes
        Fx <- seq(nx)/nx                       # rank divided by
          number to get fraction
        plot(vx, Fx, xlab="x", ylab="F(x)")    # plot and label axis
        mx <- mean(vx); sdx <- sd(vx)          # mean and stdev
        vnx <- seq(vx[1],vx[nx],length=100)    # sequence with
          increased res
        Fxn <- pnorm(vnx,mx,sdx)               # cdf values
        lines(vnx,Fxn)                         # theoretical
          cumulative plot cdf
}
```

Here, we declared object `cdf.plot` as a function and then collect the statements within curly brackets "{" to mark the start point and "}" to mark the end point. Remarks or comments are followed by the # sign and are not executed.

First, save the script as a text file with an extension **.R**, say as **cdfplot.R**. Then we can use **File|Source R Code** from R GUI menu to source it. Alternatively, we can apply the `source` function to this file name.

```
> source("lab4/cdfplot.R")
```

Note: if you are using the notepad to write scripts, then make sure to use option "all files" when saving the file from the notepad, so that the file is not saved as ***.R.txt** because this wrong name will produce failure of the source command. Once your function is stored, then all you need to do to apply it is to type its name with its argument. For example,

```
>cdf.plot(x)
```

yields results as in Figure 4.4a but with added plot of the theoretical normal. In addition, it does not include the short lines after each data point.

4.8.7   More on the Standard Normal

Recall that we obtain a standardized variable by subtracting the mean and dividing by the standard deviation. Let us use `xd$V1` as example. With `xd` as active dataset you can use Rcmdr **Data|Manage Variables in active dataset |Standardize variables ....** Select V1 and examine the result by using **View Dataset**. You will see an additional column labeled Z.V1.

Recall from Chapter 3, that to obtain the pdf of the $N(0, 1)$, we can use the Rcmdr **Distributions|Normal distribution|Plot Normal distribution**. We can also use the pop-up window selection to distribution to obtain the cumulative.

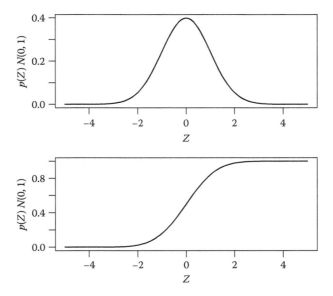

FIGURE 4.17   Standard normal pdf and cdf.

As we did in the previous two sections, using the Rconsole, to obtain the pdf of the $N(0, 1)$, generate a sequence of numbers from negative to positive. These will be in units of σ, say from -5σ, to $+5\sigma$ (every 0.1). Then apply functions dnorm and pnorm with this sequence and mean = 0, $\sigma = 1$. Also, we use the graphics function par (mfrow = c(2,1)) to establish a two-panel graph. We put the pdf in the top panel and the cdf in the bottom panel (Figure 4.17)

```
> Z <- seq(-5,+5,0.1)
> par(mfrow=c(2,1))
> plot(Z,dnorm(Z,0,1),type="l",ylab="p(Z) N(0,1)")
> plot(Z,pnorm(Z,0,1),type="l",ylab="p(Z) N(0,1)")
```

We can use the probabilities 0.1, 0.2, 0.3, …, 0.9 to calculate the quantiles

```
> Zp <- seq(0.1,0.9,0.1)
> Zq <- qnorm(Zp,0,1)
> Zq
[1] -1.2815516 -0.8416212 -0.5244005 -0.2533471 0.0000000 0.2533471
  0.5244005
[8] 0.8416212 1.2815516
>
```

That is to say we get a cumulative probability of 0.1 at $z = -1.28$, a cumulative probability of 0.2 at -0.84, and so on. These values are the 10, 20, 90 percentiles. For finer resolution, we can calculate the all the 1, 2, 3, …, 99 percentiles

```
> Zp <- seq(0.01,0.99,0.01)
> Zp
 [1]  0.01 0.02 0.03 0.04 0.05 0.06 0.07 0.08 0.09 0.10 0.11 0.12 0.13
      0.14 0.15
[16]  0.16 0.17 0.18 0.19 0.20 0.21 0.22 0.23 0.24 0.25 0.26 0.27
      0.28 0.29 0.30
[31]  0.31 0.32 0.33 0.34 0.35 0.36 0.37 0.38 0.39 0.40 0.41 0.42
      0.43 0.44 0.45
[46]  0.46 0.47 0.48 0.49 0.50 0.51 0.52 0.53 0.54 0.55 0.56 0.57
      0.58 0.59 0.60
[61]  0.61 0.62 0.63 0.64 0.65 0.66 0.67 0.68 0.69 0.70 0.71 0.72
      0.73 0.74 0.75
[76]  0.76 0.77 0.78 0.79 0.80 0.81 0.82 0.83 0.84 0.85 0.86 0.87
      0.88 0.89 0.90
[91]  0.91 0.92 0.93 0.94 0.95 0.96 0.97 0.98 0.99
> Zq <- qnorm(Zp,0,1)
> Zq
 [1]  -2.32634787 -2.05374891 -1.88079361 -1.75068607 -1.64485363
      -1.55477359
 [7]  -1.47579103 -1.40507156 -1.34075503 -1.28155157 -1.22652812
      -1.17498679

and so on ..

[97]  1.88079361 2.05374891 2.32634787
```

We can plot the probabilities vs. the quantiles

```
> plot(Zq,Zp,ylab="F(Z) N(0,1)", xlab="Z", ylim=c(0,1))
```

4.8.8 Quantile–Quantile (q–q) Plots

We can obtain q–q plots as shown in Figure 4.5 using the function qqnorm (top panel)

```
>qqnorm(x)
```

followed by the function qqline to draw the line (bottom panel).

```
>qqnorm(x);qqline(x)
```

A convenient function is qqPlot, which is part of package **car,** allows to obtain a Q–Q plot against the standard normal (among other options), and include confidence intervals. You can use it directly from the RConsole

```
>qqPlot(x, id.n=6)
[1] 38 10 74 53 52 57
```

You would load package car in case it happens to not be loaded. The identify option is invoked by using id.n; in this case id.n = 6 will identify the 6 points that depart the most from the mean (Figure 4.15, top panel). The two dashed lines around the straight line correspond to the confidence intervals. We can conclude that the data look normally distributed because it follows the straight line, except at the high and low ends where some observations depart and wander outside the confidence interval. As another illustration, we can generate a sample of 100 numbers from a normal with mean 30 and $\sigma = 1$ and produce a q–q plot similar to the one in Figure 4.15 (bottom panel)

```
>x2 <- rnorm(100,30,1)
>qqPlot(x2, id.n=6)
[1] 93 44 28 39 20 55
```

In this case, the data should be contained in the envelope generated by the confidence intervals.

We can also obtain Q–Q plots using the Rcmdr **Graphs|Quantile-comparison plot** and following it with a selection of V1, checking the normal, and using **identify observations with mouse**. In the graphics window, place cross hair at particular observations to identify outliers by number. Once you select all desired points, press **Stop** to finalize the identification (Figure 4.15, top panel).

4.8.9 FUNCTION TO PLOT EXPLORATORY DATA ANALYSIS (EDA) GRAPHS

We now practice how to put together functions by combining all the EDA graphs studied so far in the computer session. Let us create eda (x, label) function with arguments x an array of data values and label a string to label the graphs. Use **File|new script** from the Rconsole and type the following function in the script editor. All statements should be familiar to you by now except layout, which we will explain a bit later.

```
eda6 <- function(x,label="x"){
  # arguments x = array, label= string to label axis
  # divide graphics window in 6 panels
  mat <- matrix(1:6,3,2,byrow=T)
  layout(mat, widths=rep(7/2,2), heights=rep(7/3,3), T)
  par(mar=c(4,4,1,.5))
  # index plot
  plot(x, ylab=label, main="Index plot",cex.main=0.7)
  # box-whisker plot
  boxplot(x,ylab=label); title("Boxplot",,cex.main=0.7)
  # histogram
  hist(x, main="Histogram",xlab=label,cex.main=0.7)
  #density approx to histogram
  plot(density(x), main="Density approximation",xlab=label,cex.
    main=0.7)
  # quantile-quantile plot require package car
  qqPlot(x,col.lines=1,lwd=1,grid=F)
  title("QQ plot",cex.main=0.7)
  # empirical cumulative plot
  # standardize observations before plot
  z <- (x-mean(x))/sqrt(var(x))
  plot(ecdf(z), xlab=paste("Standardized ", label,sep=""),
          main= "ecdf vs Std. normal",cex.main=0.7)
  # generate standard normal and plot
  Z <- seq(-4,+4,0.1); lines(Z, pnorm(Z,0,1))
}
```

Save the script as **eda6.R** and use **File|Source R code** to source it. Alternatively, you can also source or load the function by applying

```
> source("lab4/eda.R").
```

Now, we apply the function by simply

```
eda6(x)
```

which should result on graphs like shown in Figure 4.6. In this example, we have not used argument label and therefore the function employed the default "x" as label for the graphs.

Now, what is layout? It allows dividing the graphics window in panels. It is similar to what we would obtain by using par(mfrow = c(3,2)) but here we customize it to use the space more efficiently by reducing the margins of each plot. We use mat to form a matrix of three rows of two panels each.

For later use, we can practice building functions to divide the panel in various arrangements, for example, two panels, four panels, and six panels

```
panel2 <- function (size,int="r"){
mat <- matrix(1:2,2,1,byrow=T)
layout(mat, widths=rep(size,2), heights=rep(size/2,2), TRUE)
par(mar=c(4,4,1,.5),xaxs=int, yaxs=int)
}

panel4 <- function (size,int="r"){
mat <- matrix(1:4,2,2,byrow=T)
layout(mat, widths=rep(size/2,2), heights=rep(size/2,2), TRUE)
par(mar=c(4,4,1,.5),xaxs=int, yaxs=int)
}

panel6 <- function (size,int="r"){
mat <- matrix(1:6,3,2,byrow=T)
layout(mat, widths=rep(size/2,2), heights=rep(size/3,3), TRUE)
par(mar=c(4,4,1,.5),xaxs=int, yaxs=int)
}
```

We can keep these functions in a file and source these functions to simplify the task of dividing graphics windows. For example, we would substitute the three lines in eda6 by simply panel6(size = 7).

4.8.10 TIME SERIES AND AUTOCORRELATION PLOTS

We can use functions ts.plot and acf. Arrange to plot in the same graphics page.

```
> panel2(size=5)
> ts.plot(x)
> acf(x)
```

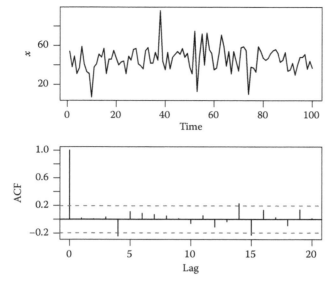

FIGURE 4.18　Exploratory analysis to check for serial correlation in the data.

We divide the window in two panels using the functions defined earlier and produce two plots as shown in Figure 4.18. The top panel has a time series plot (the values ordered as a function of time) generated using `ts.plot` and the second panel has the autocorrelation graph produced by function `acf`. The data do not show serial correlation because the spikes of the autocorrelation for nonzero lags are small. The results of this figure suggest that there is no major serial structure in the data and therefore it would be acceptable to apply parametric methods.

4.8.11　Additional Functions for the Rconsole and the R Commander

As explained in Chapter 2, one nice feature of the Rcmdr is that one can customize the menu and create a plug-in. For the purposes of this book, I have extended the Rcmdr to perform more functions. This plug-in is **RcmdrPlugin.Seeg,** which you can install, and load. Further instructions are provided in the package information.

　　RcmdrPlugin.Seeg includes modifications to the Rcmdr GUI Graph menu to facilitate building additional graphs. The following are additional items under **Graph** menu: empirical cumulative plots, time series plot, autocorrelation plot, exploratory analysis EDA plots. In the following, use the dataset `xd` already created in Rcmdr. Activate this dataset.

　　RcmdrPlugin.Seeg includes the **Graphs|Empirical Cumulative plot** menu. Select V1 and get the ecdf graph of Figure 4.4. This modified Rcmdr option invokes the **ecdf** function in R to get the empirical cumulative plot. We can look at the serial structure of the data using the modified Rcmdr **Graph|Time series plot** and then select V1 to obtain Figure 4.18 (top panel). This graph shows the values ordered as a function of time. The serial structure is not necessarily time but could be space, for example, distance. The graph is conceptually the same but we will study spatial structure later in the book. Now to examine autocorrelation using **Graph|Autocorrelation plot,** then select V1 and obtain Figure 4.18 (bottom panel), that is the correlation between values separated by a time lag.

　　Also included in the plug-in is the function eda6 that combines all the EDA graphs studied earlier. Use **Graphs|Exploratory Analysis EDA plots** item, then select V1 and type x for the variable name (this will be used as label for the graphs). We obtain graphs as given in Figure 4.6. We know how to interpret all of these plots already. This function allows us to see all of them at a glance.

4.8.12 PARAMETRIC: ONE SAMPLE *t*-TEST OR MEANS TEST

We will use a *t*-test to check if the value of the mean of 10 observations of soil porosity (in %) corresponds to a given value. This dataset is available as the first column of file **lab4/soil-porosity.txt**.

```
46  41
45  42
44  44
47  40
46  43
48  46
48  42
43  43
46  47
44  41
```

Using the Rconsole read data using `read.table`, which converts into a data frame and thus obtain the same result

```
> poro <- read.table("lab4/soil-porosity.txt")
Read 20 items
```

For the 10 samples of the first column, the null hypothesis is H0: the mean is equal to or less than 44%. However, before applying the test, use EDA to check validity of the *t*-test assumptions. Using the console by

```
eda6 (poro$V1, "porosity 1")
panel2(size=5)
ts.plot(poro$V1);acf(poro$V1)
```

Figures 4.19 and 4.20 show the resulting graphs. Then apply the `summary` by

```
> summary(poro$V1)
   Min.  1st Qu.  Median   Mean  3rd Qu.   Max.
  43.00    44.25   46.00  45.70    46.75  48.00>
```

We can see that the median is higher than mean but not by much. From the boxplot and density, the data look fairly symmetrical and normal, but not completely. There are no outliers. The q–q plot and ecdf suggest normality and the serial plots indicate that the data are uncorrelated. Therefore, we conclude that we can proceed with the *t*-test.

For `poro$V1` (the first column), the null hypothesis is H0: the mean is equal to or less than 44%. Apply `t.test` selecting alternative "`greater`" from the three options: `two.sided`, `greater`, `less`

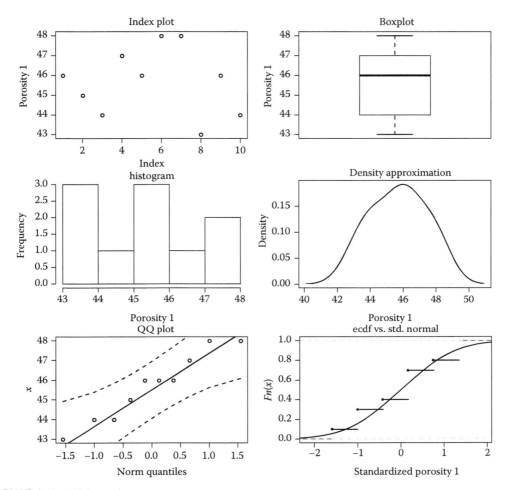

FIGURE 4.19 EDA graphs.

```
> t.test(poro$V1, alternative="greater", mu=44)

        One Sample t-test

data:  poro$V1
t = 3.1568, df = 9, p-value = 0.005805
alternative hypothesis: true mean is greater than 44
95 percent confidence interval:
 44.71284      Inf
sample estimates:
mean of x
  45.7

>
```

Thus, we have nine degrees of freedom ($df = 9$), because $n - 1 = 10 - 1 = 9$, we subtract one because one parameter (mean) was estimated. The p-value 0.0058 indicates that the t-value 3.156 exceeds the t critical value at an $\alpha = 0.01$, and we can **reject** the null hypothesis at an $\alpha = 0.01$. See a similar example with rock porosity in pages 70 and 71 of Davis (2002).

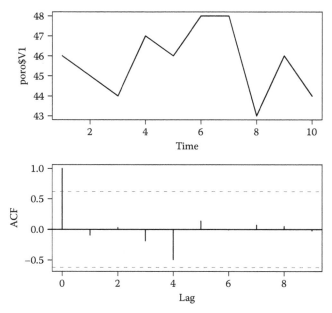

FIGURE 4.20 EDA serial check graphs.

We can do all these by using the Rcmdr. First import using Rcmdr **Data|Import data|from text file** with name `poro`, unmark variable name in file and click ok, browse to **lab4/soil-porosity.txt**. Then use the view dataset to examine it (Figure 4.21). Then obtain EDA plots using **Graphs|Exploratory Analysis EDA plots, Graphs|Time series plot, and Graphs|Autocorrelation plot** items. Select V1 for all and obtain the same graphs already discussed (Figures 4.19 and 4.20). For example, use EDA as shown in Figure 4.21. Then to run the *t*-test use the Rcmdr **Statistics|Means|One sample t-test** (Figure 4.22, top). At the dialog window select V1 and enter the null mu = 44 and mark the alternative population mean > mu0 value (Figure 4.22, bottom). In the output window, we get the same result shown earlier.

4.8.13 Power Analysis: One Sample *t*-Test

Now let us calculate power of this *t*-test. We use function `power.t.test` in the following manner

```
>power.t.test(n=10, delta = 1, sd=sd(poro$V1), type="one.sample",
  alternative = "one.sided", sig.level=0.01)
```

where we have n = 10 observations, and want to detect an effect size of `delta = 1`, given the standard deviation, only one sample, alternative is `one-sided`, and $\alpha = 0.01$. The result is

```
        One-sample t test power calculation

              n = 10
          delta = 1
             sd = 1.702939
      sig.level = 0.01
          power = 0.229953
    alternative = one.sided
```

We see that the power is low. Note from the previous section that the sample mean is 45.7.

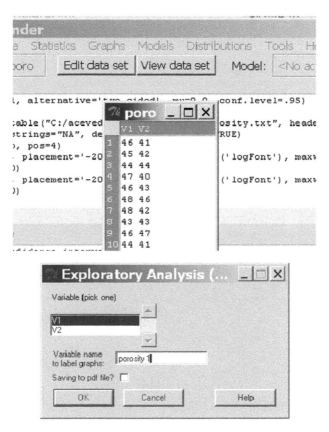

FIGURE 4.21 Porosity data as data frame imported and viewed with Rcmdr. Selecting the first column and assigning label.

If we relax the effect size to the difference 45.7 − 44, i.e., a delta ~2, the power goes up to 0.79. We conclude that for this sample the *t*-test would have relatively low power given α = 0.01.

We can now determine how many more observations would be required to raise the power to 0.9 assuming the same variability, same α, and same effect size of delta = 2.

```
>power.t.test(power=0.9, delta = 2, sd=sd(poro$V1), type="one.sample",
        alternative = "one.sided", sig.level=0.01)
```

We obtain

```
One-sample t test power calculation

          n = 12.30839
      delta = 2
         sd = 1.702939
  sig.level = 0.01
      power = 0.9
alternative = one.sided
```

Indicating that we require *n* = 13 observations to achieve this power level given these conditions.

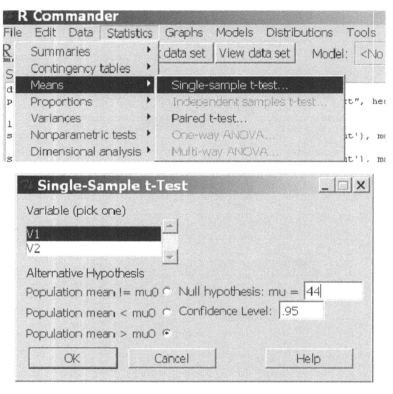

FIGURE 4.22   Rcmdr one sample *t*-test and setting one sample test.

4.8.14   PARAMETRIC: TWO-SAMPLE *t*-TEST

For the **porosity.txt** data: Is the mean porosity of column 1 the same as mean porosity of column 2? We already applied EDA on first column. Let us repeat for second column. Run from the Rcmdr or the console

```
>eda6 (poro$V2,"porosity 1")
>panel2(size=5)
>ts.plot(poro$V2);acf(poro$V2)
```

The resulting graphs indicate that the second sample is less symmetric, but it does not have outliers or serial correlation (graphs not shown here for the sake of space). Thus, we decide to proceed with parametric testing.

First, as an exploratory tool we can check if both boxplots show same spread and different means using a side-by-side graph of boxplots to compare the samples

```
> boxplot(poro, ylab="Porosity")
```

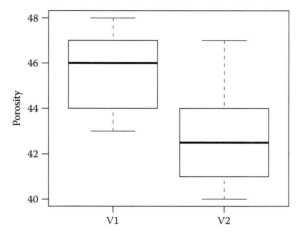

FIGURE 4.23 Boxplots side by side.

resulting in the graphs of Figure 4.23. These boxplots show that the first sample has a larger mean, and that both samples may have dissimilar variability, since the second sample has a larger maximum. We calculate both standard deviations using

```
> sapply(poro,sd)
     V1       V2
1.702939 2.233582
>
```

With the *t*-test we want to answer the question: Are the means significantly different? Before applying a two-sample *t*-test we have to check whether the sample variances are not significantly different applying the *F*-test. The null hypothesis is that the ratio *F* of variances is equal to 1 (equal variances).

```
> var.test(poro$V1,poro$V2)

        F test to compare two variances

data:  poro$V1 and poro$V2
F = 0.5813, num df = 9, denom df = 9, p-value = 0.4314
alternative hypothesis: true ratio of variances is not equal to 1
95 percent confidence interval:
 0.1443847 2.3402772
sample estimates:
ratio of variances
       0.5812918

>
```

The *F* value is 0.58, suggesting dissimilar variances as expected. However, the *p*-value is large 0.43; therefore, if we reject the null hypothesis, we still have a high probability of being wrong. Unfortunately, this is a very inconclusive result. Let us decide not to reject the null, and therefore assume similar variances.

Now run t.test on the two samples

```
> t.test(poro$V1,poro$V2)

      Welch Two Sample t-test

data: poro$V1 and poro$V2
t = 3.1525, df = 16.821, p-value = 0.005871
alternative hypothesis: true difference in means is not equal to 0
95 percent confidence interval:
 0.9245509 4.6754491
sample estimates:
mean of x mean of y
      45.7      42.9
```

The p-value is 0.58%, which indicates that we can reject the null hypothesis of equal means with $\alpha = 1\%$. Therefore, we conclude that the two samples have different means (with unknown error beta).

4.8.15   POWER ANALYSIS: TWO SAMPLE t-TEST

Now let us calculate power of this t-test. Note that the sample means 45.7 and 42.9 seem to be different by 3 units. We will use this as effect size. As we know, the sample standard deviations are 1.7 and 2.2; we will use a trade-off value of 2 for the power calculation. We use function power.t.test in the following manner

```
>power.t.test(n=10, delta = 3, sd=2, type="two.sample", alternative
  = "one.sided", sig.level=0.01)
```

Note we used type for two samples. The result is

```
      Two-sample t test power calculation

              n = 10
          delta = 3
             sd = 2
      sig.level = 0.01
          power = 0.7798381
    alternative = one.sided

 NOTE: n is number in *each* group
```

We see that the power is moderately high, almost 80%.

We can now determine how many more observations would be required to raise the power to 0.9 assuming the same variability, same α, and same effect size of delta = 3.

```
>power.t.test(power=0.9, delta = 3, sd=2, type="two.sample",
  alternative = "one.sided", sig.level=0.01)
```

We obtain

```
        Two-sample t test power calculation

              n = 13.02299
          delta = 3
             sd = 2
      sig.level = 0.01
          power = 0.9
    alternative = one.sided

  NOTE: n is number in *each* group
```

Indicating that we require $n = 13$ observations to achieve this power level given these conditions.

4.8.16 USING DATA SETS FROM PACKAGES

One advantage of the R system is a substantial number of datasets associated with many packages. We can obtain a list of datasets in a package applying function data to the package name. For example, for package **car**

```
>data(package ="car")
```

which will generate a list of datasets in package **car** in a pop-up text window. To obtain a list for all packages installed

```
>data(package = .packages(all.available = TRUE))
```

which will generate a list by package in a pop-up text window. There are arguments to data for further control of searching for the packages.

To use a dataset simply apply function data and the name of the dataset. For example, there is a dataset named CO2 in package **datasets**. We can use it by

```
> data(CO2)
> CO2
   Plant       Type    Treatment conc uptake
1   Qn1      Quebec  nonchilled   95   16.0
2   Qn1      Quebec  nonchilled  175   30.4
3   Qn1      Quebec  nonchilled  250   34.8
4   Qn1      Quebec  nonchilled  350   37.2
5   Qn1      Quebec  nonchilled  500   35.3
6   Qn1      Quebec  nonchilled  675   39.2

etc. up to 84 records
```

To do the same with Rcmdr, go to **Data|Data in packages|List datasets in packages** to obtain lists and **Data|Data in packages|Read data from** package to select a dataset and load it (Figure 4.24).

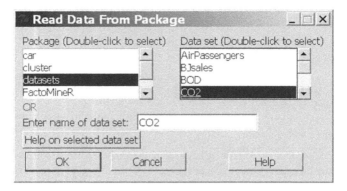

FIGURE 4.24   Selecting dataset from package.

4.8.17   NONPARAMETRIC: WILCOXON TEST

Consider the data from a very famous physics experiment by Albert Michelson to measure the speed of light with an interferometer. The dataset `michelson` is part of package **MASS**. The `Speed` values are departures from a previously reported speed of light of 299,990 km/s. There are a certain number of measurements (`Run`) for each experiment number (`Expt`)

```
> data(package="MASS")
> data(michelson)
> michelson
   Speed  Run  Expt
1   850    1    1
2   740    2    1
3   900    3    1
```

Let us explore part of the dataset; say `Expt 1` runs `1–20`.

```
> mich <- michelson$Speed[1:20]
```

Verify the recently created object. Use either the Rconsole or the view dataset in Rcmdr.

```
> mich
 [1]  850  740  900 1070  930  850  950  980  980  880 1000  980  930  650  760
[16]  810 1000 1000  960  960
>
```

You can see that `mich` has 20 entries. Now apply the EDA using the console

```
>eda6(mich, "Speed of Light")
```

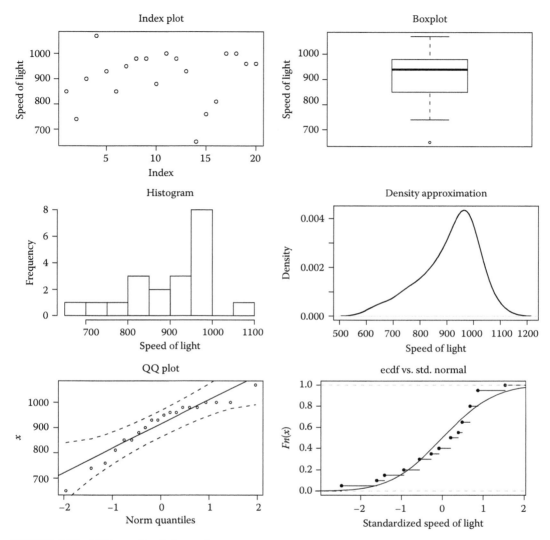

FIGURE 4.25 EDA results of observations.

and obtain results in Figure 4.25. From the median line within box in boxplot, we can see that 650 is an outlier. From the histogram and density, we see that the data are not symmetrical; from the qq plot and the ecdf we see patterns in the departure from the theoretical normal distribution. The mich data do not show much serial correlation (plots not shown for the sake of space). We conclude that mich does not seem to be normally distributed, and that it has an outlier.

Therefore, we should not apply the parametric test to these data but a nonparametric one. First, determine the ranks of data mich

```
> rank(mich)
 [1]    5.5    2.0    8.0   20.0    9.5    5.5   11.0   15.0   15.0    7.0   18.0   15.0
        9.5    1.0    3.0
[16]    4.0   18.0   18.0   12.5   12.5
```

We can see how ties are averaged. For example, the two 850 values were given rank 5.5 instead of 5 and 6. Let us look at the ranks using eda6.

```
> eda6(rank(mich),"Rank Speed of Light")
```

We can see that the ranks look more normal than the observations. The mean rank is about 10. Run the one-sample Wilcoxon Signed Rank Test to see whether the mean is equal to 990. For intuition look at the rank that the value 990 would have

```
> rank(c(mich,990))
 [1]   5.5   2.0   8.0  21.0   9.5   5.5  11.0  15.0  15.0   7.0  19.0  15.0
       9.5   1.0   3.0
[16]   4.0  19.0  19.0  12.5  12.5  17.0
>
```

990 would be rank 17, which is higher than the mean. Thus, we expect the test to tell us that the sample mean is different from 990. Now for the test, we obtain

```
> wilcox.test(mich, mu=990)
Warning in wilcox.test.default(mich, mu = 990) :
  cannot compute exact p-value with ties

        Wilcoxon signed rank test with continuity correction

data: mich
V = 22.5, p-value = 0.00213
alternative hypothesis: true location is not equal to 990

>
```

There are tied values. The low p-value of 0.21% allows to reject H0 with low probability of error ($\alpha = 0.5\%$). Therefore, we conclude that the mean is not equal to 990.

Using the Rcmdr we can compute ranks with **Data|Manage variables in dataset|compute new variable**. Use function rank in the text box to include expression to compute.

4.8.18   BIVARIATE DATA AND CORRELATION TEST

To practice correlation concepts we will generate arbitrary data using random number generation and then evaluate test procedures. Generate two independent samples x and y.u of size $n = 30$ using rnorm. These samples should be uncorrelated. For illustration then generate a third sample y.r dependent linearly on x and with some random normal variability. That is to say use script

```
# base sample
x <- rnorm(30,10,1)
# uncorrelated sample
y.u <- rnorm(30,20,2)
# correlated sample
y.r <- 2*x + rnorm(30,0,0.4)
```

Then produce scatter plots of both pairs of samples, y.u vs.x and y.r vs. x. In addition, we draw dotted lines at the mean of each sample

```
>panel2(size=5)
>plot(x,y.u);abline(v=mean(x),h=mean(y.u),lty=2)
>plot(x,y.r);abline(v=mean(x),h=mean(y.r),lty=2)
```

See Figure 4.8 already discussed.

```
>eda6(x,"x")
>eda6(y.u,"y.u")
>eda6(y.r,"y.r")
```

The resulting graphs for x and y.r suggest normality and no outliers; not reproduced here for the sake of space. However, EDA plots for y.u suggest outliers and lack of normality (Figures 4.26 and 4.27).

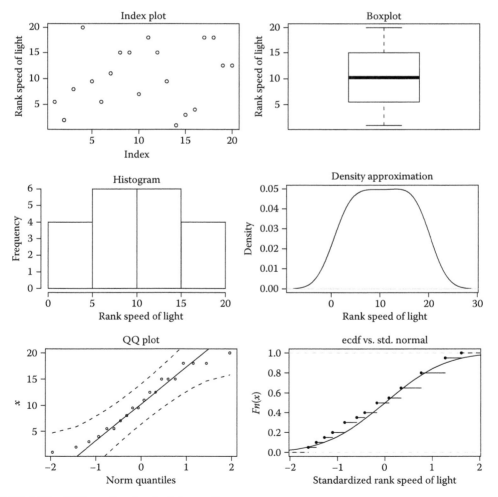

FIGURE 4.26 EDA results of ranked observations.

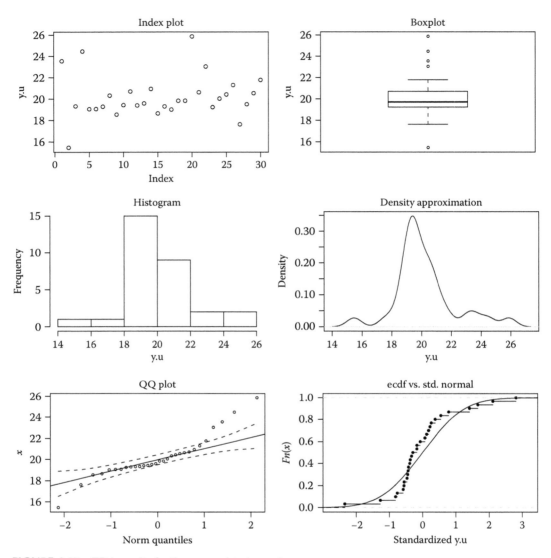

FIGURE 4.27   EDA results for the uncorrelated sample.

We decide then to apply Pearson's test to x and y.r, and Spearman's test for correlation between x and y.u.

```
> cor.test(x,y.u, method="s")
> cor.test(x,y.r, method="p")
```

Note: many times you can abbreviate arguments to first few characters; here for example "s" is Spearman's, "p" is Pearson's.

We obtain

```
> cor.test(x,y.u, method="s")

        Spearman's rank correlation rho

data:   x and y.u
S = 3510, p-value = 0.2436
alternative hypothesis: true rho is not equal to 0
sample estimates:
     rho
 0.2191324
```

With p-value 0.24 we cannot reject H0 at a reasonable α. Therefore, we cannot assure that variables are correlated.

```
> cor.test(x,y.r, method="p")

        Pearson's product-moment correlation

data: x and y.r
t = 25.6957, df = 28, p-value < 2.2e-16
alternative hypothesis: true correlation is not equal to 0
95 percent confidence interval:
 0.9567998 0.9902815
sample estimates:
    cor
 0.9794479
```

With this very low p-value 2.2×10^{-6} we can establish a low alpha error when saying that variables are correlated.

Alternatively, we can produce the scatter plot using Rcmdr. First, build a dataset containing x, y.u, and y.r using data.frame

```
X.Y <- data.frame(x,y.u,y.r)
```

Now let us go to menu **Graphs|Scatterplot**. Select variables x, y.u for x and y axis, click identify and marginal boxplots, unclick the rest. The marginal boxplots are very useful giving you an idea of the spread along each axis and the location of the mean and median of each. The result is not shown here for the sake of space.

4.8.19 COMPUTER EXERCISES

Exercise 4.5
Listed next are grades on a paper for a class of 14 students. The list is in the same sequence in which they were graded.

```
90 80 100 90 77 96 88 83 81 93 97 95 72 85
```

Calculate descriptive statistics; check these grades for normality, outliers, and serial correlation. Apply functions `eda6`, `ts.plot`, and `acf`. Note that serial correlation could exist if the grader gets either more generous or tougher during the grading process.

Exercise 4.6
Use the data in file **lab4/example2.txt**. Apply EDA methods to this dataset. Is this sample normally distributed? Are there outliers? Is there serial correlation? Use Z test to check that the population mean and standard deviation are 30 and 1.

Exercise 4.7

1. Generate 100 random numbers from a normal distribution with $\mu = 300$, $\sigma = 30$, use EDA and discuss whether the sample does indeed look normal. Calculate sample mean, variance, and standard deviation.
2. Generate another sample of the same size ($n = 100$) and standard deviation ($\sigma = 30$) but with different mean $\mu = 350$. Repeat the EDA, calculations, and discussion.
3. Select an appropriate test to see whether there is indeed a difference in the sample means of the two samples. Justify your selection. Run the test, report your results, and discuss. Calculate power and number of observations required to increase power to 0.9 if power is low.

Exercise 4.8
Use data on number of Atlantic hurricanes per season. File **lab4/hnumber.txt**. Question: Is the mean number of hurricanes per season greater than 3? Apply most appropriate test (parametric or nonparametric) to determine this. Justify your reasoning and provide conclusive answer to the question. Calculate power and number of observations required to increase power to 0.9 if power is low. Hint: Import data from **hnumber.txt** into a dataset. Use `read.table` from the R console or Rcmdr **read as table from text file**.

Exercise 4.9
Study the following question: Are there fewer Atlantic hurricanes during El Niño year as compared to La Niña? Use the file **lab4/enso-yrs.txt** provided to generate two sets of values: one for El Niño and another for La Niña. Determine the most appropriate test and provide answer to the question. Explain null and alternative hypotheses. Calculate power and number of observations required to increase power to 0.9 if power is low.

Exercise 4.10
Ground-level ozone is an important urban air pollution problem related to emissions and photochemistry. Use dataset `airquality` of package **datasets**. Is ground-level ozone in this sample correlated to solar radiation? Determine appropriate tests to run. Provide an answer to the question.

SUPPLEMENTARY READING

I have already suggested some further reading in some of the preceding sections. Supplemental reading to most of the material of this chapter can be found in Davis, 2002, Chapter 2, pp. 40–112; Rogerson, 2001, Chapter 3, pp. 42–64, Chapter 5, pp. 86–103; Carr, 1995, Chapter 4, pp. 70–85 and Chapter 2, pp. 17–26; MathSoft, 1999, Chapter 3, pp. 3-1–3-24 and 3-30–3-39. Qian, 2010, Chapter 4, pp. 49–87.

5 More on Inferential Statistics
Goodness of Fit, Contingency Analysis, and Analysis of Variance

5.1 GOODNESS OF FIT (GOF)

This important class of tests helps to determine whether a sample is drawn from a hypothetical distribution or whether two samples come from the same distribution (Davis, 2002, pp. 92–96). As in other chapters, we will use cdf to refer to the cumulative distribution function $F(x)$. **Goodness of fit (GOF)** tests can be applied to one sample or to two samples depending on the question that we seek to answer. We will cover the χ^2 or chi-square test, the Shapiro–Wilk test, and the Kolmogorov–Smirnov or **K–S test**.

5.1.1 QUALITATIVE: EXPLORATORY ANALYSIS

Before one applies a GOF test, it is a good practice to look at the data together with the theoretical or together with the second sample. A convenient plot is the ecdf plot and the hypothesized cdf. We already saw how this works with the normal distribution in the previous chapter. The concept is more general and we can apply it to other distributions such as the exponential, uniform, and others. We can add a plot of the differences between the ecdf of each observation and the corresponding theoretical cdf to visualize the magnitude of the difference and any patterns with reference to the sampled values.

Figure 5.1 illustrates an example. The sample seems to have a good fit to a standard normal distribution; the maximum absolute difference is about 0.13 and the largest differences correspond to the largest values of x. Figure 5.2 shows another example. In this case, we see a good fit to a uniform distribution $U[0, 1]$; the maximum absolute difference is about 0.17.

5.1.2 χ^2 (CHI-SQUARE) TEST

The Pearson's χ^2 (chi-square) statistic is the squared difference between observed counts (in bins or intervals) and theoretical counts (in the same bins) from the hypothesized distribution (Davis, 2002, pp. 92–96). Denote n = sample size, k = number of bins or classes, then the chi-square statistic is calculated as

$$\chi^2 = \sum_{j=1}^{k} \frac{(c_j - e_j)^2}{e_j} \tag{5.1}$$

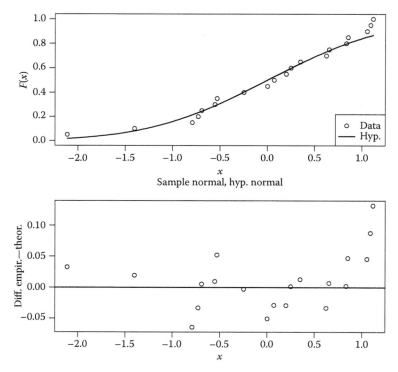

FIGURE 5.1   Example of visualizing the ecdf of a sample together with the cdf of a normal hypothetical distribution.

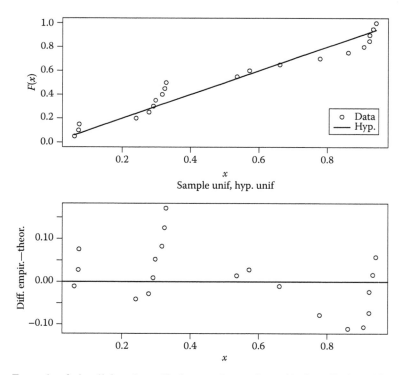

FIGURE 5.2   Example of visualizing the ecdf of a sample together with the cdf of a uniform hypothetical distribution.

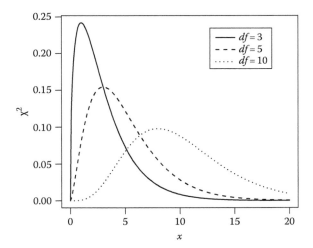

FIGURE 5.3 χ^2 pdf for $df = 3$, 5, and 10.

the sum is over $j = 1, ..., k$ where k is the number of classes or bins, c_j is number of observations counted in the jth bin, e_j is number of observations or counts expected in bin j. See Figure 5.3 which illustrates the shape of the χ^2 density for various degrees of freedom.

The test consists of checking if the test statistic is too large (for a given level of significance) because this indicates substantial departure from the hypothesized distribution. The null hypothesis H0 is: the sample is drawn from a given distribution. When you reject the null, you conclude that the sample is not drawn from the hypothesized distribution (with given alpha error). When you fail to reject, the sample may come from the hypothesized distribution.

The degrees of freedom (df) in a χ^2 test correspond to the **number of bins minus one ($k - 1$)**, and not to the number of observations minus one ($n - 1$). Increasing the number of bins may increase the df, but decrease the number of points per bin, which must be kept above five counts per bin. If additional parameters are estimated, then df decreases; for example when testing for a fit to the standard normal $N(0, 1)$, $df = k - 3$, both mean and variance are estimated from the sample to generate standard values.

Example: Suppose that we want to check if 100 values come from a uniform distribution using a GOF test. We got the following counts in $k = 5$ categories (bins or intervals) 22, 13, 18, 27, and 20. The expected value in each bin is $100/5 = 20$. Here $df = 5 - 1 = 4$. Now we calculate the statistic

$$\chi^2 = \sum_{j=1}^{k} \frac{(c_j - e_j)^2}{e_j} =$$

$$= \frac{(22 - 20)^2 + (13 - 20)^2 + (18 - 20)^2 + (27 - 20)^2 + (20 - 20)^2}{20}$$

$$= \frac{4 + 49 + 4 + 49 + 0}{20} = \frac{106}{20} = 5.30$$

The probability of getting this value or higher is the tail of the χ^2 for a $df = 4$, which is calculated as $1 - F(5.30) = 0.258$, where F is the cdf of the χ^2 for $df = 4$. Therefore, the p-value is 0.258 and we cannot reject the null with a reasonable α that the sample is drawn from a uniform distribution.

5.1.3   Kolmogorov–Smirnov (K–S)

The K–S test is a nonparametric GOF test that works with the cumulative distributions (Davis, 2002, pp. 107–112). For the one sample test: $F_1(x)$ is the ecdf of the sample $F_2(x)$ is the hypothetical cdf. For the two sample test: $F_2(x)$ is the ecdf of sample 2. The test statistic obtained as the **largest departure** between the ecdf and the cdf (sample and the hypothesized distribution), or between the two ecdf (two samples). For illustration, refer to the bottom panel of Figures 5.1 and 5.2. The largest departure in the first is 0.13 and in the second is 0.17. These would be the values of the test statistic for each case.

The test can be two-sided or one-sided. In a two-sided test, the null H0: $F_1(x) = F_2(x)$, the sample is equal to hypothetical for all x (most common). The statistic is

$$T = \max_x \left| F_1(x) - F_2(x) \right| \tag{5.2}$$

where bars represent absolute value and max is the maximum.

Coming back to the examples in Figures 5.1 and 5.2, the largest departure in the first is 0.13 and in the second is 0.17. These would be the values of the test statistic T for each case. The corresponding p-values (easily calculated with R as we will see in the computer session) are 0.83 and 0.54. We decide against rejecting the null H0 in both cases, therefore concluding that the samples may have come from the normal and uniform distributions, respectively.

In a one-sided test, we can have two cases: greater and less. The "greater" case (the name "greater" refers to the alternative) H0: $F_1(x) \leq F_2(x)$ for all x, for example, the sample values are equal to or less than the hypothetical values, the statistic is the maximum positive difference

$$T = \max_x (F_1(x) - F_2(x)) \tag{5.3}$$

We do not need the absolute value because $F_1(x)$ is greater than $F_2(x)$.

Conversely, the "less" case H0: $F_1(x) > F_2(x)$ for all x, the sample is greater than the hypothetical, the statistic is the maximum negative difference

$$T = \max_x (F_2(x) - F_1(x)) \tag{5.4}$$

Note that we inverted the order in the subtraction. Rejecting H0 (at an alpha error level) means that the samples are not drawn from the hypothetical distribution, or in other words, that the samples do not come from the same distribution. When we fail to reject it means that there is no evidence indicating that the data could not come from a given distribution.

5.1.4   Shapiro–Wilk Test

The null is that the sample comes from a normal distribution. The test statistic is W calculated by

$$W = \frac{\left(\sum_{i=1}^{n} a_i x_{(i)} \right)^2}{\sum_{j=1}^{n} (x_i - \overline{X})^2} \tag{5.5}$$

where $x_{(i)}$ (with parentheses enclosing the subscript index i) is the ith order statistic. The constants a_i are calculated from the means, variances, and covariances of the ith order statistics. The best way to present this material is using linear algebra and using the concept of covariance matrix. We will not cover these subjects until Chapter 9. For now, we will use the test using its R function implementation as you will see in the computer session.

For the example shown in Figure 5.1, the test yields $W = 0.94$ with a p-value of 0.25. We decide against rejecting the null H0, therefore concluding that the sample may have come from the normal distribution.

5.2   COUNTS AND PROPORTIONS

Recall from Chapter 3 the definition of the event "success" with a given probability of occurrence denoted here as p_s to avoid confusion with the p-value. An estimate of p_s is the ratio of number of successes to number of trials. For one sample, we formulate the following question: Is P[success] different from a presumed value? For two samples, is the P[success] different between the samples? We should also recall from Chapter 3 that the probability of obtaining r successes in n trials follows the binomial distribution

$$P(X = r) = \binom{n}{r} p_s^r (1 - p_s)^{n-r} \tag{5.6}$$

The mean is $\mu_X = np_s$ and the variance is $\sigma_X^2 = np_s(1 - p_s)$. The binomial test uses the binomial distribution as a test statistic to determine the hypothetical value of the probability of obtaining the observed ratio r/n. For one sample, the null H0 is that $p_s = p_0$, where p_0 is the hypothetical probability of success, and two-sided alternative $p \neq p_0$. A p-value is the sum of cumulative probabilities of obtaining less than r successes and more than r successes. These calculations are easily conducted in R as we will see in the computer session.

For example, suppose the hypothetical p_0 is 0.5 and that we observe a ratio of $r = 3$ successes in $n = 10$ trials. The p-value is 0.34 and we decide against rejection of the null and allowing that p_s may be 0.5. However, if we were to observe a ratio of $r = 1$ in $n = 10$ trials, the p-value is 0.02 and we can reject H0 at $\alpha = 5\%$.

5.3   CONTINGENCY TABLES AND CROSS-TABULATION

Another application of the χ^2 is the analysis of **contingency** tables. For bivariate cases, we count the number of occurrences of observations falling in joint **categories** and thus we develop a contingency table that can help answer the question: Is there association between the two variables? This is an **independence test**. Do not confuse it with a χ^2 GOF test, although the calculations are similar.

Contingency analysis applies to categorical data. Use factors with several levels and count the number of cases in each level. Then calculate frequency of two variables occurring in joint or combined levels. To build a contingency table, we use all the levels of one factor (categorical variable) as **rows** and the levels of the other variable as **columns**. We then count occurrences in each cell and find the frequency of joint levels or **cell frequency** for each cell in the table. This is a **cross-tabulation**.

Now to perform a test, first we sum the cell frequencies across both rows and columns and place the sums in the margins, which are the **marginal** frequencies. The lower-right-hand corner value contains the sum of either the row or the column marginal frequencies; both must be equal to the total number of observations n.

The test assesses whether or not one factor has effects on the other. Effects are relationships between the row and column factors; for example, are the levels of the row variable distributed at different levels according to the levels of the column variables? Rejection of the null hypothesis (independence of factors) means that there may be effects between the factors. Nonrejection means that chance could explain any differences in cell frequencies and thus factors may be independent of each other.

For example, there are three possible land cover types in a map (crop field, grassland, and forest). Using a topographic map, we classify the area in two elevation classes: lowlands and uplands. We divide the total area so that we have 80 squares of equal size and classify each square in one of the cover types and terrain classes. Then we count the number of squares in each joint class. The resulting cross-tabulation is Table 5.1.

The first step in computing the χ^2 statistic is the calculation of the row totals and column totals of the contingency table as shown in Table 5.2. The next step is the calculation of the expected cell count for each cell. This is accomplished by multiplying the marginal frequencies for the row and column (row and column totals) of the desired cell and then dividing by the total number of observations

$$e_{ij} = \frac{(Row_i Total \times Column_j Total)}{n} \tag{5.7}$$

TABLE 5.1
Cross-Tabulation of Topographic Class and Vegetation Cover

	Crop Field	Forest	Grassland
Lowlands	12	20	8
Uplands	8	10	22

TABLE 5.2
Row and Column Totals

	Crop Field	Forest	Grassland	Row Total	Total of Rows
Lowlands	12	20	8	40	
Uplands	8	10	22	40	
Column total	20	30	30		80
Total of columns				80	

TABLE 5.3
Expected Frequencies

	Crop Field	Forest	Grassland	Row Total	Total of Rows
Lowlands	12	20	8	40	
e_{1j}	10	15	15	40	
Uplands	8	10	22	40	
e_{2j}	10	15	15	40	
Column total	20	30	30		80
	20	30	30		
Total of columns				80	

TABLE 5.4
Deviation from Expected

	Crop Field	Forest	Grassland	Row Total	Total of Rows
Lowlands	12	20	8	40	
e_{1j}	10	15	15	40	
$o_{1j} - e_{1j}$	+2	+5	−7	0	
Uplands	8	10	22	40	
e_{2j}	10	15	15	40	
$o_{2j} - e_{2j}$	−2	−5	+7	0	
Column total	20	30	30		80
	20	30	30		
	0	0	0		
Total of columns				80	

For example, computation of the expected cell frequency for cell in row 1 and column 1 is $e_{11} = (20 \times 40)/80 = 10$. Use this expression to calculate all the expected cell frequencies and including these as additional rows in each cell (Table 5.3). Then we subtract the expected cell frequency from the observed cell frequency for each cell, to obtain the deviation or error for each cell. Including these in Table 5.3, we obtain the next table shown in Table 5.4.

Note that the sum of the row total for expected is the same as the sum of the observed row totals; the same is true for the column totals. Note also that the sum of the Observed − Expected for both the rows and columns equals zero. Now, to get the final table we square the difference and divide by the expected cell frequency, resulting in the chi-square values (Table 5.5).

The χ^2 statistic is computed by summing the last row of each cell; in Table 5.5 this would result in $\chi^2 = 0.4 + 0.4 + 1.66 + 1.66 + 3.26 + 3.26 = 10.64$. The number of degrees of freedom is obtained by multiplying the number of rows minus one, times the number of columns minus one; this is to say $df = (\text{Rows} - 1) \times (\text{Columns} - 1)$. In this case, $df = (2 - 1) \times (3 - 1) = 2$. We would then determine the p-value for 10.64 and compare to the desired significance level. If lower, the rows and columns of the contingency may be dependent. In this particular case, $1 - F(10.64) = 0.00489$ for $df = 2$. Therefore, the p-value ≈ 0.005 suggests to reject H0 and that vegetation cover is not distributed evenly across the different elevations.

TABLE 5.5

Final Table with Chi-Square Values

	Crop Field	Forest	Grassland	Row Total	Total of Rows
Lowlands	12	20	8	40	
e_{1j}	10	15	15	40	
$o_{1j} - e_{1j}$	+2	+5	−7	0	
$(o_{1j} - e_{1j})^2/e_{1j}$	4/10 = 0.4	25/15 = 1.66	49/15 = 3.26	5.32	
Uplands	8	10	22	40	
e_{2j}	10	15	15	40	
$o_{2j} - e_{2j}$	−2	−5	+7	0	
$(o_{2j} - e_{2j})^2/e_{2j}$	4/10 = 0.4	25/15 = 1.66	49/15 = 3.26	5.32	
Column total	20	30	30		80
	20	30	30		
	0	0	0		
	0.8	3.32	6.52	**10.64**	
Total of columns				80	

One special case of this method is to compare proportions in several groups or samples using the χ^2 distribution. In this case, c_j are the counts observed and e_j are the proportions expected in each sample or group.

For example, suppose that we wanted to know if the proportions of stakeholders willing to pay for ecosystems services in three different sites are the same. We asked 80, 90, 85 residents of each area and obtained the following positive responses 10, 12, 13. Are these proportions 0.125, 0.133, 0.153 from the same population? First, we calculate the counts of negative responses to be 70, 78, 72 in order to complete a 2 row, 3 column contingency table. Then, expected counts e_j are obtained using Equations 5.7, 10.9, 12.4, and 11.7 for positive responses and 69, 77.6, 73.3 for negative responses. Applying Equation 5.1 we obtain $\chi^2 = 0.289$, $df = (3 - 1) \times (2 - 1) = 2$, and p-value = 0.86, therefore, the proportions are not significantly different.

5.4   ANALYSIS OF VARIANCE

Analysis of variance (AOV or ANOVA) methods include a large collection of techniques and applications (Davis, 2002, pp 78–92; MathSoft, 1999, Chap. 11, Chap. 12; Manly, 2009). In this book, we emphasize three major applications: (1) comparing sample means of **response** variables to various levels of independent **factors** or groups, as we will cover in this chapter; (2) **predictor** of the expected value of each sample from the sample mean; and (3) inferential tool when performing regression analysis. We cover (2) and (3) in the next chapter along with regression. The situation described in this chapter is applied often to **experiment design** and analysis.

In the context of analyzing the response of one dependent variable to various levels of independent factors or groups, a typical application is to **compare the sample means of more than two samples**. The simplest case is that for **one factor** with m levels and one sample of size n per level. In this analysis, we compare the sample means using ratio of variances, hence the name "analysis of variance." Situations that are more complicated are **two factors**, three factors, and replicated measurements. We will cover some of these in this chapter. Then, when we have more than one response, we will have analysis of multiple variables, which we will cover later in the book together with other multivariate methods.

5.4.1   ANOVA ONE-WAY

The simplest case is **one-way** or one factor with m levels, one sample of size n per level. A response variable X is measured n times for m levels of a factor or m groups. Consider the sample mean and variance of the response X for each one of the groups or levels

$$\bar{X}_1, ..., \bar{X}_m$$
$$s_1^2, ..., s_m^2 \tag{5.8}$$

Note that each sample mean and sample standard deviation are calculated over n observations x_{ij} where $j = 1, ..., n$

$$\bar{X}_i = \frac{\sum_{j=1}^{n} x_{ji}}{n} \quad s_i^2 = \frac{\sum_{j=1}^{n} (x_{ij} - \bar{X}_i)^2}{(n-1)} \tag{5.9}$$

These variances are **within** sample variances because they are calculated using only the values in the sample. The overall sample mean can be estimated from the average of the sample means

$$\bar{\bar{X}} = \frac{\sum_{i=1}^{m} \bar{X}_i}{m}$$

and the overall variance can be estimated from the average of sample variances, that is to say from the **within** variances

$$\overline{s_p^2} = \frac{\sum_{i=1}^{m} s_i^2}{m} \tag{5.10}$$

Now note that using Equation 5.9 to expand the variance s_i^2 of each sample in Equation 5.10 we get

$$\overline{s_p^2} = \frac{\sum_{i=1}^{m} \left(\frac{\sum_{j=1}^{n} (x_{ij} - \bar{X}_i)^2}{(n-1)} \right)}{m} = \frac{\sum_{i=1}^{m} \sum_{j=1}^{n} (x_{ij} - \bar{X}_i)^2}{m \times (n-1)} = \frac{\sum_{i=1}^{m} \sum_{j=1}^{n} (x_{ij} - \bar{X}_i)^2}{N - m} \tag{5.11}$$

where $N = m \times n$ is the total number of observations. The expression on the right-hand side is the sum of square of the **within** differences or errors (SSw) divided by the number degrees of freedom (dfw) $N - m$. We refer to the right hand side of Equation 5.11 as the mean square of the errors (MSw). Therefore, the overall variance is the MSw.

$$\overline{s_p^2} = MSw = \frac{SSw}{dfw} \tag{5.12}$$

Recall that using the central limit theorem the mean of the sample means has variance

$$s_{\bar{X}}^2 = \frac{s_X^2}{n}$$

And note that the variance **among** the sample means is

$$s_{\bar{X}}^2 = \frac{\sum_{i=1}^{m}\left(\bar{X}_i - \bar{\bar{X}}\right)^2}{m-1} \tag{5.13}$$

Therefore an estimate of the variance from the variance **among** the sample means is

$$s_X^2 = n s_{\bar{X}}^2 = n\left(\frac{\sum_{i=1}^{m}\left(\bar{X}_i - \bar{\bar{X}}\right)^2}{m-1}\right) = \frac{\sum_{i=1}^{m} n\left(\bar{X}_i - \bar{\bar{X}}\right)^2}{m-1} \tag{5.14}$$

Thinking of the numerator as a sum of **among** square differences and the denominator as degrees of freedom, this is the mean square differences **among** samples

$$s_X^2 = \frac{SSa}{m-1} = MSa = \frac{SSa}{dfa} \tag{5.15}$$

Now when we ratio among variance given in Equation 5.15 to within variance given in Equation 5.12 we obtain the F value.

$$F = \frac{MSa}{MSw} = \frac{(SSa/dfa)}{(SSw/dfw)} = \frac{dfw}{dfa}\frac{SSa}{SSa} = \frac{N-m}{m-1}\frac{\sum_{i=1}^{m} n\left(\bar{X}_i - \bar{\bar{X}}\right)^2}{\sum_{i=1}^{m}\sum_{j=1}^{n}(x_{ij} - \bar{X}_i)^2} \tag{5.16}$$

As you can see these expressions are in terms of sum of squares.

The typical ANOVA one-way table is as follows. The first row is for **Among** samples, and the second row is for **Within** samples or **Residual** error. SS denotes sum of squares of differences, and MS denotes mean (average) of squares of differences. SS are divided by the corresponding df to obtain MS. For m samples and n observations the sources of variations are as shown in Table 5.6.

The F-test is used here for the ratio **among/within** variability. The question asked is: Can we detect differences **among** the samples above the "noise" or **within** variability of each sample? The H0 is $F = 1.00$ or among variance = within variance. A high F implies that H0 can be rejected with high significance (the p-value would be very low), and therefore you conclude that the difference among samples is much higher than the sample "noise." A graph of side-by-side boxplots is a helpful exploratory visualization to see if in general the sample means are different and that their difference is larger than the spread of each sample. As an example, we will see two contrasting cases of soil moisture (response variable) at four locations (factor). The example Figure 5.4a shows clear differences among means and exceeding the variability within each sample. In this case, we expect

TABLE 5.6

ANOVA One-Way Sources of Variations

Source of Variation	df	SumSq	MeanSq	F	p
Among samples	$m-1$	SSa	MSa = SSa/dfa	MSa/MSw	Pr(F)
Within samples	$N-m$	SSw	MSw = SSw/dfw		

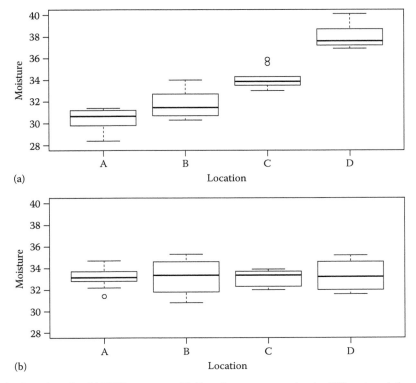

(a)

(b)

FIGURE 5.4 Boxplots for ANOVA one-way. (a) Sample means are clearly different and the among variability is above the within variability. (b) Sample means are slightly different and the among variability is not above the within variability.

a high value of F. In contrast, the example in Figure 5.4b shows little difference among means and not exceeding the within variability. In this case, we expect to get a value of F closer to 1.

When sample size varies by sample, we have an unbalanced ANOVA. The overall variance (within) can be calculated from the **pooled** sample variances. The pooled sample variance is a weighted average of the sample variances where the weights are the degrees of freedom

$$\overline{s_p^2} = \frac{\sum_{i=1}^{m}(n_i-1)s_i^2}{\sum_{i=1}^{m}(n_i-1)} = \frac{\sum_{i=1}^{m}(n_i-1)s_i^2}{\sum_{i=1}^{m}n_i-m} = \frac{\sum_{i=1}^{m}(n_i-1)s_i^2}{N-m}p \tag{5.17}$$

where
   n_i is size of the ith sample
   $N = \sum_{i=1}^{m}n_i$ is the total number of observations

Note that when $n_i = n$ for all samples, then $N = n \times m$ and that in this case the overall variance is just the average of the sample variances that we used before

$$\overline{s_p^2} = \frac{\sum_{i=1}^{m}(n_i-1)s_i^2}{N-m} = \frac{(n-1)\sum_{i=1}^{m}s_i^2}{m \times n - m} = \frac{(n-1)\sum_{i=1}^{m}s_i^2}{m \times (n-1)} = \frac{\sum_{i=1}^{m}s_i^2}{m} \tag{5.18}$$

Now note that expanding the variance s_i^2 of each sample in Equation 5.17

$$s_p^2 = \frac{\sum_{i=1}^{m} \left(\dfrac{(n_i - 1) \sum_{j=1}^{n_i} (x_{ji} - \overline{X}_i)^2}{(n_i - 1)} \right)}{N - m} = \frac{\sum_{i=1}^{m} \sum_{j=1}^{n_i} (x_{ji} - \overline{X}_i)^2}{N - m} \tag{5.19}$$

which is the same as the balanced case given in Equation 5.11. Therefore, the MSw is not affected and $s_p^2 = MSw = SSw/dfw$. However, the mean of the sample means has variance

$$s_{\overline{X}i}^2 = \frac{\sum_{i=1}^{m} \left(s_X^2 / n_i \right)}{m} = \frac{s_X^2 \sum_{i=1}^{m} (1/n_i)}{m} = \frac{s_X^2}{\dfrac{m}{\sum_{i=1}^{m} (1/n_i)}} = \frac{s_X^2}{n_h}$$

where n_h is the harmonic mean of the sample sizes. Therefore, an estimate of the variance from the variance **among** the sample means is

$$s_{\overline{X}}^2 = n_h s_{\overline{X}}^2 = n_h \left(\frac{\sum_{i=1}^{m} \left(\overline{X}_i - \overline{\overline{X}} \right)^2}{m - 1} \right) = \frac{\sum_{i=1}^{m} n_h \left(\overline{X}_i - \overline{\overline{X}} \right)^2}{m - 1} \tag{5.20}$$

Now when we ratio **among** variance given in Equation 5.20 to the **within** variance given in Equation 5.19, we obtain the F-value as before.

$$F = \frac{MSa}{MSw} = \frac{(SSa/dfa)}{(SSw/dfw)} = \frac{dfw}{dfa} \frac{SSa}{SSw} = \frac{N - m}{m - 1} \frac{\sum_{i=1}^{m} n_h \left(\overline{X}_i - \overline{\overline{X}} \right)^2}{\sum_{i=1}^{m} \sum_{j=1}^{n} (x_{ij} - \overline{X}_i)^2}$$

If the outcome is high F, then you could **predict** the expected value of each sample from the sample mean. We discuss this case in the next chapter along with regression.

5.4.2 ANOVA Two-Way

As we introduce a second factor, observations correspond to two categories or levels, one for each factor. For example, in an agronomic experiment, one factor could be crop variety and the other fertilizer treatment. There are two situations: with and without replicates. In the **nonreplicated** type, there is only one observation for each level combination of the two factors. For example, we have m samples (one for each level of factor 1) and n observations per sample (one for each level of factor 2). Thus, we have a total of $m \times n$ observations arranged as a table of n rows and m columns. It is common to use the term **block** for the rows and **treatment** for the columns. In the **replicated** case, there are several observations (say k) for each level combination of the two factors. Thus, we have a total of $m \times n \times k$ observations. In this chapter, we will only cover the nonreplicated case.

TABLE 5.7
ANOVA Two-Way Sources of Variations

Source of Variation	df	SumSq	MeanSq	F	p
Within samples	$n-1$	SSw	MSw	MSw/MSe	$\Pr(F)$
Among samples	$m-1$	SSa	MSa	MSa/MSe	$\Pr(F)$
Error (residual)	$(m-1)\times(n-1)$	SSe	MSe		

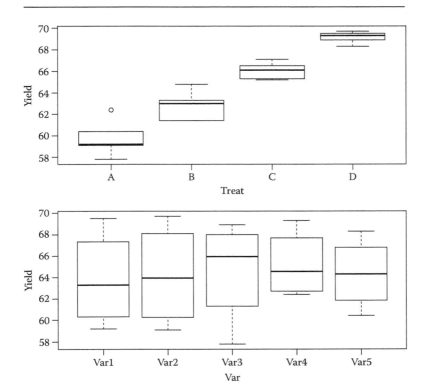

FIGURE 5.5 Boxplots for ANOVA two-way. Sample means are different above noise for one factor but not for the other.

The typical ANOVA two-way table is as follows. The first row is for **Within** samples, second row is for **Among** samples, and third row is for **Residual** error. When we have m samples and n observations per sample, then the table is as shown in Table 5.7.

The F-test is used twice. Once for the **within/error** variance ratio and again for the **among/error** variance ratio. We ask the question: Can we detect differences **among** levels of factor 2 above the "noise" of the samples? The H0 has two parts: $F = 1.00$ for the "within/error" and $F = 1.00$ for the "among/error."

In Figure 5.5, we see an example of agronomic experiment where crop yield is the response variable as a function of two factors, crop variety and fertilizer treatment. The top panel shows a clear response to treatment; however, the bottom panel shows lack of response to variety.

5.4.3 FACTOR INTERACTION IN ANOVA TWO-WAY

An interaction plot helps diagnose interaction between factors. To obtain an interaction plot, arrange categories or levels of one factor (say treatment) along the horizontal axis. Then, plot the average

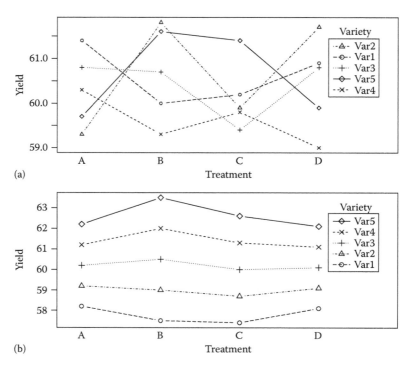

FIGURE 5.6 ANOVA: factor interaction plots. Two cases of no response to treatment. (a) No response to variety and (b) response to variety.

response of each group for the other factor (say variety). When the first factor (treatment) has an effect, we will see lines different from horizontal lines. When the other factor (variety) has an effect, we will see lines differing one from the other. If the lines are parallel to each other, then we can assume no interaction. However, significant crossings indicate interaction. Figures 5.6 and 5.7 summarize all possible four situations. In Figure 5.6a, we see an example of factor interaction and no clear effect of both factors. In the same figure (Figure 5.6b), we note a clear effect of variety but no effect of treatment. In contrast, Figure 5.7a shows effect of both factors, and no interaction. Whereas the same figure (Figure 5.7b) illustrates factor interaction, clear effect of treatment but not of variety.

5.4.4 Nonparametric Analysis of Variance

The **Kruskal–Wallis test** is a nonparametric version of the one-way ANOVA, whereas the **Friedman test** is a nonparametric version of the two-way ANOVA (nonreplicated case) (Davis, 2002; Dalgaard, 2008; MathSoft, 1999). Both nonparametric tests are based on ranks.

In Kruskal–Wallis, the procedure is similar to the Mann–Whitney test. All observations are ranked regardless of group or sample, then for each sample we add the ranks to obtain R_i, or the **sum of ranks** for each sample i. Same as in the parametric ANOVA n_i is size of the ith sample and $N = \sum_{i=1}^{m} n_i$ is the total number of observations. The statistic is

$$K = \frac{12}{N(N+1)} \sum_i n_i \left(\frac{R_i}{n_i} - \frac{N+1}{2} \right)^2 = \frac{12}{N(N+1)} \sum_i n_i \left(\frac{R_i}{n_i} \right)^2 - 3(N+1) \tag{5.21}$$

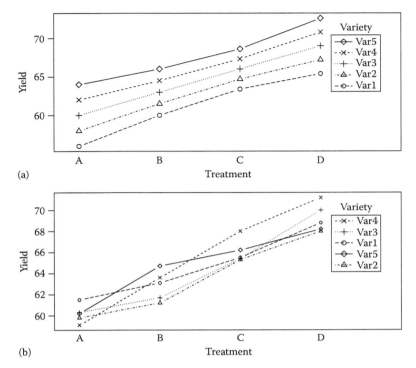

FIGURE 5.7 ANOVA: factor interaction plots. Two cases of response to treatment. (a) Response to variety and (b) no response to variety.

For small sample sizes there are tables for this statistic. However, for large sample size, K is approximately distributed as χ^2 with $m - 1$ degrees of freedom and thus it is easier to find the p-value (Davis, 2002; MathSoft, 1999). The null H0 is that all samples come from the same distribution. Note that the term (R_i/n_i) is the average rank of each sample and therefore the statistic is based on the sum of the squares of the average ranks (Dalgaart, 2008).

For the Friedman test, we rank observations within the samples (not for all observations) so that we have ranks r_{ij} for sample i and observation j, and then take the average rank $\bar{R}_i$ of each sample and the average $\bar{R}$ of all ranks. The test statistic is

$$Q = \frac{n \sum_i \left(\bar{R}_i - \bar{R} \right)^2}{\dfrac{1}{n(m-1)} \sum_i \sum_j \left(r_{ij} - \bar{R} \right)^2} \tag{5.22}$$

for which there are tables for small sample sizes. However, Q approximated by a χ^2 with $m - 1$ degrees of freedom for large sample sizes and thus it is easier to find the p-value (Dalgaard, 2008).

5.5 EXERCISES

Exercise 5.1
We want to check if 100 values come from a uniform distribution using a GOF test. We got the following counts in five categories (bins or intervals): 21, 19, 18, 22, 20. What test would you use? Calculate the statistic (should be 0.5). How many degrees of freedom do we have? The p-value for

this value of the statistic is 0.973. What is your conclusion? Repeat for counts 30, 25, 20, 15, 10. The statistic should now be 12.5 and p-value = 0.013. What is your conclusion now?

Exercise 5.2

Three forest cover types (palms, pines, and hardwoods) and three terrain types (valley, slope, and ridge). We want to see if forest cover is associated with terrain? You classify all available sites or regions and organize in a table. What test would you use? Suppose you run this test and get value for the statistic 10.64 and p-value = 0.0309. What is H0? What is your conclusion?

Exercise 5.3

Three zones (A, B, and C) are distinguished in a lake. Ten values of nitrate concentration are taken at random in each zone. Is nitrate concentration different among zones? What analysis do you run to answer this question? Suppose you get statistic = 5.8, p-value = 0.007. What is the statistic? What is H0? What is your conclusion?

Exercise 5.4

We want to check whether samples (of size n = 100) come from a uniform distribution using a GOF test. We have three different variables.

1. What statistic do you use?
2. We use counts per quartile. How many counts do we expect in each quartile. How many degrees of freedom do we have?
3. First variable: We got the following counts in the four quartiles: 26, 24, 23, 27.
   a. Calculate the statistic.
   b. The p-value for this value of the statistic is 0.94. What is your conclusion?
4. Second variable: we obtained counts 28, 22, 23, 27
   a. Calculate the statistic
   b. The p-value = 0.79. What is your conclusion now?
5. Third variable: counts 35, 35, 15, 15
   a. Calculate the statistic
   b. Suppose the p-value = 0.001. What is your conclusion now?

Exercise 5.5

Suppose we have two different community types—rural and urban-industrial—and three levels of health problems—low, medium, and high. We want to see if community type is associated with health?

1. Classify communities and health problems in a contingency table.
2. What test would you use? Then suppose you run the test and get a value for the statistic = 10.64 and p-value = 0.0309.
3. What is H0?
4. What is your conclusion?

Exercise 5.6

Four archeological sites (A, B, C, and D) are being analyzed. There is an artifact type in common among all sites. Ten values of artifact length are taken at random in each site. Is artifact length different among zones?

1. What analysis do you run to answer this question?
2. What graphs do you use to visually compare the magnitude of variability among sites to variances within each site? Illustrate with an example
3. What is the statistic?
4. What is H0?
5. Suppose you get statistic = 5.8, p-value = 0.007. What is your conclusion?

5.6 COMPUTER SESSION: MORE ON INFERENTIAL STATISTICS

5.6.1 GOF: EXPLORATORY ANALYSIS

We can modify our cdf.plot function of the previous chapter to include other distributions by changing pnorm to **p**name, where name is specific of the distribution; for example, pexp for the exponential, and punif for the uniform. For flexibility, we use an argument dist to declare the distribution. We define normal as default so that even in the absence of an argument value the function will go ahead and compare the sample to the normal.

The function cdf.plot.gof is in **cdf-plot-gof.R** file and reproduced here for easy reference. You can edit your own version of cdf.plot from the last chapter so that you become more aware of how the function works.

```
cdf.plot.gof <- function(x,dist="normal",
  mu=0,sd=1,rate=1,min=0,max=1){
nx <- length(x)            # number of observations
vx <- sort(x)              # data sorted
Fx <- seq(nx)/nx           # ranks as fractions (quantiles)
panel2(size=7)
plot(vx, Fx, xlab="x", ylab="F(x)", ylim=c(0,1)) # empirical
 cumulative plot
vnx <- seq(vx[1],vx[nx],length=100) # sequence with increased res
# cdf values
if(dist=="normal") {Fxh <- pnorm(vnx,mu,sd);Fxe <- pnorm(vx,mu,sd)}
if(dist=="exp") {Fxh <- pexp(vnx,rate);Fxe <- pexp(vx,rate)}
if(dist=="unif") {Fxh <- punif(vnx,min,max);Fxe
 <- punif(vx,min,max)}
lines(vnx,Fxh)             # theoretical cumulative plot cdf
legend("bottomright",leg=c("Data","Hyp"),pch=c(1,-1), lty=c(-1,1),
 merge=T)
plot(vx,Fx-Fxe,xlab="x", ylab="Diff Empir - Theor"); abline(h=0)
}
```

Use source to load the function in your workspace.

```
> source("lab5/cdf-plot-gof.R")
```

Now for some applications we can generate three samples of size $n = 20$: one from a normal, one from a uniform, and one from an exponential; and then use cdf.plot.gof for exploratory analysis. First, we generate a sample (using rnorm, runif, and rexp) and then call our newly developed function cdf.plot.gof (using normal, unif, and exp for argument dist). In the following sections of code, we assume a script when the prompt symbol > is not explicitly indicated.

```
xn <- rnorm(20)
cdf.plot.gof(xn,dist="normal")
mtext(side=3,line=2,"Sample Normal, Hyp Normal",cex=0.7)

#xu <- runif(20,0,1)
cdf.plot.gof(xu,dist="unif")
mtext(side=3,line=2,"Sample Unif, Hyp Unif",cex=0.7)
```

We get Figures 5.1 and 5.2 already discussed.

5.6.2   GOF: Chi-Square Test

We can get a feel for the shape of the χ^2 density using simple R functions. For example for degrees of freedom 3, 5, and 10,

```
x <- seq(0,20,0.1)
plot(x,dchisq(x,3),type="l",ylab="Chi-square")
lines(x,dchisq(x,5),lty=2)
lines(x,dchisq(x,10),lty=3)
legend("topright",leg=paste("df=",c(3,5,10)),lty=1:3)
```

To obtain Figure 5.3 already discussed. We could also do this easily from the Rcmdr using **Distributions|Chi-squared distribution|Plot chi-square distribution** but obtaining one plot at a time for each *df* value.

It is very easy to calculate the *p*-value from a chi-square from the console by typing

```
> 1- pchisq(5.30,4)
[1] 0.2578769
>
```

or using the Rcmdr **Distributions|Chi-squared distribution|Chi-square probabilities**. In the dialog window, enter 5.30 for variable value, 4 for degrees of freedom, and check Lower tail option.

Package stats include a function chisq.test we can use for GOF or for independence tests. We will use it later for contingency tables. We can build a simple χ^2 (chi-square) GOF test function to test whether a sample comes from a standard normal distribution. The χ^2 is a very important statistic, so it is worth your while to build it from scratch to become familiar with it. See Davis, 2002, pp. 94–96. Look at the **chisq-gof-norm.R** function

```
# chisq GOF test with respect to standard normal
# user gives number of equiprobable classes as argument nclass

chisq.gof.norm <- function(x, nclass, param.est){
 prob <- rep(1/nclass, nclass)         # equal prob of each class
 expec <- length(x) * prob             # expected number in each class
 numb <- floor(1 + nclass * pnorm(x))  # prob from normal scaled to
   nclass
 obser <- tabulate(numb, nclass)       # count observed in each class
 X2 <- sum((expec - obser)^2/expec)    # calc X2
 df <- nclass - 1 - param.est          # df, subtract num par
   estimated
 p.value <- 1 - pchisq(X2, df)         # pvalue from chisq
 # return result in a list
 ret.val <- list(X2=X2, df=df, p.value=p.value, observed=obser)
}
```

Then source it

```
> source("lab5/chisq-gof-norm.R")
```

Now let us use it to check normality of z generated randomly from $N(0, 1)$

```
> z <- rnorm(100)
> x2z <- chisq.gof.norm(z,4,0)
> x2z
$X2
[1]  0.56
$df
[1]  3
$p.value
[1]  0.9055252
$observed
[1]  25 22 26 27
```

We can see that the p-value is too high, and we cannot reject the hypothesis that the sample came from a standard normal distribution. We also see that all classes have large enough number of observations (none less than 5). We could increase the number of classes to have more degrees of freedom. With nclass = 8

```
> x2z <- chisqgof.norm(z,8,0)
> x2z
> x2z
$X2
[1]  3.84
$df
[1]  7
$p.value
[1]  0.7980109
$observed
[1]  13 12 9 13 17 9 13 14
```

We still have enough observations in each class (none less than 5) and we still get a high p-value and cannot reject the null. Thus, z may come from a standard normal.

5.6.3   GOF: KOLMOGOROV–SMIRNOV TEST

As examples, we use a sample drawn from an exponential distribution. First generate a random sample from an exponential with rate = 1 and test it.

```
> x <- rexp(20,rate=1)
> ks.test(x, "pexp", rate=1)

     One-sample Kolmogorov-Smirnov test

data:  x
D = 0.1801, p-value = 0.4805
alternative hypothesis: two-sided
```

As expected, with this high *p*-value, we cannot reject the null hypothesis and therefore there is no evidence that sample is not distributed according to an exponential with this rate. For comparison, test against an exponential but with rate = 2

```
> ks.test(x, "pexp", rate=2)

      One-sample Kolmogorov-Smirnov test

data:  x
D = 0.3099, p-value = 0.03334
alternative hypothesis: two-sided

>
```

As expected, the *p*-value suggests now that we can reject the null with $\alpha = 0.05$.

5.6.4   GOF: SHAPIRO–WILK

The basic stats package contains the Shapiro–Wilk test for normality. Let us apply it to test for a sample drawn from a normal like the *z* given in section 5.6.2

```
> shapiro.test(z)
      Shapiro-Wilk normality test
data: z
W = 0.9837, p-value = 0.2568
```

This high *p*-value indicates that there is no evidence to reject the null that the sample is normally distributed.

For comparison, take the x generated earlier from an exponential and apply the test

```
> shapiro.test(x)

      Shapiro-Wilk normality test

data: x
W = 0.8878, p-value = 0.02451
```

As expected, the *p*-value suggests to reject H0 to $\alpha = 0.05$ and we conclude that there is evidence against the sample being normally distributed.

5.6.5   COUNT TESTS AND THE BINOMIAL

Suppose we hypothesize that 20% of an area is covered by grass. We sample coordinates at random to select 20 locations, then survey these and count the number of times we find grass. We find grass in five sites. Is the hypothetical 20% correct? Apply the binomial test using function `binom.test`

```
> binom.test(5,20,p=0.2)

        Exact binomial test

data: 5 and 20
number of successes = 5, number of trials = 20, p-value = 0.5764
alternative hypothesis: true probability of success is not equal to
   0.2
95 percent confidence interval:
 0.08657147 0.49104587
sample estimates:
probability of success
              0.25

>
```

We get a high *p*-value indicating that we should not reject a 20% grass cover. In addition, the 95% confidence interval indicates that 20% is within the interval for 25% estimate in cover. In this case, we conclude that the cover could be 20%.

For comparison, suppose we count eight sample sites covered with grass. Now

```
> binom.test(8,20,p=0.2)

        Exact binomial test

data: 8 and 20
number of successes = 8, number of trials = 20, p-value = 0.04367
alternative hypothesis: true probability of success is not equal to
   0.2
95 percent confidence interval:
 0.1911901 0.6394574
sample estimates:
probability of success
              0.4

>

>
```

With this *p*-value we can reject the null of 20% grass cover at $\alpha = 5\%$. The lower bound of the 95% confidence interval for an estimated cover of 0.4 barely includes the hypothetical 20%. In this case, we would conclude that the cover is not 20%.

5.6.6 Obtaining a Single Element of a Test Result

Many times, you want to address only a component of the object resulting from the test. The results returned from the function are in a list. Therefore, you can do this using the dollar sign and the name of the component. For example,

```
> binom.test(8,20,p=0.2)$p.value
[1] 0.04367188
```

gives you only the *p*-value. The online help tells you about the components produced for each type of function. For example for the `binom.test`, you find this is in the help

```
statistic     the number of successes.
parameter     the number of trials.
p.value       the p-value of the test.
conf.int      a confidence interval for the probability of success.
estimate      the estimated probability of success.
null.value    the probability of success under the null, p.
alternative   a character string describing the alternative
  hypothesis.
method        the character string "Exact binomial test".
data.name     a character string giving the names of the data.
```

5.6.7 COMPARING PROPORTIONS: `prop.test`

We hypothesize that the proportion of residents of an area are willing to pay for ecosystem services is the same regardless of location. We sample three locations and obtain 10, 12, 13 positive responses out of 80, 90, 85 respondents. We discuss this example in section 5.3. We can use function `prop.test` with the positive counts, number of respondents

```
> xo <- c(10,12,13); xt <- c(80,90,85)
> prop.test(xo,xt)

        3-sample test for equality of proportions without continuity
          correction
data: xo out of xt
X-squared = 0.2898, df = 2, p-value = 0.8651
alternative hypothesis: two.sided
sample estimates:
   prop 1     prop 2     prop 3
0.1250000 0.1333333 0.1529412

>
```

We conclude that we should not reject H0 and the proportions may be the same.

Alternatively, we obtain the same result using `chisq.test` on the contingency table of positive and negative responses

```
> table <- rbind(xo,xc=(xt-xo))
> table
     [,1] [,2] [,3]
xo    10   12   13
xc    70   78   72
> chisq.test(table)

        Pearson's Chi-squared test

data:  table
X-squared = 0.2898, df = 2, p-value = 0.8651

>
```

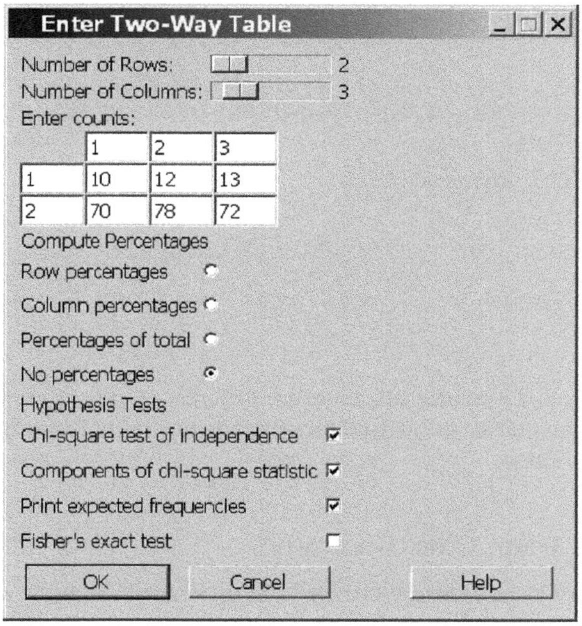

FIGURE 5.8   Using Rcmdr for contingency tables.

Using Rcmdr you can enter tables using **Statistics|Contigency table|Enter two way table** (Figure 5.8). Part of the results given in the output window are reproduced here:

```
> .Table  # Counts
   1  2  3
1 10 12 13
2 70 78 72
> .Test <- chisq.test(.Table, correct=FALSE)
> .Test
        Pearson's Chi-squared test
data:  .Table
X-squared = 0.2898, df = 2, p-value = 0.8651
> .Test$expected # Expected Counts
          1         2         3
1 10.98039 12.35294 11.66667
2 69.01961 77.64706 73.33333
> round(.Test$residuals^2, 2) # Chi-square Components
     1    2    3
1 0.09 0.01 0.15
2 0.01 0.00 0.02
```

It is typically easier to use the Rconsole directly to organize the table.

5.6.8   CONTINGENCY TABLES: DIRECT INPUT

Let us function `chisq.test` to analyze the example in Table 5.1 already discussed in the theoretical section on contingency tables (section 5.3). The degrees of freedom are calculated with (rows − 1) × (cols − 1) = (2 − 1)(3 − 1) = 2.

Enter values as a matrix object and apply the function

```
> table <- matrix(c(12,20,8,8,10,22),ncol=3,byrow=T)
> chisq.test(table)

        Pearson's Chi-squared test

data: table
X-squared = 10.6667, df = 2, p-value = 0.004828

>
```

We get the same results and conclude that there are differences in vegetation cover with elevation. Using Rcmdr you can enter tables using **Statistics|Contigency table|Enter two way table** as shown earlier in the previous section.

5.6.9 CONTINGENCY TABLES: CROSS-TABULATION

We can cross-tabulate in R use function `table`. As an example let us use dataset `airquality` which we employed for the exercises in the previous chapter.

```
> data(airquality)
> attach(airquality)
```

and check its contents

```
> airquality
    Ozone  Solar.R  Wind  Temp  Month  Day
1      41      190   7.4    67      5    1
2      36      118   8.0    72      5    2
3      12      149  12.6    74      5    3
4      18      313  11.5    62      5    4
5      NA       NA  14.3    56      5    5
6      28       NA  14.9    66      5    6
7      23      299   8.6    65      5    7
8      19       99  13.8    59      5    8
9       8       19  20.1    61      5    9
10     NA      194   8.6    69      5   10
```

Shown here are just the first 10 records of the dataset which has 153 records as seen from a `dim` command. There are six variables or factors.

```
> dim(airquality)
[1]    153    6
```

Suppose we want to build a two-way table by `Temp` and `Month`. First, let us see how many different values these factors have

```
> unique(Month)
[1] 5 6 7 8 9
> unique(Temp)
 [1] 67 72 74 62 56 66 65 59 61 69 68
[12] 58 64 57 73 81 79 76 78 84 85 82
[23] 87 90 93 92 80 77 75 83 88 89 91
[34] 86 97 94 96 71 63 70
```

We can see that `Temp` has 40 different values. These may be too many to cross-tabulate. Therefore, we need to select fewer intervals. We do this with the `cut` function using breaks based on the quantiles

```
> cut(Temp, quantile(Temp))
[1] (56,72] (56,72] (72,79] (56,72]
4 Levels: (56,72] ... (85,97]
```

Then we apply `table` with these four levels for `Temp` and 5 months.

```
> table(cut(Temp, quantile(Temp)),Month)
         Month
           5  6  7  8  9
  (56,72] 24  3  0  1 10
  (72,79]  5 15  2  9 10
  (79,85]  1  7 19  7  5
  (85,97]  0  5 10 14  5
```

We can note the higher frequency of high level of `Temp` as we increase the `Month`. Can verify with `chisq.test`

```
> chisq.test(table(cut(Temp, quantile(Temp)),Month))

        Pearson's Chi-squared test

data: table(cut(Temp, quantile(Temp)), Month)
X-squared = 104.3078, df = 12, p-value = < 2.2e-16
```

Indeed, the low p-value means that we can reject the null of no association and conclude that `Temp` is associated with `Month`.

Let us try Temp and `Ozone`. In this case, `Ozone` has many NA, so we have to remove observations that have NA with option `na.rm = T`

```
> table(cut(Temp, quantile(Temp)),cut(Ozone, quantile(Ozone,
  na.rm=T)))

                (1,18]  (18,31.5]  (31.5,63.3]  (63.3,168]
     (56,72]      17        10           5            0
     (72,79]      10        10           5            1
     (79,85]       4         6          14            6
     (85,97]       0         0           5           22
>
```

We can see that higher Ozone is associated with higher Temp, which we can verify with chisq.test

```
> chisq.test(table(cut(Temp, quantile(Temp)),cut(Ozone,
  quantile(Ozone, na.rm=T))))
      Pearson's Chi-squared test

data:  table(cut(Temp, quantile(Temp)), cut(Ozone, quantile(Ozone,
  na.rm = T)))
X-squared = 83.3787, df = 9, p-value = 3.436e-14
```

Indeed, this very low *p*-value means we can reject the null of no association and conclude that high Ozone is associated with high Temp.

5.6.10   ANOVA ONE-WAY

Let us start with data in file **lab5/maximumDO.txt**. It has 10 values of the response variable, i.e., maximum dissolved oxygen (DO) in the water for each level of a factor, i.e., zones A, B, C, D in a small estuary. Concentration of DO varies during the day; the values reported here are the daily maxima observed.

```
Dissolved Oxygen (mg/l) for zones A,B,C,D
A        B      C      D
5.2      8.5   10.1   10.7
5.4      7.7    9.5   13.1
4.1      7.4    9.4   12.5
3.5      9.3    8.1   13.4
6.5      6.9   10.8   11.0
5.8      9.3   10.6   12.3
4.9     10.6   10.7   12.2
2.5      8.1    9.9   11.4
6.1      7.2    9.6   11.7
3.3      8.9    9.4   14.2
```

First, read the data as a matrix and then convert to a column vector

```
> maxDO <- matrix(scan("lab5/maximumDO.txt",skip=2),ncol=4,byrow=T)
Read 40 items
> maxDO <- c(maxDO)
>
```

Second generate the factor, which is composed of a character vector and a levels attribute. This attribute determines what character strings may be included in the vector. The function `factor` creates a factor object and allows one to set the levels attribute.

```
> zone <- factor(rep(LETTERS[1:4], c(5,5,5,5)))
> zone
[1] A A A A A B B B B B C C C C C D D D D D
```

Next, we assemble a data frame with the data and the factor,

```
> zone.maxDO <- data.frame(zone, maxDO)
```

Function `data.frame` puts together two objects: `zone` (a factor, or vector with categorical levels) and `maxDO` (a vector with numeric entries). Double check by typing `zone.maxDO`

```
> zone.maxDO
   zone maxDO
1     A    5.2
2     A    5.4
3     A    4.1
4     A    3.5
5     A    6.5
6     A    5.8
7     A    4.9
8     A    2.5
9     A    6.1
10    A    3.3
11    B    8.5
12    B    7.7
13    B    7.4
14    B    9.3
15    B    6.9
16    B    9.3
17    B   10.6
18    B    8.1
19    B    7.2
20    B    8.9
21    C   10.1
22    C    9.5
23    C    9.4
24    C    8.1
25    C   10.8
26    C   10.6
27    C   10.7
28    C    9.9
29    C    9.6
30    C    9.4
```

(continued)

```
(continued)
31   D    10.7
32   D    13.1
33   D    12.5
34   D    13.4
35   D    11.0
36   D    12.3
37   D    12.2
38   D    11.4
39   D    11.7
40   D    14.2
>
```

Note that once you apply the data.frame function and create the dataset or data frame zone. maxDO, then you can select it using the Rcmdr. Go to **Data|Select active data set**, browse to zone.maxDO, and then **View dataset** to confirm. At this point, you can use either the Rconsole or the Rcmdr. At the Rconsole, you can use the boxplot function to obtain Figure 5.9.

```
boxplot(maxDO~zone, data=zone.maxDO,ylab="Max DO (mg/l)",
  xlab="Zone")
```

You can see that the sample means seem to be different among the samples and that these differences look greater than the variability within each sample. In addition, we see no outliers. From this diagnostic, we feel confident about proceeding to apply a parametric one-way ANOVA test.

Now we run the ANOVA with function aov and query the results using summary or anova. We obtain the typical ANOVA one-way table: first row is for **Among** samples (zone in this case), and the second row is for **Within** samples or **Residual** error. Degrees of freedom are $m - 1 = 4 - 1 = 3$ samples (zone factor) and $N - m = 4 \times 10 - 4 = 36$ observations.

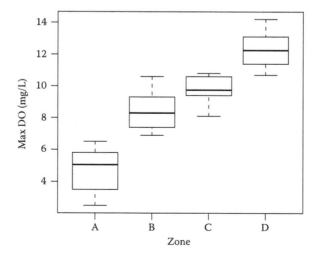

FIGURE 5.9   Boxplots for ANOVA one-way.

```
> summary(aov(maxDO~zone, data=do.zone))
            Df  Sum Sq  Mean Sq  F value    Pr(>F)
zone         3  296.56    98.85    79.86  5.62e-16 ***
Residuals   36   44.56     1.24
---
Signif. codes:  0 '***' 0.001 '**' 0.01 '*' 0.05 '.' 0.1 ' ' 1
>
```

The $F = 79.8$ obtained here implies that H0 can be rejected with high significance level (p-value is very small 5.6×10^{-16}). Therefore, you conclude that the sample means are different.

To perform the same analysis using the Rcmdr, you first use **Graphs|Boxplot**, pick maxDO variable and select identify and plot by groups you will get a second dialog window, select **zone**, click okay and obtain graphics. Then set the model using **Statitics|Fit models|Linear model**, then in the dialog select the variables to make the formula (Figure 5.10). Then go to menu **Models|Hypothesis test|ANOVA table** and mark type I to set the table (Figure 5.11) and get the result in the output window

```
> anova(anova.maxDO)
Analysis of Variance Table

Response: maxDO
            Df   Sum Sq  Mean Sq  F value    Pr(>F)
zone         3  296.555   98.852   79.855  5.619e-16 ***
Residuals   36   44.564    1.238
---
Signif. codes:  0 '***' 0.001 '**' 0.01 '*' 0.05 '.' 0.1 ' ' 1
```

Note that it is the same result as before and that it invokes function anova.

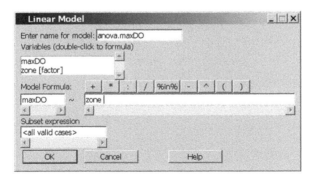

FIGURE 5.10   Using the Rcmdr to set anova model.

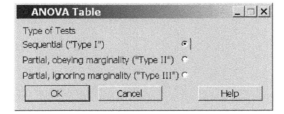

FIGURE 5.11   Selecting ANOVA table.

5.6.11 ANOVA TWO-WAY

Make a **two-way** data frame from the data in file **lab5/yield-treat-var.txt**. The data are not from a specific experiment but represent a typical experiment in agronomy and other fields. We read the response as a matrix delete the first column and put in an array named `yield`.

```
yield.raw <- matrix(scan("lab5/yield-treat-var.txt",skip=2), ncol=5,
  byrow=T)[,-1]
yield <- c(yield.raw)
> yield
 [1] 59.2 59.1 57.8 62.4 60.4 61.4 61.4 64.8 63.0 63.3 65.2 66.5 67.1
    66.1 65.3
[16]  69.5 69.7 68.9 69.3 68.3
>
```

Start by defining names for two factors; note that these are Treatment or "`Treat`" for short and Variety or "`Var`" for short.

```
treat <- factor(rep(LETTERS[1:4], rep(5,4)))
> treat
 [1] A A A A A B B B B B C C C C C D D D D D
Levels: A B C D
>
>var <- factor(rep(paste("Var", 1:5,sep=""),4))
> var
 [1] Var1 Var2 Var3 Var4 Var5 Var1 Var2 Var3 Var4 Var5 Var1 Var2
    Var3 Var4 Var5
[16] Var1 Var2 Var3 Var4 Var5
Levels: Var1 Var2 Var3 Var4 Var5
>
```

Now assemble with `data.frame` of factors and data

```
> yield.tv <- data.frame(treat, var, yield)
> yield.tv
   treat   var  yield
1      A  Var1   59.2
2      A  Var2   59.1
3      A  Var3   57.8
4      A  Var4   62.4
5      A  Var5   60.4
6      B  Var1   61.4
7      B  Var2   61.4
8      B  Var3   64.8
9      B  Var4   63.0
10     B  Var5   63.3
11     C  Var1   65.2
12     C  Var2   66.5
13     C  Var3   67.1
```

```
(continued)
14      C   Var4    66.1
15      C   Var5    65.3
16      D   Var1    69.5
17      D   Var2    69.7
18      D   Var3    68.9
19      D   Var4    69.3
20      D   Var5    68.3
> attach(yield.tv)
```

To simplify repeated use in the next few statements, we attach this data frame. For visual inspection, graph boxplots for factors. We do not need to declare the data set in the `boxplot` function because we have attached `yield.tv`

```
> par(mfrow=c(2,1))
> boxplot(yield ~ var, ylab="Yield", xlab="Var")
> boxplot(yield ~ treat, ylab="Yield", xlab="Treat")
```

We obtain results as in Figure 5.5 already discussed. Then graph an interaction plot to check for possible factor interactions

```
> interaction.plot(treat, var, yield, xlab="Treatment",ylab="Yield",
        type="b", pch=1:5, trace.label="Variety")
```

The result is in Figure 5.12. Now set the `aov` model and use `summary`

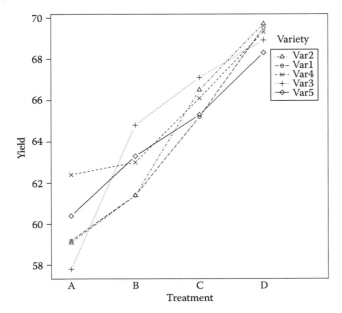

FIGURE 5.12 ANOVA: factor interaction plot for the example given earlier. We see a clear trend in factor 1 (treatment) but not for factor 2 (variety).

```
> anova.yield <- aov(yield ~ treat+ var, data=yield.tv)
> summary(anova.yield)
            Df  Sum Sq  Mean Sq  F value   Pr(>F)
treat        3  240.57    80.19  133.208  1.78e-09 ***
var          4    2.99     0.75    1.241     0.345
Residuals   12    7.22     0.60
---
Signif. codes:  0 '***' 0.001 '**' 0.01 '*' 0.05 '.' 0.1 ' ' 1
>
>
```

The results of summary(anova.yield) are similar to the typical ANOVA two-way table: first row is for **treat**, second row is for **Var**, and third row is for **Residual** error. The degrees of freedom are $4 - 1$ (4 treatment levels) and $5 - 1$ (5 variety levels). The H0 has two parts: F for the treatments (resulting $F = 133.2$ and can be rejected in this case with low α error rate) and F for the varieties (resulting $F = 1.24$ and cannot be rejected at a reasonable α error rate). You conclude that there is no difference among varieties but highly significant differences among treatments.

Once we have the data frame yield.tv, we can select it as active dataset, and then we can use Rcmdr. Go to **Data|Selectactive data set** and select yield.tv. Then view just to confirm success. Then go to **Statistics|Fit models|Linear model** and set up your model (Figure 5.13). Then go to **Models|Hypothesis test|Anova** to set the table as type I, and obtain results as before

```
anova(anova.yield)
Analysis of Variance Table

Response: yield
            Df  Sum Sq  Mean Sq  F value    Pr(>F)
treat        3  240.574   80.191 133.2079  1.778e-09 ***
var          4    2.988    0.747   1.2409     0.3453
Residuals   12    7.224    0.602
---
Signif. codes:  0 '***' 0.001 '**' 0.01 '*' 0.05 '.' 0.1 ' ' 1
```

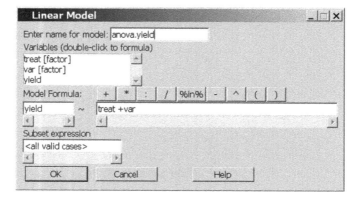

FIGURE 5.13   Using Rcmdr to set ANOVA two-way model using yield.tv data set.

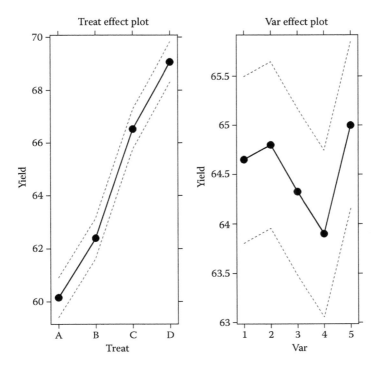

FIGURE 5.14   Effects plots.

Also can go to **Models|Graphs|Effects plot** to visualize the effect of each factor (Figure 5.14). We can see a clear increase in treatment but not a trend in variety.

5.6.12   ANOVA NONPARAMETRIC: KRUSKAL–WALLIS

Let us start with data in file **lab5/moist-loc.txt**. It has five values of the response variable (soil moisture) for each level of a factor (location A, B, C, D)

```
Soil moisture content for locations A,B,C,D
A       B       C       D
59.86   58.86   63.44   63.96
60.07   64.60   62.69   59.75
58.53   57.71   62.42   61.77
59.65   62.56   67.56   64.32
60.32   62.62   63.76   59.38
```

First, read the data as a matrix and then convert to a column vector

```
> moist <- matrix(scan("lab5/moist-loc.txt",skip=2),ncol=4,byrow=T)
Read 20 items
> moist <- c(moist)
```

Second, generate a factor `loc`,

```
> loc <- factor(rep(LETTERS[1:4], c(5,5,5,5)))
> loc
[1] A A A A A B B B B B C C C C C D D D D D
```

Next, we assemble a data frame with the data and the factor,

```
> loc.moist <- data.frame(loc,moist)
```

Double check by typing.

```
> loc.moist
   loc moist
1    A 59.86
2    A 60.07
3    A 58.53
4    A 59.65
5    A 60.32
6    B 58.86
7    B 64.60
8    B 57.71
9    B 62.56
10   B 62.62
11   C 63.44
12   C 62.69
13   C 62.42
14   C 67.56
15   C 63.76
16   D 63.96
17   D 59.75
18   D 61.77
19   D 64.32
20   D 59.38
>
```

Note that once you apply the data.frame function and create the dataset or data frame `loc.moist`, then you can select it using the Rcmdr. Go to **Data|Select active data set**, browse to **loc.moist**, and then **View dataset** to confirm. At this point, you can use either the Rconsole or the Rcmdr. At the Rconsole, you can use the `boxplot` function and `identify` to obtain Figure 5.15. We use `identify` for loc 1 and for loc 3 because we observe outliers for those locations. Once identified, we learn that outliers correspond to observations 3 and 14.

```
boxplot(moist~loc, data=loc.moist,ylab="Moisture", xlab="Location")
identify(rep(1,length(moist)),moist,cex=0.7)
identify(rep(3,length(moist)),moist,cex=0.7)
```

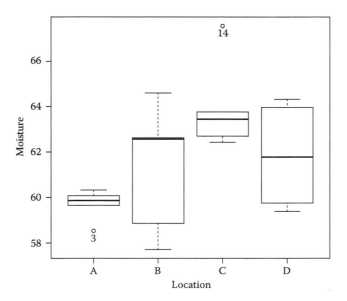

FIGURE 5.15 Boxplots for location and soil moisture data.

Because we have outliers and some of the samples do not seem to follow a normal distribution, we decide to run a nonparametric test.

Before we do so note that a parametric ANOVA results in degrees of freedom $m - 1 = 4 - 1 = 3$ samples (loc factor) and $N - m = 4 \times 5 - 4 = 20 - 4 = 16$ observations, $F = 3.43$ (p-value = 0.042), suggesting that H0 can be rejected with 5% significance level.

```
> anova.moist <- aov(moist ~ loc, loc.moist)
> summary(anova.moist)
            Df Sum Sq Mean Sq F value  Pr(>F)
loc          3 47.152  15.717  3.4309 0.04245 *
Residuals   16 73.298   4.581
---
Signif. codes: 0 '***' 0.001 '**' 0.01 '*' 0.05 '.' 0.1 ' ' 1
>
```

Now, we will run one-way analysis using Kruskal–Wallis test on moist by loc,

```
> kruskal.test(moist~loc, data=loc.moist)

      Kruskal-Wallis rank sum test

data: moist by loc
Kruskal-Wallis chi-squared = 5.9486, df = 3, p-value = 0.1141

>
```

In contrast with the parametric ANOVA, this result suggests that we cannot reject H0 unless we allow for higher error rate (15%). This result illustrates how a nonparametric test may lead to a more conservative result in the presence of outliers.

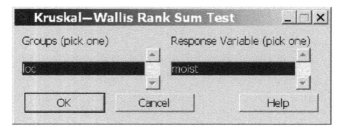

FIGURE 5.16   Dialog box for Kruskal–Wallis.

Alternatively, at the Rcmdr you can use **Graphs|Boxplot**, pick `moist` variable and select identify and plot by groups you will get a second dialog window, select `loc` and then once you get graphics can use mouse to identify outliers. Then, use **Statistics|Non parametric test|Kruskal-Wallis** and set variable a factor in the dialog box (Figure 5.16).

5.6.13   ANOVA NONPARAMETRIC: FRIEDMAN

Next, we will run the Friedman test on the same set of yield by treatment and variety. Recall that the parametric indicated very low *p*-value for treatment.

```
> friedman.test(yield, treat, var)

      Friedman rank sum test

data: yield, treat and var
Friedman chi-squared = 15, df = 3, p-value = 0.001817

>
```

In this case, the nonparametric ANOVA still allows for rejecting H0 with high confidence.

5.6.14   ANOVA: GENERATING FICTIONAL DATA FOR FURTHER LEARNING

In this section, we will generate fictional data instead of using real data values. For this purpose, we use random number generation with given mean and variances for one-way ANOVA and systematic change with respect to a factor for two-way ANOVA. This will allow us to gain an understanding of ANOVA by generating datasets for various conditions, including extreme cases. We will use the following function to generate hypothetical datasets of $m \times n$ observations, divided in m groups (by factor level) and n observations per sample.

```
invent.mxn <- function(m,n=5,d=1,p,f2="random"){
x <- matrix(nrow=n,ncol=m)
for(i in 1:m){
 if(f2=="random") x[,i] <- rnorm(n,p[i,1],p[i,2])
 if(f2=="step"){
  x[,i] <-  seq(p[i,1],p[i,2],(p[i,2]-p[i,1])/(n-1))
 }
}
return(round(x,d))
}
```

For example, use this function to generate datasets using four different means (30, 32, 34, 38) and equal standard deviations (1, 1, 1, 1). In this case, the observations within a sample will be random.

```
p <- matrix(c(30,1,32,1,34,1,38,1),byrow=T,ncol=2)
Xr <- invent.mxn(m=4,n=5,d=1,p,f2="random")
y <- c(Xr)
```

Then generate a factor, build the data frame, and plot

```
f <- factor(rep(LETTERS[1:m], rep(n,m)))
f.y <- data.frame(f, y)
boxplot(y~f, data=f.y,ylab="y", xlab="f")
```

See plots in Figure 5.17. We can see that the one-way ANOVA should result in low p-value. Let us then run the analysis

```
> summary(aov(y~f, data=f.y))
            Df  Sum Sq  Mean Sq  F value    Pr(>F)
f            3  193.54    64.51    66.47  2.99e-09 ***
Residuals   16   15.53     0.97
---
Signif. codes:  0 '***' 0.001 '**' 0.01 '*' 0.05 '.' 0.1 ' ' 1
```

We confirm that indeed we can reject H0 with high significance.

In the case of two-way analysis, the random option of argument f2 of invent.mxn will generate no significant response to f2. To obtain a fictional dataset for a two-way analysis, use option "step" in factor f2. For example, first group from 30 to 32, second from 32 to 34, third from 34 to 36, and fourth from 38 to 40.

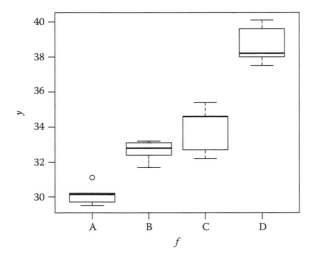

FIGURE 5.17 Boxplots for one-way ANOVA of fictional data.

```
p <- matrix(c(30,32,32,34,34,36,38,40),byrow=T,ncol=2)
Xs <- invent.mxn(m=4,n=5,d=1,p,f2="step")
y <- c(Xs)
```

Then create two factors and build a data frame

```
f1 <- factor(rep(LETTERS[1:m], rep(n,m)))
f2 <- factor(rep(paste("V", 1:n,sep=""),m))
f.y <- data.frame(f1, f2, y)
```

For boxplots and interaction plots use

```
boxplot(y~f1, data=f.y,ylab="y", xlab="f1")
boxplot(y~f2, data=f.y,ylab="y", xlab="f2")
interaction.plot(f.y$f1, f.y$f2, f.y$y,xlab="f1",ylab="y",
        type="b", pch=1:n, trace.label="f2")
```

See plots in Figures 5.18 and 5.19. As we can see this diagnostic indicates that there should be a significant difference among means with no factor interaction. Run the anova

```
> summary(aov(y~f1+f2, data=f.y))
            Df  Sum Sq  Mean Sq    F value   Pr(>F)
f1           3     175    58.33   2.393e+32   <2e-16 ***
f2           4      10     2.50   1.025e+31   <2e-16 ***
Residuals   12       0     0.00
---
Signif. codes:  0 '***' 0.001 '**' 0.01 '*' 0.05 '.' 0.1 ' ' 1
```

and yes indeed we confirm that the *p*-value is very small for both factors.

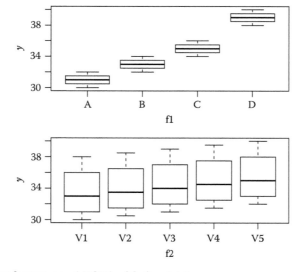

FIGURE 5.18 Boxplots for two-way ANOVA of fictional data.

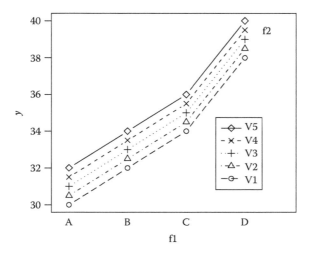

FIGURE 5.19   Interaction plot for two-way ANOVA of fictional data.

For one-way analysis, you can use function `invent.mxn` to generate different datasets and see how the results are less significant as you use means that are more similar and increase within sample standard deviation. For two-way analysis, you can use function `invent.mxn` to generate different datasets and see how the results are less significant as you cluster the groups and increase within range.

For each trial, repeat all plots and the ANOVA table. As you play with different values in invent. mxn, you will gain a better understanding of ANOVA.

5.6.15   COMPUTER EXERCISES

Exercise 5.7
Generate 20 random numbers from an exponential distribution with rate = 1. Run function `cdf. plot.gof` with proper arguments to visualize the potential fit of this sample to an exponential with rate = 1. Repeat the `cdf.plot.gof` with rate = 2 for the same sample. Are there sufficient large differences in each case to suspect lack of fit? Use the K–S test to see if the sample fits an exponential with rate 1, then repeat for rate = 2. Compare and discuss.

Exercise 5.8
Generate 100 random numbers from a normal distribution with $\mu = 300$, $\sigma = 30$. Produce visual GOF exploratory graphs to see if this sample comes from normal distribution. Apply the chi-square and the Shapiro–Wilks test. Discuss.

Exercise 5.9
Use data set airquality. Perform a cross-tabulation of ozone and wind. Perform a contingency analysis of ozone and wind. Use chi-square. Discuss whether there is indication of relation between ozone and wind. Perform a cross-tabulation of ozone and solar radiation. Perform a contingency analysis of ozone and solar radiation. Use chi-square. Discuss whether there is indication of relation between ozone and solar radiation.

Exercise 5.10
Use dataset `immer` of package MASS. Produce boxplots, interaction plot, and run appropriate two-way ANOVA of yield `Y1` as a function of factors `Loc` and `Var`. Discuss results. Are differences in `Loc` significant? Are differences in `Var` significant? Is there factor interaction?

Exercise 5.11

Use function `invent.mxn` to generate a dataset of four groups, such that all group means are separated by 3 units and the standard deviation is 1 unit for all groups. Produce boxplots and one-way ANOVA table. Now, increase standard deviation to 3 units for all groups. Produce boxplots and one-way ANOVA table. Compare results and discuss.

Exercise 5.12

Use function `invent.mxn` with `f2 = "step"` to generate a dataset of four groups, such that group pairs A-B, B-C, C-D are separated by 1 unit and that the range of all groups is 3 units. Produce boxplots, interaction plot, and a two-way ANOVA table. Now, decrease separation of pairs A-B, B-C, C-D such that there is overlap of 2 units between all three pairs. Produce boxplots and two-way ANOVA table. Compare results and discuss.

SUPPLEMENTARY READING

Davis, 2002, Chapter 2, pp. 78–112; Rogerson, 2001, Chapter 4, pp. 65–85; Carr, 1995, Chapter 2, pp. 27–30; MathSoft, 1999, Chapter 4, pp. 4-1–4-16; Chapters 5 and 6; Chapter 11, pp. 11-1–11-21; Chapter 12, pp. 12-25–12-27. Dalgaard, 2008, Chapters 7 and 8; Qian, 2010 Chapter 4, pp. 87–113.

6 Regression

6.1 SIMPLE LINEAR LEAST SQUARES REGRESSION

Let Y be a random variable defined as the "dependent" or "response" variable, and X another random variable defined as the "independent" or "factor" variable. Assume we have a joint sample x_i, y_i, $i = 1, ..., n$ or a set of n paired values of the two variables. This is a **bivariate** or two-variable situation. We already encountered this situation in Chapter 4 when we defined the covariance and correlation coefficient. Let us start this chapter by considering the scatter plot or diagram, which shows data pairs from a sample as markers on the x–y plane. As an example, Figure 6.1 shows pairs of x–y values for air temperature and ground-level ozone. Note that some pairs of points have much larger values of ozone concentration for the same temperature, as the trend would indicate. These are probably outliers and are identified by the observation number next to the marker.

Denote by $\widehat{Y}$ a **linear least squares** (LLS) estimator of Y from X

$$\widehat{Y} = b_0 + b_1 X \tag{6.1}$$

This is the equation of a straight line with **intercept** b_0 and **slope** b_1. For each data point i, we have the **estimated** value of Y at the specific values x_i

$$\widehat{y_i} = b_0 + b_1 x_i \tag{6.2}$$

The **error** (**residual**) for data point i is

$$e_i = y_i - \widehat{y_i} \tag{6.3}$$

And thus another way of writing the relationship of x_i and y_i observations is

$$y_i = b_0 + b_1 x_i + e_i \tag{6.4}$$

Take the square and sum over all observations to obtain the **total squared error**

$$q = \sum_{i=1}^{n} e_i^2 = \sum_{i=1}^{n} (y_i - \widehat{y_i})^2 \tag{6.5}$$

We want to find the value of the **coefficients** (intercept and slope) b_0, b_1 which minimize the sum of squared errors (over all $i = 1, ..., n$). That is to say we want to find b_0, b_1 such that

$$\min_{b_0, b_1} q = \min_{b_0, b_1} \sum_{i=1}^{n} e_i^2 = \min_{b_0, b_1} \sum_{i=1}^{n} (y_i - \widehat{y_i})^2 \tag{6.6}$$

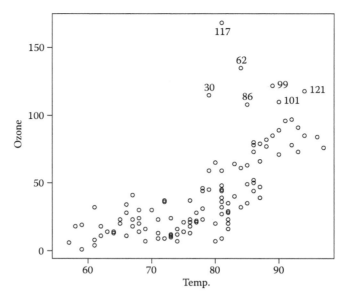

FIGURE 6.1   Scatter plot of x = air temperature and y = ozone concentration with identification of outliers.

How do we find which values of b_0, b_1 minimize q? We express q as a function of b_0, b_1 and then we find the values of b_0, b_1 that make the **gradient** of q equal to zero. The gradient is the **partial derivative** of q with respect to each coefficient. At this point, some of you may need a review of derivatives and optimization. You can skip the following section if you are familiar with the topic.

6.1.1 DERIVATIVES AND OPTIMIZATION

The **derivative** of a function $f(x)$ is denoted by df/dx. It represents a **rate of change** of f with x and is equal to the **gradient** or **slope** of f with respect to x. This assumes that $f(x)$ varies continuously along x. You can think of a derivative as a ratio of very small changes of two variables. For example, a very small change Δf divided by a very small change Δx. The derivative is approximately equal to the slope obtained as the ratio $\Delta f/\Delta x$ (Figure 6.2). Therefore, $(df/dx) \sim (\Delta f/\Delta x)$ when the deltas Δf, Δx are very small or **infinitesimal** and can be referred to as **differentials** df and dt.

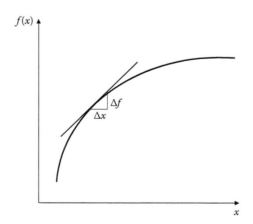

FIGURE 6.2   Concept of derivative of a function.

There are rules to calculate the derivative of a function. The simplest ones are for polynomials or sums of power functions as the following

$$f(x) = ax^c \qquad (6.7)$$

where the independent variable has been raised to a power c. The derivative is the product of the exponent and the variable raised to the power minus one. Therefore, the derivative of f with respect to x is

$$\frac{df}{dx} = acx^{c-1} \qquad (6.8)$$

When $c = 2$, Equation 6.7 is a quadratic or **parabola**. Assume $a = 1$. The derivative is $df/dx = x$, which is a linear function with respect to x.

One application of derivatives is finding an **optimum** (minimum or maximum) of a function with respect to a variable. At an optimum, the derivative takes a value of zero.

For example, the following quadratic function or parabola $f(x) = a + (x - c)^2$ has a derivative $df/dx = 2(x - c)$. At an optimum this derivative must be zero, and this will happen when the term in parenthesis becomes zero, which occurs when $x = c$. Substituting in f we see that the function takes the value $f(c) = a + (c - c)^2 = a$, which is the minimum of f.

In many instances, we have functions of more than just one variable. For example, $f(x, z)$

$$f(x,z) = a + (x - c_1)^2 + (z - c_2)^2 \qquad (6.9)$$

In this case, we can look at two-dimensional (2D) graphs to visualize the function. For example, with $c_1 = c_2 = 1$, Equation 6.9 yields the results shown in Figure 6.3 when represented as isolines

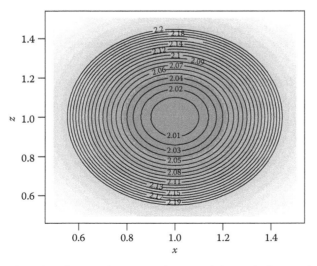

FIGURE 6.3 Parabolic function of two variables showing the minimum $f = 2$ at $x = 1$ and $z = 1$.

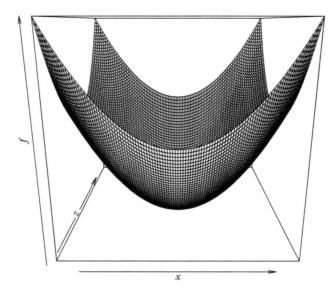

FIGURE 6.4   Perspective plot of a parabolic function of two variables showing the minimum at $x = 1$ and $z = 1$.

or contour lines (or lines of equal value of f) overlaid on an image representation and shown as a perspective plot in Figure 6.4.

When we have more than one variable, we can take the derivative of the function with respect to each one of the variables. These are the **partial derivatives** of the function. In the example given earlier, we can have two partial derivatives, one with respect to x and another with respect to z

$$\frac{\partial f}{\partial x} = 2(x - c_1) \qquad \frac{\partial f}{\partial z} = 2(z - c_2) \tag{6.10}$$

Note that the symbol ∂/∂ is used for partial derivative which is different to the one used for ordinary derivatives d/d.

One application of functions of multiple variable and partial derivatives is finding an **optimum** (minimum or maximum) of a function. At an optimum, all the partial derivatives take a value of zero. From Equation 6.10, we see that this occurs when $x = c_1$ and $z = c_2$. Substituting these values in Equation 6.9 we see that the minimum is a.

We will use this concept in the next section to calculate the regression coefficients.

6.1.2   Calculating Regression Coefficients

Coming back to the total error given in Equation 6.5, to obtain the sum of squared errors as a function of the coefficients b_0, b_1, substitute Equation 6.2 in Equation 6.5

$$q = \sum_{i=1}^{n} e_i^2 = \sum_{i=1}^{n} (y_i - \widehat{y}_i)^2 = \sum_{i=1}^{n} (y_i - b_0 - b_1 x_i)^2 \tag{6.11}$$

Expand the square of the sum of three terms in the summand

$$q = \sum_{i=1}^{n} (y_i - b_0 - b_1 x_i)^2 = \sum_{i=1}^{n} \left(y_i^2 + b_0^2 + b_1^2 x_i^2 + 2b_0 b_1 x_i - 2b_1 x_i y_i - 2b_0 y_i \right) \tag{6.12}$$

And now we find partial derivatives of q with respect to b_0, b_1. Start with b_0

$$\frac{\partial q}{\partial b_0} = \frac{\partial}{\partial b_0} \sum_{i=1}^{n} \left(y_i^2 + b_0^2 + b_1^2 x_i^2 + 2b_0 b_1 x_i - 2b_1 x_i y_i - 2b_0 y_i \right)$$

$$= \sum_{i=1}^{n} (2b_0 + 2b_1 x_i - 2y_i) = 2\left(nb_0 + b_1 \sum_{i=1}^{n} x_i - \sum_{i=1}^{n} y_i \right) \tag{6.13}$$

Set this derivative to zero

$$\frac{\partial q}{\partial b_0} = 2\left(nb_0 + b_1 \sum_{i=1}^{n} x_i - \sum_{i=1}^{n} y_i \right) = 0 \tag{6.14}$$

Note that for the left-hand side to be zero the term in parenthesis must be equal to zero. Then separate terms in b_0 and b_1

$$nb_0 = -b_1 \sum_{i=1}^{n} x_i + \sum_{i=1}^{n} y_i \tag{6.15}$$

and solve for b_0

$$b_0 = -\frac{b_1}{n} \sum_{i=1}^{n} x_i + \frac{1}{n} \sum_{i=1}^{n} y_i \tag{6.16}$$

Now take derivative with respect to b_1 and repeat the process

$$\frac{\partial q}{\partial b_1} = \frac{\partial}{\partial b_1} \sum_{i=1}^{n} \left(y_i^2 + b_0^2 + b_1^2 x_i^2 + 2b_0 b_1 x_i - 2b_1 x_i y_i - 2b_0 y_i \right)$$

$$= \sum_{i=1}^{n} \left(2b_1 x_i^2 + 2b_0 x_i - 2x_i y_i \right) = 2\left(b_1 \sum_{i=1}^{n} x_i^2 + b_0 \sum_{i=1}^{n} x_i - \sum_{i=1}^{n} x_i y_i \right) \tag{6.17}$$

Set to zero

$$\frac{\partial q}{\partial b_1} = 2\left(b_1 \sum_{i=1}^{n} x_i^2 + b_0 \sum_{i=1}^{n} x_i - \sum_{i=1}^{n} x_i y_i \right) = 0 \tag{6.18}$$

Note that the term multiplying 2 must be zero

$$b_1 \sum_{i=1}^{n} x_i^2 + b_0 \sum_{i=1}^{n} x_i - \sum_{i=1}^{n} x_i y_i = 0 \tag{6.19}$$

We can substitute our prior result given in Equation 6.16 for b_0 to obtain

$$b_1 \sum_{i=1}^{n} x_i^2 + \left[-\frac{b_1}{n} \sum_{i=1}^{n} x_i + \frac{1}{n} \sum_{i=1}^{n} y_i \right] \sum_{i=1}^{n} x_i - \sum_{i=1}^{n} x_i y_i = 0 \tag{6.20}$$

Now separate terms in b_1

$$b_1 \left[\sum_{i=1}^{n} x_i^2 - \frac{1}{n} \left(\sum_{i=1}^{n} x_i \right)^2 \right] = -\frac{1}{n} \sum_{i=1}^{n} y_i \sum_{i=1}^{n} x_i + \sum_{i=1}^{n} x_i y_i \tag{6.21}$$

Finally solve for b_1

$$b_1 = \frac{\sum_{i=1}^{n} x_i y_i - \frac{1}{n} \sum_{i=1}^{n} y_i \sum_{i=1}^{n} x_i}{\sum_{i=1}^{n} x_i^2 - \frac{1}{n} \left(\sum_{i=1}^{n} x_i \right)^2} \tag{6.22}$$

Once b_1 is calculated using Equation 6.22, then we can calculate b_0 using Equation 6.16 repeated now for easy reference

$$b_0 = -\frac{b_1}{n} \sum_{i=1}^{n} x_i + \frac{1}{n} \sum_{i=1}^{n} y_i \tag{6.23}$$

Once we have values for slope and intercept, then we can draw the regression line on the same graph as the scatter plot as illustrated in Figure 6.5.

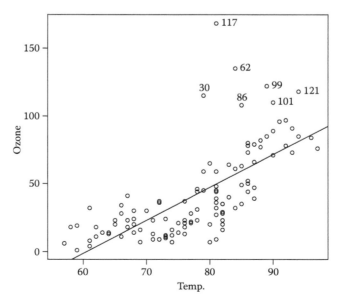

FIGURE 6.5 Regression line plotted on top of scatter plot.

6.1.3 INTERPRETING THE COEFFICIENTS USING SAMPLE MEANS, VARIANCES, AND COVARIANCE

We can re-arrange Equation 6.22 in terms of sample means of X and Y:

$$b_1 = \frac{\sum_{i=1}^{n}(x_i - \bar{X})(y_i - \bar{Y})}{\sum_{i=1}^{n}(x_i - \bar{X})^2} = \frac{s_{\text{cov}}(X,Y)}{s_x^2} \tag{6.24}$$

Here, the numerator is the sample covariance of X and Y, whereas the denominator is the sample variance of X.

We repeat Equation 6.16 here for easy reference and take note that the components are the sample means of Y and X

$$b_0 = \frac{1}{n}\sum_{i=1}^{n}y_i - \frac{b_1}{n}\sum_{i=1}^{n}x_i = \bar{Y} - b_1\bar{X} \tag{6.25}$$

In summary, Equations 6.22 or 6.24 and 6.25 are used to calculate the coefficients b_0, b_1.

Rewriting Equation 6.25 as $\bar{Y} = b_0 + b_1\bar{X}$, we note that the regression line goes through the sample means of X and Y. Using the correlation coefficient in Equation 6.24, we can rewrite as

$$b_1 = \frac{s_{\text{cov}}(X,Y)}{s_X^2} = \frac{rs_Xs_Y}{s_X^2} = r\frac{s_Y}{s_X} \tag{6.26}$$

In other words, the slope is the sample correlation coefficient multiplied by the ratio of sample standard deviations of Y over X.

Let us work out a numerical example with just a few values to illustrate the ideas. Suppose we have $n = 5$ pairs of values for X and Y: (2, 9.00), (4, 9.88), (6, 17.04), (8, 12.46), (10, 25.07). Calculate the sample means $X = (2+4+6+8+10)/5 = 6$ and $\bar{Y} = (9.00+9.88+17.04+12.46+25.07)/5 = 14.69$. Calculate the sample variances

$$s_X^2 = \frac{1}{4}\left((4+16+36+64+100) - \frac{1}{5}30^2\right) = \frac{1}{4}(220-180) = 10$$

and

$$s_Y^2 = \frac{1}{4}\left((9.00^2 + 9.88^2 + 17.04^2 + 12.46^2 + 25.07^2) - \frac{1}{5}73.45^2\right)$$

$$= \frac{1}{4}(1252.73 - 1078.98) = 43.44$$

(*continued*)

(continued)

Then the sample standard deviations are $s_X = \sqrt{10} = 3.16$ and $s_Y = \sqrt{43.44} = 6.59$. To get the sample covariance of X and Y, first calculate the sum of products $s_{XY} = 2 \times 9 + 4 \times 9.88 + 6 \times 17.04 + 8 \times 12.46 + 10 \times 25.07) = 510.14$, and then subtract the product of the sums

$$s_{\text{cov}}(X,Y) = \frac{1}{4}\left(510.14 - \frac{1}{5}(30 \times 73.45)\right) = 17.32$$

The slope is $b_1 = s_{\text{cov}}(X,Y)/s_X^2 = 17.32/10 = 1.732$ and the intercept is $b_0 = \bar{Y} - b_1\bar{X} = 14.69 - 1.73 \times 6 = 4.31$. The equation for the linear predictor is $Y = 4.31 + 1.73X$. Apply this equation to one of the x values as an illustration, say $x_2 = 4$, $y_2 = 4.31 + 1.73 \times 4 = 11.23$ the squared error for this value of y is $e_2 = (9.88 - 11.23)^2 = 1.82$.

6.1.4 REGRESSION COEFFICIENTS FROM EXPECTED VALUES

The equations of the previous section can also be developed using theoretical population and expected values. Those equations are then sample estimates of the population regression coefficients as derived here.

The mean square error is given by the expected value

$$q = E\left[(Y - b_0 - b_1 X)^2\right] \tag{6.27}$$

Expand the square of the sum of three terms in the summand

$$q = E\left[Y^2 + b_0^2 + b_1^2 X^2 + 2b_0 b_1 X - 2b_1 XY - 2b_0 Y\right] \tag{6.28}$$

Find partial derivatives of q with respect to the coefficients. Start with b_0 and set equal to zero

$$\frac{\partial q}{\partial b_0} = 2b_0 + 2b_1 E(X) - 2E(Y) = 2b_0 + 2b_1 \mu_X - 2\mu_Y = 0 \tag{6.29}$$

Solve for b_0

$$b_0 = \mu_Y - b_1\mu_X \tag{6.30}$$

Now take derivative with respect to b_1 and set equal to zero

$$\frac{\partial q}{\partial b_1} = E(2b_1 X^2 + 2b_0 X - 2XY) = 2(b_1 E(X^2) + b_0 E(X) - E(XY)) = 0 \tag{6.31}$$

We can substitute our prior result in Equation 6.30 for b_0 to obtain

$$b_1 E(X^2) + (\mu_Y - b_1\mu_X)\mu_X - E(XY) = 0 \tag{6.32}$$

Now separate terms in b_1

$$b_1\left[E(X^2)-\mu_X^2\right] = E(XY)-\mu_X\mu_Y \tag{6.33}$$

Finally solve for b_1

$$b_1 = \frac{E(XY)-\mu_X\mu_Y}{E(X^2)-\mu_X^2} = \frac{\text{cov}(X,Y)}{\sigma_X^2} \tag{6.34}$$

Once b_1 is calculated using Equation 6.34, we can calculate b_0 using Equation 6.30. Equation 6.34 says that the slope is the covariance divided by the variance of X, whereas Equation 6.30 says that the intercept is obtained by subtracting the mean of X scaled by the slope from the mean of Y.

We repeat Equation 6.34 in terms of the correlation coefficient

$$b_1 = \frac{\rho_{XY}\sigma_X\sigma_Y}{\sigma_X^2} = \rho_{XY}\frac{\sigma_Y}{\sigma_X} \tag{6.35}$$

In other words, the slope is the correlation coefficient scaled by the ratio of standard deviations of Y over X. In summary, Equations 6.34 and 6.30 are the population coefficients b_0, b_1. The corresponding equations in the previous section are sample estimates of these coefficients.

6.1.5   INTERPRETATION OF THE ERROR TERMS

There are three important error terms for Y in the regression: the residual $e_i = y_i - \widehat{y}_i$, the difference with respect to the mean $y_i - \bar{Y}$, and the error of the estimate with respect to the mean $\widehat{y}_i - \bar{Y}$. We can relate them by formulating an identity in the following manner:

$$(y_i - \bar{Y}) = (y_i - \widehat{y}_i) + (\widehat{y}_i - \bar{Y}) \tag{6.36}$$

See Figure 6.6 (see also Davis, 2002).

Now if we square both sides of Equation 6.36, we get

$$(y_i - \bar{Y})^2 = (y_i - \widehat{y}_i)^2 + (\widehat{y}_i - \bar{Y})^2 + 2(y_i - \widehat{y}_i)(\widehat{y}_i - \bar{Y}) \tag{6.37}$$

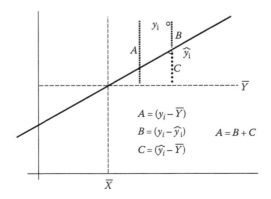

FIGURE 6.6   Error terms in regression.

And then sum over all observations, we get

$$\sum (y_i - \bar{Y})^2 = \sum (y_i - \widehat{y_i})^2 + \sum (\widehat{y_i} - \bar{Y})^2 + 2 \sum (y_i - \widehat{y_i})(\widehat{y_i} - \bar{Y}) \tag{6.38}$$

The cross-product term is zero. Therefore,

$$\sum (y_i - \bar{Y})^2 = \sum (y_i - \widehat{y_i})^2 + \sum (\widehat{y_i} - \bar{Y})^2 \tag{6.39}$$

In the following, SS denotes "sum of squares." The "total error" SS_T in Y is the sum of squared differences of observations minus the mean

$$SS_T = \sum (y_i - \bar{Y})^2 \tag{6.40}$$

The "residual" or "unexplained" error SS_E is the sum of the squares of the difference between observations and estimated values

$$SS_E = \sum (y_i - \widehat{y_i})^2 \tag{6.41}$$

The "model" or "explained" error SS_M is the sum of squared differences of estimated points minus the mean

$$SS_M = \sum (\widehat{y_i} - \bar{Y})^2 \tag{6.42}$$

Now using Equation 6.39 we write that total error is the sum of model error and the residual, that is to say

$$SS_T = SS_E + SS_M \tag{6.43}$$

And equivalently the residual is the total error minus model error

$$SS_E = SS_T - SS_M \tag{6.44}$$

Please note that when these quantities are divided by n, the number of observations, then we get the mean squares (MS). The $MS_T = s_Y^2$ sample variance of Y, for SS_M the average or mean squared model error is MS_M, and for SS_E the average or mean squared residual error (MS_E).

$$s_Y^2 = MS_E + MS_M \tag{6.45}$$

A common measure of goodness of fit is R^2, which is the ratio of the model error to the total error

$$R^2 = \frac{MS_M}{MS_T} = \frac{MS_M}{MS_E + MS_M} = \frac{MS_M}{s_Y^2} \tag{6.46}$$

When MS_E (which is minimized by the least squares procedure) is very small, then R^2 approaches 1. Note that R^2 is the fraction (or percent) of variance of Y explained by the regression model. Note that

$$1 - R^2 = \frac{MS_E}{MS_T} = \frac{MS_E}{MS_E + MS_M} = \frac{MS_E}{s_Y^2} \tag{6.47}$$

The difference of R^2 with respect to 1 is the fraction of variance of Y unexplained by the regression. In addition, it is important to realize that

$$R^2 = \frac{MS_M}{s_Y^2} = \frac{(1/n)\sum\left(b_0 + b_1 X - \bar{Y}\right)^2}{s_Y^2} = \frac{(1/n)\sum\left(\bar{Y} - b_1\bar{X} + b_1 X - \bar{Y}\right)^2}{s_Y^2} \tag{6.48}$$

The sample mean of Y cancels, b_1 is a common factor, and recognizing the sample variance of X we obtain

$$R^2 = \frac{(1/n)\sum b_1^2 (X - \bar{X})^2}{s_Y^2} = \frac{b_1^2 (1/n)\sum (X - \bar{X})^2}{s_Y^2} = \frac{b_1^2 s_X^2}{s_Y^2} \tag{6.49}$$

And by recalling the expression for b_1

$$R^2 = \frac{\left(\dfrac{s_{\text{cov}}(X,Y)}{s_X^2}\right)^2 s_X^2}{s_Y^2} \tag{6.50}$$

Therefore,

$$R^2 = \left(\frac{s_{\text{cov}}(X,Y)}{s_X s_Y}\right)^2 = r^2 \tag{6.51}$$

The square root of R^2 is equal to r, which is the correlation coefficient.

As a numerical example, let us use the data of the example given earlier. Suppose we have $n = 5$ pairs of values for x and y: (2, 9.00), (4, 9.88), (6, 17.04), (8, 12.46), (10, 25.07). Recall the sample mean of y is $\bar{Y} = 14.69$. The total error is

$$SS_T = (9.00 - 14.69)^2 + (9.88 - 14.69)^2 + (17.04 - 14.69)^2$$

$$+ (12.46 - 14.69)^2 + (25.07 - 14.69)^2 = 173.75$$

(continued)

(continued)

Then calculate all the predicted values

$$y_1 = 4.31 + 1.73 \times 2 = 7.77$$

$$y_2 = 4.31 + 1.73 \times 4 = 11.23$$

$$y_3 = 4.31 + 1.73 \times 6 = 14.69$$

$$y_4 = 4.31 + 1.73 \times 8 = 18.15$$

$$y_5 = 4.31 + 1.73 \times 10 = 21.61$$

The total residual error is

$$SS_E = (9.00 - 7.77)^2 + (9.88 - 11.23)^2 + (17.04 - 14.69)^2$$

$$+ (12.46 - 18.15)^2 + (25.07 - 21.61)^2 = 53.20$$

The explained error is $SS_M = SS_T - SS_E = 173.75 - 53.20 = 120.55$. Now $R^2 = MS_M / MS_T = 120.55/173.75 = 0.69$, which should be equivalent to

$$R^2 = \left(\frac{s_{\mathrm{cov}}(X,Y)}{s_X s_Y} \right)^2 = \left(\frac{17.32}{3.16 \times 6.59} \right)^2 = 0.832^2 = 0.69$$

The correlation coefficient is $r = 0.832$.

6.1.6 EVALUATING REGRESSION MODELS

Once we have the coefficient values of a linear regression model, we write the predictor equation. For example, for the linear regression illustrated in Figure 6.5, we obtain

$$y = -147 + 2.43x \tag{6.52}$$

where
y is the ozone concentration
x is the air temperature

This means that ozone increases with temperature and that when $x = 0$ we have a value of negative 147. This value does not make sense in terms of concentration. This coefficient value is just the one that minimizes the residual error in the range of temperature and ozone of interest. For this model, the R^2 is 0.48, which is a relatively low value, meaning that less than 50% of the variance in ozone can be explained by this model. Therefore, other factors must be important; hence, the common expression "something else must be going on."

When evaluating a regression model, we should look at other diagnostics besides the R^2. It is necessary to check the assumptions of the method and to examine the significance of the regression, the confidence interval, and the trend in the unexplained or residual error. The following are the main aspects to analyze.

First, we should examine how sound is the assumption of **linearity**. We do this by looking at the graph of Y vs. X (scatter diagrams). If the y_i points seem to follow a definite nonstraight pattern or curve, then linearity is suspicious even when getting a good R^2 (Davis, 2002, p. 192). For example,

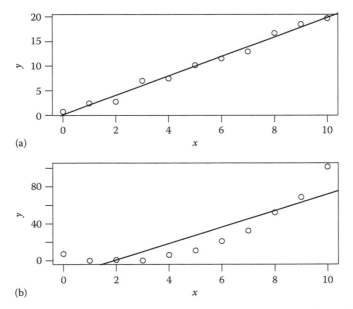

FIGURE 6.7 Contrasting examples of satisfying the linearity assumption. (a) Data follow linear relationship. (b) Data follow a curve or nonlinear pattern.

the scatter plot in Figure 6.5 may look straight disregarding the outliers, but a closer inspection indicates a curve because ozone seems to increase more rapidly as temperature increases. For contrasting hypothetical examples, look at Figure 6.7. Figure 6.7a shows one case in which the data suggest a linear relation, but Figure 6.7b illustrates a nonlinear pattern and lack of fit to the straight line obtained by linear regression.

Second, we look at the statistical significance of the **slope**, or relative magnitude of the MS_T and the MS_E. This we do using ANOVA to test whether the slope is zero. In this case, we need large F (ratio MS_M/MS_E) to reject H0, telling us that there is a nonzero slope (Davis, 2002, pp. 197–200). In the example of Figure 6.5, the F value is 108.5 with 1 and 114 degrees of freedom, which yields a p-value of 2.2×10^{-16}. The low p-value suggests that we reject H0 and therefore the slope is different from zero; that is, there is a highly significant relationship between the two variables.

Third, we evaluate the **residual error** using residual diagnostic plots like the ones shown in Figure 6.8 corresponding to the ozone vs. temperature example. At the top left, residual vs. fitted is a plot of the residuals as a function of the estimated or predicted Y. The residuals should be scattered up and down around zero (i.e., "random noise"), telling us that the error is independent of the position in the regression line. If the errors or residuals follow a pattern, then the error changes with the predicted Y. This is called **heteroscedasticity**. In the example of the figure, the errors show a pattern because instead of following a uniform or **homogeneous** spread along the horizontal axis, they dip at the midrange of fitted values. In addition, there are large positive values that seem to be outliers. At the bottom-left, we have the square root of standardized residuals as a function of fitted values. This confirms the pattern in residuals.

At the top-right, we have Q–Q plot of residuals to check for **normality** of the residuals. In this example, we note that the residual error exhibits near normal behavior only for low values of error. It departs significantly for large negative and especially for large positive errors; in addition, several observations seem to be outliers (30, 62, and 117).

At the bottom-right panel, we have a more complicated graph. It is a plot of residual vs. leverage. The term **leverage** denotes influence of outliers on the regression line. **Cook's distance** measures the effect of removing an observation. The plot displays the influence of each observation on the regression by using leverage on the horizontal axes and contour lines of Cook's distance. We can

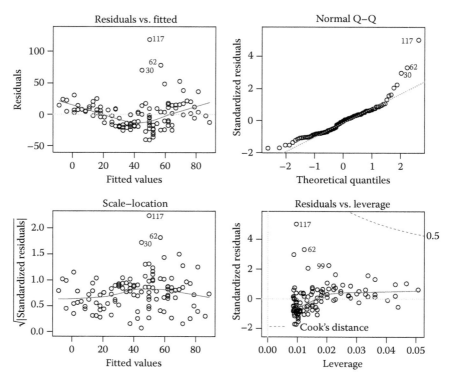

FIGURE 6.8 Residual diagnostic plots.

use these metrics to determine the influence of an observation. The higher the residual, leverage, and Cook's distance, the most influential the observation. High-leverage occurs roughly when its value exceeds $2 \times m/n$ or twice the number of coefficients (m) divided by number of observations (n). Values of Cook's distance larger than 1 are considered to have large influence.

In this example, high leverage would occur for $2 \times 2/153 = 0.03$. Here, the points with high leverage have relatively low value of residual and Cook's distance (these are <0.5 contour line). However, some points with value of leverage about 0.01–0.02 have large residual. These are observations numbered 99, 62, and 117 and these are most influential.

Fourth, we examine the **significance** of each coefficient b_0 and b_1 being different from zero using t-tests. Denote standard error of b_0 and b_1 as se_{b0} and se_{b1}

$$t_0 = \frac{b_0 - 0}{se_{b_0}} \quad \text{and} \quad t_1 = \frac{b_1 - 0}{se_{b_1}} \tag{6.53}$$

See expressions in Davis (2002, p. 201) for the standard error of the coefficients. The first t-test is often meaningless because there is no natural interpretation of the intercept. However, one case of interest would be if we have reason to believe we should consider a zero-intercept model (or regression through the origin). We will discuss this issue further in the next section. The second t-test is equivalent to the F-test performed earlier; it would tell us if the slope is different from zero. For the example at hand, we get $se_{b0} = 18.28$, $t_0 = 8.04$ with p-value 9.4×10^{-13} and $se_{b1} = 0.23$, $t_1 = 10.41$ with p-value 2×10^{-16}. We conclude that both the slope and intercept are nonzero with high significance.

Fifth, we calculate the **residual standard error** or standard deviation of the residuals

$$s_e = \sqrt{\frac{\sum (y_i - \hat{y}_i)^2}{n - 2}} \tag{6.54}$$

where we have used $n - 2$ because two parameters were estimated. In the ozone vs. temperature example, we calculate the residual standard error to be 23.71.

Sixth, calculate **confidence intervals** to express the reliability of the estimates given by the regression and of the regression coefficients (Davis, 2002, pp. 200–204). Recall from Chapter 4 that we defined the confidence level $(1 - \alpha)$ of an interval [a, b] as the probability that the interval [a,b] includes the parameter θ. The error rate α is the probability that the interval [a,b] fails to include the parameter value θ; common values for α are 0.01, 0.05, 0.1 or 1%, 5%, and 10%.

The confidence interval for a given α can be calculated using the standard error. Recall that the standard error of the estimate of the mean is the standard deviation of the sample mean $\bar{X}$. In addition, this is the same as the standard deviation of X divided by the square root of the sample size

$$\sigma_e = \sqrt{\sigma_{\bar{X}}^2} = \sqrt{\frac{\sigma_X^2}{n}} = \frac{\sigma_X}{\sqrt{n}} \qquad (6.55)$$

For a standard normal variable, we know that 95% of the probability is within two standard deviations of the mean. Therefore, with 95% probability the sample mean $\bar{X}$ is within $\bar{X} \pm 2\sigma_e$. Because of Equation 6.55 we see that this confidence interval depends on the sample size. Therefore, we can narrow the confidence interval at a given level α by increasing the number of observations.

When the number of observations is large, the standard normal is a good approximation to the distribution of the sample means. However, for small sample size it is better to use the t distribution (Davis, 2002, p. 72). In this case, the interval is given by the value of t that yields $\alpha/2$ probability in each tail

$$\bar{X} \pm t_{\alpha/2,n-1}\sigma_e \qquad (6.56)$$

Therefore, for the regression coefficients, we use a t statistic and obtain confidence intervals for both coefficients (Davis, 2002, p. 201). For the estimate of Y, we obtain a band composed of confidence intervals at each observation (Davis, 2002, p. 202).

The standard error of the estimate $\hat{Y}$ with respect to Y is the mean square error (MS_E). The standard error of estimate of observation y_i is

$$s_{\hat{y}i}^2 = MS_E\left(1 + \frac{(x_i - \bar{X})^2}{s_X^2}\right) = \frac{s_e^2}{n}\left(1 + \frac{(x_i - \bar{X})^2}{s_X^2}\right) \qquad (6.57)$$

And taking the square root

$$s_{\hat{y}i} = \frac{s_e}{\sqrt{n}}\sqrt{1 + \frac{(x_i - \bar{X})^2}{s_X^2}} \qquad (6.58)$$

which we can use to build confidence intervals using a t value for a given α. Note that the expression is quadratic and that it increases with departures of x_i from the sample mean. See Figure 6.9 for the ozone example where we note that many observations in the midrange fall outside the confidence interval reinforcing the suggestion of lack of fit to a straight line.

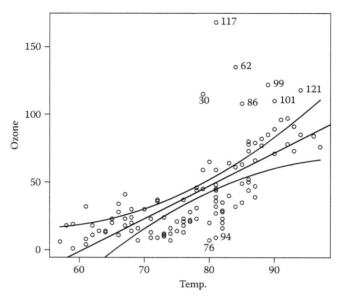

FIGURE 6.9   Scatter plot, regression line, and 90% confidence interval.

6.1.7 REGRESSION THROUGH THE ORIGIN

Now, you may wonder about the meaning of the intercept. As shown earlier, we can use a t value given by Equation 6.53 to determine that the intercept is significant. In many cases, interpreting the value of Y when $X = 0$ is meaningless since we cannot extrapolate outside the range for which we derived the regression equation. However, there are cases where it makes sense to claim that the intercept coefficient to be zero, that is to say, for the regression line to go through the origin. Some of these cases are those for which we know that $y = 0$ for $x = 0$. In these cases, we formulate a different linear regression model

$$\widehat{Y} = bX \tag{6.59}$$

where b is the slope. To estimate b, go back to Equation 6.11 and rewrite using Equation 6.59:

$$q = \sum_{i=1}^{n} e_i^2 = \sum_{i=1}^{n} (y_i - \widehat{y_i})^2 = \sum_{i=1}^{n} (y_i - bx_i)^2 \tag{6.60}$$

And expand

$$q = \sum_{i=1}^{n} \left(y_i^2 + b^2 x_i^2 - 2bx_i y_i \right) \tag{6.61}$$

Find derivative of q with respect to b

$$\frac{dq}{db} = \sum_{i=1}^{n} \left(2bx_i^2 - 2x_i y_i \right) = 2 \left(b \sum_{i=1}^{n} x_i^2 - \sum_{i=1}^{n} x_i y_i \right) \tag{6.62}$$

Setting this derivative to zero we obtain that the right-hand-side term in parenthesis must be 0

$$\left(b \sum_{i=1}^{n} x_i^2 - \sum_{i=1}^{n} x_i y_i \right) = 0 \tag{6.63}$$

Therefore, b can be solved to be

$$b = \frac{\sum_{i=1}^{n} x_i y_i}{\sum_{i=1}^{n} x_i^2} \tag{6.64}$$

A problem is that the regression line may not go through the point given by the means $(\bar{X}, \bar{Y})$. There are different ways of evaluating a model without intercept (Eisenhauer, 2003). For example, the SS_T given by Equation 6.40 is not the best way to evaluate this model, and instead we can use the deviations with respect to $Y = 0$.

As mentioned earlier, there are three important error terms for Y in the regression: the residual $e_i = y_i - \hat{y}_i$, the difference with respect to the zero $y_i - 0$, and the error of the estimate with respect to zero $\hat{y}_i - 0$. We relate these by formulating an identity in the following manner

$$(y_i - 0) = (y_i - \hat{y}_i) + (\hat{y}_i - 0) \tag{6.65}$$

See Figure 6.10.

Now if we square both sides of Equation 6.65, we get

$$(y_i)^2 = (y_i - \hat{y}_i)^2 + (\hat{y}_i)^2 + 2(y_i - \hat{y}_i)(\hat{y}_i) \tag{6.66}$$

And then sum over all observations, we get

$$\sum (y_i)^2 = \sum (y_i - \hat{y}_i)^2 + \sum (\hat{y}_i)^2 + 2 \sum (y_i - \hat{y}_i)(\hat{y}_i) \tag{6.67}$$

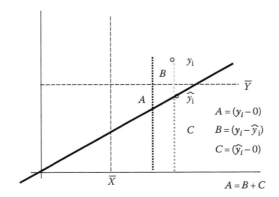

FIGURE 6.10 Error terms in regression through the origin.

As mentioned earlier, the cross-product term is zero. Therefore,

$$\sum (y_i)^2 = \sum (y_i - \widehat{y_i})^2 + \sum (\widehat{y_i})^2 \qquad (6.68)$$

In the following, SS denotes "sum of squares." We redefine terms. The "total error" SS_T in Y is the sum of squared differences of sample points

$$SS_T = \sum y_i^2 \qquad (6.69)$$

The "residual" or "unexplained" error SS_E does not change. It still is the sum of the squares of the difference between estimated and observations

$$SS_E = \sum (y_i - \widehat{y_i})^2 \qquad (6.70)$$

The "model" or "explained" error SS_M is the sum of squared differences of estimated points (see Equation 4.20, Davis, 2002)

$$SS_M = \sum \widehat{y_i}^2 \qquad (6.71)$$

Now using Equation 6.68 we write that total error is the sum of model error and the residual, that is to say

$$SS_T = SS_E + SS_M \qquad (6.72)$$

Therefore, goodness of fit or ratio of the model error to the total error gives a new R^2, which is

$$R^2 = \frac{MS_M}{MS_T} = \frac{\sum \widehat{y_i}^2}{\sum y_i^2} \qquad (6.73)$$

Note that this new R^2 is always greater (or equal) than R^2 given in Equation 6.46 for the model with intercept. Therefore, it is not fair to compare fit of the no-intercept model to the fit of the intercept model.

The SS_E and consequently the standard error formulation do not change. However, the degrees of freedom of SS_E are $n-1$ instead of $n-2$, since we have estimated only one parameter. Consequently, Equation 6.54 changes to

$$s_e = \sqrt{\frac{\sum (y_i - \hat{y}_i)^2}{n-1}} \qquad (6.74)$$

Consequently, comparing standard errors is an option to compare fit of the no-intercept model to the fit of the intercept model. Another option is to use the sample correlation coefficient between observed y_i and predicted $\widehat{y}_i$ values.

6.2   ANOVA AS PREDICTIVE TOOL

Let us look again at ANOVA but with a different perspective. For this purpose, we start by recalling the linear regression predictor given in Equation 6.4

$$y_i = b_0 + b_1 x_i + e_i \tag{6.75}$$

The ANOVA model is also a predictor, but of observations of X based on the sample means $\overline{X}_i$ (average of observations in each sample or group or level i) and the overall or grand mean $\overline{X}$ (average of all observations). To see this use the **among** variation $\delta_i = \overline{X}_i - \overline{X}$ of the sample mean from the grand mean, and the **within** variation $e_{ij} = x_{ij} - \overline{X}_i$ of the observation x_{ij} from the sample mean. We write the predictor as

$$x_{ij} = \overline{X} + \delta_i + e_{ij} \tag{6.76}$$

In similar fashion as Equation 6.75. In other words, we could estimate values of X from the grand mean plus the among variation, with a residual error:

$$\widehat{x_{ij}} = \overline{X} + \delta_i \tag{6.77}$$

However, realize this is just an identity

$$x_{ij} = \overline{X} + \delta_i + e_{ij} = \overline{X} + (\overline{X}_i - \overline{X}) + (x_{ij} - \overline{X}_i) = x_{ij} \tag{6.78}$$

In addition, we see that we assume that the outcome of ANOVA table indicates significant differences among groups, so that e_{ij} are small compared to δ_i. We also see that we assume that the residuals e_{ij} should come from an independent normally distributed variable with zero mean and constant variance. This assumption implies that we should check the residuals for normality. The example of Figures 6.11 and 6.12 illustrates one case when the residuals seem to be normally distributed.

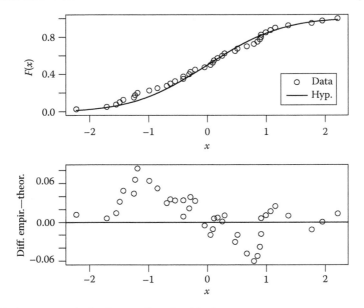

FIGURE 6.11   Exploratory analysis of normality of residuals.

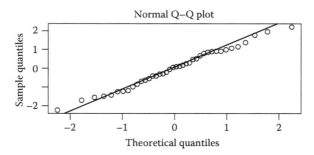

FIGURE 6.12   Q–Q plot for exploratory analysis of normality of residuals.

6.3   NONLINEAR REGRESSION

Many times the linear regression model

$$\hat{Y} = b_0 + b_1 X \qquad (6.79)$$

does not yield the best predictor model for Y as illustrated in the example of Figure 6.13a (solid line). This is an example of the attenuation of light with depth in the water column of a small estuary.

However, when we can write a sum

$$\hat{Y} = b_0 + b_1 g(X) \qquad (6.80)$$

we can still use linear regression after calculating $g(X)$ and using it instead of X in Equation 6.79. For example, when $g(X) = X^2$, the predictor

$$\hat{Y} = b_0 + b_1 X^2 \qquad (6.81)$$

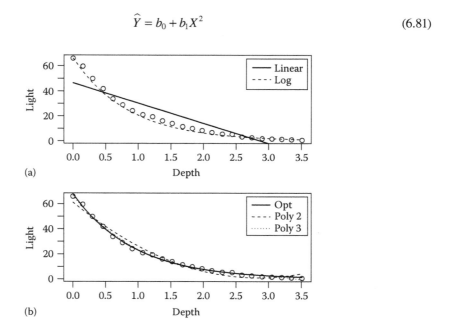

FIGURE 6.13   (a) A linear regression yields poor fit; a log-transformation is a better fit. (b) Nonlinear regression by polynomial regression and by optimization.

can be treated by linear regression after substituting X^2 for X and using it in Equation 6.79. This is a special case of polynomial regression explained later in this chapter.

However, often we need to use a nonlinear equation relating Y to X of the general form

$$\widehat{Y} = f(X, p) \tag{6.82}$$

where
  $f()$ is a function
  p are parameters

For example, an exponential model with parameters k and Y_0 of the form is a good explanatory model for light attenuation

$$\widehat{Y} = Y_0 \exp(-kX) \tag{6.83}$$

known as Beer–Lambert law. Here, k is extinction coefficient and Y_0 is the light just below the surface which is the measure reported at $X = 0$.

6.3.1 LOG TRANSFORM

In cases like the one in Equation 6.83, we can transform the nonlinear model into a linear model by use of logarithms. See Davis (2002). Take the natural log of both sides of Equation 6.83; recall that log of a product is the sum of the logs, that is, $ln(AB) = \ln(A) + \ln(B)$ and that log of an exponential is the variable itself because log is the inverse function of exp, that is, $\ln(\exp(A)) = A$ to get

$$\ln(\widehat{Y}) = \ln(Y_0) - kX \tag{6.84}$$

Now use linear regression substituting $\ln(\widehat{Y})$ for $\widehat{Y}$, $\ln(Y_0)$ for b_0 and $-k$ for b_1. Once we determine Y_0 and k, we can use the predictor Equation 6.83 as illustrated by one of the curves in the example of Figure 6.13a (dashed line). We can appreciate in this figure a better fit than the one obtained by linear regression illustrated by the solid line.

6.3.2 NONLINEAR OPTIMIZATION

Whenever we cannot transform Equation 6.82 into a linear regression problem, we can apply the process of optimization to solve the nonlinear regression problem. This consists of postulating the function that may fit the data, for example, an exponential curve, and then use an optimization algorithm to minimize the square error (residuals) with respect to the coefficients. In other words, find the value of the coefficients that would yield a minimum square error.

The **error (residual)** for data point i is

$$e_i = y_i - \widehat{y}_i = y_i - f(x_i, p) \tag{6.85}$$

Take the square and sum over all observations to obtain the **total squared error**

$$q = \sum_{i=1}^{n} e_i^{\,2} = \sum_{i=1}^{n} \left(y_i - f(xi, p) \right)^2 \tag{6.86}$$

We want to find the values of the **coefficients** p which minimize the sum of squared errors (over all $i = 1, ..., n$), that is to say, find p such that

$$\min_p q = \min_p \sum_{i=1}^{n} e_i^2 = \min_p \sum_{i=1}^{n} (y_i - f(x_i, p))^2 \qquad (6.87)$$

An optimization algorithm works in the following manner. It reads an initial guess of the values of the coefficients. Then recursively changes the parameter values in a small amount and moving down gradient (derivative) in the q surface until changes in parameter values no longer yield a decrease in q. Sometimes we can obtain the initial guess of the parameter values by means of a linear regression performed on some approximation or transform of the nonlinear function. One of the curves (solid line) of the example in Figure 6.13b corresponds to the result of optimization applied to the exponential decay example treated earlier. Comparing to the result of the log transformation (Figure 6.13a), we observe a slightly improved fit for mid-values of depth.

A key part of nonlinear regression is to define the functional relationship or model. This is usually possible by knowing or postulating how the system works. In this example, if we are trying to find a coefficient of light attenuation in the water column of a lake, we know that light attenuation follows an exponential law, because the rate of decay is linear with depth.

6.3.3 Polynomial Regression

A special case of nonlinear regression is polynomial regression (Davis, 2002, pp. 207–213). The predictor is a linear combination of increasing powers of the X. In this case, we formulate the nonlinear relation as a polynomial instead of a functional relationship. It is useful when you do not know what model to apply. Although a solution is always found, we may not know the meaning or interpretation attached to the coefficients.

We already saw an example earlier when we wrote the predictor as a quadratic equation. In general,

$$\widehat{Y} = b_0 + b_1 X + b_2 X^2 + \cdots + b_m X^m \qquad (6.88)$$

where the polynomial is mth order. Figure 6.13b (dashed and dotted lines) shows examples for second and third order polynomial regression applied to the exponential decay problem. In the case of $m = 3$, we estimated four coefficients

$$\widehat{Y} = b_0 + b_1 X + b_2 X^2 + b_3 X^3 \qquad (6.89)$$

As we see in the figure, we have achieved as good a fit as the exponential model solved by optimization. However, because we do not know the meaning of the coefficients, we cannot claim that we have a better understanding of a generic response of light to depth than the exponential model. One practical application of polynomial regression is the calibration of sensors and instruments.

By transforming each power of X into a variable, the polynomial regression is converted to a multivariable linear regression problem that we will discuss later in the book after covering matrices.

6.3.4 Predicted vs. Observed Plots

An interesting graph to evaluate regression results is the fitted vs. observed plot together with a 1:1 line or of 45° slope. An indication of good fit is how close the points follow the 1:1 line. In Figure 6.14, we can observe several instances for the exponential law example given earlier and analyzed by linear,

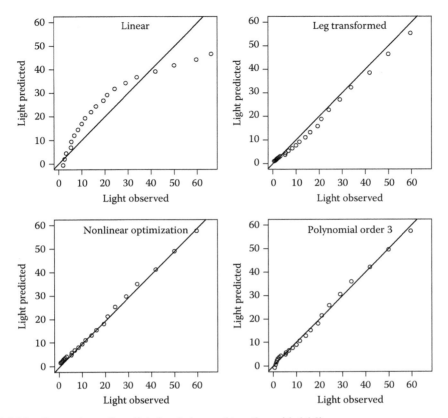

FIGURE 6.14 Comparison of predicted and observed together with 1:1 line.

log-transformed, optimization, and polynomial methods. We can easily visualize how in this example the nonlinear optimization model is better at following the 1:1 line.

Exercise 6.1
We have 5 observations of two variables y and x

```
x        y
1.0      1.67
3.0      7.90
5.0      9.03
7.0     17.84
9.0     13.60
```

Using a calculator, determine sample means, sample variances, and sample standard deviations of X and of Y. Calculate sample covariance of X and Y. Then, using these results, calculate the regression coefficients. Write the equation for the linear predictor. Calculate the predicted $\widehat{y}_4$ for $x_4 = 7$. Calculate the residual error for $x_4 = 7$. Sketch the scatter plot and predicted line.

Exercise 6.2
Using results of exercise 6.1, calculate the explained, residual, and total error for $x_4 = 7$. Illustrate on the scatter plot. Calculate the explained, residual, and total error for all values of x and the overall total. Calculate R^2. Calculate the correlation coefficient r. How much variance does the regression model explains? Calculate the residual standard error.

6.4   COMPUTER SESSION: SIMPLE REGRESSION

6.4.1   SCATTER PLOTS

We will use the `airquality` sample dataset available in package **datasets**. We already used this dataset when we built contingency tables in Chapter 5. Ozone is an important urban air pollution problem when it occurs in excessive amount in the lower troposphere. Its concentration relates to meteorological variables.

Let us look at the contents of `airquality` dataset,

```
> airquality
     Ozone  Solar.R  Wind  Temp  Month  Day
1       41      190   7.4    67      5    1
2       36      118   8.0    72      5    2
3       12      149  12.6    74      5    3
4       18      313  11.5    62      5    4
5       NA       NA  14.3    56      5    5
6       28       NA  14.9    66      5    6

and so on …

151     14      191  14.3    75      9   28
152     18      131   8.0    76      9   29
153     20      223  11.5    68      9   30
>
```

It has 153 records. Note that this dataset includes NA for some records. Recall that you can address a component of the dataset by using the $ sign. For example, `airquality$Temp`.

First, attach the dataset so that it becomes easier to use the component variables

```
>attach(airquality)
```

And then we can simply use

```
> Temp
  [1]  67 72 74 62 56 66 65 59 61 69 74 69 66 68 58 64 66 57 68 62 59
       73 61 61 57
 [26]  58 57 67 81 79 76 78 74 67 84 85 79 82 87 90 87 93 92 82 80 79
       77 72 65 73
 [51]  76 77 76 76 76 75 78 73 80 77 83 84 85 81 84 83 83 88 92 92 89
       82 73 81 91
 [76]  80 81 82 84 87 85 74 81 82 86 85 82 86 88 86 83 81 81 81 82 86
       85 87 89 90
[101]  90 92 86 86 82 80 79 77 79 76 78 78 77 72 75 79 81 86 88 97 94
       96 94 91 92
[126]  93 93 87 84 80 78 75 73 81 76 77 71 71 78 67 76 68 82 64 71 81
       69 63 70 77
[151]  75 76 68
```

Alternatively, from the Rcmdr use the Data menu and then select package datasets and then data airquality.

We will focus on ozone as a function of one of the meteorological variables; for example, air temperature. First, do a scatter plot using `plot(x, y)` function to get

```
> plot(Temp, Ozone)
```

Note that some pairs of points have larger values of ozone concentration. As we have done in previous chapters, to check these points we can run the function `identify` and use the cursor to identify points

```
> identify(Temp, Ozone)
```

Then move cursor (crosshair) to each outlier and click. The observation number will appear next to the point. Terminate the process pressing Stop. Upon termination, we have the observation number for each one of these points and the plot (Figure 6.1).

```
[[1] 30 62 86 99 101 117 121
```

Alternatively, using the Rcmdr we can plot using menu item **Graphs|Scatter plot**. Select Temp, Ozone, identify, and smooth line (Figure 6.14 Comparison of predicted and observed together with 1:1 line Figure 6.15). Then move cursor (crosshair) to each outlier and click. The observation number will appear next to the point. The process can be terminated with menu item Stop. Upon termination, we have the observation number for each one of these points (Figure 6.16).

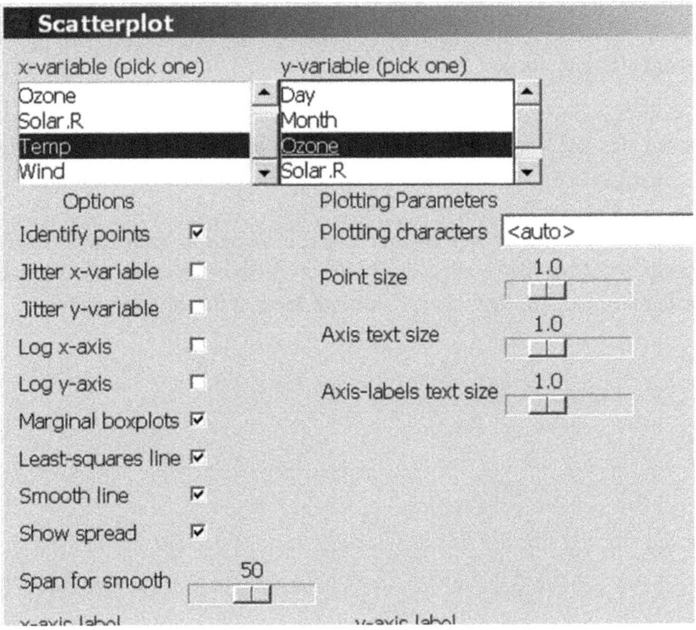

FIGURE 6.15 Rcmdr dialog window for scatter plots.

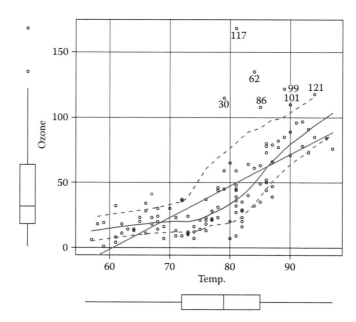

FIGURE 6.16   Scatter plots with barplots, smooth line, and spread.

```
[1]  30 62 86 99 101 117 121
```

An important thing to note is that a linear regression may not be the best choice. We will proceed with a linear model for now and later see how to do nonlinear.

6.4.2 SIMPLE LINEAR REGRESSION

Simple linear regression is done using function lm(y ~ x), where y is the dependent variable and x is the independent variable. Recall we use symbol ~ to denote a functional relationship. A nonintercept regression, or regression through the origin, is done by adding character 0 in the expression, that is, using lm(y ~ 0 + x).

Now we will use simple linear regression to explain ozone concentration from temperature lm(Ozone ~ Temp). Store the results in an object ozone.lm. We use extension lm to remind ourselves that this is an object of type lm. Then we look at the object.

```
> ozone.lm <- lm(Ozone ~ Temp)
> ozone.lm
Call:
lm(formula = Ozone ~ Temp)
Coefficients:
(Intercept)         Temp
  -146.995         2.429
```

These are the regression coefficients b_0 (intercept) and b_1 (slope) yielding Equation 6.52 already discussed. For this model, the R^2 is 0.48, which is a relatively low value, meaning that less than 50% of the variance in ozone can be explained by this model. Therefore, other factors must be important; hence, the common expression "something else must be going on." Using these values, we can plot the regression line on top of the scatter plot using the function abline. First, use plot to get the scatter plot, then identify outliers, press stop when done, and then add abline

```
> plot(Temp, Ozone)
> identify(
> abline(ozone.lm$coef)
```

We obtain Figure 6.5 already described. We get more information about the regression by looking at the summary

```
>summary(ozone.lm)
Call:
lm(formula = Ozone ~ Temp)
Residuals:
      Min        1Q    Median        3Q       Max
 -40.7295  -17.4086   -0.5869   11.3062  118.2705
Coefficients:
                Estimate Std. Error t value Pr(>|t|)
(Intercept) -146.9955     18.2872   -8.038 9.37e-13 ***
Temp           2.4287      0.2331   10.418  < 2e-16 ***
---
Signif. codes:  0 `***' 0.001 `**' 0.01 `*' 0.05 `.' 0.1 ` ' 1
Residual standard error: 23.71 on 114 degrees of freedom
Multiple R-Squared: 0.4877,    Adjusted R-squared: 0.4832
F-statistic: 108.5 on 1 and 114 DF,  p-value: < 2.2e-16
```

As discussed earlier, each coefficient has a standard error (standard deviation of its estimate) and a p-value of a *t*-test to evaluate its significance. For the example at hand, we get $se_{b0} = 18.28$, $t_0 = 8.04$ with p-value 9.4×10^{-13} and $se_{b1} = 0.23$, $t_1 = 10.41$ with p-value 2×10^{-16}. These p-values are very low; therefore, we can reject H0, namely that the coefficients are zero. Therefore, there is a non-zero slope and nonzero intercept. The residual standard error is 23.71. The degrees of freedom are df = 114 because two parameters were estimated (116 − 2 = 114). Note that 153 − 116 = 37 values were no data (NA) and thus removed. The R^2 is 0.48. The F value is 108.5 with 1 and 114 degrees of freedom, which yields a p-value of 2.2×10^{-16}. The low p-value suggests that we reject H0 and therefore the slope is different from zero.

Now let us produce diagnostic plots with pre-built function for lm-type objects. We get four graphs. Therefore, use par of four panels. Since we have the Rcmdr loaded, the plots correspond to those of the Rcmdr (Figure 6.8)

```
>par(mfrow=c(2,2));plot(ozone.lm)
```

You could also obtain individual displays of each plot by using

```
> par(mfrow=c(1,1));plot(ozone.lm, ask=T)
Hit <Return> to see next plot:
Hit <Return> to see next plot:
Hit <Return> to see next plot:
Hit <Return> to see next plot:
>
```

Use the following function conf.int.lm that calculates and plots the confidence intervals (MathSoft, 1999, 7-26–7-29). The function is available from the seeg package.

```
conf.int.lm <- function (dat.lm, alpha){
        # variables and names
        lm.x <- dat.lm$model[,2]                    # extract x variable
        lm.y <- dat.lm$model[,1]                    # extract y variable
        name.x <- names(dat.lm$model[2])            # name of x
        name.y <- names(dat.lm$model[1])            # name of y

          # calculate stderr and conf int
        rmse <- sqrt(sum(dat.lm$resid^2)/dat.lm$df)  # calculates
          residual standard error
        more <- ((lm.x - mean(lm.x))^2)/var(lm.x)    # square dev
          over variance
        stderr <- (rmse/sqrt(dat.lm$df+2))*(1+ more) # std error of
          estimates
        t.value <- qt(1 - alpha/2, dat.lm$df)        # calculates
          t value for given alpha
        lower <- dat.lm$fitted - stderr*t.value      # confidence
          interval low end
        upper <- dat.lm$fitted + stderr*t.value      # confidence
          interval high end

          # graphics
        plot(lm.x, lm.y, xlab=name.x, ylab=name.y)   # scatter plot
        abline(dat.lm$coef) # regression line
        ord <- order(lm.x)  # sort
        lines(lm.x[ord], lower[ord])                 # plot lower
        lines(lm.x[ord], upper[ord])                 # plot high
        identify(lm.x,lm.y,labels=row.names(dat.lm$model))
          # return
          invisible(list(lower = lower, upper = upper))
  }
```

Apply to ozone.lm object with alpha = 0.05

```
>conf.int.lm(ozone.lm, 0.05)
```

to obtain Figure 6.9. We have used alpha 0.05 for a 90% confidence interval (recall we have two tails). Input a different value and observe changes in the interval as you change the value of alpha.

Here is a recapitulation of all the commands we have applied to perform the regression analysis

```
> attach(airquality)

> plot(Temp, Ozone)            # get scatter plot
> identify(Temp, Ozone)        # identify outliers

> ozone.lm <- lm(Ozone ~ Temp) # regression object
> ozone.lm                     # look at coeffs

> plot(Temp, Ozone)            # get scatter plot
> abline(ozone.lm$coef)        # add regression line to scatter plot

> summary(ozone.lm)            # get more info on regression object
> par(mfrow=c(2,2));plot(ozone.lm) # diagnostic plots residual error
>conf.int.lm(ozone.lm, 0.05) # regression plot with confidence interval
```

We can perform all this using the Rcmdr by going to menu item **Data|Select Active data set** and select airquality. Then go to **Statistics|Fit Models|Linear Regression**. We get a dialog window as shown in Figure 6.17. Enter a name (say ozone.lm), select Ozone for response and Temp for explanatory variable. Click ok and then we get the results on the Output window of the Rcmdr.

Another way is to use **Statistics|Fit Models|Linear model**. We get a different dialog window where we can compose the function. For either one of dialog windows, we obtain the same result in the Rcmdr output window as the one obtained using the Rconsole.

```
Call:
lm(formula = Ozone ~ Temp, data = airquality)

Residuals:
    Min       1Q   Median      3Q     Max
-40.7295  -17.4086  -0.5869  11.3062  118.2705

Coefficients:
            Estimate  Std. Error  t value  Pr(>|t|)
(Intercept)  -146.9955    18.2872   -8.038  9.37e-13 ***
Temp            2.4287     0.2331   10.418   < 2e-16 ***
---
Signif. codes:  0 '***' 0.001 '**' 0.01 '*' 0.05 '.' 0.1 ' ' 1

Residual standard error: 23.71 on 114 degrees of freedom
  (37 observations deleted due to missingness)
Multiple R-Squared: 0.4877,      Adjusted R-squared: 0.4832
F-statistic: 108.5 on 1 and 114 DF,  p-value: < 2.2e-16
```

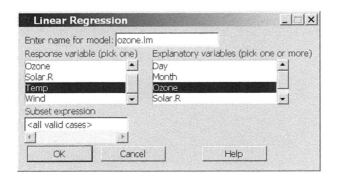

FIGURE 6.17   Dialog window for linear regression.

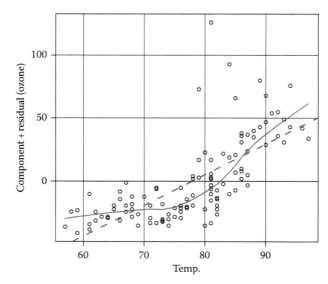

FIGURE 6.18   Additional diagnostic plot.

Now for diagnostics, go to **Models|Select active models**, scroll to `ozone.lm`. We can use **Models|Summarize model** to obtain a summary in the output window. We can also use **Models|Graphs|Basic Diagnostic plots** to get Figure 6.8. In addition, **Models|Graphs|Componen t+residuals** produces a useful diagnostic plot (Figure 6.18) illustrating the behavior of the residual in conjunction with the scatter plot and the regression line. We conclude that the error behavior is not uniform with Temp and therefore the linear model may not be a good model.

6.4.3   Nonintercept Model or Regression through the Origin

Now, you may wonder about the intercept coefficient which would imply a negative value of ozone concentration for `Temp` = 0. What happens is that the predictor may be valid only in the temperature interval of 56°F–97°F, since our data values do not extend outside this interval. So, extrapolating to `Temp` = 0 would yield erroneous result. Moreover, the equation would still predict a negative value in the extreme of the Temp range as we can see by substitution in Equation 6.52, we would get about −10 ppb, which is not a valid concentration.

Not very often, but sometimes we do have a reason to force the regression line through the origin. For the sake of illustration, consider the data in file **test-RTO.txt**.

```
Exp Resp
0  0
3.11  8.27
3.19  6.24
4.66  7.76
5.83  12.63
5.47  11.44
7.97  14.41
7.79  15.03
9.38  17.12
9.75  19.34
```

where `Exp` is an organism exposure to a toxicant and `Resp` is the response to this exposure. We have determined that there is no response for zero exposure.

We can do a nonintercept regression by adding 0 or −1 to the expression y ~ x to read y ~ 0 + x or y ~ x − 1. Let us try

```
tox <- read.table(file="lab6/test-RTO.txt",header=T)
attach(tox)
tox.lm <- lm(Resp ~ 0 + Exp)      # regression with 0 intercept
plot(Exp,Resp)                    # scatter plot
abline(a=0, b=tox.lm$coef)        # add regression line to scatter plot
summary(tox.lm)                   # get more info on regression object
par(mfrow=c(2,2));plot(tox.lm)    # diagnostic plots residual error
```

to get graphs in Figures 6.19 and 6.20.

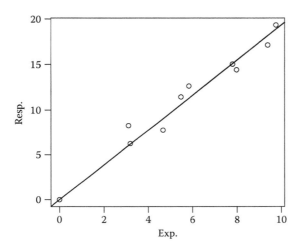

FIGURE 6.19   Regression through the origin.

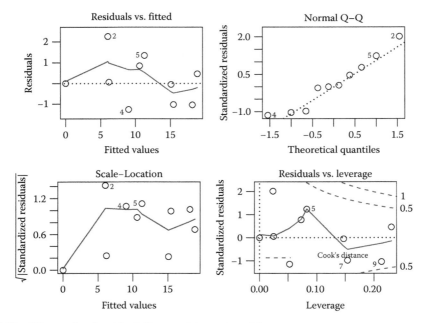

FIGURE 6.20   Diagnostic of residuals for zero intercept model.

The summary indicates a slope estimate of 1.94 and a R^2 of 0.99. This R^2 is defined differently and **cannot** be compared to the one obtained for an intercept model.

```
> summary(tox.lm)

Call:
lm(formula = Resp ~ 0 + Exp)

Residuals:
     Min       1Q   Median       3Q      Max
-1.26223 -0.77861  0.03192  0.75289  2.24872

Coefficients:
    Estimate Std. Error t value Pr(>|t|)
Exp  1.93610    0.05569   34.77 6.66e-11 ***
---
Signif. codes:  0 '***' 0.001 '**' 0.01 '*' 0.05 '.' 0.1 ' ' 1

Residual standard error: 1.131 on 9 degrees of freedom
Multiple R-squared: 0.9926,    Adjusted R-squared: 0.9918
F-statistic:  1209 on 1 and 9 DF,  p-value: 6.658e-11

>
```

Also, note that degrees of freedom are $n - 1$, or $10 - 1 = 9$ and that the standard error is 1.13. The predictor is $y = 1.94x$, where y is Resp and x is Exp.

We can use Rcmdr to perform regression through the origin. We can use the **Statistics|Linear model** menu item and enter the formula for the lm using $\sim 0 + x$, which yields the same results as when using the console.

6.4.4 ANOVA ONE WAY: AS LINEAR MODEL

We will use the example of maximum DO from the last chapter. Using the console you can get predicted and residuals

```
> fitted.values(anova.maxDO)
    1     2     3     4     5     6     7     8     9    10    11
   12    13
 4.73  4.73  4.73  4.73  4.73  4.73  4.73  4.73  4.73  4.73  8.39
 8.39  8.39
   14    15    16    17    18    19    20    21    22    23    24
   25    26
 8.39  8.39  8.39  8.39  8.39  8.39  8.39  9.81  9.81  9.81  9.81
 9.81  9.81
   27    28    29    30    31    32    33    34    35    36    37
   38    39
 9.81  9.81  9.81  9.81 12.25 12.25 12.25 12.25 12.25 12.25 12.25
12.25 12.25
   40
12.25
```

```
(continued)
> resid(anova.maxDO)
    1     2     3     4     5     6     7     8     9    10    11
   12    13
 0.47  0.67 -0.63 -1.23  1.77  1.07  0.17 -2.23  1.37 -1.43  0.11
-0.69 -0.99
   14    15    16    17    18    19    20    21    22    23    24
   25    26
 0.91 -1.49  0.91  2.21 -0.29 -1.19  0.51  0.29 -0.31 -0.41 -1.71
 0.99  0.79
   27    28    29    30    31    32    33    34    35    36    37
   38    39
 0.89  0.09 -0.21 -0.41 -1.55  0.85  0.25  1.15 -1.25  0.05 -0.05
-0.85 -0.55
   40
 1.95
>
>
```

Note that the predicted values are identical for each group, but the residuals are not. The **predicted** are actually the sample means 4.73, 8.39, 9.81, and 12.25. The error or **residual** of each observation is the difference between the predicted value and the observed value.

Now let us check residuals for normality

```
cdf.plot.gof(resid(anova.maxDO))
qqnorm(resid(anova.maxDO)); qqline(resid(anova.maxDO))
```

To obtain Figures 6.11 and 6.12. Now we run the Shapiro–Wilk test

```
> shapiro.test(resid(anova.maxDO))

     Shapiro-Wilk normality test

data: resid(anova.maxDO)
W = 0.9863, p-value = 0.9027
```

indicating that there is no evidence that the residuals are not normally distributed.

To get fitted values and residuals using the Rcmdr, use **Models|Add observations to Data** and select fitted, residuals and observation indices as shown in Figure 6.21. Now if you use view data set you will see the desired values. Now we can go to **Graphs** menu and select some useful features of the Cmdr. For example, we can produce a plot of means using **Graphs|plot of means** menu (Figure 6.22) to obtain Figure 6.23.

For two-way ANOVA, we proceed in a similar way.

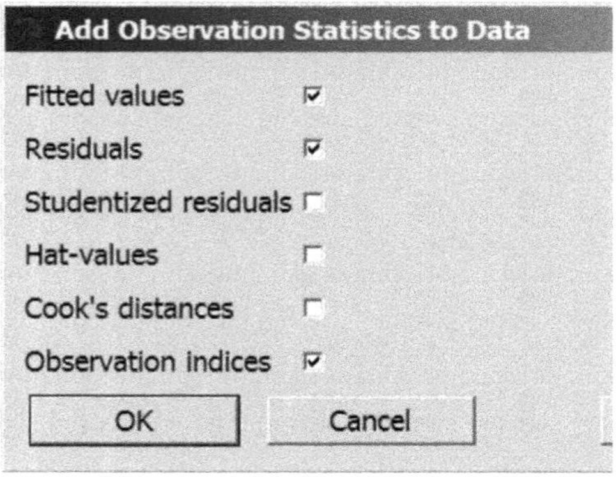

Add Observation Statistics to Data

Fitted values   ☑
Residuals   ☑
Studentized residuals ☐
Hat-values   ☐
Cook's distances   ☐
Observation indices   ☑

OK   Cancel

FIGURE 6.21   Using Rcmdr to add statistics to datasets.

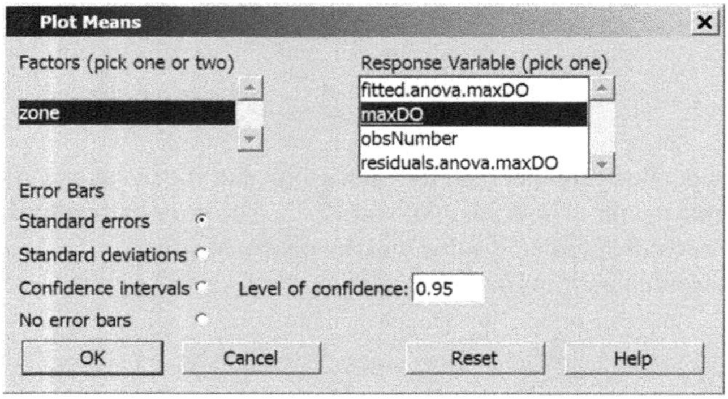

Plot Means   ✕

Factors (pick one or two)   Response Variable (pick one)
                                                          fitted.anova.maxDO
zone   maxDO
                                                          obsNumber
                                                          residuals.anova.maxDO

Error Bars
Standard errors   ⦿
Standard deviations   ○
Confidence intervals   ○   Level of confidence: 0.95
No error bars   ○

OK   Cancel   Reset   Help

FIGURE 6.22   Using Rcmdr for plot of means.

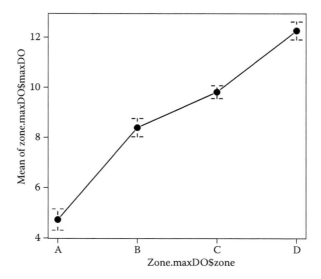

FIGURE 6.23   Plot of means.

6.4.5   Linear Regression: Lack-of-Fit to Nonlinear Data

Assume a sample of light measurements (W/m^2) measured at different depths in the water column of an estuary and given in file **lab6/light-depth.csv** provided in the seeg package. Light attenuation can increase with concentration of suspended sediments and enhanced algae production due to increase in nutrients. Estuaries are important receptors of water contaminants resulting from point sources and from freshwater runoff. Runoff carries contaminants generated upstream in the watershed.

In this file we have three columns, the first is depth in m, second and third are down welling and upwelling light (in W/m^2).

```
Depth,Light,Upwell

0,66.02,4.19
0.15,59.58,4.03
0.3,49.92,3.86
0.46,41.87,3.38
0.61,33.82,3.22
0.76,28.99,2.58
0.91,24.15,2.25
1.07,20.93,1.93
1.22,19.32,1.77
1.37,16.1,1.61
1.52,14.01,1.38
1.68,11.43,1.22
1.83,9.98,1.08
1.98,8.37,0.9
2.13,6.76,0.71
2.29,5.48,0.58
2.44,5.31,0.43
2.59,3.22,0.34
2.74,2.58,0.27
2.9,1.93,0.21
3.05,1.61,0.16
3.2,1.32,0.14
3.35,0.98,0.09
3.51,0.58,0
```

Create a data frame and store object light.depth

```
light.depth <- read.table(file="lab6/light-depth.
  csv",sep=",",header=T)
attach(light.depth)
```

Now the following statements are performed for regression of Light vs. Depth and we extract information by using `summary`

```
> light.lm <- lm(Light ~ Depth) # regression object
> summary(light.lm)
Call:
lm(formula = Light ~ Depth)
```

(*continued*)

```
(continued)

Residuals:
    Min     1Q  Median     3Q     Max
 -8.251 -6.997  -2.225  5.129 19.459

Coefficients:
             Estimate Std. Error t value Pr(>|t|)
(Intercept)    46.561      3.269   14.24 1.39e-12 ***
Depth         -16.244      1.598  -10.16 8.96e-10 ***
---
Signif. codes:  0 '***' 0.001 '**' 0.01 '*' 0.05 '.' 0.1 ' ' 1

Residual standard error: 8.265 on 22 degrees of freedom
Multiple R-squared: 0.8245,    Adjusted R-squared: 0.8165
F-statistic: 103.3 on 1 and 22 DF,  p-value: 8.958e-10
>
```

Examining the summary, we see that we have a relatively high R^2 of ~0.8; therefore, more than ¾ of the variance is explained by regression. The large value of F = 103 with a very small p-value ~8.9×10^{-11} indicate that H0 can be rejected. Therefore, there is a trend (slope different from zero). The coefficients are significantly different from zero (p-values very low in *t*-tests).

Now we perform more diagnostics, such as scatter plot, add a regression line, and do confidence interval at alpha 0.05

```
light.lm <- lm(Light ~ Depth) # regression object
par(mfrow=c(2,1))
plot(Depth, Light) # get scatter plot
abline(light.lm$coef) # add regression line to scatter plot
conf.int.lm(light.lm, 0.05)
```

You could put all of these statements in a ***.R** script and use Edit|Run all. We get Figure 6.24. Note that many observations are outside the confidence interval. We can also do the residual diagnostic plots

```
par(mfrow=c(2,2))
plot(light.lm) # diagnostic plots residual error
```

We can see in Figure 6.25 that the residuals have pattern (heteroscedastic), and that several observations are outliers. The residuals go down and then up with predicted light. They do not seem to be normally distributed (also indicated by the qq plot). In this example, high leverage would occur for 2 × 2/23 = 0.17. The residual vs. leverage plot indicates that observations 1, 2, and 24 most influential since they have high values of residual, Cook's distance, and almost high leverage.

We have learned that the *Y vs. X* plot does not indicate a linear relation, but curvilinear. In conclusion, we need to try a nonlinear regression. The message of this exercise is that of alert: your R^2 and p-value looked fine, but the linear regression is a poor model for this dataset.

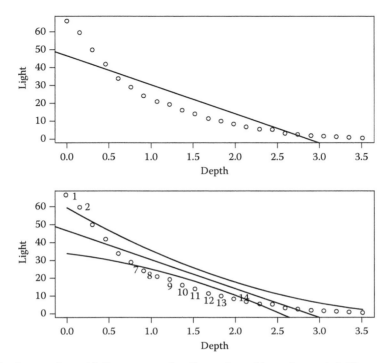

FIGURE 6.24 Scatter plot, with linear regression line and confidence interval; it illustrates lack-of-fit of regression line to nonlinear data.

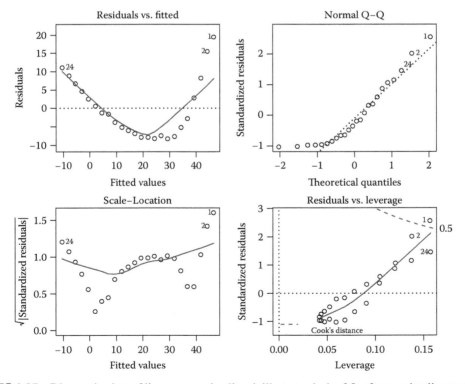

FIGURE 6.25 Diagnostic plots of linear regression line; it illustrates lack-of-fit of regression line to nonlinear data.

6.4.6 Nonlinear Regression by Transformation

The examples we have worked so far, particularly the last one, illustrate the need for nonlinear regression. Let us work the estuarine light–depth example $x =$ depth, $y =$ light (down welling) which we attempted to solve by linear regression in the previous section. First, we need an equation that may describe the nonlinear nature of the data. We could start with an exponential

$$y = y_0 \exp(-kx) \tag{6.90}$$

which is a natural model because it prescribes a linear rate of decrease of light with depth. In this case, we can transform by taking natural log of both sides

$$\ln(y) = \ln(y_0) - kx \tag{6.91}$$

Move the $\ln(y_0)$ to the left and because subtracting logs is the same as the log of the ratio

$$\frac{\ln(y)}{\ln(y_0)} = -kx \tag{6.92}$$

Now we can perform a linear regression of $\ln(y/y_0) \sim x$ to obtain a value of coefficient k

```
> lnlight.lm <- lm(log(Light/Light[1])~ 0+ Depth)
> summary(lnlight.lm)

Call:
lm(formula = log(Light/Light[1]) ~ 0 + Depth)

Residuals:
      Min        1Q    Median        3Q       Max
-0.61212  -0.00600   0.07318   0.20794   0.34546

Coefficients:
      Estimate Std. Error t value Pr(>|t|)
Depth  -1.1745     0.0216  -54.37   <2e-16 ***
---
Signif. codes:  0 '***' 0.001 '**' 0.01 '*' 0.05 '.' 0.1 ' ' 1

Residual standard error: 0.2165 on 23 degrees of freedom
Multiple R-squared: 0.9923,      Adjusted R-squared: 0.9919
F-statistic:  2956 on 1 and 23 DF,   p-value: < 2.2e-16
```

To do these operations in the Rcmdr: go to **Data|Manage variables ... | Compute new variable**. Then calculate new variable with name lnLight and formula log(Light/Light[1]) as in Figure 6.26 and view dataset to double check that new variable is there. Now go to **Statistics|Fit Models|Linear regression** and setup regression of ln Light vs. Depth as in Figure 6.27. We can see that we obtained same results as before when using the console.

Notice that we cannot compare the R-square and standard error to the nontransformed linear model because the units are different. How do we know if we really improved the fit of the model by using the log transform. One option is to convert the fitted transformed values to the original by using

$$\exp(fitted) = \exp\left(\ln\left(\frac{y}{y_0}\right)\right) = \frac{y}{y_0}$$

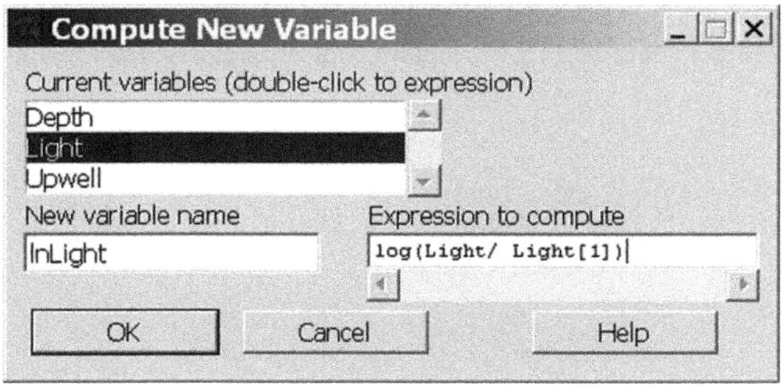

FIGURE 6.26   Use Rcmdr to compute new variable.

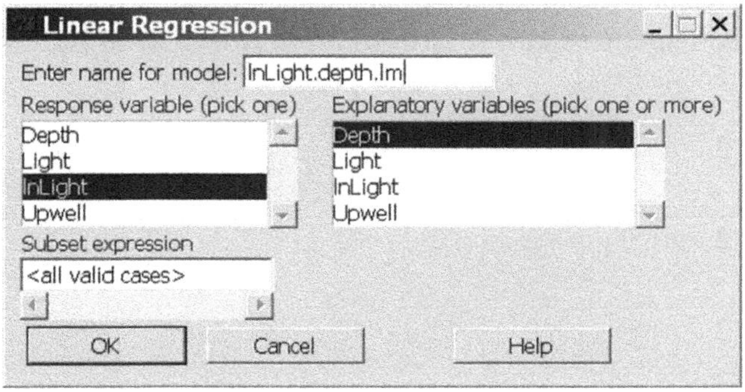

FIGURE 6.27   Use Rcmdr to setup regression.

$$\widehat{yt} = y_0 \exp(\mathit{fitted})$$

Then compare the SS of residuals $\sum_{i=1}^{n} (y_i - \widehat{y_i})^2$ of the linear to the transformed SS of residuals $\sum_{i=1}^{n} (y_i - \widehat{yt_i})^2$. The following lines of code make the comparison.

```
> # fitted values by log transform
> yt.est <- Light[1]*exp(lnlight.lm$fitted)
> # fitted values of linear model
> y.est <- light.lm$fitted
>
> # sum of square residual errors
> SSEl <- sum((Light - y.est)^2)
> SSEt <- sum((Light - yt.est)^2)
> SSEl; SSEt
[1] 1502.805
[1] 104.1913
>
```

Note that we have referred to the fitted values of the regression object using or lnLight. lm$fitted. This is equivalent to using fitted(lnLight.lm). We can see a definite improvement by a decrease of SSE from 1502.80 to 104.19. Also, we can do this in terms of standard error of both, which is obtained by the square root of SSE divided by degrees of freedom

```
> # Standard errors of residual
> Se.l <- sqrt(SSEl/light.lm$df)
> Se.t <- sqrt(SSEt/lnlight.lm$df)
> Se.l; Se.t
[1] 8.264944
[1] 2.128393
>
>
```

We can see a definite improvement by a decrease of Se from 8.26 to 2.12. Another option is to compare correlation coefficients

```
> # correlation coefficients
> r.l <- cor(Light, y.est)
> r.t <- cor(Light, yt.est)
> r.l ; r.t
[1] 0.9079952
[1] 0.9977884
>
```

which clearly shows an improvement from 0.90 to 0.99.

We can visualize the difference between the linear and log-transformed fit to the data by plotting the scatter plot and superimposing the estimate from both regression procedures

```
plot(Depth,Light)
lines(Depth,y.est,lty=1)
lines(Depth,yt.est,lty=2)
legend("topright",lty=1:2,leg=c("Linear","Log"))
```

as shown in Figure 6.13a.

6.4.7 Nonlinear Regression by Optimization

In this section, we will employ function nls to perform nonlinear regression. We need to start the optimization algorithm from an initial guess for the coefficients. Generally, this requires some additional knowledge of the system. In the exponential example treated earlier, we use the measured value (66) at zero depth as y_0 and the value ($k = -1.17$) obtained by transformation as first guess for the algorithm. The initial guess is declared in a start = list(...) argument to nls

```
> light.nls <- nls(Light ~ y0*exp(-k*Depth), start=list(k=1.1,
  y0=66))
> light.nls
Nonlinear regression model
  model: Light ~ y0 * exp(-k * Depth)
   data: parent.frame()
      k      y0
  1.082 67.800
 residual sum-of-squares: 21.31

Number of iterations to convergence: 3
Achieved convergence tolerance: 5.265e-06
>
```

We now have a residual sum of squares of 21.31. This is less than 104 obtained by transformation. The standard error and correlation coefficient of fitted values are calculated using

```
> yn.est <- fitted(light.nls)
> SSEn <- sum((Light - yn.est)^2)
> Se.n <- sqrt(SSEn/(length(yn.est)-2)); Se.n
[1] 0.9842852
> r.n <- cor(Light, yn.est); r.n
[1] 0.9988176
>
```

We have referred to the contents of the regression object with `fitted(light.nls)`. With respect to the transformed method, we have a decrease of standard error from 2.12 to 0.98 and a very small improvement of correlation coefficient from 0.997 to 0.998.

With the fitted values, we can do graph like the one shown in Figure 6.13a (solid line) discussed earlier.

```
> plot(Depth,Light)
> lines(Depth, yn.est)
```

We can also refer to the regression coefficients and the residuals using `coefficients` and `residuals` functions.

```
> residuals(light.nls)
[1] 5.125242 -5.041343 -4.009494 -5.523206 1.692044 1.225148
  8.186048
[8] 8.350103
attr(,"label")
[1] "Residuals"
```

It is a good idea to do diagnose the residuals for uniformity and normality. We will do two graphics pages of two panels each. The first-page panels are one to plot residuals vs. predicted (adding a line at zero) and the other for a q–q plot. The second page is for comparison to a normal cdf using function `cdf.plot.gof` created in the last chapter.

```
panel2(size=5)
plot(yn.est, residuals(light.nls))
abline(h=0)
qqnorm(residuals(light.nls)); qqline(residuals(light.nls))
cdf.plot.gof(residuals(light.nls))
```

We get the two-panel graph shown in Figures 6.28 and 6.29. These graphs reveal that we have reduced the errors with respect to the linear model but that we still have some nonuniformity pattern in the errors and deviations from the normal. A Shapiro–Wilk test of the residuals

```
> shapiro.test(residuals(light.nls))

      Shapiro-Wilk normality test

data: residuals(light.nls)
W = 0.9396, p-value = 0.1599

>
```

tells us that we could reject H0 of non-normality but with a higher 15% error rate. We conclude that the residuals are nearly normally distributed.

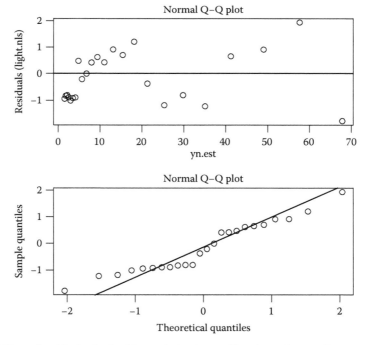

FIGURE 6.28 Plots of residuals obtained by optimization: uniformity and normality.

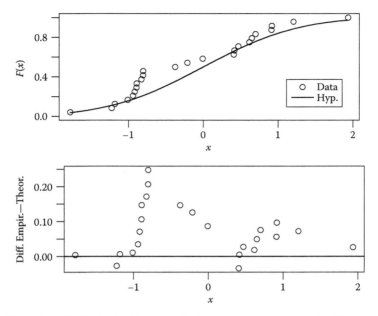

FIGURE 6.29 Plots of residuals obtained by optimization: comparing to normal cdf.

6.4.8 POLYNOMIAL REGRESSION

This is another way of performing nonlinear regression. We use a polynomial to approximate the nonlinear functional relationship. It is useful when you do not know what model to apply. However, there may not be a physical meaning attached to the coefficients.

Let us apply it to the same light vs. depth example. Let us try a second-order polynomial (a parabola) or up to power 2, that is, up to a X^2 term

```
> light.poly <- lm(Light ~ poly(Depth, 2), data=light.depth)
> summary(light.poly)

Call:
lm(formula = Light ~ poly(Depth, 2), data = light.depth)

Residuals:
    Min      1Q  Median      3Q     Max
-4.5961 -1.5265  0.4013  1.8560  4.7877

Coefficients:
                 Estimate Std. Error t value Pr(>|t|)
(Intercept)       18.0942     0.5784   31.29  < 2e-16 ***
poly(Depth, 2)1  -84.0119     2.8334  -29.65  < 2e-16 ***
poly(Depth, 2)2   36.5269     2.8334   12.89 1.92e-11 ***
---
Signif. codes:  0 '***' 0.001 '**' 0.01 '*' 0.05 '.' 0.1 ' ' 1

Residual standard error: 2.833 on 21 degrees of freedom
Multiple R-squared: 0.9803,     Adjusted R-squared: 0.9784
F-statistic: 522.7 on 2 and 21 DF,  p-value: < 2.2e-16

> yp.est2 <- light.poly$fitted
>
```

We can see that all coefficients are significant (low p-value of t-test) as well as the F statistic. The R^2 is also substantially high. Now let us graph the scatter plot and superimpose the fitted line

```
yp.est2 <- light.poly$fitted
plot(Depth,Light)
lines(Depth, yp.est2)
```

As we see in Figure 6.13a (dashed line), we have achieved a relatively good fit for this dataset. We can get an improved fit by increasing the order of the polynomial to the third power.

```
> light.poly <- lm(Light ~ poly(Depth, 3), data=light.depth)
> summary(light.poly)

Call:
lm(formula = Light ~ poly(Depth, 3), data = light.depth)

Residuals:
     Min       1Q    Median       3Q      Max
-2.16349 -0.83575   0.05711  0.87719  2.15683

Coefficients:
                 Estimate Std. Error t value Pr(>|t|)
(Intercept)        18.094      0.242  74.779  < 2e-16 ***
poly(Depth, 3)1   -84.012      1.185 -70.872  < 2e-16 ***
poly(Depth, 3)2    36.527      1.185  30.814  < 2e-16 ***
poly(Depth, 3)3   -11.853      1.185  -9.999 3.17e-09 ***
---
Signif. codes:  0 '***' 0.001 '**' 0.01 '*' 0.05 '.' 0.1 ' ' 1

Residual standard error: 1.185 on 20 degrees of freedom
Multiple R-squared: 0.9967,     Adjusted R-squared: 0.9962
F-statistic:  2024 on 3 and 20 DF,  p-value: < 2.2e-16

> yp.est3 <- light.poly$fitted
```

As with second order, we can see that all coefficients are significant (low p-value of t-test) as well as the F statistic. The R^2 is now higher. Now let us graph the scatter plot and superimpose the fitted line as before

```
yp.est3 <- light.poly$fitted
plot(Depth,Light)
lines(Depth, yp.est3)
```

As we see in Figure 6.13b dotted line), we have achieved a better fit for this dataset. We may have done as good a job fitting data points with a curve, but we cannot claim that we have a better understanding of a generic response of light to depth.

To perform polynomial regression in Rcmdr, go to **Statistics|Fit Models|Linear Model**, then enter the poly expression in the function to obtain the same results as with the console.

6.4.9 PREDICTED VS. OBSERVED PLOTS

We can produce graph of fitted vs. observed together with a 1:1 line for all our four models—linear, log-transformed, optimization, and polynomial—with the following code

```
panel4(size=7)
xlabel = "Light Observed"; ylabel = "Light Predicted"
plot(Light,y.est,ylim=c(0,60),xlab=xlabel,ylab=ylabel)
abline(a=0,b=1); mtext(side=3,line=-1,"Linear",cex=0.8)
plot(Light,yt.est,ylim=c(0,60),xlab=xlabel,ylab=ylabel)
abline(a=0,b=1); mtext(side=3,line=-1,"Log transformed",cex=0.8)
plot(Light,yn.est,ylim=c(0,60),xlab=xlabel,ylab=ylabel)
abline(a=0,b=1);mtext(side=3,line=-1,"Nonlinear
Optimization",cex=0.8)
plot(Light,yp.est3,ylim=c(0,60),xlab=xlabel,ylab=ylabel)
abline(a=0,b=1);mtext(side=3,line=-1,"Polynomial order 3",cex=0.8)
```

yielding the plots of Figure 6.14 discussed earlier.

6.4.10 COMPUTER EXERCISES

Exercise 6.3
Perform linear regression on 10 observations of two variables y and x (file **lab6/exercise6-3.txt**)

	x	y
[1,]	1	1.674
[2,]	2	8.997
[3,]	3	7.904
[4,]	4	9.877
[5,]	5	9.034
[6,]	6	17.037
[7,]	7	17.836
[8,]	8	12.462
[9,]	9	13.599
[10,]	10	25.067

Produce scatter plots, regression line, confidence interval lines, and diagnostic (e.g., residual error plots). Write the predictor. Discuss these results. Calculate the explained, residual, and total error. How much variance does the model explain?

Exercise 6.4
Assume the following data for X (first column) and Y (second column). File **lab6/exercise6-4.txt**)

0.97	3.31
0.25	1.15
0.07	1.21
2.60	5.48
0.77	0.87
0.28	0.91

(continued)

```
(continued)
1.96    4.47
1.32    4.13
1.78    5.01
0.06    1.14
0.81    2.57
1.87    6.43
0.31    2.37
2.85    6.88
1.01    4.08
1.39    3.60
2.41    5.73
1.73    4.07
2.83    7.90
0.43    1.67
```

Perform exploratory data analysis and descriptive statistics for each variable. Run a correlation test. Build a linear predictor of Y from X using linear regression. Evaluate the regression. Discuss thoroughly.

Exercise 6.5

Use the upwelling part of light measurements in the **lab6/light-depth.csv** file. Assume that upwelling light extinction follows an exponential law same as down welling. Produce a scatter plot of **upwell** light vs. depth. Demonstrate graphically that a linear regression would not yield a good model. Obtain estimates of k and y_0 from a linear regression applied to log-transformed data. Plot and compare. Use these estimates as initial guess for a nonlinear regression optimization applied to an exponential model. Graph the exponential curve and compare to the log-transformed. Diagnose the residual error. Use polynomial regression and compare to nonlinear regression. Draw predicted vs. observed plots for all four models.

Exercise 6.6

Estuarine sediments are important receptors of water contaminants resulting from point sources and from freshwater runoff. Assume a sample of moisture content (g water/100 g dried solids) measured at different depths (m) in the sediments of an estuary (data from Davis, 2002, Chapter 4). The depth and moisture pairs are (0,124), (5,78), (10,54), (15,35), (20,30), (25,21), (30,22), (35,18). Demonstrate graphically that a linear regression would not yield a good model. Obtain estimates of k and y_0 from a linear regression applied to log-transformed data. Plot and compare. Use these estimates as initial guess for a nonlinear regression optimization applied to an exponential model. Graph the exponential curve and compare to the log-transformed. Diagnose the residual error. Use polynomial regression and compare to nonlinear regression. Draw predicted vs. observed plots for all four models.

Exercise 6.7

Work the ozone vs. temperature example from `airquality` dataset using nonlinear optimization and polynomial regression. For an equation that may describe the nonlinear nature of the data, assume an exponential

$$y = y_0 \exp\big(k(x - x_0)\big) \tag{6.93}$$

where
  y is ozone
  x is temperature

Note that $x_0 \sim 50$ from the ozone vs. temperature scatter plot. Start the optimization algorithm from an initial guess for the coefficients from a log-transformed linear regression. Select the best order of the polynomial by trial and error. Diagnose the residual error. Use polynomial regression and compare to nonlinear regression. Draw predicted vs. observed plots for all four models.

SUPPLEMENTARY READING

Chapter 3, pp. 123–158, Chapter 4, pp. 191–228 (Davis, 2002); Chapter 6, pp. 104–122 (Rogerson, 2001); Chapter 2, pp. 27–30, Chapter 3, pp. 44–63 (Carr, 1995); Chapter 7, pp. 7-1–7-9 and 7-26–7-29, 7-38–7-43, Chapter 24, pp. 24-2–24-14, Chapter 24, pp. 24-33–24-35 (MathSoft, 1999); (Eisenhauer, 2003).

7 Stochastic or Random Processes and Time Series

7.1 STOCHASTIC PROCESSES AND TIME SERIES: BASICS

A **stochastic or random process** is a sequence of random variables $X(t)$ with a distribution that may vary with time t and a joint distribution for the entire sequence. For example, a Gaussian process is a sequence of normally distributed variables with a joint distribution that is also normal. The process is **stationary** if the distribution is constant with time t. To be less restrictive, we can make a weaker statement considering the process stationary if the mean and variance are constant. For example, a Gaussian process defined by identical independent normal variables with $N(0, \sigma)$ is stationary. This definition is similar to the one of a stationary spatial random variable, as we will discuss in Chapter 8.

We will consider two types of random processes. One is a **regular** process with RV defined at all times t and the interval between two successive time points is constant (say Δt), such as a Gaussian process. When Δt is very small, we have a time-continuous process. The other is a jump or **point** process that does not have a value at all t but only in a set of time values separated by a variable time interval, such as a Poisson process. We will study both Gaussian and Poisson processes in this chapter.

A **time series** $x(t)$ is a collection of values $\{x(t_i)\}$ at times t_i or, in other words, a sample or realization of an stochastic process indexed by time. Same as before, we can have a **regular** time series, a series such that all successive time instants are separated by the same amount of time. In other words, $t_i - t_j = \Delta t$ for all pairs t_i, t_j of successive points. In other cases, the series is **irregular**, which means that the separation between successive time instants is not constant; pairs of successive points have $t_i - t_j$ different from other points.

Figure 7.1 illustrates two examples. The left panels correspond to a case when every day there is a value for the variable drawn from a distribution and the value is independent of the previous values. The histogram at the bottom corresponds to all values and it suggests that they may come from a symmetrical distribution. Whereas the right-hand-side panels show an example of a point process, such that not all days have an event and the interval between days with nonzero values varies. The interval for a particular pair of successive values is independent of previous intervals. At the bottom, the histogram corresponds only to nonzero values and it suggests that the values come from a skewed distribution.

A characteristic of both examples is that the value at time t does not depend on the history of when events occur or values at previous times. However, there are cases when the values depend on the history; we will treat those with more detail in later chapters.

An important case for time series and random processes is one where there is periodicity. For a concrete example, consider the yearly number of sunspots in the period 1700–2011 (Figure 7.2). These values are from the Royal Observatory of Belgium (SIDC-team, 2012). The number of sunspots seems to be related to magnetic storms on the Sun and therefore to solar wind and the auroras. This example corresponds to a **regular** time series.

7.2 GAUSSIAN

Consider a Gaussian process formed by a sequence of identical independent normal variables with distribution $N(\mu, \sigma)$ at all times t. A special case is that for $\mu = 0$, which corresponds to "white noise." As example consider Figure 7.3, which depicts nine realizations of a stationary Gaussian process with $\mu = 30$ and $\sigma = 5$ for 30 day period of time. For each run, at each simulation day we

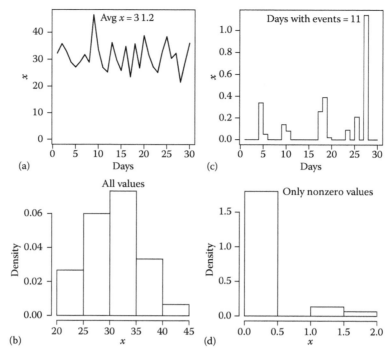

FIGURE 7.1   Examples of time series. (a) Time series for maximum daily temperature in a given month; (b) histogram of all values in the time series. (c) Time series for daily rainfall; (d) histogram of nonzero values in the time series.

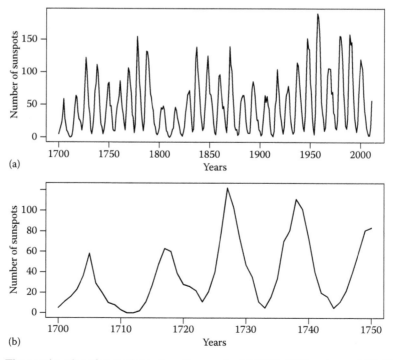

FIGURE 7.2   Time series plot of annual number of sunspots. (a) 1700–2011 series and (b) 1700–1750 time window.

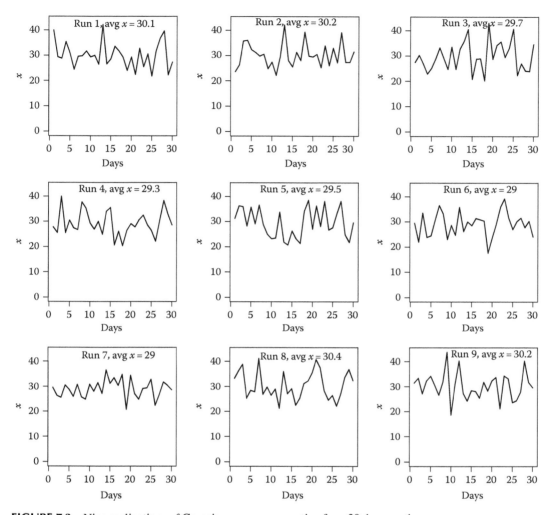

FIGURE 7.3   Nine realizations of Gaussian process generation for a 30 day month.

draw a random number from a normal distribution with this mean and standard deviation. As we can appreciate in this figure, the average of each run is near 30. The histograms shown in Figure 7.4 suggest symmetry and possibly normality in some cases.

As we increase the time span or duration of the series, the histograms should show more normality. For example, now we simulate white noise of unit variance, a Gaussian process with mean 0 and standard deviation 1 for 300 units of time. Figure 7.5 shows the time series, and Figure 7.6 illustrates how the histograms suggest a normal distribution.

7.3   AUTOCOVARIANCE AND AUTOCORRELATION

Of great interest is the relationship of values separated by various time differences or time **lags**. For example, if the time series represents a periodic process, that is, a pattern that repeats itself every so many units T of time, then we expect a strong correlation for values separated by T units. For example, examination of the plot in Figure 7.2a indicates about five peaks in a 50 year period, suggesting a ~10 year cycle. In the narrower time window (Figure 7.2b), you should be able to see that it is more like a ~11 year cycle; this is the well-known sunspot 11-year cycle. However, in the time series shown in Figure 7.7, we cannot appreciate any periodicity and may assume that the process is independent.

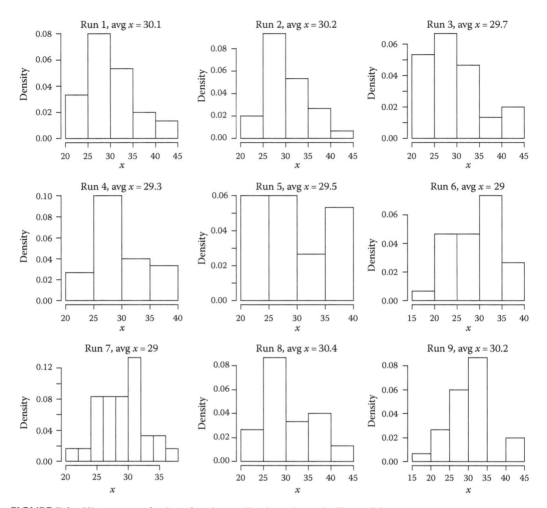

FIGURE 7.4   Histograms of values for nine realizations shown in Figure 7.3.

Lagged scatter plots are scatter plots of pairs of values in the time series separated by various time lags. In the example of Figure 7.8, we have scatter plots for values lagged by 1–4 time units for the white noise time series given earlier. We see no indication of potential correlation for values lagged by 1–4 time units. Next, we will discuss the calculation of correlation and covariance between values separated by all possible lags.

The autocovariance for lag h, assuming a stationary process, is

$$c(h) = E((X(t) - \mu)(X(t + h) - \mu)) \qquad (7.1)$$

which does not depend on t, but only on the lag h. The autocorrelation for lag h is the autocovariance scaled by dividing the $c(h)$ by the covariance at zero lag $c(0)$. Note that $c(0)$ is the maximum value of the autocovariance and happens to be just the variance σ^2,

$$\rho(h) = \frac{c(h)}{c(0)} = \frac{c(h)}{\sigma^2} \qquad (7.2)$$

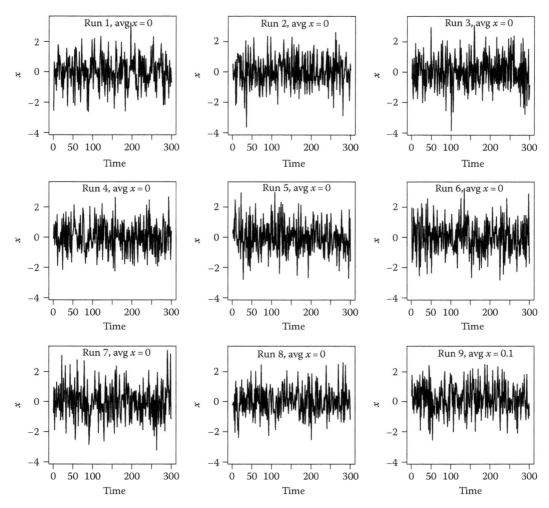

FIGURE 7.5   Nine realizations of a stationary zero-mean Gaussian process for 300 time units.

Therefore, $\rho(h)$ is always less or equal than 1. Both $c(h)$ and $\rho(h)$ are even functions, that is to say they have the same values for h and $-h$. For example, a Gaussian process with variance $c(0) = \sigma^2$ and distribution $N(0,\sigma)$ has $c(h) = 0$ $h \neq 0$ for all other lags because they are independent random variables; therefore, both autocovariance and autocorrelation should be a simple spike at $h = 0$. This is confirmed by Figure 7.9, which corresponds to the white noise process with unit variance given earlier.

Going back to the annual number of sunspots example, in Figures 7.10 and 7.11, we have scatter plots for values lagged by 1–12 years. We can see some potential positive correlation for values lagged by 1, 10, and 11 years. In addition, we can examine the autocorrelation and autocovariance shown in Figure 7.12. Note that high positive correlation occur for short lags ($h = 1, 2, 3$) and then again at lags around 11 years ($h = 10, 11, 12$). The maximum correlation ($=1$) is obtained at zero lag, and peak correlations are obtained at 5 years (negative) and 10–11 years (positive). The horizontal dashed lines are for the 95% confidence interval for the autocorrelation estimate at each lag, that is for null H0: rho = 0. For the plot of autocovariance, the maximum value will be the variance ($\sim$1600 in the example).

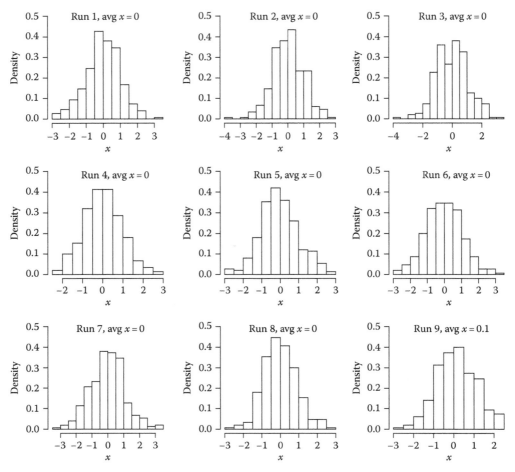

FIGURE 7.6 Histograms of values for nine realizations shown in Figure 7.5.

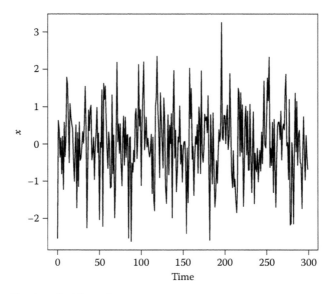

FIGURE 7.7 One realization of white noise.

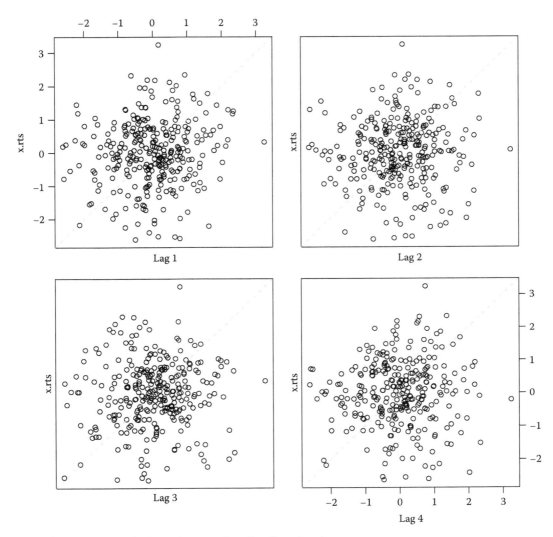

FIGURE 7.8   White noise lagged scatter plots. Lag from 1 to 4.

7.4   PERIODIC SERIES, FILTERING, AND SPECTRAL ANALYSIS

We will now focus on periodic time series. Consider the example in Figure 7.13 which contains hypothetical monthly data for 25 years. The units of the plots are in years. Looking at the time series plot,
we suspect a periodicity of 1 year in the underlying process (i.e., the "signal"). In addition, there is
variability or noise superimposed on the signal. The autocorrelation plot shows that there is a maximum positive correlation for a lag of 1 year (and it repeats for lag = 2 years) and a maximum negative
correlation for half a year (repeats for lag = 1.5 years). This confirms the periodicity of 1 year.

It is often of interest to separate the signal from the noise, or to **filter** out the noise and leave
the signal. There are several filtering methods including moving average filter, running median
filter, and exponential smoothing filter. We will focus here on moving average filters based on
convolution.

A moving average filter works by assigning to the value at each t a weighted average of values
around the calculation point. Options are to include equal number of points before and after the target point, or to include points only after or only before the target point. The calculation is based on

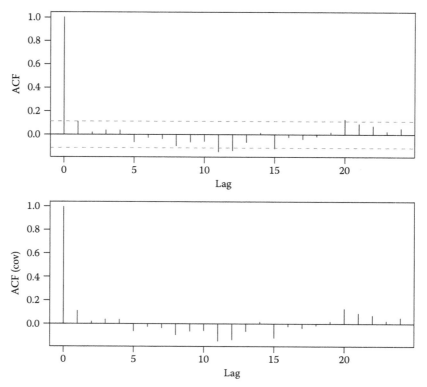

FIGURE 7.9   White noise autocorrelation plots.

convolution. A typical structure is even or equal number of points k in each side of the target point; it requires an odd number of coefficients in order to include the target point itself

$$\hat{x}(t) = a_{-k}x(t - k\Delta t) + \cdots + a_{-1}x(t - \Delta t) + a_0 x(t) + a_1 x(t + \Delta t) + \cdots + a_k x(t + k\Delta t) \qquad (7.3)$$

The coefficients determine the filtering operation. Thus, the filter is $g(t) = \sum_{i=-k}^{k} a_i \delta(t - i\Delta t)$ and the delta functions represent spikes or impulses at the times indicated. When all coefficients add up to 1, we obtain a **low-pass** filter or **smoother**, which reduces the variability from point to point. Whereas when all coefficients add up to zero, we obtain a **high-pass** filter, which enhances the variability of the series.

Figure 7.14 shows an example. The top panel is the series given earlier, and the mid-panel is the smoothed series by a five-point smoother with all coefficients equal to 1/5. This is to say $a_i = \{1/5, 1/5, 1/5, 1/5, 1/5\}$. In a case like this, when all coefficients are the same, we call the filter **rectangular**. The bottom panel is the residual or noise that was filtered out. The filtered signal is the slow component or **trend**. For contrast, Figure 7.15 shows the result of high-pass filtering with coefficients equal to −1/5 for all offset points and equal to 4/5 at the target point. This is to say $\{-1/5, -1/5, 4/5, -1/5, -1/5\}$. Note that in this case the residual is the trend.

Convolution consists of calculating Equation 7.3 at all times and summing over the entire time series. Denote the convolution operator by the symbol * and $g(t)$ as the filter, then the filtered series is $\hat{x}(t) = g(t) * x(t)$.

Spectral analysis is a useful method to study periodic time series. The **spectrum** is based on expressing the time series as a sum of sine and cosine components of multiple frequencies.

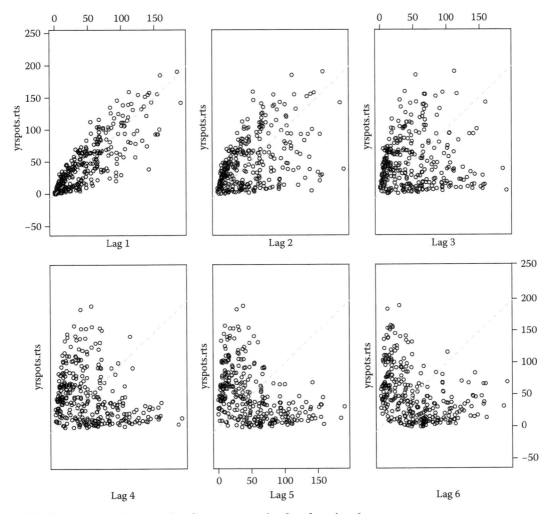

FIGURE 7.10 Lagged scatter plots for sunspots series. Lag from 1 to 6.

There are several ways of estimating the spectrum of a time series. One is the periodogram obtained by the **Discrete Fourier Transform** (DFT). The other is by Fourier transform of the truncated and smoothed autocovariance function. We will focus here on the periodogram method. Computations can be performed rapidly on a computer using the Fast Fourier Transform (FFT) algorithm. The length should be a power of 2, for example, 256.

Suppose the series is of length n, then

$$x(t) = \sum_{i=1}^{(n-1)/2} G_x(f_i)\exp(j2\pi f_i t) \tag{7.4}$$

where
f_i is frequency
j is $\sqrt{-1}$ for imaginary numbers
$G_x(f_i)$ is the DFT at frequency f_i

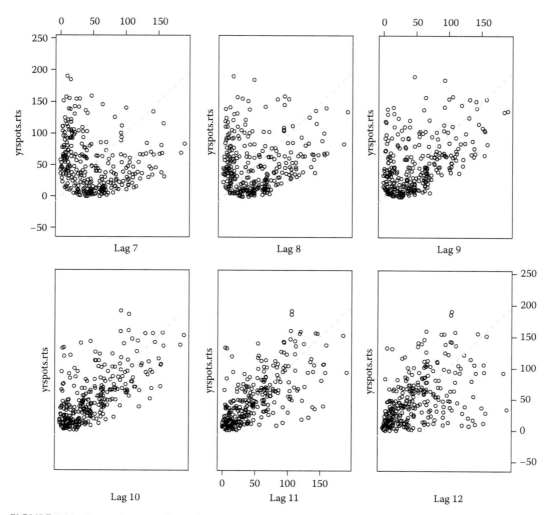

FIGURE 7.11   Lagged scatter plots of sunspots series. Lag from 7 to 12.

Here, we need to recall the Euler relationship $\exp(jy) = \cos(y) + j\sin(y)$. Thus, the summation of the exponential terms implies a summation of sinusoids. In other words, the time series can be decomposed in sinusoidal terms of multiple frequencies $f_i = i/n$ where $i = 1, 2,\ldots, (n-1)/2$.

The DFT $G_x(f)$ is

$$G_x(f) = \frac{1}{n}\sum_{m=0}^{n-1} x(t_m)\exp(-j2\pi ft_m) \tag{7.5}$$

where t_m are the values of time. The $G_x(f_i)$ are complex numbers and therefore have a magnitude $|G_x(f_i)|$ and a phase angle. The magnitude represents the strength of the periodicity at this frequency f_i. Once we formulate the DFT, we can obtain the time series by **inverse DFT** given by Equation 7.4.

The **periodogram** is a graph of the frequency components of a time series. See examples in Figure 7.16. The horizontal axis is frequency (inverse of period) and the vertical axis is n times the square of the magnitude $n|G_x(f_i)|^2$ at the period. To facilitate visualization, the periodogram is

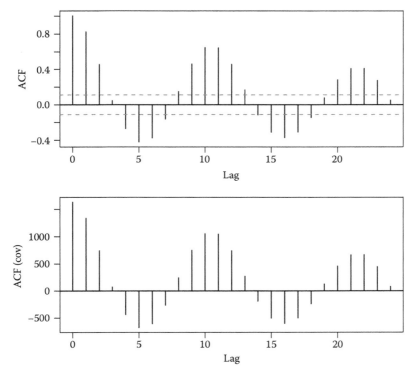

FIGURE 7.12 Autocorrelation plot of sunspots series.

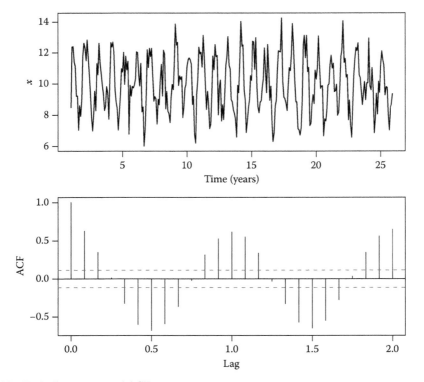

FIGURE 7.13 Periodic process and ACF.

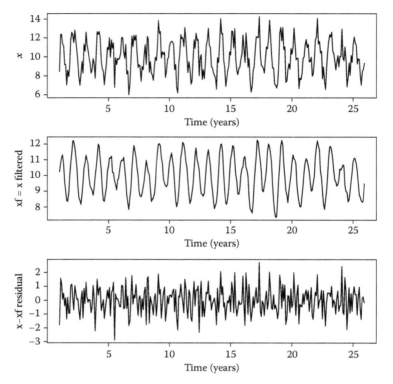

FIGURE 7.14   Low-pass filtering of periodic process.

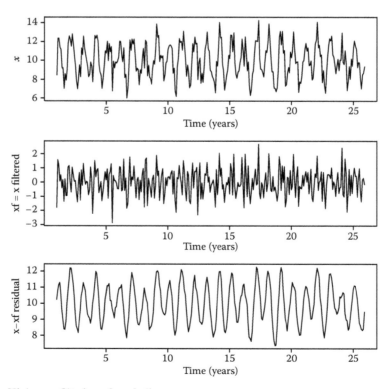

FIGURE 7.15   High-pass filtering of periodic process.

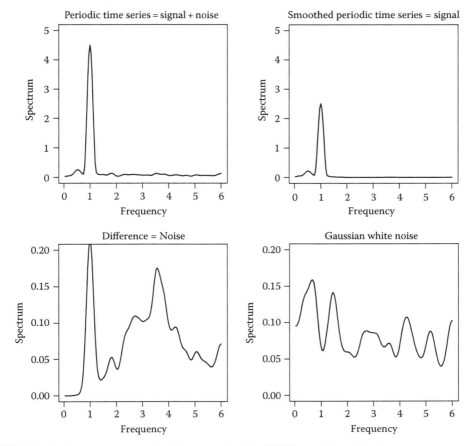

FIGURE 7.16 Periodograms. Smoothed by successive Daniell filters of 5 and 7 points.

smoothed by a **Daniell filter**. This type of low-pass filter differs from a rectangular filter in that the first and last coefficients are half as large $1/2(k-1)$ as the other ones $1/(k-1)$. For example, a five-point smoother will have {1/8, 1/4, 1/4, 1/4, 1/8}. In Figure 7.16 (top left), we have the periodogram of the periodic time series discussed at the beginning of section 7.4 and shown in Figure 7.13. Note that there is large value at a frequency close to 1 which is the inverse of 1 year. The top-right panel corresponds to the periodogram of the low-pass filtered series smoothed by a five-point rectangular filter. Note how the peak at frequency = 1 remains and the higher frequency bumps are eliminated. At the bottom-left panel, we can see the spectrum of the noise removed by filtering suggesting abundance of higher frequencies. Lastly, for comparison, the spectrum of Gaussian time series in the bottom-right panel shows abundance of many higher frequencies.

An alternative to perform filtering is to use DFT and the frequency domain because the Fourier transform of the convolution of two series is the product of their transforms.

$$G_{x(t)*g(t)}(f) = G_{x(t)}(f)G_{g(t)}(f) \tag{7.6}$$

Therefore, for filtering, simply multiply the series' DFT by the filter's DFT and then apply inverse DFT to get the filtered series.

7.5 POISSON PROCESS

A common phenomenon is the "arrival" or occurrence of an event at a time t independently of the time of previous occurrence of the events. Saying it in different words, events on nonoverlapping time intervals are mutually independent. In addition, the average rate of arrivals is constant. The Poisson pmf is a good model for the number of arrivals in an interval. Typical applications include occurrence of earthquakes. As we increase the rate, the pmf would be more and more as a normal distribution (Figure 7.17).

Let us now derive the Poisson pmf. Suppose we have a rate of λ occurrences or arrivals per unit time. Therefore, the probability of an arrival in a small interval delta t is $\lambda\Delta t$. However, the probability of having more than one arrival in that small interval is zero. Therefore, the probability of no-arrival is $1 - \lambda\Delta t$. We can summarize as

$$P(k,\Delta t) = \begin{cases} 1 - \lambda\Delta t & \text{for } k = 0 \\ \lambda\Delta t & \text{for } k = 1 \\ 0 & \text{for } k > 1 \end{cases} \tag{7.7}$$

where $P(k, t)$ is the probability of k arrivals in interval t. We are interested in determining the pmf of the number arrivals in a time interval t, the pdf of the arrival time of the kth occurrence, and the pdf of the time interval between arrivals of successive occurrences (interarrival time).

At this point, it is worthwhile to note that this process refers to arrivals on a continuous line. For many applications, this line is time, but for others it may be considered a spatial domain of dimension one; for example, a transect along an ecosystem, or the midline of a river, or a road.

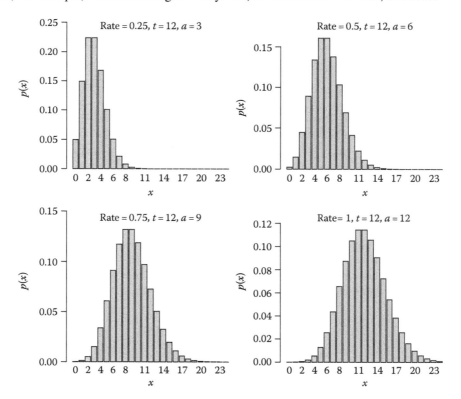

FIGURE 7.17 pmf of Poisson RV for increasing values of rate in a fixed interval.

To determine the pmf of the number arrivals in a time interval t, we write a recursive equation on $P(k, t)$

$$P(k, t + \Delta t) = P(k, t)P(0, \Delta t) + P(k-1, t)P(1, \Delta t) \tag{7.8}$$

In other words, the arrival count was k at time t and there were no new arrivals plus the probability that the count was $k-1$ and we get one new arrival. Now use Equation 7.7

$$P(k, t + \Delta t) = P(k, t)(1 - \lambda \Delta t) + P(k-1, t)\lambda \Delta t \tag{7.9}$$

Rearranging, dividing by Δt, using infinitesimal Δt and the concept of derivative, we obtain

$$\frac{P(k, t + \Delta t) - P(k, t)}{\Delta t} = \frac{dP(k, t)}{dt} = -\lambda P(k, t) + \lambda P(k-1, t) \tag{7.10}$$

which we can show has solution (Drake, 1967)

$$P(k, t) = \frac{(\lambda t)^k \exp(-\lambda t)}{k!} \tag{7.11}$$

For $t \geq 0$, $k = 0, 1, 2, \ldots$ which is the Poisson pmf when $a = \lambda t$ (see Figure 7.17). The Poisson RV has mean and variance equal to a as we described in Chapter 3. The cdf is shown in Figure 7.18.

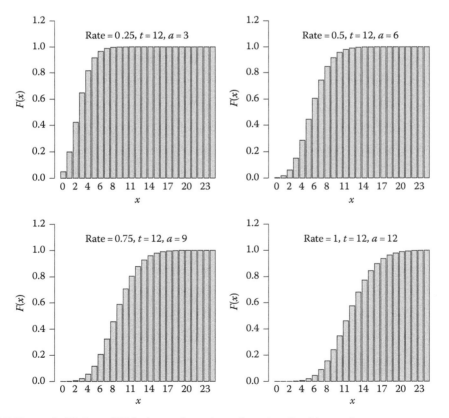

FIGURE 7.18 cmf of Poisson RV for increasing values of rate in a fixed interval.

Using a similar argument, we can calculate the pdf of arrival times as RV T_k taking values τ. Say, the density

$$P(\tau < \tau_k \le \tau + \Delta\tau) = \Delta\tau p(k,\tau) = P(k-1,\tau)\lambda\Delta\tau \tag{7.12}$$

Now use Equation 7.11

$$p(k,\tau) = \frac{(\lambda\tau)^{k-1}\exp(-\lambda\tau)}{(k-1)!}\lambda = \frac{\lambda^k\tau^{k-1}\exp(-\lambda\tau)}{(k-1)!} \tag{7.13}$$

This is the pdf of an **Erlang** RV of order k. The cdf is

$$F(k,\tau) = 1 - \exp(-\lambda\tau)\sum_{k=0}^{k-1}\frac{(\lambda\tau)^k}{k!} \tag{7.14}$$

Figure 7.19 shows both pdf and cdf. As you can see, the curve looks more and more as a normal as we increase the order k (Acevedo, 1980; Acevedo et al. 1996; Hennessey, 1980; Lewis, 1977).

When you look carefully at the curve for $k = 1$, you realize it is an exponential. Indeed, the first-order arrival time, $k = 1$, T_1, is the exponential

$$p(1,\tau) = \frac{\lambda^k\tau^{k-1}\exp(-\lambda\tau)}{(k-1)!} = \lambda\exp(-\lambda\tau) \tag{7.15}$$

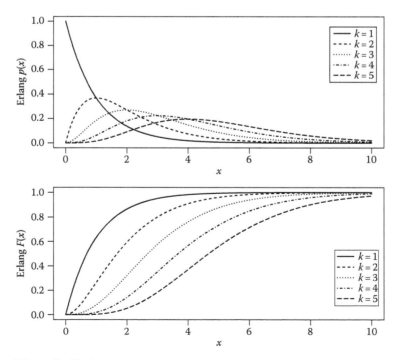

FIGURE 7.19   Erlang distribution.

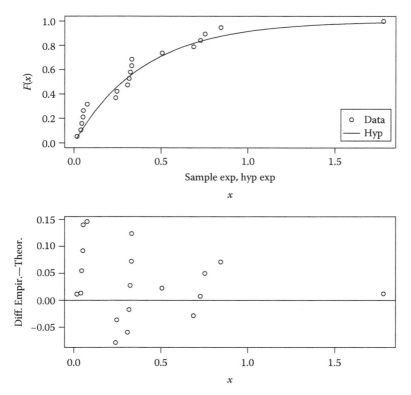

FIGURE 7.20   ecdf plots of Poisson interarrival times.

This is the interarrival time or time between successive arrivals. The interarrival time has mean $1/\lambda$ and variance $1/\lambda^2$. Random variable T_k is the sum of k independent random variables T_1 or exponential RV. Thus, we have that T_k has mean k/λ and variance k/λ^2 (Drake, 1967).

In summary, the arrival times T_k follow an Erlang of order k, the first-order arrival or interarrival times of a Poisson process are exponentially distributed independent random variables. In fact, we can determine whether it is reasonable to model a process as a Poisson process is to see if the interarrival times are independent exponential variables (Drake, 1967). Figure 7.20 illustrates one case where the empirical cdf seems to follow the theoretical cdf for an exponential distribution. Figure 7.21 shows an exploratory analysis of this particular example of interarrival times; as you can see the graphs also suggest a skewed distribution as it would be achieved with an exponential distribution.

7.6   MARKED POISSON PROCESS

It is of interest also to include some quantity (a **mark**) to the occurrence of the event at time t. For earthquakes, this quantity may be intensity, magnitude, and energy. For rain events, the quantity may be rainfall intensity. Associating a quantity y_i to the time t_i we have a **marked** Poisson process. We assume that the RV describing quantity is independent from the RV describing arrival times.

Suppose $Y(t)$ is a Poisson process with rate λ and that at each arrival time t_i we have a mark or value from a RV X_i, with given pdf. The sum of all marks for arrivals occurring in the interval t

$$Z(k, t) = \sum_{i=1}^{Y(t)} X_i \tag{7.16}$$

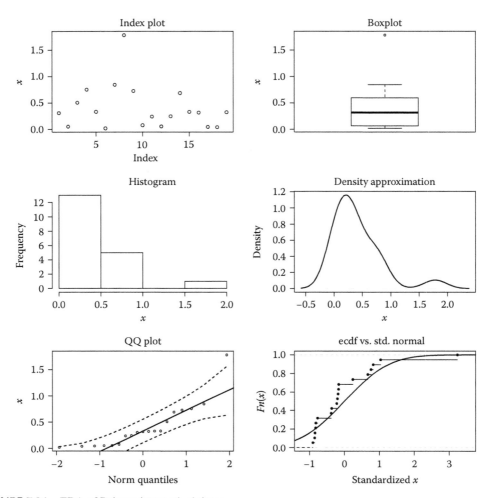

FIGURE 7.21 EDA of Poisson interarrival times.

This is called a **compound** Poisson process and has mean $\mu_Z = \lambda\mu_X t$ and variance $\sigma_Z^2 = \lambda(\mu_X^2 + \sigma_X^2)t$.

As an example, think about modeling rainfall for every day of a month. A rainy or wet day be decided upon a Poisson process, and the mark would be the amount of rain for that day if it is a wet day. The frequency distribution of rainfall in rainy days at a site determines the amount of rain, once a day is selected as wet (Richardson and Nicks, 1990). Daily rainfall distribution is skewed toward low values and it varies month to month according to climatic records. We will study four distributions to generate rainfall amount: exponential, Weibull, gamma, and skewed normal (Figure 7.22).

The simplest model is an exponential RV,

$$p(x) = a\exp(-ax)$$
$$F(x) = 1 - \exp(-ax)$$

(7.17)

where x is the daily rainfall amount and has only one parameter, the rate a. Recall that both the mean and standard deviation are equal to $1/a$, which is the average rainfall for wet days in the month.

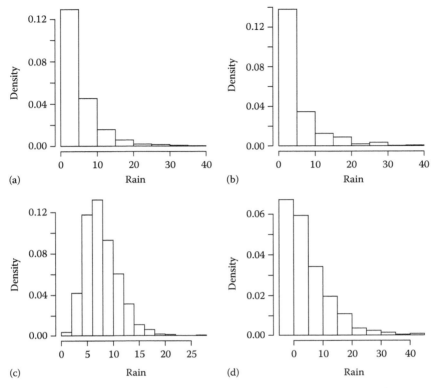

FIGURE 7.22 Examples of samples (size = 1000) generated by four different distributions: (a) exponential with μ = scale = 5, (b) Weibull with μ = 5, shape = 1.3, (c) gamma with μ = 5, shape = 0.8, and (d) skewed with μ = 5, σ = 8, γ = 2 (skew).

A more complicated model is the Weibull, which has the following pdf and cdf

$$p(x) = \left(\frac{c}{b}\right)\left(\frac{x}{b}\right)^{c-1} \exp\left(-\left(\frac{x}{b}\right)^{c}\right)$$

$$F(x) = 1 - \exp\left(-\left(\frac{x}{b}\right)^{c}\right)$$

(7.18)

with parameters shape = c and scale = b, and mean and variance

$$\mu_X = b\Gamma\left(\frac{(1+1)}{c}\right) = \left(\frac{b}{c}\right)\Gamma\left(\frac{1}{c}\right)$$

$$\sigma_X^2 = b^2\left(\Gamma\left(\frac{(1+2)}{c}\right) - \left(\Gamma\left(\frac{(1+1)}{c}\right)\right)^2\right) = \left(\frac{b^2}{c}\right)\left(2\Gamma\left(\frac{2}{c}\right) - \left(\frac{1}{c}\right)\left(\Gamma\left(\frac{1}{c}\right)\right)^2\right)$$

(7.19)

where Γ is the gamma function. Another way to write the Weibull is to use parameters rate and power coefficients equal to the inverse of scale and shape, $a = 1/b$, $d = 1/c$

$$p(x) = \left(\frac{a}{d}\right)(ax)^{1/d-1} \exp\left(-(ax)^{1/d}\right)$$

$$F(x) = 1 - \exp\left(-(ax)^{1/d}\right)$$

We can see that when shape $c = 1$ or $d = 1$, this expression reduces to the exponential. A value of shape c lower than one (that is $d > 1$) will produce a pdf with higher values to the left than the exponential pdf. Whereas a value of c higher than 1 ($d < 1$) will produce higher values to the right (Figure 7.23). The case $d > 1$ is of more interest for modeling rainfall amount. However, for $d > 1$, $p(0)$ does not a have a finite value.

For example when $a = 10$, $d = 1.5$,

$$\mu_X = \left(\frac{1}{a}\right)\Gamma(1+d) = 10\Gamma(2.5) = 10 \times 1.33 = 13.3$$

and for $a = 10$, $d = 0.5$

$$\mu_X = \left(\frac{1}{a}\right)\Gamma(1+d) = 10\Gamma(1.5) = 10 \times 0.89 = 8.9$$

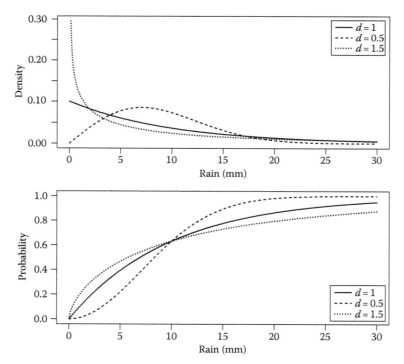

FIGURE 7.23   Weibull distribution for scale = 10 ($a = 0.1$) and three different values of shape ($1/d$). The solid line for $d = 1$ reduces to an exponential. The other two values of d are for shape greater than 1 and shape less than 1.

Another pdf typically employed to model rainfall amount is the Gamma pdf

$$p(x) = \frac{(x/b)^{c-1} \exp(-x/b)}{b\Gamma(c)} \tag{7.20}$$

where the parameters c and b are shape and scale, respectively. The scale is the inverse of the rate $a = 1/b$. The mean and variance are

$$\mu_X = cb \quad \sigma_X^2 = cb^2 \tag{7.21}$$

For the gamma, it is relatively easy to generate rainfall with a given mean and variance of rain since we can solve for c and b from Equation 7.21

$$\frac{\mu_X}{b} = c \quad \text{and} \quad c = \frac{\sigma_X^2}{b^2}$$
$$b = \frac{\sigma_X^2}{\mu_X} \quad \text{and} \quad c = \frac{\mu_X^2}{\sigma_X^2} \tag{7.22}$$

There is no closed form equation for the cdf unless the shape parameter c is an integer and in this case $\Gamma(c) = (c - 1)!$ and the pdf reduces to the Erlang pdf

$$p(x) = \frac{(ax)^{c-1} \exp(-ax)}{(c-1)!}$$
$$F(x) = 1 - \exp(-ax) \sum_{c=0}^{c-1} \frac{(ax)^c}{c!}$$

Another pdf used to generate rainfall is based on a skewed distribution generated from a normal (Neitsch et al., 2002)

$$x = \mu_X + 2\frac{\sigma_X}{\gamma_X} \left[\left[\frac{\gamma_X}{6} \left(z - \frac{\gamma_X}{6} \right) + 1 \right]^3 - 1 \right] \tag{7.23}$$

where, in addition to the mean daily rainfall for the month, we use the σ_X = standard deviation daily rainfall for the month and γ_X = skew coefficient of daily rainfall. In this case, z is a value from a standard normal random variable (mean equal 0 and standard deviation equal 1). The distribution parameters—mean, standard deviation, and skewness—vary month to month according to climatic records.

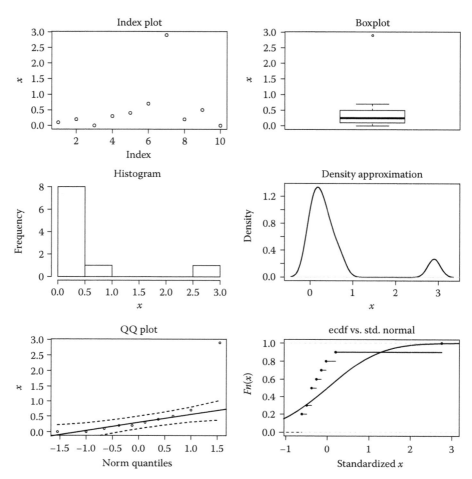

FIGURE 7.24 EDA of daily rainfall.

As an example, consider the EDA graphs in Figure 7.24, which pertain to the daily rainfall in a given month, and in Figure 7.25, which corresponds to interarrival time. In this case, the interarrival time is the number of days elapsed in between wet days. We can see that these sets do not exhibit normality and that they could correspond to an exponential or Weibull RV. Furthermore, if we look at the graphs in Figure 7.26, we see that the ecdf for the rainfall amount can be modeled as a Weibull RV (with shape $c = 0.7$ and scale $b = 0.4$), whereas Figure 7.27 shows that the ecdf for the days in between wet days can be modeled simply by an exponential with rate $\lambda = 0.5$.

We conclude that in this example rainfall for this month could be modeled as a marked Poisson process Y with rate $\lambda = 0.5$ and Weibull distributed marks X with shape $c = 0.7$ and scale $b = 0.4$. The mean and variance of X calculated by Equation 7.19 are 0.51 and 0.54. The empirical mean and variance of X are 0.53 and 0.74. The mean and variance of T1 are $1/0.5 = 2$ and $1/(0.5)^2 = 4$, the empirical mean of T1 is 2.2 days and its variance 2.4. The rainfall for the month of $t = 30$ days is a compound Poisson process Z and has mean

$$\mu_Z = \lambda \mu_X t = 0.5 \times 0.51 \times 30 = 7.65$$

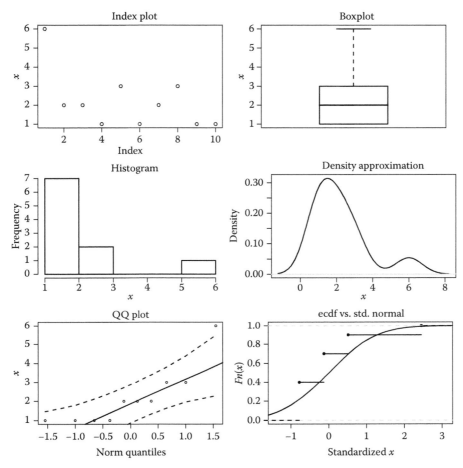

FIGURE 7.25 EDA of days in between wet days.

and variance

$$\sigma_Z^2 = \lambda\left(\mu_X^2 + \sigma_X^2\right)t = 0.5 \times (0.51^2 + 0.54^2) \times 30 = 8.27$$

7.7 SIMULATION

Once we have a suitable model for a random process, we can generate hypothetical time series as possible realizations of the random process. This is a Monte Carlo simulation. For example, we can generate multiple realizations of the rainfall time series for the month to use as input to crop and ecosystem simulation models and to explore possible rainfall sequences due to scenarios of climate change.

Looking at this example further, consider rainfall for the month modeled as a Poisson with rate $\lambda = 0.5$ and Weibull distributed marks X with shape $c = 0.7$ and scale $b = 0.4$. Figure 7.28 shows nine rainfall time series or realizations of this process. We can see how we have 10 or more days with rain for most realizations. Some realizations have days with intense rain (3–4 cm) and in some realizations most of the wet days have low amount of rain. Figure 7.29 shows the histograms of the marks or rain amount generated by Weibull distribution.

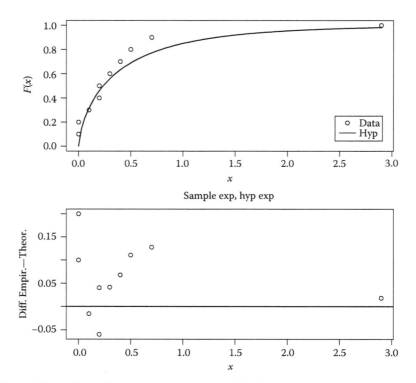

FIGURE 7.26   ecdf of daily rainfall and comparison to a Weibull cdf.

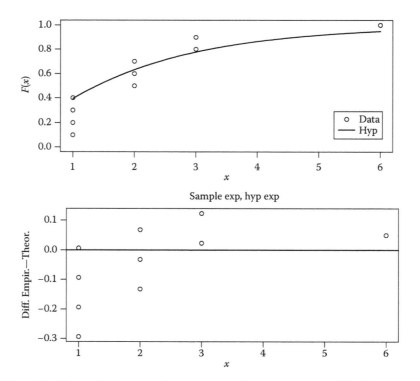

FIGURE 7.27   ecdf of days in between wet days and comparison to an exponential cdf.

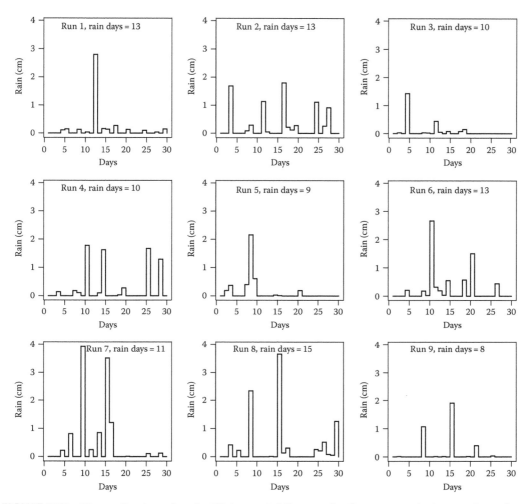

FIGURE 7.28 Nine realizations of marked Poisson rainfall generation for a wet month. Amount drawn from Weibull density.

7.8 EXERCISES

Exercise 7.1

Time series modeling of tidal height is important to many environmental models. Tidal heights are referenced to a datum and are specified at equally spaced intervals, for example, 0.5 h. Data are obtained from tidal stage recorders or from the U.S. Coast and Geodetic Survey Tide Tables. Consider the following model of tidal height

$$x(t) = A_0 + \sum_{k=1}^{3} A_k \sin(k 2\pi f t) + \sum_{k=1}^{3} B_k \cos(k 2\pi f t)$$

where
  $x(t)$ is the tidal elevation
  A_k, B_k are coefficients
  f is frequency, h^{-1}
  t is time, h

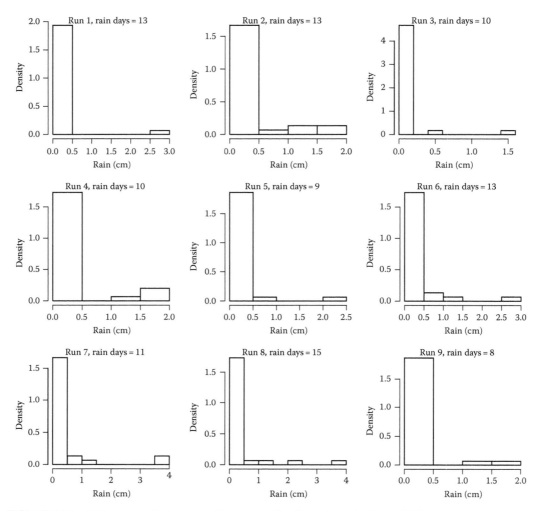

FIGURE 7.29 Histograms of rain amount for nine realizations shown in Figure 7.28.

Assume all coefficients equal to 1. Sketch each one of the tidal time series model components for one period. Then graphically add all the components. Sketch the periodogram for each component and for the total tidal elevation.

Exercise 7.2

Consider a longitudinal transect of 100 m along a direction N-S, establish a starting point as distance $d = 0$ m and that we record the distance from this datum to each one of the points where we detect the presence of a plant species. Assuming that detections are independent and that the expected rate is 0.1 plants/m. Use a Poisson pdf to model this spatial process; determine the probability of observing $k = 1, 2, 3, 4$ plants in the transect. Sketch the pdf of distances to the detection of the fifth plant.

7.9   COMPUTER SESSION: RANDOM PROCESSES AND TIME SERIES

7.9.1   Gaussian Random Processes

We start by studying how to simulate a Gaussian process. The following lines of code simulate nine runs of 30 days (1 month) of a Gaussian process with mean 30 and standard deviation 5. We loop through the runs or realizations drawing a sample of size 30 from a normal distribution using `rnorm`. Within the loop we calculate the average so that we can report in the plots to be produced later.

```
#gaussian simulation
muT=30;sdT=5;ndays= 30; nruns=9
# define array
zg <- matrix(ncol=nruns, nrow=ndays);tavg <- array()
# loop realizations
for(i in 1:nruns){
zg[,i] <- round(rnorm(ndays,muT,sdT),2); tavg[i]
  <- round(mean(zg[,i]),1)
}# end of realization loop
```

Then we plot the series for all realizations. First, design a layout for nine panels (3 × 3), plot one series in each panel (Figure 7.3), and then in a second-page plot the histogram for each run (Figure 7.4).

```
z <- zg
# plot
 mat<- matrix(1:9,3,3,byrow=T)
 layout(mat,rep(7/3,3),rep(7/3,3),res=TRUE)
 par(mar=c(4,4,1,.5), xaxs="r", yaxs="r")

for(i in 1:nruns){
 plot(z[,i], type="l", ylab="x", xlab="Days", ylim=c(0,max(z)))
 mtext(side=3,line=-1,paste("run",i,", avg x=",tavg[i]),cex=0.7)
}
for(i in 1:nruns){
 hist(z[,i],prob=T,main="",xlab="x")
 mtext(side=3,line=0,paste("run",i,", avg x=",tavg[i]),cex=0.7)
}
```

Now increase the time span or duration of the series. For example, simulate noise of unit variance, a Gaussian process, with mean 0 and standard deviation 1, for 300 units of time.

```
muT=0;sdT=1;ndays= 300; nruns=9
# define array
zg <- matrix(ncol=nruns, nrow=ndays);tavg <- array()
# loop realizations
for(i in 1:nruns){
zg[,i] <- round(rnorm(ndays,muT,sdT),2); tavg[i]
  <- round(mean(zg[,i]),1)
}# end of realization loop
z <- zg
```

We repeat the plots as shown earlier. Figure 7.5 shows the time series, and Figure 7.6 illustrates how the histograms suggest a normal distribution.

7.9.2   Autocorrelation

We will use one of the realizations (run 1) of the Gaussian process with mean 0 and variance 1. First, convert to regular time series using function **ts**. Then plot using ts.plot

```
x <- z[,1]
x.rts <- ts(x, start=0, deltat=1)
ts.plot(x.rts, ylab="x",xlab="Time")
```

The result is Figure 7.7. Next we do scatter plots with four lags (Figure 7.8) and autocorrelation plots (Figure 7.9).

```
lag.plot(x.rts, lags=4, layout=c(2,2))
panel2(size=7)
acf(x.rts)
acf(x.rts, type="covariance")
```

7.9.3   Periodic Process

File **lab7/periodic.csv** contains 25 years of monthly data, for a total of $25 \times 12 = 300$ records. The reported period is 1981–2005. The columns are year, month (1–12), and data value.

```
Yr,Mo,x
1981,1,8.47
1981,2,12.362
1981,3,12.38
1981,4,11.382
1981,5,11.16
1981,6,9.2
1981,7,9.2
1981,8,7.048
1981,9,8.58
1981,10,7.938
1981,11,9.07
1981,12,11.78
1982,1,12.6
1982,2,12.292
```

We read this file as a data frame, create regular time series with function **ts** applied to third column x, with a time interval of 1 year (delta $t = 1/12$ year) starting in 1981; then plot the time series with ts.plot and the autocorrelation function with acf to get results as in Figure 7.13.

```
x <- read.table("lab7/periodic.csv",sep=",",header=T)
x.rts <- ts(x$x, start=1981, deltat=1/12)
panel2(size=7)
ts.plot(x.rts, ylab="x",xlab="Time")
acf(x.rts)
```

7.9.4 FILTERING AND SPECTRUM

Apply function `filter` using a rectangular low-pass filter of 5 points, subtract the filtered signal `x.f` from the original series `x`, and plot all three time series (Figure 7.14)

```
x.f <- filter(x,filter=rep(1/5,5), method="convolution", sides=2,
  circular=T)
y <- x-x.f
panel3(size=5)
plot(t,x, type="l")
plot(t,x.f, type="l")
plot(t,y, type="l")
acf(y)
```

Apply function `filter` using a high-pass filter of 5 points, subtract the filtered signal `x.f` from the original series `x`, and plot all three time series (Figure 7.15)

```
x.f <- filter(x.rts,filter=c(-1/5,-1/5,4/5,-1/5,-1/5),
  method="convolution", sides=2, circular=T)
y <- x-x.f
panel3(size=7)
ts.plot(x.rts, type="l",ylab="x")
ts.plot(x.f, type="l",ylab="xf = x filtered")
ts.plot(y, type="l",ylab="x-xf residual")
```

Function `spec.pgram` estimates the periodogram, argument `demean = T` removes the mean, and `span` declares the number of points of successive Daniell filters to smooth the periodogram. Here, we use `plot = F` and then employ a separate plot command to draw the periodogram. If we had used `plot = T`, the function would have produced the periodogram in decibels (dB), which is logarithm scale corresponding to $20 \times \log(x)$. The result of the following lines of code is Figure 7.16.

```
panel4(size=7)
x.spec <- spec.pgram(x, demean=T,span=c(5,7),circular=T,plot=F)
plot(x.spec$freq,x.spec$spec,type="l",xlab="Frequency",ylab=
  "Spectrum")
title("Periodic time series = signal + noise",cex.main=0.7)

xf.spec <- spec.pgram(x.f, demean=T, span=c(5,7),plot=F)
plot(xf.spec$freq,xf.spec$spec,type="l",xlab="Frequency",ylab=
"Spectrum")
title("Smoothed periodic time series = signal",cex.main=0.7)

xy.spec <- spec.pgram(y, span=rep(5,7),circular=T,plot=F)
plot(xy.spec$freq,xy.spec$spec,type="l",xlab="Frequency",ylab=
  "Spectrum")
title("Difference = Noise",cex.main=0.7)

x.wn <- z[,1]
xwn.spec <- spec.pgram(x.wn, demean=T,span=c(5,7,9),circular=T,plot=F)
plot(xwn.spec$freq,xwn.spec$spec,type="l",xlab="Frequency",ylab=
  "Spectrum")
title("Gaussian white noise",cex.main=0.7)
```

7.9.5 SUNSPOTS EXAMPLE

As another example, we will use the **lab7/year-spot1700-2011.txt** dataset, which contains yearly number of sunspots in the period 1700–2011. These values are from SIDC-team (2012) World Data Center for the Sunspot Index, Royal Observatory of Belgium.

First, we read the file as a matrix object and check

```
> yrspots <- matrix(scan("lab7/year-spot1700-2011.
  txt",skip=6),ncol=2,byrow=T)
> yrspots
        [,1] [,2]
 [1,] 1700    5
 [2,] 1701   11
 [3,] 1702   16
```

Next create regular time series with function `ts` applied to second column of `yrspots`, with a time interval of 1 year `deltat=1` starting in 1700; next do a plot of the time series as shown in Figure 7.2a discussed earlier.

```
> yrspots.rts <- ts(yrspots[,2], start=1700, deltat=1)
> ts.plot(yrspots.rts, ylab="Number Sunspots",xlab="Years")
```

We can use `window` to zoom in a 50 year period (Figure 7.2b).

```
> winspots.rts <- window(yrspots.rts, start= 1700, end=1750)
> ts.plot(winspots.rts)
```

Lagged scatter plots shown in Figures 7.10 and 7.11 are obtained by

```
> lag.plot(yrspots.rts, lags=11, layout=c(2,3))
```

You can use any other `layout`, `panel`, or `par` function that we have used in previous computer sessions.

Another visual exploration is an autocorrelation plot as we have done before with function `acf` to obtain Figure 7.12 discussed previously. A plot of autocovariance can be obtained by using `type = "autocovariance"` in the `acf` function call.

```
> acf(yrspots.rts)
> acf(spots.rts, type="covariance")
```

The maximum value of the covariance, or covariance at lag 0, should be equal to the variance. It seems to be approximately 1500. We can confirm by calculating the variance of the entire set of values and the maximum autocovariance

```
> var(yrspots[,2])
[1] 1631.299
> max(acf(yrspots.rts, type="covariance")$acf)
[1] 1626.071
>
```

7.9.6 POISSON PROCESS

As an example, we will use the **lab7/bird-arrival.txt** dataset, which contains arrival times (in accumulated hours) of birds to a site in a 12h period starting from hour 0. The first value is how long it took for the first individual to arrive, the second value is how long for the second, and so forth.

We scan the file and calculate the time difference in between successive arrivals to obtain interarrival time `tau`. Round to three decimals, since this is the number of decimals of the values given in the file. We check mean and variance and whether the interarrival time has mean $1/\lambda$ and variance $1/\lambda^2$. Or, in other words, whether λ estimated by $1/\overline{T}$ and $1/\sqrt{s_T^2}$ are similar.

```
>ctau <- scan(file="lab7/bird-arrival.txt")
>tau <- round(c(ctau[1],diff(ctau)),3)
>1/mean(tau);1/sqrt(var(tau))
[1] 2.475248
[1] 2.350483
```

We see that the last two numbers are similar and around 2.41. We estimate that $\lambda = 2.41$.

We use the `eda6` function and `cdf.plot.gof` function developed in Chapter 5 to visualize if the interarrival time may be exponentially distributed with rate equal to 2.41. Label with `mtext`.

```
eda6(tau)
cdf.plot.gof(tau,dist="exp",rate=2.41)
mtext(side=3,line=2,"Sample Exp, Hyp Exp",cex=0.7)
```

The graphics are shown in Figures 7.20 and 7.21, which suggest exponentially distributed times with this rate. Next, perform a K–S test to confirm

```
> ks.test(tau,"pexp", rate=2.41)

      One-sample Kolmogorov-Smirnov test

data: tau
D = 0.1504, p-value = 0.7284
alternative hypothesis: two-sided
```

We conclude that the sample may be exponentially distributed with that mean.

7.9.7 POISSON PROCESS SIMULATION

First, generate a Poisson distributed number k using `rpois` and then for each integer up to k generate an interarrival time using `rexp`. The variable `ctau` is the cumulative time to each arrival.

```
# generate intearrival time
a <- 2; tf<- 12
tau <- array(); ctau <- array()
k <- rpois(1,lambda=a*tf)
for(i in 1:k){
 tau[i] <- rexp(1,rate=a)
 ctau[i] <- sum(tau[1:i])
}
```

You can check the results visually and by K–S as we did in the previous section.

```
eda6(tau)
cdf.plot.gof(tau,dist="exp",rate=a)
mtext(side=3,line=2,"Sample Exp, Hyp Exp",cex=0.7)
ks.test(tau,"pexp", rate=a)
```

7.9.8   Marked Poisson Process Simulation: Rainfall

Apply seeg's function `poisson.rain` reproduced here for easy reference.

```
poisson.rain <- function(rate=1, ndays=30, shape=1, scale=1,plot.
  out=T){
 nk <- rpois(1,lambda=rate*ndays)
 tau1 <- array(); ctau1 <- array()
 for(i in 1:nk){
 tau1[i] <- ceiling(rexp(1,rate))
 ctau1[i] <- sum(tau1[1:i])
}
if (ctau1[nk]<=ndays) k <- nk else
k <- min(which(ctau1>ndays))-1
tau <- tau1[1:k]; ctau <- ctau1[1:k]

# amount of rain
x <- round(rweibull(k, shape,scale),2)
y <- cbind(tau,ctau,x)
days <- seq(1,ndays); rain <- days; rain[] <- 0
for(i in 2:ndays){
 ab <- which(ctau==days[i])
 if(length(ab)>0) rain[i] <- x[ab]
}
z <- cbind(days,rain)
 if(plot.out == T){
 panel4(size=7)
 plot(days,rain,type="s",xlab="Day",ylab="Rain (cm/day)")
 hist(rain,prob=T,main="",xlab="Rain (cm/day)")
 hist(tau,prob=T,main="",xlab="Interarrival time")
}
return(list(y=y,z=z))
}
```

Now employ this function within a loop for nine realizations in the same manner as we did for the Gaussian process.

```
#marked poisson simulation
ndays= 30;nruns= 9;rate=0.5;shape=0.7; scale=0.4
# define array
zp <- matrix(ncol=nruns, nrow=ndays);nwet <- array()
# loop realizations
for(j in 1:nruns) {
 rainy <- poisson.rain(rate,ndays,shape,scale,plot.out=F)
 zp[,j] <- rainy$z[,2]; nwet[j] <- length(rainy$y[,3])
} # end of realization loop
```

Then plot

```
z <- zp
# plot
 mat<- matrix(1:9,3,3,byrow=T)
 layout(mat,rep(7/3,3),rep(7/3,3),res=TRUE)
 par(mar=c(4,4,1,.5), xaxs="r", yaxs="r")

for(i in 1:nruns) {
 plot(z[,i], type="s", ylab="Rain(cm)", xlab="Days",
   ylim=c(0,max(z)))
 mtext(side=3,line=-1,paste("run",i,", rain days=",nwet[i]),cex=0.8)
}

for(i in 1:nruns){
 hist(z[,i],prob=T,main="",xlab="Rain (cm)")
 mtext(side=3,line=-1,paste("run",i,", rain days=",nwet[i]),cex=0.8)
}
```

The result should be like Figures 7.28 and 7.29.

7.9.9 COMPUTER EXERCISES

Exercise 7.3
Sea surface temperatures of the Pacific Ocean in several regions are considered indicative of El Niño. File **lab7/sstoi_pa.txt** contains monthly data for years 1950–1997 and for regions Niño1 + 2, Niño3, Niño4, and Niño3.4 (Acevedo et al., 1999). Read the file, convert the columns to time series, plot and analyze the time series for regions 1 + 2 and 3.4 using autocorrelation and periodograms.

Exercise 7.4
Streamflow is an important variable of watersheds measured as a volume discharge per unit time. The U.S. Geological Survey operates monitoring stations throughout the country. Consider, for example, flow of the Neches River in Texas at station USGS 08040600 near Town Bluff, TX. File **lab7/TB-flow.csv** contains daily flow data 1952–2010. Read the file, convert the flow to time series, plot and analyze the time series using autocorrelation and periodograms. Hint: when applying `ts` use `freq = 365, start = 1952, end = 2010`.

Exercise 7.5

Simulate daily rainfall for the month modeled as a Poisson with rate $\lambda = 0.5$ and Weibull distributed marks X with shape $c = 0.9$ and scale $b = 0.5$.

SUPPLEMENTARY READING

Chapters 16 and 17 (MathSoft, 1999). Chapters 1–5 (Box and Jenkins, 1976). Chapter 8 ((Manly, 2009). Carr (1995) covers spectral analysis in Chapter 9. Davis (2002) covers stochastic process (pp. 150–176), and autocorrelation (pp. 217–234).

8 Spatial Point Patterns

8.1 TYPES OF SPATIALLY EXPLICIT DATA

In this book, we will look at two main types of spatial data: one type corresponds to **point** patterns and the other type to **lattice** arrangements. The difference is whether we consider points or regions placed over a spatial domain. We will focus on **two-dimensional** (2D) domains because they encompass many important geographical and environmental situations.

Point patterns are typically a collection of points placed over a spatial domain. They may be regular as in a grid or irregular. We may have values of variables associated at each point. Lattice data correspond to regions or polygons. Regions are regularly arranged when they occur in a regular pattern, say a rectangular grid. For example, a remote sensed image composed of pixels. Regions are irregularly arranged when they do not follow a regular spatial pattern, as it happens for example when we have political divisions, such as counties.

We will cover three types of spatial analysis: **analysis of point** patterns, that is, to examine the spatial distribution of points, for example, check whether points are clustered or uniformly distributed; **geostatistics**, to predict values of variables in nonsampled points using a collection of sampled points; **spatial regression**, to predict values of variables in regions from values at the neighboring regions. In this book, we will cover the fundamentals of these methods. Nowadays many of these methods are also implemented in Geographical Information Systems (GIS) software. In this chapter, we will cover spatial point patterns. We will wait to cover the other two methods until we learn about matrices and linear algebra.

8.2 TYPES OF SPATIAL POINT PATTERNS

A spatial point pattern is a collection of points located within a bounded region or domain. Points in a **two-dimensional domain** have coordinates x, y within the domain where x is displayed horizontally and y vertically. Coordinate x increases from left to right and coordinate y from bottom to top. The coordinate system can be geographical; for example x increasing towards the east and y increasing towards the north.

Figure 8.1 illustrates four examples. At the top left, points are regularly located in the domain as a grid; all three other patterns are irregular in distribution. At the top right, points spread uniformly, but both patterns at the bottom are nonuniform. The one on the left follows a gradient with x, and the one on the right shows aggregation. In this chapter, we will consider distribution in detail.

Applications are many and varied. For example, spatial distribution of plants and trees on the landscape, sampling sites in a waste field, and sampling sites in a lake, locations of intense events such as quakes or tornadoes. The nature of the data could be **location only** (just the pair of coordinates for each point), or **marked point** (pair of coordinates plus data values associated with each location). A spatial point pattern is a starting point for **kriging** analysis as discussed later in the book.

8.3 SPATIAL DISTRIBUTION

We now consider more details of the distribution of points over space. As illustrated by Figure 8.1, major types of distributions are uniform or homogeneous and nonuniform or nonhomogeneous. These terms refer to the uniformity or homogeneity of the **density of points**, that is, spatial variability of the number of points per unit area.

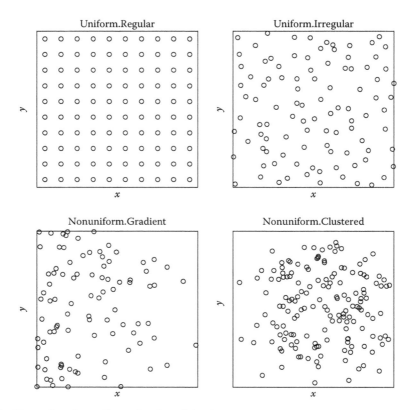

FIGURE 8.1 Examples of two-dimensional spatial point patterns.

Uniform or homogeneous patterns correspond to those when density of points is the same over the spatial domain (both panels at the top in Figure 8.1). In turn, these patterns can be **regular** when the distance between points is the same (as in a grid, top-left panel) or **random** when the location of points independent of other points and the density is the same (top-right panel). The pattern on the top right was generated by assigning a location randomly to one point per square of the grid shown on the left. Note that a regular pattern is uniform, but a uniform pattern is not necessarily regular.

Nonhomogeneous patterns correspond to those with changing density over the domain as it occurs when the point-density follows a **gradient** (bottom left) or when they are **aggregated or clustered** (bottom right). Gradients are common in geographical and environmental analysis. They can occur because of elevation, substrate conditions, and many other factors. Aggregation can occur due to facilitation processes, lack of dispersion, bias sampling, and many other factors.

8.4 TESTING SPATIAL PATTERNS: CELL COUNT METHODS

In many instances, it is important to determine whether the point distribution follow one of the categories mentioned earlier. In this section, we cover grid-based methods that are amenable to analysis using the chi-square test.

8.4.1 TESTING UNIFORM PATTERNS

Divide the domain in a grid of T cells or tiles of equal size. Count the total number of points, m. If the distribution is uniform, then the expected number e_i of points per cell is $E = m/T$. We now count the number o_i of observed points per cell and use a chi-square test with $df = T - 2$.

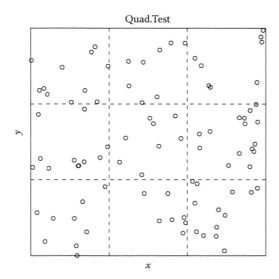

FIGURE 8.2   Testing uniform pattern of 90 points with 3×3 grid, this is $T = 9$ cells.

TABLE 8.1
Points per Tile in Figure 8.2

11	8	8
11	8	15
9	7	13

$$\chi^2 = \sum_{i=1}^{T} \frac{(o_i - e_i)^2}{e_i} \qquad (8.1)$$

We formulate the null H0 to be that the distribution is uniform. When there are large departures from the expected, the chi-square value is sufficiently large and we can reject the null and conclude that the pattern is nonuniform.

For example, Figure 8.2 shows a pattern of 90 points distributed over a domain divided in a grid of $3 \times 3 = 9$ square cells. We would expect $e_i = 90/9 = 10$ points per cell for all cells i. We obtain observed values by counting how many points we have in each cell (Table 8.1) and then calculating the chi-square value for each cell using Equation 8.1. We get chi-square of 5.8, which for $df = 9 - 2 = 7$ has a p-value of 0.56. Therefore, we should not reject H0 and we conclude that the pattern may be uniform.

8.4.2   TESTING FOR SPATIAL RANDOMNESS

Spatially random patterns should follow a Poisson distributed RV. Recall that the Poisson pdf is a binomial with very small probability of the event "success." Therefore, the Poisson is an RV model for rare events. Therefore, we test for random pattern by testing whether the spatial pattern follow a Poisson distribution.

Define λ density (number of points per unit area) by

$$\lambda = \frac{m}{A} \qquad (8.2)$$

where A is total area of the domain. Divide the total area in T cells or tiles, then the mean number of points per cell is m/T and should be equal to the density multiplied by the area a of a cell, therefore

$$a\lambda = \frac{A}{T}\frac{m}{A} = \frac{m}{T} \tag{8.3}$$

Use a Poisson RV with rate $a\lambda = mT$ to calculate probability of having $r = 0, 1, 2,\dots$ points in a cell

$$P(r) = \frac{\exp(-a\lambda)(a\lambda)^r}{r!} \tag{8.4}$$

then the expected number of cells with $r = 0, 1, 2,\dots$ points in a cell

$$e_r = TP(r) \tag{8.5}$$

Then count how many cells o_r are observed to have $r = 0, 1, 2,\dots$ points in a cell. Then use chi-square (χ^2) to compare to observed values.

Denote by r_i the points in cell i, then the sample variance is

$$s^2 = \frac{\sum_{i=1}^{T} \left(r_i - \frac{m}{T} \right)^2}{T-1} \tag{8.6}$$

Recall from Chapter 3 that for a Poisson the mean is equal to the variance and equal to the rate. Therefore, we can compare the variance s^2 of number of points per cell to the mean m/T to determine whether the pattern is uniform, random, or clustered. For this purpose, we use the ratio of mean to variance. If the ratio is less than 1, then the pattern is not random and is clustered; if the ratio is larger than 1, then the pattern is not random and is uniform. If the ratio is equal to 1, then the pattern is random. Therefore, we formulate a t-test with t value based on the departure of the ratio from unity

$$t = \frac{\dfrac{m/T}{s^2} - 1}{s_e} = \frac{\dfrac{m/T}{s^2} - 1}{\sqrt{2/(T-1)}} \tag{8.7}$$

with $df = T - 1$ (Manly, 2009).

For example, take the pattern in Figure 8.3. We got $m = 8$ points in a grid of $5 \times 5 = 25$ cells over a domain of unit area $A = 1$. Each cell has a side of $1/5 = 0.2$ units of distance. The area of each cell is $a = (0.2)^2 = 0.04$. The rate is

$$a\lambda = \frac{m}{T} = \frac{8}{25} = 0.32$$

The probabilities of a Poisson with this rate are 0.73, 0.23, 0.04 for $r = 0, 1, 2$. The expected values of number of cells that have r points are 18, 6, 1. The observed are 17 cells with 0 points, and 8 cells with 1 point, 0 cells with 2 points. The chi-square values are 0.055, 0.666, and 1.00. These add up to 1.72 and we have $df = 3 - 2 = 1$ because we used three categories or bins. The p-value is 0.19 and can reject only to 20%. Therefore, the pattern could be random.

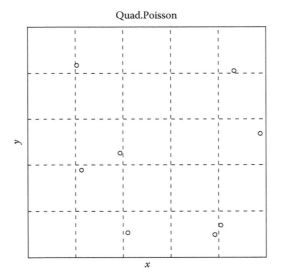

FIGURE 8.3   Testing for randomness.

The sample variance s^2 is calculated from Equation 8.6. First, subtract 0.32 from 0 or 1 depending on whether the cells had zero or one point, then square and add over all cells to get 5.44. Then divide by $T - 1 = 24$ to obtain $s^2 = 0.23$. The ratio of the mean m/T to the variance s^2 is

$$\frac{m/T}{s^2} = \frac{0.32}{0.23} = 1.41$$

Because this ratio is larger than 1, then pattern could be more uniform than random. To test

$$t = \frac{\dfrac{m/T}{s^2} - 1}{\sqrt{2/(T-1)}} = \frac{1.41 - 1.00}{\sqrt{2/24}} = \frac{0.41}{0.288} = 1.42$$

with $df = T - 1 = 24$. The p-value is 0.08 and we could reject that the pattern is random with 10% error rate.

8.4.3   CLUSTERED PATTERNS

A clustered pattern is a set of points closer together than expected by chance alone. One way of modeling clustered patterns is with the negative binomial.

$$P(r) = \binom{k+r-1}{r} \left(\frac{p}{1+p}\right)^r \left(\frac{1}{1+p}\right)^k \tag{8.8}$$

where
    r is the number of points in a cell
    k is a clustering or shape parameter (Davis, 2002)

$$k = \frac{(m/T)^2}{s^2 - (m/T)} \tag{8.9}$$

and p is the probability that a cell contains a point

$$p = \frac{\lambda}{k} = \frac{m/T}{k} \tag{8.10}$$

The mean is the rate m/T. We then use chi-square to test if pattern distributed according to a negative binomial.

For example, suppose we have $T = 3 \times 3 = 9$ cells and that we observe $m = 13$ points. The rate is $m/T = 13/9 = 1.44$. Suppose we observe the following counts in each cell 1, 0, 1, 0, 1, 5, 2, 2, 1. Therefore, the observed values of number of cells with $r = 0, 1, 2, 3, 4, 5$ are 2, 4, 2, 0, 0, 1. The variance is calculated to be $s^2 = 2.27$. Therefore, the clustering parameter is $k = 2.50$. Consequently, parameter $p = 1.44/2.50 = 0.576$. A theoretical negative binomial with this p and k would yield the following expected values 2.88, 2.63, 1.69, 0.93, 0.47, and 0.22. The chi-square is 5.14 with $df = 9 - 2 = 7$. The p-value is 0.64. Therefore, we cannot reject H0 and we conclude that the pattern may follow a negative binomial and therefore it may be clustered.

8.5 NEAREST-NEIGHBOR ANALYSIS

The methods presented in the previous section work on a grid and the number of points in each cell. In this section, we cover methods based on **distance** to the nearest-neighbor point (Davis, 2002; Manly, 2009). Since the points occur in 2D, the distance between points with coordinates (x_1, y_1) and (x_2, y_2) is the Euclidian distance

$$d_{12} = \sqrt{\frac{(x_1 - x_2)^2 + (y_1 - y_2)^2}{2}} \tag{8.11}$$

8.5.1 FIRST-ORDER ANALYSIS

Consider the nearest-neighbor distance d_i for each point i. If the points follow a Poisson distribution, with density λ, then the pdf of distances is

$$p(d) = 2\lambda\pi d \exp(-\lambda\pi d^2) \tag{8.12}$$

which has mean $\mu_d = 1/2/(\sqrt{\lambda})$ and variance $\sigma_d^2 = (4 - \pi)/4\pi\lambda$. We perform the analysis by comparing an empirical cdf of d with the theoretical cdf, which is

$$G(d) = 1 - \exp(-\lambda\pi d^2) \tag{8.13}$$

To calculate the empirical cdf, for each d we count of all distances to nearest neighbors less or equal than the value d

$$\widehat{G}(d) = \frac{1}{m} \sum_{1}^{n} \delta_i(d_i, d) \tag{8.14}$$

Here, we use an indicator function

$$\delta_i(d_i, d) = \begin{cases} 1 & \text{when } d_i \leq d \\ 0 & \text{otherwise} \end{cases} \qquad (8.15)$$

Now we can compare the ecdf $\widehat{G}(d)$ with the theoretical cdf $G(d)$.

For example, consider the point pattern in Figure 8.4, which looks clustered. The process mentioned earlier would yield a plot of empirical and theoretical cdfs for visual comparison as in Figure 8.5. The circles correspond to the ecdf raw or uncorrected for edge effects. There are several schemes to correct for edge effect such as the Kaplan–Meier estimates. When we include this correction for this example (Figure 8.5), we visually appreciate that the corrected ecdf departs only slightly from the ecdf and mostly for larger values of d. In conclusion, this visual exploration tells us that the ecdf departs substantially from the theoretical cdf, reinforcing the suspicion of a clustered pattern of the points.

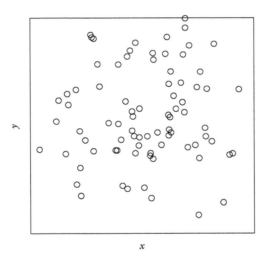

FIGURE 8.4   Spatial point pattern for distance analysis.

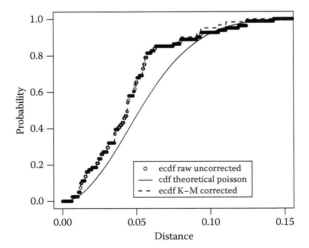

FIGURE 8.5   Empirical (raw and corrected) and theoretical cdf of distance.

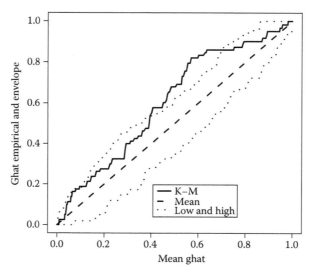

FIGURE 8.6   ecdf departing from band obtained by Monte Carlo simulation.

For a quantitative assessment, we could apply simple GOF tests, like chi-square and K–S, to test for the fit of the ecdf to the theoretical cdf. Alternatively, we can use Monte Carlo simulation and generate a region where the cdf would fall if the distribution were to be Poisson. The extremes of all the simulation runs define the envelope of the region (Kaluzny et al., 1996). For this purpose, it is usually better to plot the ecdf vs. the cdf in the same manner as quantile–quantile plots. For example, Figure 8.6 illustrates one case where the ecdf departs from the region generated from random runs. We plot the G for the mean of the Monte Carlo runs (horizontal axis) vs. the G for the data (vertical axis) and statistics of the simulated pattern. If the data had a perfect random pattern, the empirical line would have slope of 1:1 and will follow the centerline for the mean.

Now we examine the graph to detect departures from this slope by excursions outside the envelope of maximum and minimum G for the Monte Carlo runs. Since the data line falls outside the envelopes of random pattern, this indicates that this point pattern is clustered.

8.5.2   Second-Order Analysis

Another more refined approach is to look at the distance to the second closest neighbor, the third neighbor, and so on until the kth neighbor. A powerful approach is the **Ripley's K** function, an estimator of the second-order properties. The K function is the cumulative distribution of points within a distance interval. The theoretical $K(d)$ is

$$K(d) = \frac{N(d)}{\lambda} \tag{8.16}$$

where $N(d)$ is the expected number of points within a distance d. The empirical cdf is

$$\widehat{K}(d) = \frac{1}{\lambda} \sum_{i}^{m} \sum_{j}^{m} \frac{\delta_{ij}(d_{ij}, d)}{m} \tag{8.17}$$

for $i \neq j$ and where d_{ij} is distance between points i and j. Here, we use an indicator function

$$\delta_{ij}(d_{ij}, d) = \begin{cases} 1 & \text{when } d_{ij} \leq d \\ 0 & \text{otherwise} \end{cases} \tag{8.18}$$

Recall that $\lambda = m/A$ and substitute in Equation 8.17 to get

$$\widehat{K}(d) = \frac{A}{m^2} \sum_i^m \sum_j^m \delta_{ij}(d_{ij}, d) \tag{8.19}$$

Again, we can compare the theoretical (Poisson) with the empirical. There are several schemes to correct for edge effect such as the border and isotropic estimates. When we include isotropic correction for this example (Figure 8.7), we visually appreciate that the corrected departs only slightly from the raw and mostly for larger values of d. In conclusion, this visual exploration tells us that the empirical K departs substantially from the theoretical K, reinforcing the suspicion of a clustered pattern of the points.

As an aid to visualization we can calculate

$$\widehat{L}(d) = \sqrt{\frac{K(d)}{\pi}} \tag{8.20}$$

and plot the $L(d)$ function vs. d and check whether it follows a straight line. The reason is that a completely random pattern will have

$$K(d) = \pi d^2 \tag{8.21}$$

and therefore if the pattern is close to random

$$\widehat{L}(d) = \sqrt{\frac{K(d)}{\pi}} = \sqrt{\frac{\pi d^2}{\pi}} = d \tag{8.22}$$

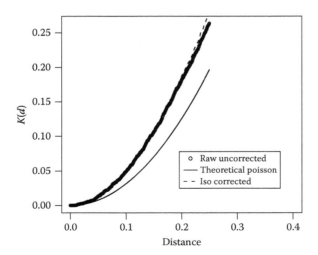

FIGURE 8.7   Empirical K raw and corrected vs. theoretical comparison.

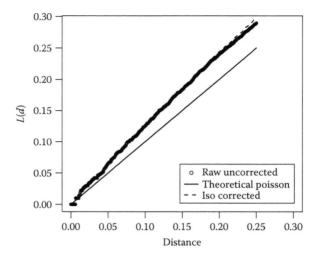

FIGURE 8.8 Empirical L raw and corrected vs. theoretical comparison.

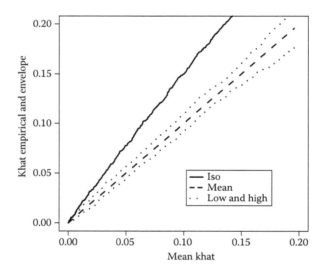

FIGURE 8.9 Empirical K departing outside the band obtained by simulation.

As illustrated in Figure 8.8, the empirical $L(d)$ departs substantially from the 1:1 line given by the theoretical Poisson (1:1 line). We compare these also with Monte Carlo simulation as we did for $G(d)$ (see Figure 8.9).

8.6 MARKED POINT PATTERNS

In addition to its location, each point may have a "mark" or associated value. In this case, we have **marked point patterns**. The mark may be a value from a continuous variable or from a categorical variable. For example, if we measure pollutant concentration of two pollutants at a set of points located in the domain, we have two marks at each point: concentration values of the pollutants. These vary continuously. For another example, consider trees of four species. Suppose we record species type and diameter for each tree. The marks are species (a categorical type) and diameter

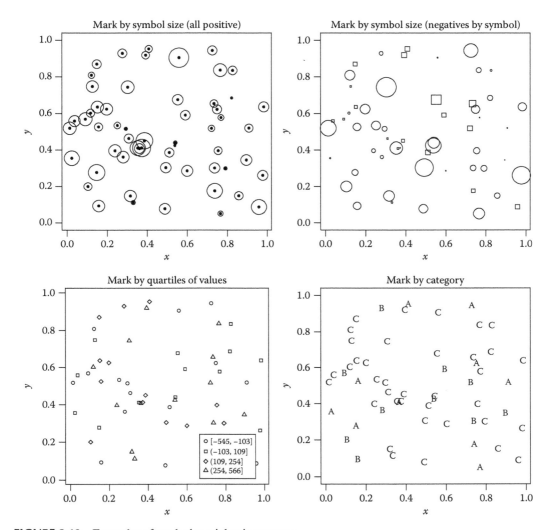

FIGURE 8.10   Examples of marked spatial point patterns.

(a continuous mark). Of great interest, in addition to the spatial distribution, is how the values or marks relate to each other over the domain given the spatial structure of the point pattern.

As illustrated in Figure 8.10, there are several ways of depicting the marks. For example, we can make the symbol size proportional to the value of a continuous variable (top panels). This suffices when all values are positive (top left), but in addition we change symbol type for negative values (top right). Alternatively, we can divide the range of the continuous variable into intervals (say quartiles) and use different symbol type for each one of the intervals (bottom left). Finally, the simplest is to add a label at the point corresponding to the level of a categorical variable (bottom right).

8.7   GEOSTATISTICS: REGIONALIZED VARIABLES

In this part of the chapter, we will start the study **geostatistics**, a technique to predict values of variables in nonsampled points of a spatial domain using a collection of sampled points. We start with **marked point patterns** that placed regularly or irregularly over the spatial domain. Being "marked," we have values of variables at each point. We calculate a measure of similarity using the **covariance** between points separated at given distances or dissimilarity using the **semivariance**.

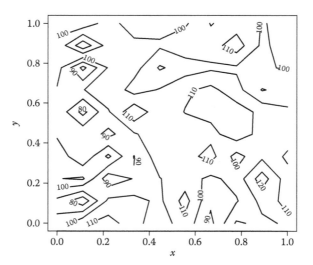

FIGURE 8.11   Example of a regionalized variable with values expressed by contour lines.

As we will see in a later chapter, we then use **kriging** to interpolate values at nonsampled points; that is, use the covariance structure to predict the values of the variables at the nonsampled points. This way, we have values for the entire domain; some of these are measured values while others are kriging-estimates.

In geostatistics, the term **regionalized variable** denotes an RV taking values at locations over a spatial domain. For example, elevation, concentration of a contaminant, soil hydraulic conductivity, and rainfall. Geostatistics assumes that the variable is continuous over the domain. The regionalized variable is a random variable $Z(x, y)$ where location x, y are coordinates of a point in 2D spatial domain. An example is in Figure 8.11, where we indicate the values $z(x, y)$ of Z at all points x, y by a contour line map or isoline map. You can think of $z(x, y)$ as a realization of $Z(x, y)$. The concepts extend to 3D but in this book we will limit the discussion to 2D.

Being a random variable, $Z(x, y)$ has a distribution (pdf and cdf). In principle, the distribution could be different at each point x, y, yielding a complicated model. An important spatial case is that of a **stationary** regionalized variable, or the spatially invariant case where the distribution of variable $Z(x, y)$ does not depend on the location x, y. A less demanding specification is to require at least that the first two moments (mean and variance) do not depend on location x, y, even though the entire distribution may vary.

Under this weak stationary condition, the expected value of $Z(x, y)$ at all points is a constant μ_Z,

$$E[Z(x, y)] = \mu_Z \tag{8.23}$$

and the central second moment or variance at all points is σ_Z^2

$$E[(Z(x, y) - \mu_Z)^2] = \sigma_Z^2 \tag{8.24}$$

8.8   VARIOGRAMS: COVARIANCE AND SEMIVARIANCE

Now, we want a measure of spatial dependence between values at different points separated by various distances. The most natural one to use is the covariance between values at different points. In geostatistics, the dissimilarity between points measured by the semivariance became prevalent. Semivariance and covariance are however related.

8.8.1 COVARIANCE

The covariance of Z between points (x_1, y_1) and (x_2, y_2) separated by distance h is

$$\text{cov}(Z(x_1, y_1), Z(x_2, y_2)) = E[Z(x_1, y_1)Z(x_2, y_2)] - \mu_Z^2 \tag{8.25}$$

When the regionalized variable is stationary, this covariance is the same for any two points separated by this distance h and therefore we can write it as a function of h only

$$\text{cov}(h) = E[Z(x_1, y_1)Z(x_2, y_2)] - \mu_Z^2 \tag{8.26}$$

In geostatistics, we will refer to the distance h as a **lag**. This expression is similar to the autocorrelation of a random process $X(t)$, where the lag is time. However, time was 1D and lag was simply a difference in time. Here, the points occur in 2D and the lag h is the Euclidian distance between points defined earlier in this chapter and repeated here for easy reference

$$d_{12} = \sqrt{\frac{(x_1 - x_2)^2 + (y_1 - y_2)^2}{2}} \tag{8.27}$$

Note that cov(h) is maximum at zero lag ($h = 0$) and equal to the variance because at $h = 0$ it is the same sample point $Z(\boldsymbol{x})$ and therefore

$$\text{cov}(0) = E[Z(x, y)Z(x, y)] - \mu_Z^2 = E[Z^2(x, y)] - \mu_Z^2 = \sigma_Z^2 \tag{8.28}$$

Covariance decreases as h increases because the similarity of values will tend to decrease as points are increasingly apart. Note that Equation 8.25 is theoretical covariance. In practice, we obtain an **empirical covariance** as the average of products of values z at all point pairs (x_i, y_i) and (x_j, y_j) separated by distance in the interval $h \le d_{ij} < h + \Delta h$ and subtracting the square of the sample mean

$$c(h) = \frac{\displaystyle\sum_{h \le d_{ij} < h + \Delta h} z(x_i, y_i)z(x_j, y_j)}{n_h} - \overline{Z}^2 \tag{8.29}$$

Here, n_h is the number of point pairs at distance h.

The top panels of Figure 8.12 show examples of a theoretical covariance (left) and an empirical covariance (right). A plot of covariance as a function of h is called a **covariogram**.

Now, divide cov(h) by $c(0)$ or σ_Z^2 to obtain the **correlation coefficient**

$$\rho(h) = \frac{\text{cov}(h)}{\text{cov}(0)} = \frac{\text{cov}(h)}{\sigma_Z^2} \tag{8.30}$$

which has a value of 1 at $h = 0$ and decays to 0 with increasing h. The empirical correlation coefficient $r(h)$ is the ratio of empirical covariance over variance. The bottom panels of Figure 8.12 show correlation coefficients corresponding to the examples in the top panels. On the left, theoretical $\rho(h)$ and on the right-hand side the empirical $r(h)$ coefficient. The plot of correlation coefficient vs. lag is the **correlogram**.

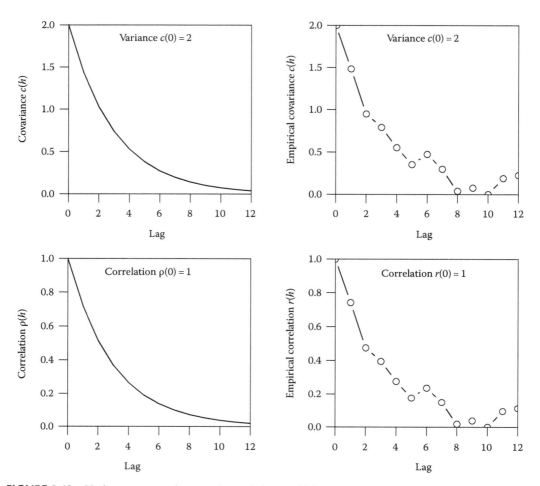

FIGURE 8.12   Variograms: covariance and correlation coefficient.

8.8.2 SEMIVARIANCE

The semivariance is a measure of dissimilarity denoted with the Greek letter gamma γ. It is defined as **half** of the average of the squares of the differences between points (x_i, y_i) and (x_j, y_j) separated by distance in the interval $h \leq d_{ij} < h + \Delta h$

$$\gamma(h) = \frac{\displaystyle\sum_{h \leq d_{ij} < h + \Delta h} \left(z(x_i, y_i) - z(x_j, y_j)\right)^2}{2n_h} \tag{8.31}$$

The prefix "semi" reminds us that this is half of the total squared difference at a given h. A plot of semivariance vs. lag is a **semivariogram**.

Note that if $h = 0$, the difference should be zero and therefore we would expect that $\gamma(0) = 0$. However, the empirical semivariance is not calculated at $h = 0$, but at a small distance representing a minimum lag in the data set. Then $\gamma(h)$ is extrapolated from the calculated values back to the $h = 0$ axis. The nonzero intercept found by this extrapolation is called the **nugget effect** and represents the fine-scale or micro-scale variation of the regionalized variable (example in Figure 8.13).

Many times, as points become more and more distant, the dissimilarity reaches a plateau because there is no additional difference in values accrued. In these cases, the semivariance has an upper bound, that is, it reaches an asymptotic value called the **sill** as distance increases (example in

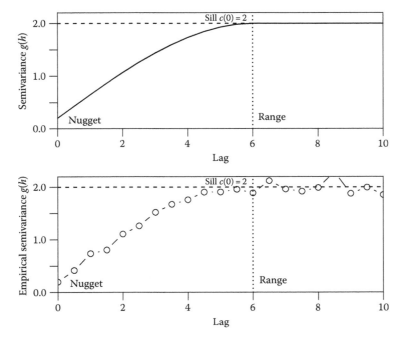

FIGURE 8.13 Semivariance: nugget, sill, and range.

Figure 8.13). In other words, the sill is the ceiling or saturation of a bounded semivariogram. The **range** is the distance at which the semivariance reaches the sill (example in Figure 8.13).

As anticipated, the semivariance and covariance are related. Start with Equation 8.31, expand the numerator, use the fact that the average sum of squares should be independent of location for a stationary variable, and then use Equation 8.29

$$\gamma(h) = \frac{\sum z(x_i, y_i)^2 + z(x_j, y_j)^2 - 2z(x_j, y_j)z(x_i, y_i)}{2n_h}$$

$$= MS_Z - \frac{\sum z(x_j, y_j)z(x_i, y_i)}{n_h} = MS_Z - c(h) - \overline{Z}^2 = s_Z^2 - c(h) \tag{8.32}$$

Therefore, the semivariance is equal to the sample variance s_Z^2 minus the covariance $c(h)$, and the variance is covariance at zero lag $c(0)$

$$\gamma(h) = s_Z^2 - c(h) = c(0) - c(h) \tag{8.33}$$

Now, because $c(h)$ goes to zero for large h, the asymptotic value of the semivariance or sill is equal to the variance

$$sill = \gamma(\infty) = c(0) = s_Z^2 \tag{8.34}$$

Figure 8.14 shows another example of semivariogram and illustrates the relationship with the covariogram. We have ignored the nugget for simplicity.

To enhance our understanding of semivariograms, consider scatter plots of the raw calculations of the square differences for each point pair and their corresponding distance (Figure 8.15, bottom-left panel). This type of plot is a variogram "cloud." The density of points indicates how many pairs

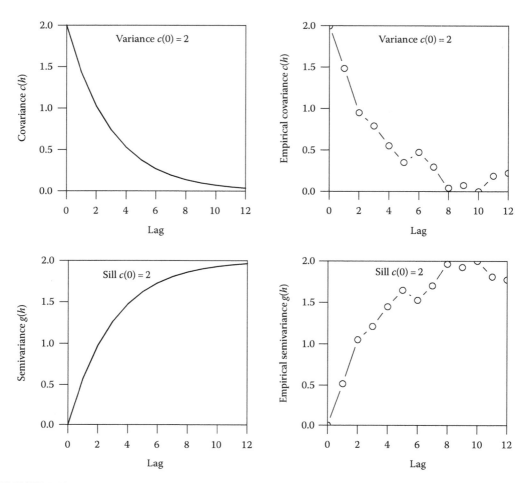

FIGURE 8.14 Variograms: covariance and semivariance.

go into the calculation for each lag interval. The bottom-right panel is a summary of the cloud using boxplots for each lag interval. Both plots are useful to detect and identify outliers.

Before we leave this section, be aware that in practice, the term **variogram** is used to refer generically to covariograms, correlograms, and semivariograms.

8.9 DIRECTIONS

When searching for points separated at given distances h, we can move along lines with given directions; for example, north to south or east to west. The angle θ of the line of search with respect to a reference defines direction. There are several ways of defining the direction angle. One is by **azimuth**, which is the angle with respect to the North (N) direction and increasing clockwise (CW) (Deutsch and Journel, 1992). Examples at 45° intervals are illustrated in Figure 8.16 (left panel).

Another one is by measuring the angle θ with respect to East (E) direction and increasing counterclockwise (CCW) (Carr, 1995). Examples at 45° intervals are illustrated in Figure 8.16 (right panel). We can calculate as many directions as we want according to desired directional resolution. For example, if we double the resolution, we would have eight directions separated by 22.5°.

When we search all directions, then we have **omnidirectional** variograms. When finding pairs to calculate the distance in a given direction, a bandwidth is established to restrict the search. The diagram in Figure 8.17 illustrates the relationship between direction and bandwidth.

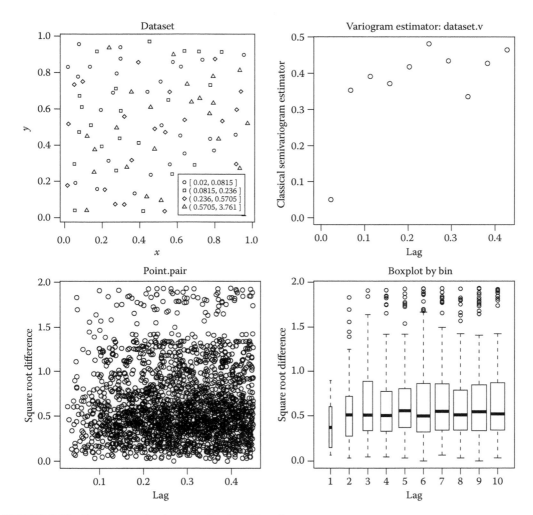

FIGURE 8.15   Dataset, semivariogram, cloud, and boxplot.

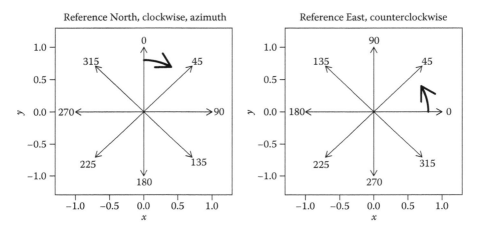

FIGURE 8.16   Theta angle measured as azimuth or as angle from the East.

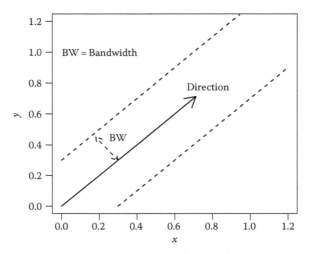

FIGURE 8.17   Direction vector and search bounds around the direction vector.

If the variograms are the same regardless of direction, then the behavior of regionalized variable $Z(x, y)$ is **isotropic**. But if the variograms change with direction, then $Z(x, y)$ is **anisotropic**. There are two types of anisotropic behavior. One is **geometric** anisotropy, which implies that the range changes with direction, and the other is **zonal** anisotropy, which means that the sill changes with direction. **Nested** variograms and linear transforms are tools to correct for anisotropic behavior.

8.10   VARIOGRAM MODELS

A variogram model is a theoretical equation defining the semivariance or the covariance. We can fit a model variogram to an empirical variogram by finding the values of the model parameters that produce the least difference between the model and the empirical variogram. One important application of model variograms is as input to the process of Kriging.

Several mathematical functions are available and employed to model variograms (Davis, 2002; Carr, 1995). In summary, we can consider two major types of models: **Bounded** models (those that reach a sill) as the exponential, gauss, and spherical; and **Unbounded** models (those that do not reach a sill and keep increasing with distance) as the linear and power models.

8.10.1   EXPONENTIAL MODEL

This model is the simplest way to describe decay of covariance with increasing lag. The equation is

$$c(h) = c(0)\exp\left(\frac{-h}{a}\right) \tag{8.35}$$

prescribing a decay of covariance from the maximum $c(0) = \sigma_Z^2$ to zero at a rate $1/a$ (see Figures 8.12 and 8.14). The parameters of this covariance model are $c(0)$ and a.

Then the semivariance is

$$\gamma(h) = c(0) - c(h) = c(0) - c(0)\exp\left(\frac{-h}{a}\right) = c(0)\left(1 - \exp\left(\frac{-h}{a}\right)\right) \tag{8.36}$$

It asymptotically reaches a maximum value (sill) of $c(0)$ when h/a is large. Recall that $c(0) = \sigma_Z^2$, and that the sill is the variance of Z. From the last equation also note that $\gamma(0) = 0$. Thus, in this model the nugget is zero.

To account for a nonzero nugget effect, we can use the following discontinuity $\gamma(0^+) = nugget$, where 0^+ is a value of lag slightly larger than 0. Therefore for values of h larger than 0, we have

$$\gamma(h) = \begin{cases} \left(c(0) - \gamma(0^+)\right)\left(1 - \exp(-h/a)\right) + \gamma(0^+) & \text{when} \quad h > 0 \\ 0 & \text{when} \quad h = 0 \end{cases} \qquad (8.37)$$

The covariance is

$$c(h) = c(0) - \gamma(h) \qquad (8.38)$$

So for $h > 0$ substitute Equation 8.37 to obtain

$$c(h) = c(0) - \gamma(h) = c(0) - [c(0) - \gamma(0^+)]\left(1 - \exp\left(\frac{-h}{a}\right)\right) + \gamma(0^+) \qquad (8.39)$$

and therefore $c(h) = [c(0) - \gamma(0^+)]\exp(-h/a)$, whereas for $h = 0$, $c(h) = c(0) - \gamma(0) = c(0)$. In summary, the covariance is

$$c(h) = \begin{cases} \left(c(0) - \gamma(0^+)\right)\exp(-h/a) & \text{when} \quad h > 0 \\ c(0) & \text{when} \quad h = 0 \end{cases} \qquad (8.40)$$

The exponential model is illustrated in Figure 8.18.

The semivariance reaches the sill asymptotically; therefore, the range is defined as that distance at which gamma is close to the sill. To select how close we can use, for example, 95%, of the sill. That is,

$$\gamma(h_{95}) = 0.95 \times c(0) = 0.95 \times \sigma_Z^2 \qquad (8.41)$$

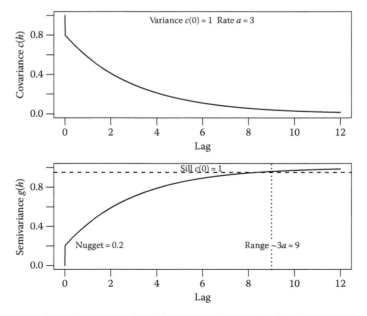

FIGURE 8.18 Exponential variogram model with nugget = 0.2, $a = 3$, $c(0) = 1$.

Solving for h_{95} assuming the nugget small compared to $c(0)$, we obtain an approximate value

$$h_{95} \approx -a \ln(0.05) \approx 3 \times a \tag{8.42}$$

In this case, $a = 3$, and the range would be $3 \times 3 \approx 9$ (see Figure 8.18).

A more precise formula taking into account the nugget is

$$h_{95} \approx -a \ln\left(1 - \frac{0.95 \times c(0) - \gamma(0+)}{c(0) - \gamma(0+)}\right) \tag{8.43}$$

8.10.2 SPHERICAL MODEL

This model arises from a Poisson distribution that yields a sphere (circle on 2D) of radius a. The covariance is proportional to the intersection of two spheres when the distance between centers is less than the radius a. In this case, the range is given exactly by parameter a.

$$\gamma(h) = \begin{cases} 0 & \text{when } h = 0 \\ \gamma(0^+) + [c(0) - \gamma(0^+)]\left(\dfrac{3h}{2a} - \dfrac{h^3}{2a^3}\right) & \text{when } 0 < h \leq a \\ \gamma(0^+) + [c(0) - \gamma(0^+)] = c(0) & \text{when } h > a \end{cases} \tag{8.44}$$

This is illustrated in Figure 8.19 for a range of $a = 6$.

8.10.3 GAUSSIAN MODEL

This model arises from a Gauss distribution.

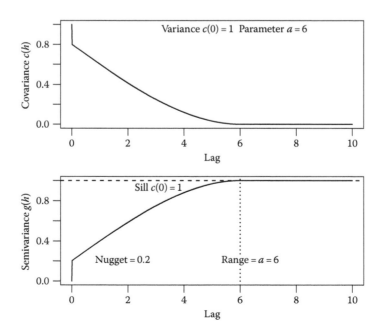

FIGURE 8.19 Spherical model with nugget = 0.2, and $a = 3$ = range and sill = $c(0) = 1$.

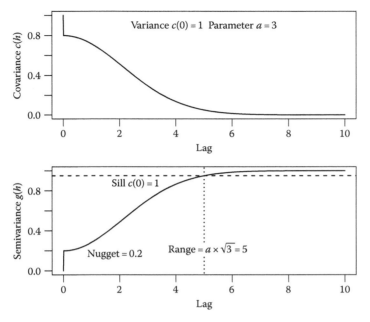

FIGURE 8.20 Gaussian model with nugget = 0.2 and $a = 3$ and sill = $c(0) = 1$.

$$\gamma(h) = \begin{cases} \gamma(0^+) + [c(0) - \gamma(0^+)]\left(1 - \exp\left(-\dfrac{h^2}{a^2}\right)\right) & \text{when} \quad h > 0 \\ 0 & \text{when} \quad h = 0 \end{cases} \qquad (8.45)$$

It reaches the sill asymptotically as the exponential model. We define the range as that distance at which gamma is 95% of the sill as in Equation 8.41. It can be calculated from parameter a using

$$h_{95} = \sqrt{3} \times a = 1.732 \times a \qquad (8.46)$$

This is illustrated in Figure 8.20 for $a = 3$, $c(0) = 1$, $\gamma(0^+) = 0.2$. In this case, the range is

$$h_{95} = \sqrt{3} \times 3 = 1.732 \times 3 \approx 5$$

8.10.4 LINEAR AND POWER MODELS

The semivariance increases linearly with h and does not reach a maximum. This indicates a nonstationary variable. This is to say the variance does not reach a constant value, at least within the scale of the maximum distance used.

$$\gamma(h) = \begin{cases} \gamma(0^+) + ah^b & \text{when} \quad h > 0 \\ 0 & \text{when} \quad h = 0 \end{cases} \qquad (8.47)$$

In this case, the concepts of sill and range are not applicable. When $b = 1$, the power law becomes linear. The parameter a determines the slope. When $b = 2$, we get a parabolic increase (see Figure 8.21).

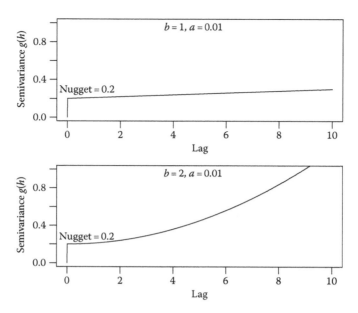

FIGURE 8.21   Power model: linear and parabolic with nugget = 0.2 and $a = 0.01$.

8.10.5 MODELING THE EMPIRICAL VARIOGRAM

Usually have to fit a model variogram to an empirical one. This means finding the parameter values of the model variogram that minimizes the difference between the empirical and the model variogram. For this purpose, we can employ a numerical fitting routine; it requires an initial guess for values of model parameters, and other conditions such as number of iterations and tolerance for convergence. A convenient procedure is nonlinear least-squares regression. A couple of examples using spherical models are shown in Figures 8.22 and 8.23.

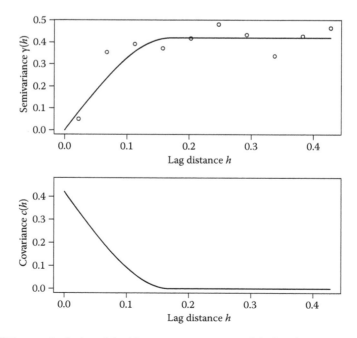

FIGURE 8.22   Fitting a spherical model with zero nugget to an empirical variogram.

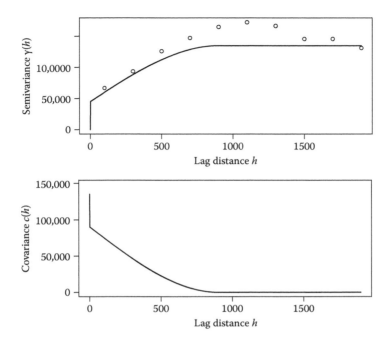

FIGURE 8.23 Fitting a spherical model with nonzero nugget to an empirical variogram.

8.11 EXERCISES

Exercise 8.1

Use the spatial point pattern of Figure 8.24 and the chi-square test to determine if the pattern is uniform (homogeneous).

1. What is the number of cells, T, the total number of points, m, and the expected number of points for each cell if the distribution were uniform?
2. What is the null hypothesis for a chi-square test? How many degrees of freedom df?
3. Calculate the chi-square value.
4. Use the 1- pchisq(x,df) expression in R to calculate the p-value.
5. Can you reject the null hypothesis? What is your conclusion?

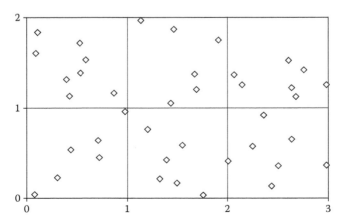

FIGURE 8.24 Spatial point pattern for quadrat analysis.

Exercise 8.2
Uniform point patterns can be either regular or random, but the χ^2 test does not specify which.

1. Does the point pattern of Exercise 8.1 appear more regular or random?
2. What distribution and test could you use to determine the nature of a uniform pattern?

Exercise 8.3
Examine the spatial point pattern of Figure 8.25 for randomness.

1. What is the total area, A, the total number of points, m, and the density, λ, the total number of cells, T, the area for each cell, a, and the rate, $\lambda a = m/T$?
2. If the points are random, they will follow the Poisson distribution given by Equation 8.4. Using the expected number of points per cell as the rate yields the following probabilities that a cell will have r number of points: $P(0) = 0.646$, $P(1) = 0.282$, $P(2) = 0.062$, $P(3) = 0.009$.
3. Calculate the expected number of cells that will have 0, 1, 2, or 3 points. How do these values compare to the point pattern?
4. Determine the number of points r_i in each cell i. Calculate the sample variance using Equation 8.6.
5. Calculate the ratio of the mean over the sample variance, $(m/T)/s^2$, and compare the value to 1. Is it greater than 1, equal to 1, or less than 1? What does this imply about the point pattern being regular, random, or clustered? What test could you use to determine if the point pattern is random?
6. The t-value (from Equation 18.7) is approximately 0.288, which has a corresponding p-value of 0.389. Can you reject the null hypothesis? What is your conclusion?

Exercise 8.4
Consider the point pattern of Figure 8.26 on the x–y plane, A = (0, 1), B = (3, 3), C = (2, 0), and D = (2, 1).

1. For each point identify its nearest neighbor and calculate that distance to complete Table 8.2.
2. If the point pattern were random, then the points would follow a Poisson distribution. Calculate what the mean nearest neighbor distance would be if the point pattern were

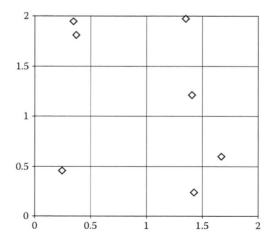

FIGURE 8.25   Spatial point pattern with rare occurrence.

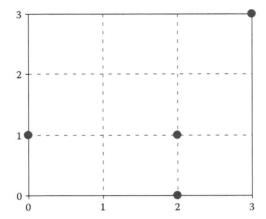

FIGURE 8.26   A few demo points to calculate distance.

TABLE 8.2
NND Table

Point	Nearest Neighbor	Nearest-Neighbor Distance (NND)
A		
B		
C		
D		
Average nearest-neighbor distance ($\overline{NND}$)		

---

random, that is, use density $\lambda = m/A$ and $\mu_d = 1/(2\sqrt{\lambda})$. How does this compare with the average NND?

3. The ratio $\overline{NND}/\mu_d$ is called the nearest-neighbor statistic, or the standardized mean, and ranges from 0 (perfectly clustered, i.e., all points are in the same location) through 1 (random) to 2.15 (perfectly dispersed, i.e., the average nearest-neighbor distance is maximized). Calculate this ratio and use it to judge if this point pattern is more clustered, random, or regular.

Exercise 8.5

1. Do you need values at each point to calculate chi-square and Ripley's K and L? Explain.
2. Do you need values at each point to calculate the semivariogram? Explain.

Exercise 8.6

Suppose a semivariogram has range of 4, sill of 20, and a nugget of 2.

1. Write an expression for the variogram using the spherical model given by Equation 8.44. Hint: substitute the values given into the equation so that it is only function of h.
2. Sketch a graph of the semivariogram.
3. Write an expression for the covariance as a function of h.
4. Draw a graph of the covariogram.

8.12   COMPUTER SESSION: SPATIAL ANALYSIS

8.12.1   PACKAGES AND FUNCTIONS

We will use the spatial analysis `spatstat` and geostatistics `sgeostat` R packages. First, install these packages from the CRAN website if you have not done this yet. Then load the packages. The command `library` offers another way of loading an installed package.

```
>library(spatstat)
>library(sgeostat)
```

Recall that the packages can be loaded automatically whenever you start the RGui by including them in your **Rprofile.site** file. You can add the following segment to your **etc/Rprofile.site** file.

```
# loading packages automatically
local({
old <- getOption("defaultPackages")
options(defaultPackages = c(old, "Rcmdr","spatstat","sgeostat"))
})
```

Package **seeg**, which you already loaded, includes several functions for spatial analysis.

8.12.2   FILE FORMAT

The simplest way to specify a point pattern is a text file containing columns for the coordinates x, y and the marks (if any). For example, see the first few lines of file **xy100.txt**

```
x       y
0.301   0.376
0.439   0.845
0.48    0.491
0.525   0.262
0.319   0.679
0.488   0.155
0.225   0.531
... etc
```

This is an unmarked point pattern with two columns, one for x and one for y. However, some formats contain more information about the data structure in the header.

A common format for spatial data is the **GeoEAS** format. This format is from the U.S. EPA, Environmental Monitoring Systems Laboratory (Englund and Sparks, 1991). It includes point and grid specifications. We can import these files into R following a sequence of commands given in the function `scan.geoeas.ppp` from seeg. This function will read a GeoEAS file and make a data frame.

As a first example, we will look at file **lab8/unif100geoEAS.txt**. It has a header with title, number of columns (ncol) to be found in each record (two of those will be coordinates), labels for the coordinates and variables (for marked patterns), and then the records (rows). In this section, we will be concerned only with x and y. This is how this file **lab8/unif100geoEAS.txt** should look like for the same data given earlier in file **xy100.txt**

```
Example of location-only data
2
x
y
0.301   0.376
0.439   0.845
0.48    0.491
0.525   0.262
0.319   0.679
0.488   0.155
0.225   0.531
.....
```

We can read it with function `scan.geoeas.ppp`

```
> unif100xy <- scan.geoeas.ppp("lab8/unif100geoEAS.txt")
Read 1 item
Read 2 items
>
```

and obtain the following data frame

```
> unif100xy
        x      y
1    0.301 0.376
2    0.439 0.845
3    0.480 0.491
4    0.525 0.262
5    0.319 0.679
6    0.488 0.155
7    0.225 0.531
```

If using the seeg add on for Rcmdr, go to **Spatial|Read geoEAS ppp**, type the name to be given to the dataset, say **unif100xy**, and then click ok, browse to lab8, and click Open. The desired dataset is active and you can view it using **View data set**.

8.12.3   CREATING A PATTERN: LOCATION-ONLY

Once we have a data frame, we can convert into a ppp object using function ppp, which specifies the coordinates

```
> unif100ppp <- ppp(unif100xy$x, unif100xy$y)
```

We can check the contents

```
> unif100ppp
 planar point pattern: 100 points
window: rectangle = [ 0, 1 ] x [ 0, 1 ]
```

We can also go backward from a ppp to a data frame simply applying the `data.frame` function. For example,

```
>pppset <- unif100ppp
>dataset <- data.frame(x=pppset$x,y=pppset$y)
```

We can use the `summary` extractor function on a ppp

```
> summary(unif100ppp)
planar point pattern: 100 points
average intensity 100 points per unit area
window: rectangle = [ 0, 1 ] x [ 0, 1 ]
Window area = 1
>
```

To plot the ppp

```
plot(unif100ppp$x,unif100ppp$y, xlab="x",ylab="y")
title("uni100xy",cex.main=0.8)
```

which yields Figure 8.27 (top-left panel).

As another example, use the data in file **lab8/pois100geoEAS.txt** which has points drawn from a Poisson distribution. Read the data using the scan geoEAS function, then apply ppp, and plot

```
pois100xy <- scan.geoeas.ppp("lab8/pois100geoEAS.txt")
pois100ppp <- ppp(pois100xy$x, pois100xy$y)
plot(pois100ppp$x,pois100ppp$y, xlab="x",ylab="y")
title("pois100xy",cex.main=0.8)
```

which yields Figure 8.27 (bottom-left panel). In the following sections, we will see how to analyze the patterns.

8.12.4 GENERATING PATTERNS WITH RANDOM NUMBERS

As we discussed earlier, we can supplement the test methods for spatial patterns by comparing the empirical statistics to the theoretical ones, using a band given by simulation of many patterns drawn at random. As a first example, let us create a randomly distributed pattern as a ppp object. This will help us understand how to test against random patterns later. We will use the [0, 1] × [0, 1] spatial domain for simplicity because it is the default for the spatstat functions.

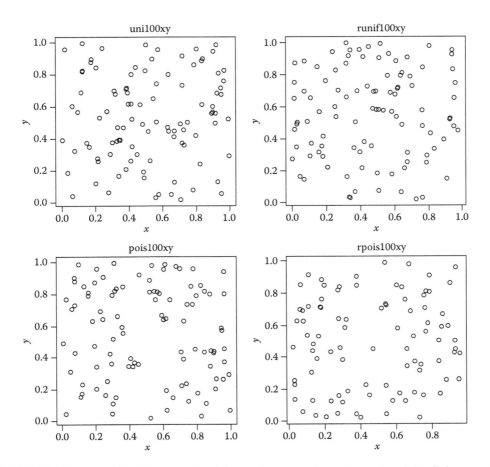

FIGURE 8.27 Plot of unifxy100 pattern (top left), random generated uniform (top right), Poisson pattern from file (bottom left), and random generated Poisson (bottom right).

First use random function for uniform patterns generated with function `runifpoint`, and then plot (Figure 8.27, top right). To do this we specify the intensity as an argument

```
> runif100ppp <- runifpoint(100)
> runif100ppp
  planar point pattern: 100 points
window: rectangle = [ 0, 1 ] x [ 0, 1 ]
> plot(runif100ppp$x,runif100ppp$y, xlab="x",ylab="y")
> title("runif100xy",cex.main=0.8)
```

Another useful function is `rmpoispp` to generate a Poisson point pattern with a given intensity

```
rpois100ppp <- rmpoispp(100,types=c("A"))
rpois100ppp
plot(rpois100ppp$x,rpois100ppp$y, xlab="x",ylab="y")
title("rpois100xy",cex.main=0.8)
```

as shown in Figure 8.27 (bottom right). Here, we have created a marked pattern for the sake of an example. Note that the same function can be used to specify multiple patterns.

8.12.5 Grid or Quadrat Analysis: Chi-Square Test for Uniformity

We will use seeg function `quad.chisq.ppp`, which requires a point pattern and a target density. We will apply to `unif100xy` given earlier. The function will define the number of cells in the grid based on the target intensity or density. Recall that chi-square requires 5 points per cell. Note that `unif100xy` complies with this minimum for a 4 × 4 grid because the expected number of points per cell is

$$\frac{m}{T} = \frac{100}{4 \times 4} = \frac{100}{16} = 6.25$$

where
   m is the number of points
   T is the number of cells

So, apply function `quad.chisq.ppp` with arguments `unif100xy` and 5

```
>quad100 <- quad.chisq.ppp(unif100xy,5)
```

And check the results

```
> quad100
$pppset
 planar point pattern: 100 points
 window: rectangle = [ 0 , 1 ] x [ 0 , 1 ]

$Xint
       [,1] [,2] [,3] [,4]
 [1,]    8    4    4    9
 [2,]    4   10    5    9
 [3,]    6   11   10    3
 [4,]    3    5    5    4

$intensity
 [1] 6.25

$chisq
 [1] 18.4

$p.value
 [1] 0.1891652
>
```

The matrix `Xint` is the number of points in each cell. The `intensity` is the grand mean = 100/16 = 6.25. The chi-square and p-value indicates that H0 cannot be rejected and there is no evidence that the pattern is not uniform (Figure 8.28). In this figure, intensity or density for each cell (as given by matrix `Xint`) is visualized as a lattice or raster image (upper-right panel). Higher intensity corresponds to darker gray.

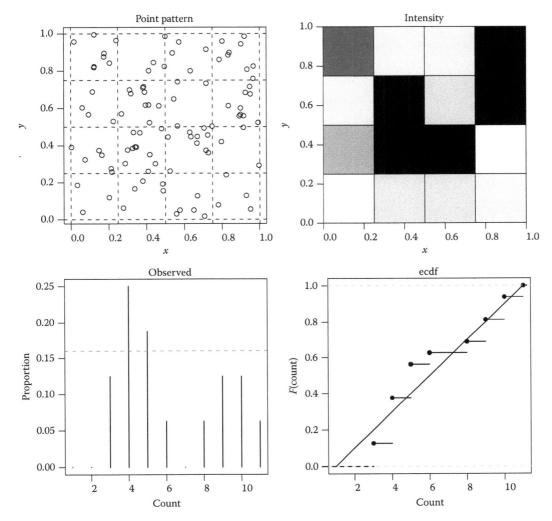

FIGURE 8.28 Quadrat analysis: grid, intensity per cell (darker gray tones are higher intensity), histogram, and comparison to cumulative uniform distribution.

This process can be done with the seeg addon to Rcmdr. With the unif100xy active, go to Spatial|quadrat, and then select Quad ChiSq. This would yield the same results as given earlier using the R console.

8.12.6 GRID OR QUADRAT ANALYSIS: RANDOMNESS, POISSON MODEL

We will use seeg's function quad.poisson.ppp, which requires a dataset with coordinates and a target density. This function will define the number of cells in the grid based on the target intensity or density. Let us use the sample data pois100xy given earlier in section 8.12.3. Recall that the Poisson model is for rare events. Let us assume a density of about 0.2. Note that this number corresponds to the pois100xy dataset with a grid of about 500 cells because the expected number of points per cell is $(m/T) = (100/500) = 0.2$. Let us apply the function with arguments pois100xy and 0.2

```
> pois100 <- quad.poisson.ppp(pois100xy,0.2)
```

And check the results

```
> pois100
$pppset
  planar point pattern: 100 points
window: rectangle = [ 0, 1 ] x [ 0, 1 ]

$num.cells
[1] 484

$Xint
[,1] [,2] [,3] [,4] [,5] [,6] [,7] [,8] [,9] [,10] [,11] [,12] [,13]
  [,14] [,15] [,16] [,17] [,18] [,19] [,20] [,21] [,22]
[1,] 0 0 1 0 1 0 1 0 0 0 0 1 0 2 1 1 0 1 0 0 0 0
[2,] 0 0 0 0 0 0 0 0 0 0 1 0 0 0 0 0 0 0 0 0 0 1
[3,] 0 2 0 0 1 1 0 0 0 0 0 1 0 0 0 0 0 0 0
etc …

$chisq
[1] 0.02195254

$df
[1] 2

$p.value
[1] 0.9890837

$intensity
[1] 0.2066116

>
```

The chi-square and p-value indicate that H0 cannot be rejected and there is no evidence that the pattern is not random (Figure 8.29). As mentioned earlier, we are visualizing intensity or density for each cell (i.e., lattice data) in the upper-right panel using a gray scale for intensity.

To do the same but using the Rcmdr, go to Spatial, select Quad Poisson, then type 0.2 and obtain same results as with the console.

8.12.7   Nearest-Neighbor Analysis: *G* and *K* Functions

The number of points should follow a Poisson with mean λA, where λ is the rate or intensity (number of points per unit area) and A is area. We will use nearest-neighbor statistics G and K. First, apply Gest function to pois100xy dataset with "none" and "km" for edge correction

```
> G.u <- Gest(pois100ppp,correction=c("none","km"))
> G.u
Entries:
id        label           description
--        -----           -----------
r         r               distance argument r
theo      G[pois](r)      theoretical Poisson G(r)
raw       hat(G)[raw](r)  uncorrected estimate of G(r)
```

```
(continued)
rs          hat(G)[bord](r)      border corrected estimate of G(r)
km          hat(G)[km](r)        Kaplan-Meier estimate of G(r)
hazard      hat(lambda)[km](r)   Kaplan-Meier estimate of hazard
function lambda(r)
-----------------------------------------
Recommended range of argument r: [0, 0.079639]
   Available range of argument r: [0, 0.19143]
>
```

As indicated r is distance, raw the uncorrected G, km the Kaplan–Meier estimate of G, rs the reduced sample correction or border corrected G, and theo the theoretical G that would be obtained if data followed Poisson.

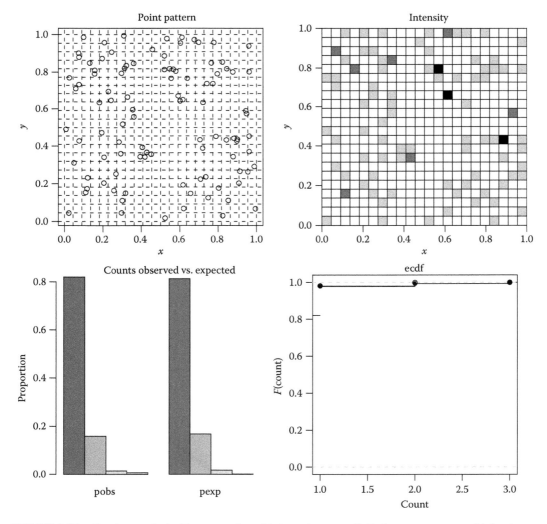

FIGURE 8.29 Quadrat analysis. Upper panels: grid, intensity per cell (darker gray tones are higher intensity); lower panels: comparison to Poisson. Histogram and ecdf.

Kest provides estimates of Ripley's *K*-function. There are several corrections: for example, isotropic and border. Apply Kest function to pois100xy dataset with "none" and "iso" for edge correction

```
> K.u <- Kest(pois100ppp,correction=c("none","iso"))
> K.u
Entries:
id        label             description
--        -----             -----------
r         r                 distance argument r
theo      K[pois](r)        theoretical Poisson K(r)
un        hat(K)[un](r)     uncorrected estimate of K(r)
iso       hat(K)[iso](r)    Ripley isotropic correction estimate of
K(r)
----------------------------------------
Recommended range of argument r: [0, 0.25]
   Available range of argument r: [0, 0.25]
```

Note that r is distance, un the uncorrected *K*, iso the isotropic corrected *K*, and theo the theoretical *K* that would be obtained if data followed Poisson.

We can produce graphics for comparison of empirical to theoretical. To simplify, in seeg we have function nnGK.ppp to calculate and plot Ghat, Khat, and Lhat. Simply apply the function nnGK.ppp to the pois100xy dataset to obtain Figure 8.30.

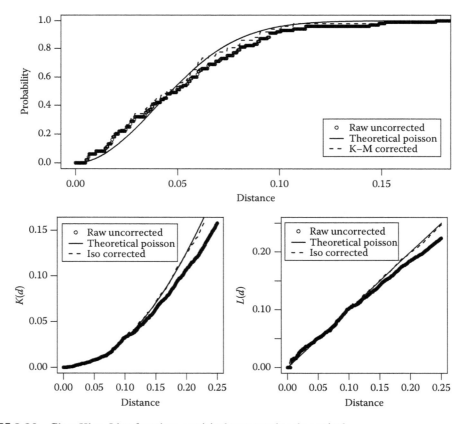

FIGURE 8.30   Ghat, Khat, Lhat function: empirical compared to theoretical.

```
>nnGK.ppp(pois100xy)
```

There seems to be a little clustering, but overall the empirical is very close to the theoretical and therefore the pattern seems to be random.

This function is based on the following code, which also illustrates how to split the screen in a more complicated mode. We plot G on one top panel, and then divide the bottom panel in two to plot K and L side by side. Recall that $L(d) = \sqrt{K(d)/\pi}$. You can easily vary this code to include other edge corrections.

```
split.screen(c(2,1))
  screen(1)
  par(mar=c(4,4,1,.5),xaxs=r, yaxs=r)
  plot(raw~r, data=G.u, type="p", xlim=c(0,r[min(which(raw==1))]),xl
    ab="Distance",ylab="Probability",cex=0.6)
  lines(theo ~r, data=G.u, lty=1)
  lines(km~r, data=G.u,lty=2)
  legend("bottomright", c("Raw Uncorrected", "Theoretical
    Poisson","K-M Corrected"),
         pch=c(1,-1,-1), lty=c(-1,1,2), merge=T,cex=0.7)

split.screen(c(1,2), screen = 2)
screen(3)
plot(un~r, data=K.u, type="p", xlim=c(0,max(r)),xlab="Distance",yl
  ab="K(d)",cex=0.6)
lines(theo ~r, data=K.u, lty=1)
lines(iso~r, data=K.u,lty=2)
legend("topleft", c("Raw Uncorrected", "Theoretical Poisson","Iso
  Corrected"),
       pch=c(1,-1,-1), lty=c(-1,1,2), merge=T,cex=0.7)
screen(4)
plot(sqrt(un/pi)~r, data=K.u, type="p", xlim=c(0,max(r)),ylim=c(0,
  max(r)),xlab="Distance",ylab="L(d)",cex=0.6)
lines(sqrt(theo/pi) ~r, data=K.u, lty=1)
lines(sqrt(iso/pi)~r, data=K.u,lty=2)
legend("topleft", c("Raw Uncorrected", "Theoretical Poisson","Iso
  Corrected"),
       pch=c(1,-1,-1), lty=c(-1,1,2), merge=T,cex=0.7)
```

When using the Rcmdr seeg addon, go to Spatial, select Nearest Neighbor, and obtain same results as with the console.

8.12.8 MONTE CARLO: NEAREST-NEIGHBOR ANALYSIS OF UNIFORMITY

In this section, we will use simulation of many random patterns and calculate the G and K metrics to determine an envelope (Kaluzny et al., 1996). The seeg function GKhat.env will perform the calculation and plots. The function will compute G or K for s Monte Carlo simulated random patterns generated with function runifpoint, then it plots the mean, low end and high end of the G or K for the simulated random pattern and compare to the empirical one. In turn the seeg function nnGKenv.ppp allows the application of the function mentioned earlier to a given spatial

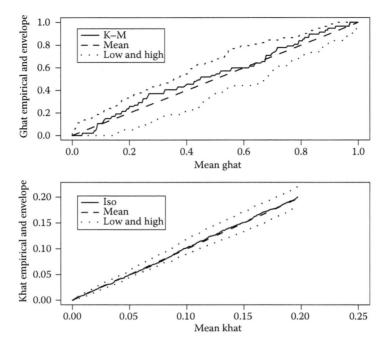

FIGURE 8.31   Monte Carlo analysis.

pattern dataset. We do not reproduce these functions here for the sake of space. You can study the functions from the console.

Apply function nnGKenv.ppp to examine the G for the pois100xy pattern and 100 simulation runs. The parameter nsim is the number of Monte Carlo realizations or simulated patterns.

```
>nnGKenv.ppp (unif100xy,nsim=100)
```

This will plot the G for the mean of the Monte Carlo runs (horizontal axis) vs. the G for the data (vertical axis) and the simulated pattern (Figure 8.31, top panel). This approach is also employed with the second-order statistic, the *K* function (Figure 8.31, bottom panel). If the data had a perfect random pattern, the empirical line would have slope 1 and will follow the centerline for the mean. We can observe that the empirical function stays inside the envelope suggesting that pattern is random.

If using the Rcmdr, go to Spatial, select Monte Carlo, and enter number of realizations and obtain same result as with the console.

8.12.9   MARKED SPATIAL PATTERNS: CATEGORICAL MARKS

We will use as an example the data frame lansing from package **spatstats**. This dataset has coordinates x, y to indicate where a tree is located and a categorical variable to indicate the tree species. This makes the pattern **marked**.

```
> data(lansing)
> lansing
marked planar point pattern: 2251 points
multitype, with levels = blackoak hickory maple misc redoak whiteoak
window: rectangle = [ 0, 1 ] x [ 0, 1 ]
```

For a summary of lansing

```
> summary(lansing)
Marked planar point pattern: 2251 points
Average intensity 2250 points per square unit (one unit = 924 feet)

*Pattern contains duplicated points*
Multitype:
          frequency proportion intensity
blackoak        135    0.0600        135
hickory         703    0.3120        703
maple           514    0.2280        514
misc            105    0.0466        105
redoak          346    0.1540        346
whiteoak        448    0.1990        448

Window: rectangle = [0, 1]x[0, 1]units
Window area =  1 square unit
Unit of length: 924 feet
```

plot the spp object

```
> plot(lansing)
```

Various markers in Figure 8.32 correspond to different species. We obtain a location-only plot by unmarking (Figure 8.33)

Lansing

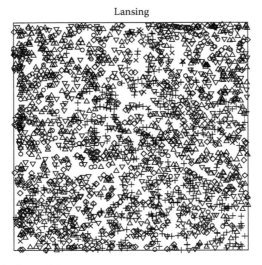

FIGURE 8.32 Plot of marked ppp object.

Unmark (lansing)

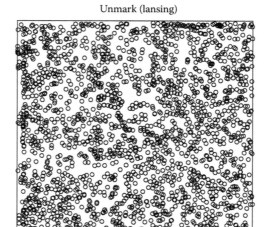

FIGURE 8.33 Location-only plot.

```
> plot(unmark(lansing))
```

Contents of the object include the marks, as well as location coordinates x and y

```
> lansing$mark
  [1] blackoak blackoak blackoak blackoak blackoak blackoak
      blackoak blackoak blackoak blackoak
 [11] blackoak blackoak blackoak blackoak blackoak blackoak
      blackoak blackoak blackoak blackoak
....
[131] blackoak blackoak blackoak blackoak blackoak hickory  hickory
      hickory  hickory  hickory
> lansing$x
  [1] 0.078 0.076 0.051 0.015 0.030 0.102 0.135 0.121 0.040 0.065
      0.091 0.139 0.146 0.128 0.030
 [16] 0.064 0.035 0.069 0.091 0.107 0.096 0.079 0.129 0.140 0.040
      0.003 0.002 0.001 0.036 0.048
 [31] 0.053 0.129 0.118 0.061 0.271 0.222 0.158 0.186 0.179 0.249
      0.265 0.242 0.195 0.205 0.203
...
> lansing$y
  [1] 0.091 0.266 0.225 0.366 0.426 0.474 0.498 0.489 0.596 0.608
      0.587 0.627 0.671 0.705 0.665
 [16] 0.739 0.776 0.769 0.748 0.728 0.767 0.776 0.815 0.839 0.844
      0.857 0.858 0.887 0.859 0.883
 [31] 0.907 0.917 0.975 0.976 0.986 0.921 0.939 0.611 0.495 0.487
      0.465 0.557 0.554 0.545 0.514
```

```
(continued)
  [46]  0.528 0.510 0.320 0.015 0.268 0.080 0.018 0.370 0.367 0.381
        0.418 0.469 0.492 0.471 0.497
  [61]  0.567 0.594 0.587 0.819 0.840 0.869 0.853 0.880 0.886 0.882
        0.878 0.869 0.900 0.922 0.975
  ...
```

You could subset the species by creating separate objects and plotting them separately, or together but with a different color. For example, we can separate the hickory and maples. Try

```
> hick.spp <- lansing[lansing$marks == "hickory", ]
> hick.spp
marked planar point pattern: 703 points
multitype, with levels = blackoak hickory maple misc redoak whiteoak
window: rectangle = [ 0, 1 ] x [ 0, 1 ]
> maple.spp <- lansing[lansing$marks=="maple",]
> maple.spp
marked planar point pattern: 514 points
multitype, with levels = blackoak hickory maple misc redoak whiteoak
window: rectangle = [ 0, 1 ] x [ 0, 1 ]
> plot(hick.spp)
blackoak hickory maple misc redoak whiteoak
    1       2      3     4     5       6
> points(maple.spp, col="gray")
```

to obtain Figure 8.34. This plot suggests that maple and hickory trees do not overlap.

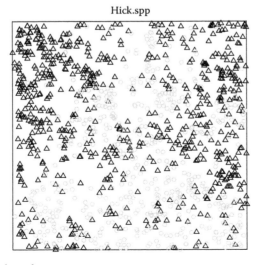

Hick.spp

FIGURE 8.34 Hickory and maple trees.

8.12.10 MARKED SPATIAL PATTERNS: CONTINUOUS VALUES

Now we will work with point patterns that have continuous values at each point. As a first example, we will use the **lab8/xyz-geoEAS.txt** data file, which has a mark z each point. The file contains a header with title, number of columns (3) in a record (row), labels for the measurements, and then 100 records (rows). The records are x coordinate, y coordinate and marks. This is how the first few records of the file should look like:

```
xyz marked point pattern
3
x
y
z
0.510   0.751   1.830
0.273   0.713   0.732
0.051   0.734   0.747
0.719   0.659   3.761
0.619   0.832   3.444
0.052   0.296   2.058
0.357   0.317   1.825
etc
```

We will use function **scan.geoeas.ppp**

```
> xyz <- scan.geoeas.ppp("lab8/xyz-geoEAS.txt")
Read 1 item
Read 3 items
```

Or use the Rcmdr seeg add on, go to **Spatial|read geoEAS**, browse to folder, select file, double check is active dataset and view it.

Plot as in Figure 8.35 by using the following code to generate marks by quartiles applying seeg function `plot.point.bw`

```
xyv <- point(xyz)
plot.point.bw(xyv,v='z',legend.pos=2,pch=c(21:24),cex=0.7)
mtext(side=1,line=2,"x")
mtext(side=2,line=2,"y")
title("Dataset",cex.main=0.7)
```

The plot indicates that we may have a uniform pattern. We can double check by quadrat analysis and nearest-neighbor analysis. We can conclude that the pattern is uniform but not random (see Exercise 8.8). Alternatively, use the Rcmdr, go to **Spatial|quadrat chisq analysis**; and **Spatial|nearest neighbor**.

More information can be gathered by looking at the statistics of the marks. Let us start with eda of variable z.

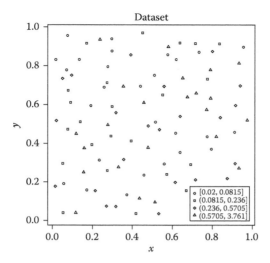

FIGURE 8.35 Point pattern read from file in GeoEAS format.

```
   panel6(size=7)
 eda6(xyz$z)
```

We get EDA plots as in Figure 8.36. We can conclude from the histogram, boxplot, and density that most points have low value of the variable, and a few have large value. From the Q–Q plot and ecdf we can confirm that the data are not normally distributed.

We will use seeg's function `vario`, which in turn uses package **sgeostat** to analyze the `xyz` pattern. Function `vario` uses functions `point` to generate a point object, plots it to visualize the point pattern, then uses function `pair` to generate a point object and a pair object. Function `pair` requires us to define number of lags and the maximum distance. The pair object contains all pairs separated at each lag up to the maximum distance. When applying command pair, we can also decide the direction. From the help of sgeostat

```
Type: either 'isotropic' or 'anisotropic'. If 'isotropic' then all
possible pairs of points are represented in the pair object. If
'anisotropic', then the arguments theta and dtheta are used to
determine which pairs of points to include.
```

Now proceed to calculate the variogram. First, the omnidirectional variogram is obtained by using maxdist = 0.45, which is slightly less than half of the side (= 1) of the point pattern

```
> xyz.v <- vario(xyz,num.lags=10,type='isotropic', maxdist=0.45)
```

Look at resulting object

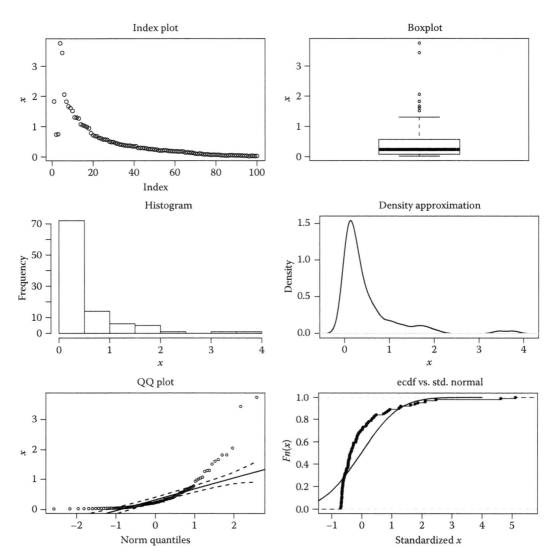

FIGURE 8.36   EDA of marks.

```
> xyz.v
   lags    bins      classic        robust          med     n
1     1  0.0225  0.09990675  0.04330418  0.03840832     12
2     2  0.0675  0.70693031  0.23132770  0.14496928     68
3     3  0.1125  0.78272027  0.33766519  0.14451244    138
4     4  0.1575  0.74364678  0.29546471  0.13872255    192
5     5  0.2025  0.83539287  0.37358955  0.20924215    217
6     6  0.2475  0.96216997  0.34356786  0.13403939    267
7     7  0.2925  0.86799281  0.35400730  0.19816677    288
8     8  0.3375  0.67231396  0.29973627  0.14735033    279
9     9  0.3825  0.85489899  0.34439950  0.19108823    327
10   10  0.4275  0.92901103  0.33775452  0.15892916    292
>
```

From the help of sgeostat, we see that a variogram object has the following components

```
A variogram object:
  lags          vector of lag identifiers
  bins          vector of midpoints of each lag
  classic        vector of classic variogram estimates for each lag
  robust         vector of robust variogram estimates for each lag
  med          vector of median variogram estimates for each lag
  n          vector of the number of pairs in each lag
```

As a rule of thumb, we usually want n to be at least 30. This is satisfied except for the first lag bin with 12 pairs. The function also plotted the dataset, the variogram estimated, cloud, and boxplots. The last two are obtained make use of **spacecloud** and **spacebox**. From the boxplots, it is easier to visualize the distribution of differences at each lag (Figure 8.15).

All of these can be done using the Rcmdr, make sure xyz is the active dataset, go to Spatial|variogram calculation to obtain same results as with the console.

We will specify a variogram model based on visual inspection of the estimate. As an example, look carefully at the variogram estimate in Figure 8.15 (top-right panel). We can guess that the model parameter values are nugget ~0, sill ~0.42, and range ~0.17. We can confirm the guess of the sill by calculating the variance of the marks

```
> var(xyz$z)
[1] 0.414
```

Next we visualize a spherical model with these parameter values using seeg's function model. semivar.cov to plot model of the semivariance, together with the empirical, and determine the covariance (Figure 8.22).

```
> m.xyz.v <- model.semivar.cov(var=xyz.v, nlags=10, n0=0, c0=0.42,
a=0.17)
```

Using the Rcmdr, make sure xyz.v is active dataset, go to Spatial|Model semi-var & Cov, enter parameters, and obtain same results as with the console.

8.12.11 MARKED PATTERNS: USE SAMPLE DATA FROM sgeostat

We will use R's **sgeostat** sample dataset maas containing x, y coordinates (two coordinates in 2D) and a value of variable zinc of Zinc measurements as groundwater quality variable. First load set and examine it

```
>data(maas)
> maas
            x         y    zinc
1      181072    333611    1022
2      181025    333558    1141
3      181165    333537     640
4      181298    333484     257
5      181307    333330     269
6      181390    333260     281
7      181165    333370     346
8      181027    333363     406
9      181060    333231     347
...
152    180201    331160     126
153    180173    331923     210
154    181118    333214     279
155    180627    330190     375
```

In this case, x, y are labels for the coordinates and zinc is the label for Z. There are 155 points. Now use function `vario` to visualize the point pattern, variogram, cloud, boxplot, and obtain the variogram

```
> maas.v <- vario(maas,num.lags=10,type='isotropic', maxdist=2000)
```

Look at resulting object

```
> maas.v
      lags     bins    classic     robust        med     n
1        1      100   133699.5    73581.9   42741.73   315
2        2      300   187012.8   105830.1   68462.36   811
3        3      500   252080.4   167882.1  138254.47   978
4        4      700   294647.7   217601.3  250478.35  1090
5        5      900   329714.7   239934.7  266252.73  1065
6        6     1100   344545.8   257999.7  265463.74   970
7        7     1300   333049.4   250315.0  254121.03   850
8        8     1500   291395.0   214845.5  205960.29   813
9        9     1700   291489.5   218285.6  190160.09   771
10      10     1900   262528.5   177701.1  140957.35   707
```

Then examine the plots shown in Figure 8.37. Note that the dataset does not fill the plotting area (top-left panel). The initial values for the model variogram parameters are determined by "eyeball-ing" the graph of the omnidirectional variogram (top right). We can guess that the model parameter values are nugget = 45,000, sill = 150,000, and range = 900. We can confirm the sill by calculating the variance of the marks

```
> var(maas$zinc)
[1] 134743.2>
```

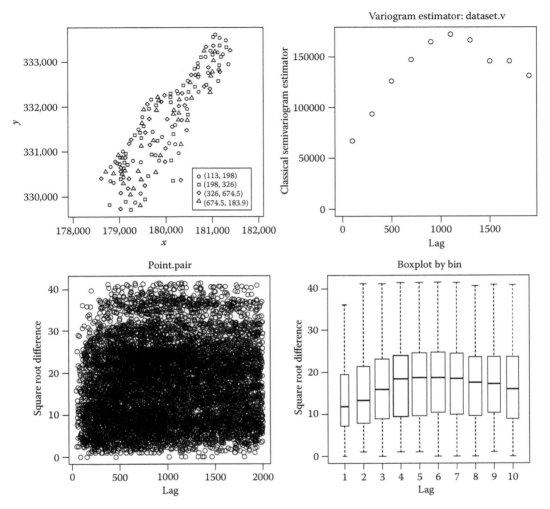

FIGURE 8.37 Point pattern of maas dataset, omnidirectional semivariogram, cloud, and boxplots for maas zin.

So refine model parameter values are nugget = 45,000, sill = 135,000, and range = 900. Use the model.semivar.cov function to visualize this model (Figure 8.23)

```
>m.maas.v <- model.semivar.cov(var = maas.v, nlags=10, n0=45000,
c0=135000, a=900)
```

We can also compute a directional variogram. Note that sgeostat refers to directional as anisotropic

```
theta: an angle, measured in degrees from the horizontal x axis,
that determines pairs of points to be included in the pair object
dtheta: a tolerance angle, around theta, measured in degrees that
determines pairs of points to be included in the pair object
```

Package sgeostat's theta angle convention is angle with respect to E and counter clockwise. So, NE would be $\theta = 45$, N would be $\theta = 90$, NW would be $\theta = 135$, SW would be $\theta = 225$, S would be $\theta = 270$, and SE would be $\theta = 315$.

Generate pairs for NE direction with narrow tolerance of 7.5°

```
>maas45.v <- vario(maas,num.lags=10,type='anisotropic', theta=45,
dtheta=7.5, maxdist=2000)
```

The variogram

```
> maas45.v
     lags         bins      classic       robust          med     n
1      1     99.95315    135180.0     68358.08    104208.72     31
2      2    299.85945    132956.6     88038.82     85371.46     72
3      3    499.76576    189284.1    101127.83     51681.92    104
4      4    699.67206    214109.3    137603.59     99414.04    135
5      5    899.57836    269251.2    198077.18    221222.48    165
6      6   1099.48467    256012.4    152654.59    142534.34    177
7      7   1299.39097    280148.9    171929.56    143049.11    170
8      8   1499.29727    268449.4    182379.69    194706.73    229
9      9   1699.20358    217030.7    148399.51    126476.07    222
10    10   1899.10988    151925.7    106604.17     88028.70    253
>
```

The plots (not shown here for the sake of space) indicate that the NE variogram is different to the omnidirectional variograms. Note that the cloud for the directional variogram would be less dense. One could calculate several variograms to see if data are isotropic. Note: do not confuse the term isotropic as property of the dataset with `isotropic` as argument of function `pair` in sgeostat.

One important application of nonlinear estimation is to model a semivariogram. The package sgeostat has a function to do this for bounded (exponential, Gaussian, spherical) and unbounded (linear) variograms models. We need to specify a variogram object, initial values (guess and educated) for model parameters, and other conditions such as number of iterations, tolerance for convergence. Let us fit a spherical model to the omnidirectional variogram of maas. The initial values for the parameters were determined by "eyeballing" the graph of the omnidirectional variogram

```
> maas.vsph <- fit.variogram(model="spherical", maas.v, nugget=45000,
sill=135000, range=900)
Initial parameter estimates:
  first: 45000 135000 900
  second: 45045 135135 900.9
Iteration: 1
Gradient vector: 3773.46 -27426 55.58101
New parameter estimates: 48773.46 107574 955.581

rse.dif = 1306198516 (rse = 1306198516 ) ; parm.dist = 27684.43

Iteration: 2
Gradient vector: 3067.826 -26857.27 46.75541
New parameter estimates: 48067.83 108142.7 946.7554

rse.dif = 7043079 (rse = 1313241595 ) ; parm.dist = 27031.96
```

and so on ....for iterations 3–10

```
Iteration: 10
Gradient vector: -2865.902 -41348.67 -45.1847
New parameter estimates: 47134.1 108651.3 954.8153
rse.dif = 9.608141 (rse = 1298420601 ) ; parm.dist = 41447.89
Convergence not achieved!
>
```

After 10 iterations (default), convergence was not achieved. Let us increase the number of iterations to 20

```
maas.vsph <- fit.variogram(model="spherical", maas.v, nugget=45000,
sill=150000, range=900, iterations=20)
```

The first iterations are the same as given earlier, but now at iteration 18 the process achieves convergence

```
Iteration: 18
Gradient vector: 3350.953 -42098.13 44.7204
New parameter estimates: 48350.95 107901.9 944.7204

rse.dif = 0 (rse = 1312807276 ) ; parm.dist = 42231.31

Convergence achieved by sums of squares.
Final parameter estimates: 48350.95 107901.9 944.7204
```

The model parameter values are nugget = 48,351, sill = 107,902, and range = 945.

We can do this in Rcmdr. First, select as active dataset. Then go to **Spatial|Fit Model semi-var**, press ok, and get same results as from the console.

8.12.12 COMPUTER EXERCISES

Exercise 8.7
Generate a uniform pattern of 100 points in a [0, 1] x [0, 1] domain. Convert to data frame. Run grid (quadrat) and nearest-neighbor analysis. Check whether the pattern is uniform. Confirm using the Monte Carlo method. Hint: use `runifpoint`, `data.frame`, `quad.chisq.ppp`, `nnGK.ppp`, and `nnGKenv.ppp`.

Exercise 8.8
Use both grid (quadrat) and nearest-neighbor analysis (plot G, *K*, and L and use envelopes from 20 simulation runs) to determine that indeed the `xyz` pattern in **lab8/xyz-geoEAS.txt** is uniform but not random. Hint: use `quad.chisq.ppp`, `nnGK.ppp`, and `nnGKenv.ppp`.

Exercise 8.9
Use `hick.spp` created in the computer session. Apply the tools we learned for quadrat and nearest-neighbor analysis. Hint: apply `data.frame`, `quad.chisq.ppp`, `nnGK.ppp`, and `nnGKenv.ppp`.

Demonstrate that pattern is clustered. Note: the iso correction estimate cannot be computed for 500 or more points. Subset a portion of the data frame, say 450 points, by using [1:450].

Exercise 8.10

Use marked pattern given in file **lab8/unif100marked-geoEAS.txt**. This pattern is uniform because the location data are the same as `unif100` already analyzed in the computer session. Run omnidirectional variogram analysis on the marks and determine the semivariance spherical model. Determine the covariance model and plot it.

Exercise 8.11

Plot model semivariance and covariance for the `zinc` variable in `maas` dataset using the directional NE variogram.

SUPPLEMENTARY READING

On point patterns: Davis, 2002: Chapter 5, pp. 299–313, Rogerson, 2001: Chapter 7 pp. 124–153, Carr, 1995: Chapter 6, pp. 150–180, Deutsch and Journel (1992): Chapters II and III, pp. 9–60, Kaluzny et al. (1996): Chapter 3, pp. 21–71, and Chapter 6, pp. 161–186. On Variograms: Davis, 2002, Chapter 4, pp. 254–265, Carr, 1995, Chapter 6, pp. 160–184, Deutsch and Journel, 1992, Chapter II, pp. 22–31, Chapter III, pp. 39–60, Kaluzny et al. (1996): Chapter 4, pp. 74–107.

Part II

Matrices, Tempral and Spatial Autoregressive Processes, and Multivariate Analysis

9 Matrices and Linear Algebra

So far, we have dealt with one variable, or a series of identical variables (stationary process), and in the case of regression with two variables. In the remainder of the book, we study methods of analysis with several or many variables, say $X_1, X_2, \ldots, X_n$, where n is the dimension of the system. Because we need to solve the equations simultaneously, we use mathematical strategies that take advantage of the relationships among the variables to find a solution. A very useful strategy is to arrange the interactions among the variables in an array called **matrix**.

More specifically, when the interactions among the variables X_i are linear, then we can use **linear algebra**. Linear algebra underlies **multivariate** analysis and statistics as well as geostatistics. Therefore, a good grasp of linear algebra is of great importance to understand the remaining chapters of this book. This chapter provides a review of linear algebra.

9.1 MATRICES

A matrix is an array of numbers organized into rows and columns. The position of each element in the array is identified by its position in row i and column j. In the following matrix $\mathbf{A}$, element a is in row 1 and column 1; element b is in row 1 and column 2; element c is in row 1 and column 3. Notice that we use bold font to denote matrices

$$\mathbf{A} = \begin{bmatrix} a & b & c \\ d & e & f \\ g & h & i \end{bmatrix} \tag{9.1}$$

Each one of the entries of a matrix is a **scalar**. The column and row numbers are used as sub-indices to write the matrix. So, matrix $\mathbf{A}$ is written as

$$\mathbf{A} = \begin{bmatrix} a_{11} & a_{12} & a_{13} \\ a_{21} & a_{22} & a_{23} \\ a_{31} & a_{32} & a_{33} \end{bmatrix} \tag{9.2}$$

9.2 DIMENSION OF A MATRIX

The number of rows and columns of a matrix represents the **dimension** of the matrix. Matrix $\mathbf{A}$ mentioned earlier has three rows and three columns and therefore is a 3×3 matrix. A single number is a matrix of dimension 1×1 and is a **scalar**. If the number of rows in a matrix is different than the number of columns, it is a **rectangular matrix**. An $n \times m$ matrix $\mathbf{B}$ has n rows and m columns:

$$\mathbf{B} = \begin{bmatrix} b_{11} & b_{12} & . & . & . & b_{1m} \\ b_{21} & b_{22} & . & . & . & b_{2m} \\ . & . & & & & . \\ . & . & & & & . \\ . & . & & & & . \\ b_{n1} & b_{n2} & . & . & . & b_{nm} \end{bmatrix} \tag{9.3}$$

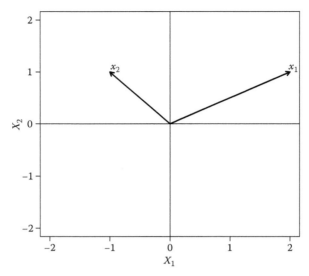

FIGURE 9.1   Geometric interpretation of two-dimensional vectors.

9.3 VECTORS

A matrix with only one column is a **column vector** and a matrix with only one row is a **row vector**. For example, a column vector with three entries has dim 3×1, a row vector with four entries has dim 1×4. Column vector $\mathbf{x}$ and row vector $\mathbf{y}$ shown next have three rows and four columns, respectively:

$$\mathbf{x} = \begin{bmatrix} x_{11} \\ x_{21} \\ x_{31} \end{bmatrix} \tag{9.4}$$

$$\mathbf{y} = \begin{bmatrix} y_{11} & y_{12} & y_{13} & y_{14} \end{bmatrix} \tag{9.5}$$

The dimensions of column vector $\mathbf{x}$ are 3×1 and of row vector $\mathbf{y}$ are 1×4.

We can think of a vector geometrically as an $n \times 1$ matrix in such a way that the entries define the coordinates of a point in n-space. The length of the vector will be the distance from the origin of coordinates to the point, whereas an arrow pointing from the origin toward the point indicates the direction of the vector. We can easily visualize vectors of dimension 2×1 on a plane by drawing an arrow from the origin of coordinates to the point with coordinates given by elements of the vector. For example, Figure 9.1 shows vectors $\mathbf{x}_1 = \begin{bmatrix} 2 \\ 1 \end{bmatrix}$ and $\mathbf{x}_2 = \begin{bmatrix} -1 \\ 1 \end{bmatrix}$ in two-dimensional space. Their lengths are $\sqrt{2^2 + 1^2} = \sqrt{5} = 2.24$ for $\mathbf{x}_1$ and $\sqrt{(-1)^2 + 1^2} = \sqrt{2} = 1.41$ for $\mathbf{x}_2$.

9.4 SQUARE MATRICES

Matrix $\mathbf{A}$ in Equation 9.2 is a square matrix because it has the same number of columns as rows, three rows and three columns, that is, a 3×3 matrix. Square matrices represent linear systems of equations that have the same number of unknown variables as linearly independent equations.

The **main diagonal** of a square matrix is composed of all of the entries for which the column number i is equal to the row number j. The main diagonal of matrix $\mathbf{A}$ would consist of the elements a_{11}, a_{22}, a_{33}. That is, all elements a_{ij} such that $i = j$. All entries above the diagonal would have $j > i$; for example in matrix $\mathbf{A}$, these elements are a_{23}, a_{12}, a_{13}. Entries below the diagonal would have i greater than j. For example in matrix $\mathbf{A}$, these are a_{21}, a_{31}, a_{32}.

9.4.1 TRACE

The trace of a square matrix is the sum of all of the elements in the main diagonal. This is to say,

$$\text{trace}(\mathbf{A}) = \sum_{i=1}^{n} a_{ii} \tag{9.6}$$

For example consider matrix $\mathbf{A} = \begin{bmatrix} 1 & 4 \\ 5 & 2 \end{bmatrix}$, the trace of A is $1 + 2 = 3$. Note that the trace is a **scalar**.

9.4.2 SYMMETRIC MATRICES: COVARIANCE MATRIX

A square matrix is said to be **symmetric** if all the entries above the main diagonal are equal to the corresponding entries below the main diagonal, $a_{ij} = a_{ji}$. Matrix $\mathbf{C}$ shown next is an example of a symmetric matrix:

$$\mathbf{C} = \begin{bmatrix} 2 & 1 & 4 \\ 1 & 2 & 3 \\ 4 & 3 & 2 \end{bmatrix} \tag{9.7}$$

Symmetric matrices are very important in multivariate analysis because the pairwise covariance among many (say n) variables can be conveniently arranged in a square $n \times n$ matrix. The covariance of pairs of variables X_i, X_j will be the entries of the matrix. When the variables are the same X_i, X_i, the covariance reduces to the variance of X_i; therefore, the main diagonal entries contain the variances. Also, since covariance of X_i, X_j is the same as covariance of X_j, X_i, the matrix will be symmetric.

This type of matrix is a **variance-covariance matrix** or simply a **covariance matrix**. For example in a bivariate case, the covariance matrix is 2×2

$$\mathbf{C} = \begin{pmatrix} \sigma_{X_1}^2 & \text{cov}(X_1, X_2) \\ \text{cov}(X_2, X_1) & \sigma_{X_2}^2 \end{pmatrix} \tag{9.8}$$

and because $\text{cov}(X_1, X_2) = \text{cov}(X_2, X_1)$ the matrix is symmetric

$$\mathbf{C} = \begin{pmatrix} \sigma_{X_1}^2 & \text{cov}(X_1, X_2) \\ \text{cov}(X_1, X_2) & \sigma_{X_2}^2 \end{pmatrix} \tag{9.9}$$

Note that the **trace of a covariance matrix is the sum of the variances**, or **total variance** because the main diagonal entries are the variances.

$$\text{trace}(\mathbf{C}) = \sum_{i=1}^{n} \sigma_{Xi}^2 \tag{9.10}$$

Example: Suppose variances of X_1 and X_2 are 2 and 3 respectively and that the covariance of X_1 and X_2 is 1. The covariance matrix would be

$$\mathbf{C} = \begin{pmatrix} 2 & 1 \\ 1 & 3 \end{pmatrix} \tag{9.11}$$

The trace is $2 + 3 = 5$, and this is equal to the total variance.

9.4.3 IDENTITY

The **identity** matrix $\mathbf{I}$ is a symmetric matrix with all the diagonal entries equal to 1, $a_{ii} = 1$, and all off diagonal entries equal to zero $a_{ij} = 0$, when $i \neq j$. The identity matrix works with matrix multiplication (to be covered subsequently) as the multiplicative identity, 1. This is to say, $\mathbf{AI} = \mathbf{A}$ and $\mathbf{IA} = \mathbf{A}$. The following matrix is an identity 3×3 matrix:

$$\mathbf{I} = \begin{bmatrix} 1 & 0 & 0 \\ 0 & 1 & 0 \\ 0 & 0 & 1 \end{bmatrix} \tag{9.12}$$

For example, take matrix $\mathbf{C}$ from Equation 9.7

$$\mathbf{CI} = \begin{bmatrix} 2 & 1 & 4 \\ 1 & 2 & 3 \\ 4 & 3 & 2 \end{bmatrix} \begin{bmatrix} 1 & 0 & 0 \\ 0 & 1 & 0 \\ 0 & 0 & 1 \end{bmatrix} = \begin{bmatrix} 2 & 1 & 4 \\ 1 & 2 & 3 \\ 4 & 3 & 2 \end{bmatrix} = \mathbf{C}$$

9.5 MATRIX OPERATIONS

Now we will learn some basic operations of linear algebra: addition, multiplication, transposition, and inversion.

9.5.1 ADDITION AND SUBTRACTION

Matrix addition (and subtraction) is the most intuitive operation on matrices. If the dimensions of each of the summand matrices match, then the matrices can be added (or subtracted) together entry by entry. $\mathbf{C} = \mathbf{A} + \mathbf{B}$ is obtained by $c_{ij} = a_{ij} + b_{ij}$ for all entries ij. The following example shows the addition of two 3×2 matrices:

$$\begin{bmatrix} 1 & 2 & 3 \\ -1 & -2 & -3 \end{bmatrix} + \begin{bmatrix} -4 & -5 & -6 \\ 4 & 5 & 6 \end{bmatrix} = \begin{bmatrix} 1-4 & 2-5 & 3-6 \\ -1+4 & -2+5 & -3+6 \end{bmatrix} = \begin{bmatrix} -3 & -3 & -3 \\ 3 & 3 & 3 \end{bmatrix} \tag{9.13}$$

9.5.2 SCALAR MULTIPLICATION

By scalar we mean a single number or a 1×1 matrix. A scalar can multiply a matrix to produce a new matrix "proportional" to the previous one. We multiply the scalar by each entry within the matrix. If $\mathbf{D} = k\mathbf{B}$, then $d_{ij} = k \times b_{ij}$ for each i and j. For example, when matrix

$$\mathbf{A} = \begin{bmatrix} a_{11} & a_{12} & a_{13} \\ a_{21} & a_{22} & a_{23} \\ a_{31} & a_{32} & a_{33} \end{bmatrix}$$

is multiplied by a scalar with value equal 3, we would get

$$3\mathbf{A} = \begin{bmatrix} 3a_{11} & 3a_{12} & 3a_{13} \\ 3a_{21} & 3a_{22} & 3a_{23} \\ 3a_{31} & 3a_{32} & 3a_{33} \end{bmatrix}.$$

As another example, $2\mathbf{C}$ will be

$$2 \times \mathbf{C} = 2 \times \begin{bmatrix} 2 & 1 & 4 \\ 1 & 2 & 3 \\ 4 & 3 & 2 \end{bmatrix} = \begin{bmatrix} 4 & 2 & 8 \\ 2 & 4 & 6 \\ 8 & 6 & 4 \end{bmatrix}$$

9.5.3 LINEAR COMBINATION

A linear combination of vectors is the sum of scalar multiplication of a set of vectors by a set of coefficients. That is to say, vector $\mathbf{y}$ of dimension $n \times 1$ is

$$\mathbf{y} = c_1\mathbf{x}_1 + c_2\mathbf{x}_2 + \cdots + c_m\mathbf{x}_m = \sum_{i=1}^{m} c_i\mathbf{x}_i \tag{9.14}$$

where scalars c_i are coefficients and $\mathbf{x}_i$ are vectors all of dimension $n \times 1$. For example, take vectors $\mathbf{x}_1 = \begin{bmatrix} 2 \\ 1 \end{bmatrix}$ and $\mathbf{x}_2 = \begin{bmatrix} -1 \\ 1 \end{bmatrix}$ that we plotted before and coefficients $c_1 = 0.5$, $c_2 = 1$

$$\mathbf{y} = 0.5 \times \begin{bmatrix} 2 \\ 1 \end{bmatrix} + 1 \times \begin{bmatrix} -1 \\ 1 \end{bmatrix} = \begin{bmatrix} 0 \\ 1.5 \end{bmatrix}$$

Figure 9.2 shows vector $\mathbf{y}$ resultant from this linear combination.

9.5.4 MATRIX MULTIPLICATION

Matrix multiplication is more complex than addition and scalar multiplication. The simplest multiplication is an entry-wise product or Hadamard product of two matrices $\mathbf{C} = \mathbf{A} \circ \mathbf{B}$, where $c_{ij} = a_{ij} \times b_{ij}$ and the two multiplier matrices $\mathbf{A}$, $\mathbf{B}$ should have the same dimension. This product has application in image and map processing; however, for multivariate analysis, we require a more complicated matrix multiplication, because the matrix elements correspond to the coefficients of equations. Therefore, in the remainder of the chapter, we will refer to **matrix multiplication** of two

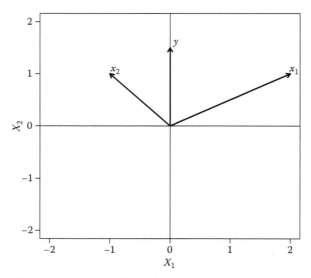

FIGURE 9.2   Linear combination of two two-dimensional vectors.

matrices as the operation of obtaining the product by summing the element-wise product of each of the rows of the first matrix with each of the columns of the second matrix. This operation is **not commutative**. Therefore, we need to distinguish between **premultiplication** and **postmultiplication**. In order to multiply two matrices, we must match number of columns in the first multiplier with the number of rows in the second

$$\mathbf{A}_{n \times r} \times \mathbf{B}_{r \times m} = \mathbf{C}_{n \times m}$$

where

$n \times r$ are the dimensions of matrix $\mathbf{A}$
$r \times m$ are the dimensions of matrix $\mathbf{B}$

Here, matrix $\mathbf{A}$ premultiplies matrix $\mathbf{B}$. The resulting matrix $\mathbf{C}$ has dimensions $n \times m$. A nonrigorous representation of this match is

$$n \times \underbrace{r \times r}_{\text{must match}} \times m \rightarrow n \times m$$

where the two values in the center must be equal.

When the dimensions of the matrices match, the two matrices are multiplied by summing the entry-wise product of each of the rows of the first matrix with each of the columns of the second matrix. For example, a 2×3 matrix premultiplies a 3×1 matrix:

$$\begin{bmatrix} 1 & 2 & 3 \\ 4 & 5 & 6 \end{bmatrix} \times \begin{bmatrix} 7 \\ 8 \\ 9 \end{bmatrix} = \begin{bmatrix} 1 \times 7 + 2 \times 8 + 3 \times 9 \\ 4 \times 7 + 5 \times 8 + 6 \times 9 \end{bmatrix} = \begin{bmatrix} 50 \\ 131 \end{bmatrix} \tag{9.15}$$

Note that $2 \times 3 \times 3 \times 1 \rightarrow 2 \times 1$, where the 3 and 3 in the center indicates match.

A special case of multiplication is very important to mention here: when we postmultiply a square matrix $n \times n$ by a column vector $n \times 1$, we obtain another vector $n \times 1$. Thus, we can see this

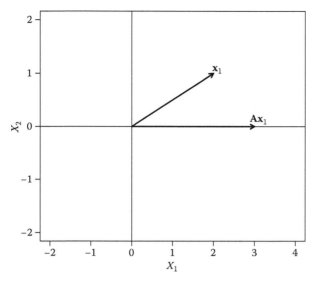

FIGURE 9.3 A vector $\mathbf{x}_1$ transformed to another vector by matrix multiplication.

matrix multiplication as a transformation of the vector into another, usually with different length and direction. For example, $\mathbf{Ax}_1 = \begin{bmatrix} 1 & 1 \\ -2 & 4 \end{bmatrix} \begin{bmatrix} 2 \\ 1 \end{bmatrix} = \begin{bmatrix} 3 \\ 0 \end{bmatrix}$ transforms vector $\mathbf{x}_1 = \begin{bmatrix} 2 \\ 1 \end{bmatrix}$ by rotating it clockwise $45°$ and lengthening it from $\sqrt{2^2 + 1^2} = \sqrt{5} = 2.24$ to $\sqrt{3^2 + 0^2} = \sqrt{9} = 3$ (Figure 9.3).

9.5.5 DETERMINANT OF A MATRIX

The **determinant** of a matrix is an operation that **assigns a scalar to a square matrix**. Matrix inversion and other complex processing of matrices require this operation. Determinants are only relevant for square matrices. The determinant of matrix $\mathbf{A}$ is usually denoted $|A|$ or $\det(\mathbf{A})$. A simple formula exists for evaluating the determinant of 2×2 or 3×3 matrices. The determinant is the sum of the products of the elements in the upper left to lower right diagonals minus the sums of products of lower left to upper right diagonals. For a 2×2 matrix this is

$$\begin{vmatrix} a_{11} & a_{12} \\ a_{21} & a_{22} \end{vmatrix} = a_{11}a_{22} - a_{21}a_{12} \tag{9.16}$$

and for a 3×3 matrix this is

$$\begin{vmatrix} a_{11} & a_{12} & a_{13} \\ a_{21} & a_{22} & a_{23} \\ a_{31} & a_{32} & a_{33} \end{vmatrix} = (a_{11}a_{22}a_{33} + a_{21}a_{32}a_{13} + a_{31}a_{12}a_{23})$$

$$- (a_{31}a_{22}a_{13} + a_{21}a_{12}a_{33} + a_{11}a_{32}a_{23}) \tag{9.17}$$

For example, the determinant of matrix $\mathbf{C}$ in Equation 9.11 would be

$$|\mathbf{C}| = \begin{vmatrix} 2 & 1 \\ 1 & 3 \end{vmatrix} = 2 \times 3 - 1 \times 1 = 5$$

and the determinant of matrix $\mathbf{C}$ from Equation 9.7

$$|\mathbf{C}| = \begin{vmatrix} 2 & 1 & 4 \\ 1 & 2 & 3 \\ 4 & 3 & 2 \end{vmatrix}$$

$$= 2 \times 2 \times 2 + 4 \times 1 \times 3 + 4 \times 1 \times 3 - 4 \times 2 \times 4 - 1 \times 1 \times 2 - 2 \times 3 \times 3$$

$$= 8 + 12 + 12 - 16 - 2 - 12 = 2$$

As you can see increasing the dimensions from 2×2 to 3×3 increased the number of calculations from 2 to 6. The increase in number of calculations for a large value of n for an $n \times n$ matrix can be significant. Matrices represent applications in n-space and evaluating the determinant of an $n \times n$ matrix is slightly more complex. Practically, we use computers to calculate determinants for matrices of large dimensions.

9.5.6   MATRIX TRANSPOSITION

Transposition consists of exchanging the rows and columns of a matrix. The transpose of matrix $\mathbf{A}$ would be denoted $[\mathbf{A}]'$ or $[\mathbf{A}]^T$. This would result in a row vector becoming a column vector. For example, a matrix with dimensions 2×3 would become a 3×2 matrix as shown next:

$$\begin{bmatrix} 6 & 5 & 4 \\ 3 & 2 & 1 \end{bmatrix}^T = \begin{bmatrix} 6 & 3 \\ 5 & 2 \\ 4 & 1 \end{bmatrix} \tag{9.18}$$

9.5.7   MAJOR PRODUCT

The **major product** of a matrix is a square symmetric matrix formed by the product of the matrix and its transpose. The major product of the matrix is useful in linear regression methods. For example, we form a two-column matrix with the data values or observations x_i of a variable X in a sample of size n in column 2 and all 1's in column 1 in the following manner:

$$\mathbf{x} = \begin{bmatrix} 1 & x_1 \\ 1 & x_2 \\ 1 & x_3 \\ \dots & \dots \\ 1 & x_n \end{bmatrix} \tag{9.19}$$

Now if we transpose and premultiply $\mathbf{x}$ by its transpose $\mathbf{x}^T$, we obtain a square matrix

$$\mathbf{x}^T\mathbf{x} = \begin{bmatrix} 1 & 1 & 1 & \dots & 1 \\ x_1 & x_2 & x_3 & \dots & x_n \end{bmatrix} \begin{bmatrix} 1 & x_1 \\ 1 & x_2 \\ 1 & x_3 \\ \dots & \dots \\ 1 & x_n \end{bmatrix} = \begin{bmatrix} n & \sum x_i \\ \sum x_i & \sum x_i^2 \end{bmatrix} \tag{9.20}$$

The entries are the sum of squares of x_i in the diagonal and sum of x_i in the off-diagonal entries. The matrix $\mathbf{S_x} = \mathbf{x^T x}$ plays an important role in regression.

For example, suppose we have five values of X, $x_i = 1, 2, 2, 1, 0$

$$\mathbf{S_x} = \mathbf{x^T x} = \begin{bmatrix} 1 & 1 & 1 & 1 & 1 \\ 1 & 2 & 2 & 1 & 0 \end{bmatrix} \begin{bmatrix} 1 & 1 \\ 1 & 2 \\ 1 & 2 \\ 1 & 1 \\ 1 & 0 \end{bmatrix} = \begin{bmatrix} 5 & 1+2+2+1+0 \\ 1+2+2+1+0 & 1+4+4+1+0 \end{bmatrix} = \begin{bmatrix} 5 & 6 \\ 6 & 10 \end{bmatrix}$$

9.5.8 MATRIX INVERSION

Division is not strictly defined for matrices. However, recall that division of two numbers a/b is defined by multiplying a by the inverse $1/b = b^{-1}$. For two matrices $\mathbf{A}$ and $\mathbf{B}$, we multiply $\mathbf{A}$ by the inverse of $\mathbf{B}$, denoted $\mathbf{B}^{-1}$, instead of dividing $\mathbf{A}$ by $\mathbf{B}$. The inverse of matrix $\mathbf{A}$ is another matrix defined as a matrix that when multiplied by $\mathbf{A}$ yields the identity matrix $\mathbf{I}$. That is to say $\mathbf{AA}^{-1} = \mathbf{I}$. The inverse only exists for square matrices with linearly independent rows. The determinant of a matrix can be used to determine if the rows of the matrix are linearly independent. Linearly independent square matrices, that is, nonsingular, have nonzero determinants. Singular matrices have determinant equal to zero. In other words for the inverse of $\mathbf{A}$ to exist, A must be nonsingular $|\mathbf{A}| \neq 0$.

The inverse of $\mathbf{A}$ is calculated from the **adjoint** matrix adj$\mathbf{A}$ divided by the determinant:

$$\mathbf{A} = \frac{\text{adj}\mathbf{A}}{|\mathbf{A}|} \tag{9.21}$$

The adjoint matrix adj$\mathbf{A}$ is formed by the cofactors of the entries of the transpose of $\mathbf{A}$ or equivalently by the transpose of the matrix of cofactors of $\mathbf{A}$:

$$\text{adj}\mathbf{A} = (\mathbf{A}^T)^c = (\mathbf{A}^c)^T \tag{9.22}$$

The cofactor of an entry is the determinant of the matrix that remains after deleting the row and column corresponding to this entry. The sign depends on whether the sum of row and column numbers for the entry position is even (assign positive sign) or odd (assign negative sign). Now form a new matrix formed by cofactors of each entry and divide by determinant.

To illustrate how to calculate adjoint and the inverse, we will apply Equations 9.21 and 9.22 to a 2×2 matrix:

$$\mathbf{A} = \begin{bmatrix} 4 & 10 \\ 10 & 30 \end{bmatrix} \tag{9.23}$$

Step 1: Calculate the transpose.

$$\mathbf{A}^T = \begin{bmatrix} 4 & 10 \\ 10 & 30 \end{bmatrix}$$

In this case, it is just the same matrix because it is symmetric.

Step 2: Calculate cofactors of $\mathbf{A}^T$. In this simple case, the matrix remaining after deleting a row and a column is simply a scalar. The determinant of a scalar is just that same scalar. Take entry a_{11}, delete row 1 and column 1, the matrix remaining is just scalar 30, the sign is positive because $1 + 1 = 2$ is even. Take entry a_{12} delete row 1, column 2, the matrix remaining is scalar 10, the sign is negative because $1 + 2 = 3$ is odd. Take entry a_{21}, delete row 2 and column 1, the matrix remaining is just scalar 10, the sign is negative because $2 + 1 = 3$ is odd. Take entry a_{22}, delete row 2 and column 2, the matrix remaining is just scalar 4, the sign is positive because $2 + 2 = 4$ is even. Using these cofactors, the adjoint matrix is

$$\text{adj}\mathbf{A} = (\mathbf{A}^T)^c = \begin{bmatrix} 30 & -10 \\ -10 & 4 \end{bmatrix}$$

Step 3: Calculate the determinant

$$|\mathbf{A}| = \begin{vmatrix} 4 & 10 \\ 10 & 30 \end{vmatrix} = 4 \times 30 - 10 \times 10 = 20 \tag{9.24}$$

Step 4: Divide the adjoint matrix by the determinant

$$\mathbf{A}^{-1} = \frac{\text{adj}\mathbf{A}}{|\mathbf{A}|} = \frac{(\mathbf{A}^T)^c}{|\mathbf{A}|} = \frac{1}{20}\begin{bmatrix} 30 & -10 \\ -10 & 4 \end{bmatrix} = \begin{bmatrix} 1.5 & -0.5 \\ -0.5 & 0.2 \end{bmatrix} \tag{9.25}$$

A practical way of inverting a matrix of low dimensions, without remembering cofactors, is to simultaneously perform operations on the matrix and the identity matrix as to convert all the diagonal elements to ones and the off-diagonal elements to zeros. For example:

Step 1: Place the matrix $\mathbf{A}$ besides an identity matrix

$$\begin{bmatrix} 4 & 10 \\ 10 & 30 \end{bmatrix} \begin{bmatrix} 1 & 0 \\ 0 & 1 \end{bmatrix}$$

Step 2: Row 1 of both matrices is divided by the element in first row, first column to produce 1 at a_{11} entry

$$\begin{bmatrix} 1 & 2.5 \\ 10 & 30 \end{bmatrix} \begin{bmatrix} .25 & 0 \\ 0 & 1 \end{bmatrix}$$

Step 3: The first row scaled by first element in row 2 is subtracted from row 2 to reduce a_{21} to zero

$$\begin{bmatrix} 1 & 2.5 \\ 0 & 5 \end{bmatrix} \begin{bmatrix} .25 & 0 \\ -2.5 & 1 \end{bmatrix}$$

Step 4: Row 2 is divided by element in second column to give $a_{22} = 1$

$$\begin{bmatrix} 1 & 2.5 \\ 0 & 1 \end{bmatrix} \begin{bmatrix} .25 & 0 \\ -0.5 & 0.2 \end{bmatrix}$$

Step 5: Row 2 is scaled by remaining nonzero off-diagonal term and subtracted from row 1 to reduce remaining off-diagonal a_{12} term to zero

$$\begin{bmatrix} 1 & 0 \\ 0 & 1 \end{bmatrix} \begin{bmatrix} 1.5 & -0.5 \\ -0.5 & 0.2 \end{bmatrix}$$

The final matrix in the right is the inverse of the matrix on the left in the first step. Multiply them to verify that their product is the identity matrix. We showed this procedure for pedagogical purposes; we use computers to calculate the inverse of a large matrix.

9.6   SOLVING SYSTEMS OF LINEAR EQUATIONS

Based upon the elements of matrices expressed earlier, you can now apply that information to solving systems of equations:

$$4x + 10y = 38$$
$$10x + 30y = 110 \tag{9.26}$$

By using the coefficients of the each of the equations as entries in a matrix, we can rewrite Equations 9.26 as follows:

$$\begin{bmatrix} 4 & 10 \\ 10 & 30 \end{bmatrix} \cdot \begin{bmatrix} x \\ y \end{bmatrix} = \begin{bmatrix} 38 \\ 110 \end{bmatrix} \tag{9.27}$$

Note that you can check your matrix multiplication skills by multiplying the first two matrices. It should yield a matrix containing the left-hand side of Equations 9.26. Denoting the vector of unknowns x, y as $\mathbf{x}$

$$\mathbf{x} = \begin{bmatrix} x \\ y \end{bmatrix} \tag{9.28}$$

The matrix is now in the form $\mathbf{Ax} = \mathbf{b}$, where

$$\mathbf{A} = \begin{bmatrix} 4 & 10 \\ 10 & 30 \end{bmatrix} \tag{9.29}$$

and

$$\mathbf{b} = \begin{bmatrix} 38 \\ 110 \end{bmatrix} \tag{9.30}$$

To solve the equation

$$\mathbf{Ax} = \mathbf{b} \tag{9.31}$$

for **x**, simply multiply both sides of the equation by the inverse of **A**. That is,

$$\mathbf{A}^{-1}\mathbf{A}\mathbf{x} = \mathbf{I}\mathbf{x} = \mathbf{x} = \mathbf{A}^{-1}\mathbf{b} \tag{9.32}$$

to obtain

$$\mathbf{x} = \mathbf{A}^{-1}\mathbf{b} \tag{9.33}$$

For example, solve for **x** in Equation 9.27. We already calculated the inverse of **A** in the previous section and is given in equation

$$\mathbf{A}^{-1} = \begin{bmatrix} 1.5 & -0.5 \\ -0.5 & 0.2 \end{bmatrix}$$

Apply Equation 9.32

$$\begin{bmatrix} 1.5 & -.5 \\ -.5 & 0.2 \end{bmatrix} \cdot \begin{bmatrix} 4 & 10 \\ 10 & 30 \end{bmatrix} \cdot \begin{bmatrix} x \\ y \end{bmatrix} = \begin{bmatrix} 1.5 & -.5 \\ -.5 & 0.2 \end{bmatrix} \cdot \begin{bmatrix} 38 \\ 110 \end{bmatrix} \tag{9.34}$$

which after multiplication is simplified to

$$\begin{bmatrix} 1 & 0 \\ 0 & 1 \end{bmatrix} \cdot \begin{bmatrix} x \\ y \end{bmatrix} = \begin{bmatrix} 2 \\ 3 \end{bmatrix} \tag{9.35}$$

Therefore, the unique solutions to the system of Equations 9.26 above is $x = 2$ and $y = 3$. If you graph both equations, the lines should intersect at the point $(2, 3)$.

A typical use of this method is to solve the **regression problem** based on solving for the slope b_1 and intercept b_0 of a line of best fit between X and Y using observations y_i and x_i. We will see that the need to solve the matrix equation

$$\begin{bmatrix} n & \sum_{i=1}^{n} x_i \\ \sum_{i=1}^{n} x_i & \sum_{i=1}^{n} x_i^2 \end{bmatrix} \begin{bmatrix} b_0 \\ b_1 \end{bmatrix} = \begin{bmatrix} \sum_{i=1}^{n} y_i \\ \sum_{i=1}^{n} x_i y_i \end{bmatrix} \tag{9.36}$$

which can abbreviated by using names $\mathbf{S_x}$ for the major product matrix in the left, **b** for the vector of slope and intercept, and $\mathbf{S_y}$ for the right-hand side,

$$\underbrace{\begin{bmatrix} n & \sum_{i=1}^{n} x_i \\ \sum_{i=1}^{n} x_i & \sum_{i=1}^{n} x_i^2 \end{bmatrix}}_{\mathbf{S_x}} \underbrace{\begin{bmatrix} b_0 \\ b_1 \end{bmatrix}}_{\mathbf{b}} = \underbrace{\begin{bmatrix} \sum_{i=1}^{n} y_i \\ \sum_{i=1}^{n} x_i y_i \end{bmatrix}}_{\mathbf{S_y}}$$

to write in matrix form the equation as

$$\mathbf{S_x b} = \mathbf{S_y} \tag{9.37}$$

Note that $\mathbf{S_x}$ is a symmetric matrix involving only the observations of X. Solving the matrix equation for unknown $\mathbf{b}$ allows calculation of intercept b_0 and slope b_1 of the regression line. The solution is

$$\mathbf{b} = \mathbf{S_x^{-1}}\,\mathbf{S_y} \tag{9.38}$$

9.7 LINEAR ALGEBRA SOLUTION OF THE REGRESSION PROBLEM

An alternative way to solve the simultaneous regression equations is to use linear algebra (Carr, 1995, pages 47–49). These equations are repeated here for easy reference:

$$nb_0 + b_1 \sum_{i=1}^{n} x_i - \sum_{i=1}^{n} y_i = 0 \tag{9.39}$$

$$b_1 \sum_{i=1}^{n} x_i^2 + b_0 \sum_{i=1}^{n} x_i - \sum_{i=1}^{n} x_i y_i = 0 \tag{9.40}$$

Moving terms independent of b_0 and b_1 to the right-hand side, we obtain

$$nb_0 + b_1 \sum_{i=1}^{n} x_i = \sum_{i=1}^{n} y_i \tag{9.41}$$

$$b_1 \sum_{i=1}^{n} x_i^2 + b_0 \sum_{i=1}^{n} x_i = \sum_{i=1}^{n} x_i y_i \tag{9.42}$$

These can be summarized in matrix form as in equation 4.17 in Davis, 2002, page 194

$$\begin{bmatrix} n & \sum_{i=1}^{n} x_i \\ \sum_{i=1}^{n} x_i & \sum_{i=1}^{n} x_i^2 \end{bmatrix} \begin{bmatrix} b_0 \\ b_1 \end{bmatrix} = \begin{bmatrix} \sum_{i=1}^{n} y_i \\ \sum_{i=1}^{n} x_i y_i \end{bmatrix} \tag{9.43}$$

and abbreviated by using new names $\mathbf{S_x}$ and $\mathbf{S_y}$.

$$\mathbf{S_x b} = \mathbf{S_y} \tag{9.44}$$

Note that $\mathbf{S_x}$ is a symmetric matrix involving only the observations of X. Solving the matrix equation for unknown $\mathbf{b}$ allows calculation of intercept b_0 and slope b_1 of the regression line. The solution is

$$\mathbf{b} = \mathbf{S_x^{-1}}\,\mathbf{S_y} \tag{9.45}$$

Upon calculation we find that the determinant of matrix $\mathbf{S_x}$ is the sample variance of X. The slope is sample covariance divided by sample variance of X:

$$\begin{bmatrix} b_0 \\ b_1 \end{bmatrix} = \begin{bmatrix} \bar{Y} - \dfrac{s_{\text{cov}}(X,Y)}{s_X^2} \bar{X} \\ \dfrac{s_{\text{cov}}(X,Y)}{s_X^2} \end{bmatrix} \tag{9.46}$$

Similarly, we can use the population approach,

$$b_0 + b_1 \mu_X = \mu_Y \tag{9.47}$$

$$b_1 E(X^2) + b_0 \mu_X = E(XY) \tag{9.48}$$

in matrix form

$$\begin{bmatrix} 1 & \mu_X \\ \mu_X & E(X^2) \end{bmatrix} \begin{bmatrix} b_0 \\ b_1 \end{bmatrix} = \begin{bmatrix} \mu_Y \\ E(XY) \end{bmatrix} \tag{9.49}$$

Note the similarity between Equation 9.43 for sample and Equation 9.49 for population. This equation can also be written in brief form as $\mathbf{Ab = c}$. Now, finding $\mathbf{A^{-1}}$ and solving for b

$$\begin{bmatrix} b_0 \\ b_1 \end{bmatrix} = \frac{\begin{bmatrix} E(X^2) & -\mu_X \\ -\mu_X & 1 \end{bmatrix} \begin{bmatrix} \mu_Y \\ E(XY) \end{bmatrix}}{\sigma_X^2} = \frac{\begin{bmatrix} E(X^2)\mu_Y - \mu_X E(XY) \\ E(XY) - \mu_Y \mu_X \end{bmatrix}}{\sigma_X^2} \tag{9.50}$$

$$\begin{bmatrix} b_0 \\ b_1 \end{bmatrix} = \begin{bmatrix} \mu_Y - \dfrac{\text{cov}(X,Y)}{\sigma_X^2} \mu_X \\ \dfrac{\text{cov}(X,Y)}{\sigma_X^2} \end{bmatrix} \tag{9.51}$$

The determinant of A is the variance of X. The slope is covariance divided by variance of X. We can illustrate numerically.

$$\begin{bmatrix} 5 & 30 \\ 30 & 220 \end{bmatrix} \begin{bmatrix} b_0 \\ b_1 \end{bmatrix} = \begin{bmatrix} 73.45 \\ 510.14 \end{bmatrix}$$

Solve for b using the inverse of S_x

$$\begin{bmatrix} b_0 \\ b_1 \end{bmatrix} = \begin{bmatrix} 1.10 & -0.15 \\ -0.15 & 0.025 \end{bmatrix} \begin{bmatrix} 73.45 \\ 510.14 \end{bmatrix} = \begin{bmatrix} 4.27 \\ 1.73 \end{bmatrix}$$

9.8   ALTERNATIVE MATRIX APPROACH TO LINEAR REGRESSION

We now show that the regression coefficients can be calculated directly from the data matrices. We know that the linear regression estimate for each observation

$$\widehat{y_i} = b_0 + b_1 x_i \tag{9.52}$$

And that this estimate has a residual error

$$e_i = y_i - \widehat{y_i} \tag{9.53}$$

So that any observation can be written as the estimate plus a residual error

$$y_i = \widehat{y_i} + e_i = b_0 + b_1 x_i + e_i \tag{9.54}$$

Using matrix algebra, we can write equations for all observations in a vector equation

$$\mathbf{y} = \mathbf{x}\mathbf{b} + \mathbf{e} \tag{9.55}$$

where matrix $\mathbf{x}$ is $n \times 2$, b is 2×1, therefore y is nx1

$$\mathbf{x} = \begin{bmatrix} 1 & x_1 \\ 1 & x_2 \\ 1 & x_3 \\ \dots & \dots \\ 1 & x_n \end{bmatrix} \tag{9.56}$$

$$\mathbf{y} = \begin{bmatrix} y_1 \\ y_2 \\ y_3 \\ \dots \\ y_n \end{bmatrix} \tag{9.57}$$

This is easy to see if we substitute in Equation 9.55

$$\begin{bmatrix} y_1 \\ y_2 \\ y_3 \\ \dots \\ y_n \end{bmatrix} = \begin{bmatrix} 1 & x_1 \\ 1 & x_2 \\ 1 & x_3 \\ \dots & \dots \\ 1 & x_n \end{bmatrix} \begin{bmatrix} b_0 \\ b_1 \end{bmatrix} + \begin{bmatrix} e_1 \\ e_2 \\ e_3 \\ \dots \\ e_n \end{bmatrix} = \begin{bmatrix} b_1 x_1 + b_0 + e_1 \\ b_1 x_2 + b_0 + e_2 \\ b_1 x_3 + b_0 + e_3 \\ \dots \\ b_1 x_n + b_0 + e_n \end{bmatrix} \tag{9.58}$$

Solving for the error in Equation 9.55

$$\mathbf{e} = \mathbf{y} - \mathbf{xb} \tag{9.59}$$

Now, we note that the sum of all square errors can be written as the major product matrix $\mathbf{e}$

$$\mathbf{e}^{\mathrm{T}}\mathbf{e} = \begin{bmatrix} e_1 & e_2 & \cdots & e_n \end{bmatrix} \begin{bmatrix} e_1 \\ e_2 \\ \cdots \\ e_n \end{bmatrix} = \sum_{i=1}^{n} e_i^2 \tag{9.60}$$

As we know we need to find $\mathbf{b}$ that minimizes this total square error:

$$\min_{\mathbf{b}}[\mathbf{e}^{\mathrm{T}}\mathbf{e}] = \min_{\mathbf{b}}[(\mathbf{y} - \mathbf{xb})^{\mathrm{T}}(\mathbf{y} - \mathbf{xb})] \tag{9.61}$$

This is achieved by the b obtained by solving the unknown $\mathbf{b}$ from the system of equations $\mathbf{y} = \mathbf{xb}$, where $\mathbf{y}$ and $\mathbf{x}$ are known. To do this premultiply both sides by $\mathbf{x}^{\mathrm{T}}$

$$\mathbf{x}^{\mathrm{T}}\mathbf{y} = \mathbf{x}^{\mathrm{T}}\mathbf{xb} \tag{9.62}$$

Note that $\mathbf{x}^{\mathrm{T}}\mathbf{x}$ is a 2×2 matrix

$$\mathbf{x}^{\mathrm{T}}\mathbf{x} = \begin{bmatrix} 1 & 1 & 1 & \cdots & 1 \\ x_1 & x_2 & x_3 & \cdots & x_n \end{bmatrix} \begin{bmatrix} 1 & x_1 \\ 1 & x_2 \\ 1 & x_3 \\ \cdots & \cdots \\ 1 & x_n \end{bmatrix} = \begin{bmatrix} n & \sum x_i \\ \sum x_i & \sum x_i^2 \end{bmatrix} \tag{9.63}$$

which is the same as matrix $\mathbf{S}_x$ in Equation 9.43. The entries are sum of squares of entries of $\mathbf{x}$ and sum of entries of $\mathbf{x}$. Now

$$\mathbf{x}^{\mathrm{T}}\mathbf{y} = \begin{bmatrix} 1 & 1 & 1 & \cdots & 1 \\ x_1 & x_2 & x_3 & \cdots & x_n \end{bmatrix} \begin{bmatrix} y_1 \\ y_2 \\ y_3 \\ \cdots \\ y_n \end{bmatrix} = \begin{bmatrix} \sum y_i \\ \sum x_i y_i \end{bmatrix} \tag{9.64}$$

which is a 2×1 matrix the same as $\mathbf{S}_y$ in Equation 9.43. The entries are sum of entries of $\mathbf{y}$ and sum of cross-products of entries of $\mathbf{y}$ with entries of $\mathbf{x}$.

Therefore, we can see that $\mathbf{S}_x = \mathbf{x}^{\mathrm{T}}\mathbf{x}$ and $\mathbf{S}_y = \mathbf{x}^{\mathrm{T}}\mathbf{y}$.

Now to find $\mathbf{b}$ we find inverse of $\mathbf{S}_x = \mathbf{x}^{\mathrm{T}}\mathbf{x}$ and premultiply both sides by $(\mathbf{x}^{\mathrm{T}}\mathbf{x})^{-1}$

$$\mathbf{b} = (\mathbf{x}^{\mathrm{T}}\mathbf{x})^{-1}(\mathbf{x}^{\mathrm{T}}\mathbf{y}) = \mathbf{S}_x^{-1}\mathbf{S}_y \tag{9.65}$$

just as we did in Equation 9.45 to obtain the final result given in Equation 9.46. As a numerical example, use the following data $x_i = \{2, 4, 6, 8, 10\}$, $y_i = \{9.00, 9.88, 17.04, 12.46, 25.07\}$,

$$\mathbf{x}^T\mathbf{x} = \begin{bmatrix} 1 & 1 & 1 & 1 & 1 \\ 2 & 4 & 6 & 8 & 10 \end{bmatrix} \begin{bmatrix} 1 & 2 \\ 1 & 4 \\ 1 & 6 \\ 1 & 8 \\ 1 & 10 \end{bmatrix} = \begin{bmatrix} 5 & 30 \\ 30 & 220 \end{bmatrix}$$

$$\mathbf{x}^T\mathbf{y} = \begin{bmatrix} 1 & 1 & 1 & 1 & 1 \\ 2 & 4 & 6 & 8 & 10 \end{bmatrix} \begin{bmatrix} 9.00 \\ 9.88 \\ 17.04 \\ 12.46 \\ 25.07 \end{bmatrix} = \begin{bmatrix} 73.45 \\ 510.14 \end{bmatrix}$$

the resulting matrices are the same as $\mathbf{S}_x$ and $\mathbf{S}_y$ in the previous section.

9.9 EXERCISES

Exercise 9.1
Identify the row and column number of the remaining elements in matrix $\mathbf{A}$ of Equation 9.1

Exercise 9.2
Write a 2×3 matrix as matrix $\mathbf{B}$ of Equation 9.3.

Exercise 9.3
What would be the dimensions of a row vector with n entries? What would be the dimensions of a column vector with n entries?

Exercise 9.4
Write a 2×2 matrix. Determine the elements above the diagonal and below the diagonal.

Exercise 9.5
Suppose $\mathbf{C}$ of Equation 9.7 is a covariance matrix. What are the various variances and covariances represented in the entries of matrix $\mathbf{C}$?

Exercise 9.6
What would the dimensions of the resulting matrix be if a 3×4 matrix is postmultiplied with a 4×5 matrix?

Exercise 9.7
Multiply the following matrices:

$$\begin{bmatrix} 1 & 4 & -7 \\ 2 & -5 & 8 \end{bmatrix} \times \begin{bmatrix} 1 & 2 \\ 3 & 4 \\ 5 & 6 \end{bmatrix}$$

Exercise 9.8

Find the determinant for the 2×2 identity matrix. Based upon this calculation what do you think the determinant of the 5×5 identity matrix would be?

Exercise 9.9

Find the transpose of square symmetric matrix $\mathbf{C}$ in Equation 9.7. Do you think the transpose of a square symmetric matrix is always the same matrix?

Exercise 9.10

Multiply the matrix and its transpose given in Equation 9.18. What is the dimension of the product? Is it symmetric?

Exercise 9.11

Consider the following matrices:

$$\mathbf{A} = \begin{bmatrix} 5 & 0 \\ 0 & 5 \end{bmatrix} \qquad \mathbf{B} = \begin{bmatrix} 4 & 1 \\ 3 & 0 \\ 2 & -1 \end{bmatrix}$$

1. Can you add these? If yes, find $\mathbf{A} + \mathbf{B}$, if not explain why.
2. Can you find $\mathbf{AB}$? $\mathbf{BA}$? If yes, complete the operation. If not, why not?
3. What is the transpose of $\mathbf{B}$?
4. Find the det $(\mathbf{B}^T\mathbf{B})$.

Exercise 9.12

Suppose we have six values of X, xi = 2, 1, 0, 0, 1, 2. Calculate $\mathbf{Sx} = \mathbf{x}^T\mathbf{x}$.

Exercise 9.13

Suppose we also have yi = 3, 2, 1, 1, 2, 5. Determine matrix $\mathbf{Sy} = \mathbf{x}^T\mathbf{y}$ Use this and $\mathbf{Sx}$ from exercise 9.12 to write a matrix equation where vector $\mathbf{b}$ of regression coefficients is the unknown. Solve this matrix equation.

9.10   COMPUTER SESSION: MATRICES AND LINEAR ALGEBRA

9.10.1   CREATING MATRICES

We use the R base matrix library. We already know how to create a matrix from scanning a file. We can also create a matrix using a sequence of numbers. For example,

```
> A <- matrix(1:12, nrow=3, ncol=4)
> A
     [,1]  [,2]  [,3]  [,4]
[1,]   1     4     7    10
[2,]   2     5     8    11
[3,]   3     6     9    12
```

Note that A is a rectangular matrix with 3 rows and 4 columns. We can obtain the dimension of matrices with `dim` function. For example,

```
> dim(A)
[1] 3 4
```

As another example

```
> B <- matrix(1:9, nrow=3, ncol=3)
> B
     [,1] [,2] [,3]
[1,]    1    4    7
[2,]    2    5    8
[3,]    3    6    9
```

Note that B is a square matrix. Confirm with dim()

```
> dim(B)
[1] 3 3
```

Another way of generating a matrix is by means of the structure function and directly declaring the dimension

```
> structure(1:9, dim=c(3,3))
     [,1] [,2] [,3]
[1,]    1    4    7
[2,]    2    5    8
[3,]    3    6    9
```

which yields the same result as before.

If we want a matrix with random numbers from the uniform distribution, we could do

```
> D <- structure(runif(9),dim=c(3,3))
> round(D,2)
     [,1] [,2] [,3]
[1,] 0.12 0.45 0.60
[2,] 0.36 0.28 0.29
[3,] 0.48 0.85 0.19
```

9.10.2 OPERATIONS

Operators + and * represent an entry-by-entry sum and multiplication (Hadamard product) and the dimensions must match. Therefore, we can perform the following

```
> A + A
     [,1] [,2] [,3] [,4]
[1,]    2    8   14   20
[2,]    4   10   16   22
[3,]    6   12   18   24

> B+B
     [,1] [,2] [,3]
[1,]    2    8   14
[2,]    4   10   16
[3,]    6   12   18
```

But we cannot use + and * on A and B because their dimensions do not match. Try to see

```
> A+B
Error in A + B : non-conformable arrays
> A*B
Error in A * B : non-conformable arrays
>
```

Operators *, /, +, - can also be used with scalars; for example,

```
> C <- 3*B
> C
     [,1]   [,2]   [,3]
[1,]    3     12     21
[2,]    6     15     24
[3,]    9     18     27

> C/3
     [,1]   [,2]   [,3]
[1,]    1      4      7
[2,]    2      5      8
[3,]    3      6      9
```

To perform matrix multiplication, we use operator %*% (which is not the same as *). Matrices need to conform for multiplication. Operator %*% is not commutative. For example, we can do BA, that is premultiply A by B because $(3 \times 3) \times (3 \times 4) = (3 \times 4)$

```
> B%*%A
     [,1]   [,2]    [,3]    [,4]
[1,]   30     66     102     138
[2,]   36     81     126     171
[3,]   42     96     150     204
```

However, we cannot do AB, that is premultiply B by A, because they do not conform; that is, $(3 \times 4) \times (3 \times 3)$ do not match. Thus, if we try to multiply we get the following error message:

```
> A%*%B
Error in A %*% B : non-conformable arguments
>
```

As another example multiply 3**M**, where **M** is the following matrix $\mathbf{M} = \begin{pmatrix} 1 & 4 \\ 2 & 5 \\ 3 & 6 \end{pmatrix}$

```
> M <- structure(1:6, dim=c(3,2))
> M
     [,1]    [,2]
[1,]    1       4
[2,]    2       5
[3,]    3       6
> 3*M
     [,1]    [,2]
[1,]    3      12
[2,]    6      15
[3,]    9      18>
```

Calculate $\mathbf{A}^3$ where matrix **A** is $\mathbf{A} = \begin{pmatrix} 17 & -6 \\ 45 & -16 \end{pmatrix}$. First, generate the matrix

```
> A <- matrix(c(17,-6,45,-16), byrow=T, ncol=2)
> A
     [,1]    [,2]
[1,]   17      -6
[2,]   45     -16
>
```

Then multiply three times

```
> A%*%A%*%A
     [,1]    [,2]
[1,]   53     -18
[2,]  135     -46
>
```

By the way, the power operation ^ does **not** yield the correct result, as you can check trying

```
> A^3
      [,1]     [,2]
[1,]  4913     -216
[2,] 91125    -4096
```

What this operation does is simply raise each entry to a power, for example, 17^3 = 4913, and that is not the result of matrix power operation.

9.10.3 OTHER OPERATIONS

To **transpose** a matrix use function t

```
> t(C)
        [,1]   [,2]   [,3]
[1,]       3      6      9
[2,]      12     15     18
[3,]      21     24     27
```

To extract **diagonal** and the **trace**

```
> diag(C)
[1]     3        15       27
> sum(diag(C))
[1]     45
```

To construct an **identity** matrix

```
> diag(4)
        [,1]   [,2]   [,3]   [,4]
[1,]       1      0      0      0
[2,]       0      1      0      0
[3,]       0      0      1      0
[4,]       0      0      0      1
```

To calculate the **determinant** of a matrix use det. The matrix must be square. We cannot calculate determinant of rectangular matrix. For example, if we try

```
> A
        [,1]   [,2]   [,3]   [,4]
[1,]       1      4      7     10
[2,]       2      5      8     11
[3,]       3      6      9     12
> det(A)
Error in det(A)  : x must be a square matrix
```

But we can calculate det of matrix B

```
> B
        [,1]   [,2]   [,3]
[1,]       1      4      7
[2,]       2      5      8
[3,]       3      6      9
> det(B)
[1] 0
```

Note that **B** is singular because det(**B**) is 0. This is because the columns depend on each other. We can also calculate the determinant of an identity matrix, and it should be 1

```
> det(diag(3))
[1] 1
```

Also, the `det` of matrix **D** above

```
> det(D)
[1] -0.01062473
```

9.10.4   SOLVING SYSTEM OF LINEAR EQUATIONS

Recall that a system of linear equations $\mathbf{Bx} = \mathbf{c}$ is solved by premultiplying by the inverse $\mathbf{B}^{-1}\mathbf{Bx} = \mathbf{B}^{-1}\mathbf{c}$ to obtain the solution $\mathbf{x} = \mathbf{B}^{-1}\mathbf{c}$. We can use function `solve`. For example,

```
> D <- matrix(c(19,2,15,8,18,19,11,17,10), nrow=3, ncol=3)
> D
      [,1]  [,2]  [,3]
[1,]    19     8    11
[2,]     2    18    17
[3,]    15    19    10
> c <- c(9,5,14)
> x <- solve(D, c)
> round(x,2)
[1]  0.45  0.58 -0.38
>
```

However, try

```
> c <- c(1,2,3)
> x <- solve(B,c)
Error in solve.default(B, c) :
  Lapack routine dgesv: system is exactly singular
>
```

This cannot be solved because **B** is singular, therefore det = 0 and it has no inverse.

9.10.5   INVERSE

To calculate the inverse of a matrix **B** also use the function `solve` but make **c** the identity matrix of the same dimension as **B**. This works because

$$\mathbf{B}^{-1}\mathbf{Bx} = \mathbf{B}^{-1}\mathbf{I}$$

$$\mathbf{x} = \mathbf{B}^{-1}$$

So, say if we use **D** above, we build an identity matrix and then apply `solve`

```
> I <- diag(3)
> I
        [,1]   [,2]   [,3]
[1,]     1      0      0
[2,]     0      1      0
[3,]     0      0      1
D.inv <- solve(D,I)
round(D.inv,2)
         [,1]    [,2]    [,3]
[1,]     0.04   -0.04    0.02
[2,]    -0.07   -0.01    0.09
[3,]     0.07    0.07   -0.10
>
```

9.10.6 COMPUTER EXERCISES

Exercise 9.14
Calculate the major product matrix $\mathbf{x}^T\mathbf{x}$ for 10 values of x drawn from a standard normal RV. Hint: use random number generation.

Exercise 9.15
Calculate the determinant of the major product matrix of the previous exercise. Calculate the inverse.

Exercise 9.16
Use the following matrices $\mathbf{A} = \begin{bmatrix} 1 & 3 \\ 2 & 4 \end{bmatrix}$, $\mathbf{B} = \begin{bmatrix} 5 & 0 \\ 0 & 3 \end{bmatrix}$ and the following vectors.
$\mathbf{c} = \begin{bmatrix} 2 \\ 0 \end{bmatrix}$ $\mathbf{x} = \begin{bmatrix} x_1 \\ x_2 \end{bmatrix}$ Calculate **AI, AB, BA, Bc, Ic** where **I** is the identity matrix. Write the equation **Bx = c**. Solve for **x**.

SUPPLEMENTARY READING

Davis, 2002, Chapter 3, pp. 123–140; Carr, 1995, Chapter 3, pp. 44–59, especially pp. 47–49, Chapter 4, pp. 77–79. MathSoft Guide, Chapter 7, pp. 7-9–7-24 and 7-29–7-38, Chapter 25, pp. 25-1–25-12, pp. 25-36–25-57.

10 Multivariate Models

10.1 MULTIPLE LINEAR REGRESSION

We will extend the simple linear regression model to more than one independent variable X. That is, we now have several (m) independent variables X_i, influencing one dependent or response variable Y.

As defined earlier, we develop a linear least squares (LLS) estimator of Y

$$\widehat{Y} = b_0 + b_1 X_1 + b_2 X_2 + \cdots + b_m X_m \tag{10.1}$$

This is the equation of a plane with **intercept** b_0 and **coefficients** b_k, $k = 1, \ldots, m$. Note that we now have $m + 1$ regression parameters to estimate; these are m coefficients and one intercept. For each observation i we have a set of data points y_i, x_{ki}, we have the **estimated** value of Y at the specific points x_{ki}

$$\widehat{y_i} = b_0 + b_1 x_{1i} + b_2 x_{2i} + \cdots + b_m x_{mi} \tag{10.2}$$

This is an extension of the regression problem stated in a previous chapter. We need to find the values of the coefficients b_k, $k = 0, 1, \ldots, m$ that minimize the square error. We define a column vector $\mathbf{b} = [b_0\ b_1\ b_2\ \ldots\ b_m]^{\mathrm{T}}$. Then find $\mathbf{b}$ such that

$$\min_{\mathbf{b}} q = \min_{\mathbf{b}} \sum_{i=1}^{n} e_i^2 = \min_{\mathbf{b}} \sum_{i=1}^{n} (y_i - \widehat{y_i})^2 \tag{10.3}$$

A starting visual point in multiple regression is to obtain pairwise scatter plots of all variables involved. This allows one to explore potential relationships among the variables. For example, Figure 10.1 shows scatter plots of ozone and several meteorological variables (wind, solar radiation, temperature). We observe potential increase of ozone with temperature and solar radiation and decrease with wind. At the same time, there is an indication of relationship between wind and temperature.

10.1.1 MATRIX APPROACH

We can proceed as in simple linear regression; first, find derivatives of q with respect to b_0 and each one of the coefficients b_k. Then set these derivatives equal to zero, so that q is at a minimum, and find equations to solve for $\mathbf{b}$. Matrix algebra will help us now; we can find a matrix equation to solve for $\mathbf{b}$ as an extension of the linear algebra process developed in the previous chapter.

We can build a vector $\mathbf{y}$ as we did in the last chapter. This is $n \times 1$ (a column vector)

$$\mathbf{y} = \begin{bmatrix} y_1 \\ y_2 \\ y_3 \\ \ldots \\ y_n \end{bmatrix} \tag{10.4}$$

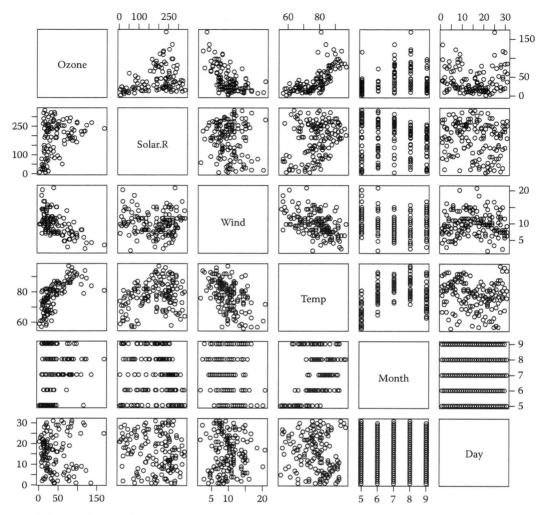

FIGURE 10.1   Scatter plot pairs.

and a matrix **x** which is rectangular $n \times (m + 1)$, n rows for observations and $m + 1$ columns for the intercept and m variables

$$\mathbf{x} = \begin{bmatrix} 1 & x_{11} & x_{21} & \dots & x_{m1} \\ 1 & x_{12} & x_{22} & \dots & x_{m2} \\ \dots & \dots & \dots & \dots & \dots \\ 1 & x_{1n} & x_{2n} & \dots & x_{mn} \end{bmatrix} \qquad (10.5)$$

Now, the unknown coefficient vector **b** is a column vector $(m + 1) \times 1$ with entries $b_0, b_1, \dots, b_m$

$$\mathbf{b} = \begin{bmatrix} b_0 \\ b_1 \\ \dots \\ b_m \end{bmatrix} \qquad (10.6)$$

as before we need to solve **b** from

$$y = xb \tag{10.7}$$

Premultiplying both sides by $\mathbf{x}^T$

$$\mathbf{x}^T\mathbf{y} = \mathbf{x}^T\mathbf{x}\mathbf{b} \tag{10.8}$$

Solve for **b** by premultiplying both sides by the inverse of $\mathbf{x}^T\mathbf{x}$

$$\mathbf{b} = (\mathbf{x}^T\mathbf{x})^{-1}(\mathbf{x}^T\mathbf{y}) \tag{10.9}$$

Now let us write $\mathbf{x}^T\mathbf{x}$ in terms of entries of **x**

$$\mathbf{x}^T\mathbf{x} = \begin{bmatrix} 1 & 1 & \cdots & 1 \\ x_{11} & x_{12} & \cdots & x_{1n} \\ x_{21} & x_{22} & \cdots & x_{2n} \\ \vdots & \vdots & \vdots & \vdots \\ x_{m1} & x_{m2} & \cdots & x_{mn} \end{bmatrix} \begin{bmatrix} 1 & x_{11} & x_{21} & \cdots & x_{m1} \\ 1 & x_{12} & x_{22} & \cdots & x_{m2} \\ \cdots & \cdots & \cdots & \cdots & \cdots \\ 1 & x_{1n} & x_{2n} & \cdots & x_{mn} \end{bmatrix} \tag{10.10}$$

After multiplication we get

$$\mathbf{x}^T\mathbf{x} = \begin{bmatrix} n & \sum x_{1i} & \sum x_{2i} & \cdots & \sum x_{mi} \\ \sum x_{1i} & \sum x_{1i}^2 & \sum x_{1i}x_{2i} & \cdots & \sum x_{1i}x_{mi} \\ \sum x_{2i} & \sum x_{2i}x_{1i} & \sum x_{2i}^2 & \cdots & \sum x_{2i}x_{mi} \\ \vdots & \vdots & \vdots & \ddots & \vdots \\ \sum x_{mi} & \sum x_{mi}x_{1i} & \sum x_{mi}x_{2i} & \cdots & \sum x_{mi}^2 \end{bmatrix} \tag{10.11}$$

Note that $\mathbf{x}^T\mathbf{x}$ is a matrix of dimension $m \times m$. Therefore, Equation 10.8 can be expanded

$$\begin{bmatrix} n & \sum x_{1i} & \sum x_{2i} & \cdots & \sum x_{mi} \\ \sum x_{1i} & \sum x_{1i}^2 & \sum x_{1i}x_{2i} & \cdots & \sum x_{1i}x_{mi} \\ \sum x_{2i} & \sum x_{2i}x_{1i} & \sum x_{2i}^2 & \cdots & \sum x_{2i}x_{mi} \\ \vdots & \vdots & \vdots & \ddots & \vdots \\ \sum x_{mi} & \sum x_{mi}x_{1i} & \sum x_{mi}x_{2i} & \cdots & \sum x_{mi}^2 \end{bmatrix} \begin{bmatrix} b_0 \\ b_1 \\ b_2 \\ \vdots \\ b_m \end{bmatrix} = \begin{bmatrix} \sum y_i \\ \sum x_{1i}y_i \\ \sum x_{2i}y_i \\ \vdots \\ \sum x_{mi}y_i \end{bmatrix} \tag{10.12}$$

This is a well-known equation (for example, see page 463 in Davis, 2002). For brevity, in all of the aforementioned equations we can use the notation $\mathbf{S_x} = \mathbf{x}^T\mathbf{x}$ and $\mathbf{S_y} = \mathbf{x}^T\mathbf{y}$ as in the previous chapter. Thus

$$\mathbf{S_y} = \mathbf{S_x}\mathbf{b} \tag{10.13}$$

$\mathbf{S_x}$ entries are the sum of squares and cross products of entries of $\mathbf{x}$, whereas $\mathbf{S_y}$ entries are sum of squares and cross products of entries of y with entries of x. So now, we solve the equation premultiplying by the inverse of $\mathbf{S_x}$ to obtain

$$\mathbf{S_x^{-1} S_y = S_x^{-1} S_x b} \tag{10.14}$$

and therefore

$$\mathbf{b = S_x^{-1} S_y} \tag{10.15}$$

This is also a well-known equation (see Davis, 2002, page 462).

To gain some insight into the solution, let us develop it for $m = 2$, so that it is easier to find the inverse

$$\begin{bmatrix} n & \sum x_{1i} & \sum x_{2i} \\ \sum x_{1i} & \sum x_{1i}^2 & \sum x_{1i}x_{2i} \\ \sum x_{2i} & \sum x_{2i}x_{1i} & \sum x_{2i}^2 \end{bmatrix} \begin{bmatrix} b_0 \\ b_1 \\ b_2 \end{bmatrix} = \begin{bmatrix} \sum y_i \\ \sum x_{1i}y_i \\ \sum x_{2i}y_i \end{bmatrix} \tag{10.16}$$

Divide both sides by n and use average or sample mean notation (i.e., a bar on top of the variable)

$$\begin{bmatrix} 1 & \overline{X}_1 & \overline{X}_2 \\ \overline{X}_1 & \overline{X_1^2} & \overline{X_1 X_2} \\ \overline{X}_2 & \overline{X_1 X_2} & \overline{X_2^2} \end{bmatrix} \begin{bmatrix} b_0 \\ b_1 \\ b_2 \end{bmatrix} = \begin{bmatrix} \overline{Y} \\ \overline{X_1 Y} \\ \overline{X_2 Y} \end{bmatrix} \tag{10.17}$$

After some algebraic work, we can find that the determinant of $\mathbf{Sx}$ is

$$\left| \mathbf{Sx} \right| = (s_{X_1})^2 (s_{X_2})^2 - (s_{\mathrm{cov}(X_1,X_2)})^2 \tag{10.18}$$

Note that in the special case of perfect correlation between X_1 and X_2, say $X_2 = X_1$, this expression reduces to

$$\left| \mathbf{Sx} \right| = (s_{X_1})^2 (s_{X_1})^2 - \left((s_{X_1})^2 \right)^2 = 0 \tag{10.19}$$

which means that we cannot find the inverse and there will be no solution. So, let us assume that X_1 and X_2 are not perfectly correlated. After performing the inverse and multiplication operations, we find

$$\begin{bmatrix} b_0 \\ b_1 \\ b_2 \end{bmatrix} = \begin{bmatrix} \overline{Y} - b_1 \overline{X}_1 - b_2 \overline{X}_2 \\ \dfrac{(s_{X_2})^2 s_{\mathrm{cov}(X_1,Y)} - s_{\mathrm{cov}(X_2,Y)} s_{\mathrm{cov}(X_1,X_2)}}{(s_{X_1})^2 (s_{X_2})^2 - (s_{\mathrm{cov}(X_1,X_2)})^2} \\ \dfrac{(s_{X_1})^2 s_{\mathrm{cov}(X_2,Y)} - s_{\mathrm{cov}(X_1,Y)} s_{\mathrm{cov}(X_1,X_2)}}{(s_{X_1})^2 (s_{X_2})^2 - (s_{\mathrm{cov}(X_1,X_2)})^2} \end{bmatrix} \tag{10.20}$$

Note that the covariance between X_1 and X_2, plays an important role here. If this covariance where to be zero, that is, if X_1 and X_2, are uncorrelated, then the solution simplifies to

$$
\begin{bmatrix} b_0 \\ b_1 \\ b_2 \end{bmatrix} = \begin{bmatrix} \bar{Y} - b_1\bar{X}_1 - b_2\bar{X}_2 \\ \dfrac{s_{\text{cov}(X_1,Y)}}{s_{X_1}^2} \\ \dfrac{s_{\text{cov}(X_2,Y)}}{s_{X_2}^2} \end{bmatrix}
\tag{10.21}
$$

Coefficients b_1, b_2 are **partial** or **marginal** coefficients, i.e., the rate of change of Y with one of the X while holding all of the other Xs constant. We can see that in the special case of uncorrelated X_1 and X_2, this marginal change of Y with X_1 or X_2 depends only on the covariance of X_1 or X_2 and Y and the variance of X_1 or X_2. However, when X_1 and X_2 are correlated, then the marginal coefficient for one variable is affected by (1) the variance of the other variable, (2) the covariance of the other variable with Y, and (3) the covariance of X_1 and X_2.

Recall that by using the definition of correlation coefficient

$$
r_{(X,Y)} = \frac{s_{\text{cov}(X,Y)}}{s_X s_Y}
\tag{10.22}
$$

we can substitute in Equation 10.29 to obtain

$$
\begin{bmatrix} b_0 \\ b_1 \\ b_2 \end{bmatrix} = \begin{bmatrix} \bar{Y} - b_1\bar{X}_1 - b_2\bar{X}_2 \\ \dfrac{r_{(X_1,Y)}s_{X_1}s_Y}{s_{X_1}^2} \\ \dfrac{r_{(X_2,Y)}s_{X_2}s_Y}{s_{X_2}^2} \end{bmatrix} = \begin{bmatrix} \bar{Y} - b_1\bar{X}_1 - b_2\bar{X}_2 \\ \dfrac{r_{(X_1,Y)}s_Y}{s_{X_1}} \\ \dfrac{r_{(X_2,Y)}s_Y}{s_{X_2}} \end{bmatrix}
\tag{10.23}
$$

In this case, the product of a correlation coefficient and a ratio of standard deviations give the marginal or partial coefficients b_1, b_2. The correlation coefficient is that between the corresponding independent variable X_i and the dependent variable Y. The ratio is that of the standard deviation of the dependent variable Y to the standard deviation of independent variable X_i.

In general, for m independent variables we get the solution in the special case as

$$
\begin{bmatrix} b_0 \\ b_1 \\ b_2 \\ \vdots \\ b_m \end{bmatrix} = \begin{bmatrix} \bar{Y} - b_1\bar{X}_1 - b_2\bar{X}_2 - \cdots - b_m\bar{X}_m \\ \dfrac{s_{\text{cov}(X_1,Y)}}{s_{X_1}^2} \\ \dfrac{s_{\text{cov}(X_2,Y)}}{s_{X_2}^2} \\ \vdots \\ \dfrac{s_{\text{cov}(X_m,Y)}}{s_{X_m}^2} \end{bmatrix}
\tag{10.24}
$$

The strength of the linear relationship between independent variables X_1 and X_2 is **collinearity**. More specifically, X_1 and X_2 are perfect collinear if all pairs of observations are linearly related

$$a_0 + a_1 X_{1i} + a_2 X_{2i} = 0 \tag{10.25}$$

Then, high correlation between X_1 and X_2 implies strong collinearity. When we have several independent variables, we can have **multicollinearity** among these variables. More specifically, there is perfect multicollinearity if there is a linear relationship among some of the independent variables

$$a_0 + a_1 X_{1i} + a_2 X_{2i} + \cdots + a_k X_{ki} = 0 \tag{10.26}$$

Thus, high correlation among variables would imply strong multicollinearity.

10.1.2 POPULATION CONCEPTS AND EXPECTED VALUES

We can derive these equations using a population approach. Equation 10.17 becomes

$$\begin{bmatrix} 1 & \mu_{X_1} & \mu_{X_2} \\ \mu_{X_1} & E(X_1^2) & \mathrm{cov}(X_1, X_2) \\ \mu_{X_2} & \mathrm{cov}(X_1, X_2) & E(X_1^2) \end{bmatrix} \begin{bmatrix} b_0 \\ b_1 \\ b_2 \end{bmatrix} = \begin{bmatrix} \mu_Y \\ \mathrm{cov}(X_1, Y) \\ \mathrm{cov}(X_2, Y) \end{bmatrix} \tag{10.27}$$

Equation 10.20 becomes

$$\begin{bmatrix} b_0 \\ b_1 \\ b_2 \end{bmatrix} = \begin{bmatrix} \mu_Y - b_1 \mu_{X_1} - b_2 \mu_{X_2} \\ \dfrac{\sigma_{X_2}^2 \, \mathrm{cov}(X_1, Y) - \mathrm{cov}(X_2, Y)\mathrm{cov}(X_1, X_2)}{\sigma_{X_1}^2 \sigma_{X_2}^2 - (\mathrm{cov}(X_1, X_2))^2} \\ \dfrac{\sigma_{X_1}^2 \, \mathrm{cov}(X_2, Y) - \mathrm{cov}(X_1, Y)\mathrm{cov}(X_1, X_2)}{\sigma_{X_1}^2 \sigma_{X_2}^2 - (\mathrm{cov}(X_1, X_2))^2} \end{bmatrix} \tag{10.28}$$

and Equation 10.21 is now

$$\begin{bmatrix} b_0 \\ b_1 \\ b_2 \end{bmatrix} = \begin{bmatrix} \mu_Y - b_1 \mu_{X_1} - b_2 \mu_{X_2} \\ \dfrac{\mathrm{cov}(X_1, Y)}{\sigma_{X_1}^2} \\ \dfrac{\mathrm{cov}(X_2, Y)}{\sigma_{X_2}^2} \end{bmatrix} \tag{10.29}$$

As given earlier, recall that by using the definition of correlation coefficient

$$\rho(X, Y) = \frac{\mathrm{cov}(X, Y)}{\sigma_X \sigma_Y} \tag{10.30}$$

we can substitute in Equation 10.29 to obtain

$$\begin{bmatrix} b_0 \\ b_1 \\ b_2 \end{bmatrix} = \begin{bmatrix} \mu_Y - b_1\mu_{X_1} - b_2\mu_{X_2} \\ \dfrac{\rho(X_1,Y)\sigma_{X_1}\sigma_Y}{\sigma_{X_1}^2} \\ \dfrac{\rho(X_2,Y)\sigma_{X_2}\sigma_Y}{\sigma_{X_2}^2} \end{bmatrix} = \begin{bmatrix} \mu_Y - b_1\mu_{X_1} - b_2\mu_{X_2} \\ \dfrac{\rho(X_1,Y)\sigma_Y}{\sigma_{X_1}} \\ \dfrac{\rho(X_2,Y)\sigma_Y}{\sigma_{X_2}} \end{bmatrix} \qquad (10.31)$$

In this case, the product of the correlation coefficient and the ratio of standard deviations give the marginal or partial coefficients b_i. The correlation coefficient is that between the corresponding independent variable X_i and the dependent variable Y. The ratio is of the standard deviation of the dependent variable to the standard deviation of independent variable X_i.

10.1.3 EVALUATION AND DIAGNOSTICS

In a similar fashion to simple regression (see Chapter 6), we use ANOVA and t-tests for each coefficient. For diagnostic of residuals and outliers, we use plots of residuals vs. fitted, Q–Q plots of residuals, and the residuals vs. leverage plot. As an example, Figure 10.2 illustrates these diagnostic plots when we perform a multiple regression of ozone vs. temperature, wind, and solar radiation. The interpretation is as in simple regression. Here we see that observations 117, 62, 30 are identified as potential outliers by most plots, and that additionally observations 9 and 48 are detected as

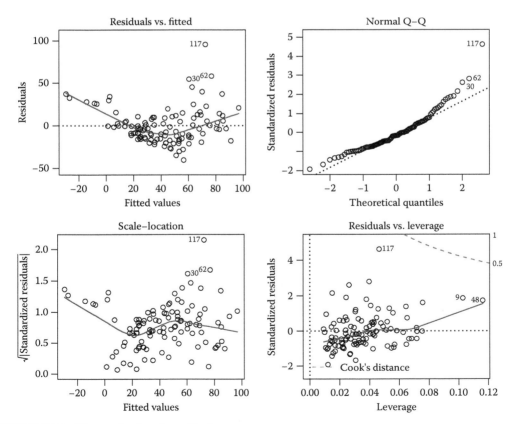

FIGURE 10.2 Diagnostic plots for residuals.

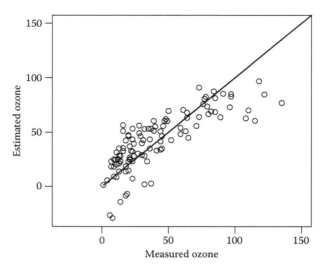

FIGURE 10.3 Ozone measure vs. fitted.

outliers by the residual vs. leverage plot. Recall from Chapter 6, that high-leverage occurs when its value exceeds $2 \times m/n$ or twice the number of coefficients (m) divided by number of observations (n), and those values of Cook's distance larger than 1 are considered having large influence. In this example, high-leverage would occur for $2 \times 4/153 = 0.052$.

A useful plot is that of ozone estimated by the regression model vs. the ozone observed in the data, together with the hypothetical line where both values would coincide, or the line with 1:1 slope. An example is in Figure 10.3 for the ozone example already described.

However, there is also the need to check for multicollinearity among independent variables. As we concluded in the previous section, correlation among the independent variables X_i makes the marginal coefficient depend on correlations among these variables and the response Y. We also saw that in extreme cases of perfect correlation there is no solution. Correlation among independent variables can make a coefficient more important than what it really is.

A practical metric of collinearity is **tolerance**, defined as the amount of variance in an independent variable not explained by the other variables. Its inverse is the **variance inflation factor** (VIF). To calculate this metric for the variable X_j use

$$Tol = 1 - R_j^2 \tag{10.32}$$

where R_j^2 is the R^2 of the regression of variable X_j on all other independent variables. A high value of R_j^2 implies that variable X_j is well explained by the others, and then the tolerance will be low, and because VIF is the reciprocal $VIF = 1/Tol$ we will have high VIF. Therefore, low tolerance or high VIF imply problems with multicollinearity. As a rule of thumb $Tol < 0.2$ or $VIF > 5$ indicates potential problems (Rogerson, 2001). A less strict rule is $Tol < 0.1$ or $VIF > 10$. Although these thresholds are arbitrary, they provide practical guidance.

One possible approach to remedy multicollinearity problems is to remove some variables from the analysis taking into account our knowledge of the underlying processes, and the VIF of the various variables. Actually, this is only one aspect of variable selection, which we consider, with more details in the next section.

10.1.4 VARIABLE SELECTION

Key issues to consider when applying multiple linear regression is how many variables and which X_i to use (Rogerson, 2001). There are several ways of proceeding. (1) **Backward** selection: start by including all variables at once, and then **drop** variables in sequence without significant reduction of R^2. (2) **Forward** selection: we start with the X_i most likely to affect the Y variable, and then **add** independent variables. (3) **Stepwise** selection: Drop and add variables as in forward and backward selection; as we add variables we check to see if we can drop a variable added before. This process can be automated using metrics that describe how good the current selection is. The **Mallows' Cp** statistic or the **Akaike Information Criterion** (AIC) is used to decide whether an X can be dropped or added and as guide to stop the trial and error process.

Mallows' Cp is calculated for a subset of p variables of all m independent variables in the following manner,

$$Cp(r) = \frac{SS_r}{SS_m} + 2r - n = \frac{\sum_{i=1}^{n} \left(y_i - y_i(r) \right)^2}{\sum_{i=1}^{n} \left(y_i - y_i(m) \right)^2} + 2r - n \tag{10.33}$$

Here SS_r and SS_m correspond to the residual mean square error obtained when the regression is calculated for r and m independent variables, respectively. Those mean square errors use $y_i(r)$ and $y_i(m)$, which are the fitted dependent variable for the ith observation. Recognize that if r were to be equal to m, Cp takes the value $1 + 2m - n$. In addition, the ratio of sum of square errors would tend to be larger than 1. Cp is used to select the set of variables by picking the set that would make Cp less than $2r$ when the sets are ordered according to increasing values of r.

Akaike's information criterion (AIC) is based on two concepts we have not explained yet. One of these is the **likelihood function** and the other is **information content**. So, first we will briefly explain these concepts.

The likelihood function applies to discrete and continuous random variables. For example, for a discrete random variable X, the likelihood function $L(\theta)$ of a parameter θ of its PMF is a continuous function formed by the products of the PMF evaluated at the observed values

$$L(\theta) = \prod_{i=1}^{n} p(\theta, x_i) \tag{10.34}$$

The symbol Π denotes product in a similar way in which $\sum$ denotes sum. Note that because PMF is evaluated at the observed values, then L is only a function of θ. A maximum likelihood estimate (MLE) is a value $\hat{\theta}$ of θ that maximizes $L(\theta)$. A MLE is obtained by calculus taking the derivative $\partial L/\partial\theta$, making it equal to zero, and finding the point where the optimum occurs as we explained in Chapter 6, or by numerical optimization methods. This concept is expanded for more than one parameter.

Information theory is founded on Shannon's **information** content or entropy of a discrete random variable X with PMF $p(x)$.

$$H(X) = -\sum_{j} p(x_j) \ln\left(p(x_j) \right) \tag{10.35}$$

where the summation is over all possible values of the RV. Do not confuse with observations x_i of X. Actually, the logarithm can be employed with any base; for example, when using base 2 we obtain binary information. Shannon's information content is a measure of uncertainty; it increases as PMF becomes more even or uniform.

Armed with these concepts, we are ready to state the definition of AIC as

$$AIC = 2r - 2\ln(L) \tag{10.36}$$

where
   r is the number of variables
   L is the maximized likelihood function of the regression model for these r variables

This expression comprises two parts: a positive cost $2r$ of increasing the number of variables and a negative benefit $2\ln(L)$ derived from the goodness of fit. Then we select the set of variables such that AIC or the balance of two terms should be as low as possible. The AIC gives a measure of information lost when using a regression model vs. another.

Cp and AIC can be used in an iterative or stepwise process. For example, to augment a model by one variable, we add the X_i for which AIC is lowest among all alternatives. To reduce a model drop the X_i for which AIC is lower than current. At each step, we run multiple regression. This stepwise regression procedure is automated by programming. It usually commences by augmentation from zero-order regression (intercept only).

When the number of observations is low or the number of variables high, we can correct AIC by adding a factor $2r(r + 1)/(n - r - 1)$ to penalize with an extra cost the use of too many variables when n is low.

10.2   MULTIVARIATE REGRESSION

We will extend the multiple linear regression models to more than one dependent variable Y. That is, now in addition to several (m) independent variables X_i, we have several (k) dependent or response variables Y_j.

We can build a $n \times k$ matrix $\mathbf{y}$ (that is one column vector per dependent variable)

$$\mathbf{y} = \begin{bmatrix} y_{11} & y_{12} & \cdots & y_{1k} \\ y_{21} & y_{22} & \cdots & y_{2k} \\ \cdots & \cdots & \cdots & \cdots \\ y_{n1} & y_{n2} & \cdots & y_{nk} \end{bmatrix} \tag{10.37}$$

and a rectangular $n \times (m + 1)$ matrix $\mathbf{x}$, that is n rows for observations and $m + 1$ columns for the intercept and m variables

$$\mathbf{x} = \begin{bmatrix} 1 & x_{11} & x_{21} & \cdots & x_{m1} \\ 1 & x_{12} & x_{22} & \cdots & x_{m2} \\ \cdots & \cdots & \cdots & \cdots & \cdots \\ 1 & x_{1n} & x_{2n} & \cdots & x_{mn} \end{bmatrix} \tag{10.38}$$

Now $\mathbf{b}$, the unknown coefficient vector, is a matrix of dimension $(m + 1) \times k$

$$\mathbf{b} = \begin{bmatrix} b_{01} & b_{02} & \cdots & b_{0k} \\ b_{11} & b_{12} & \cdots & b_{1k} \\ \cdots & \cdots & \cdots & \cdots \\ b_{m1} & b_{m2} & \cdots & b_{mk} \end{bmatrix} \tag{10.39}$$

as before we need to solve $\mathbf{b}$ from

$$\mathbf{y} = \mathbf{xb} \tag{10.40}$$

As before, we can do this by premultiplying both sides by $\mathbf{x^T}$

$$\mathbf{x^T y} = \mathbf{x^T xb} \tag{10.41}$$

and premultiplying by the inverse of $\mathbf{x}^T\mathbf{x}$

$$\mathbf{b} = (\mathbf{x}^T\mathbf{x})^{-1}(\mathbf{x}^T\mathbf{y}) = \mathbf{S}_x^{-1}\,\mathbf{S}_y \qquad (10.42)$$

For example if you had two dependent variables Y_1 and Y_2 and three independent variables X_1, X_2, X_3, the matrix $\mathbf{x}$ is $n \times 4$, the $\mathbf{y}$ matrix is $n \times 2$, and the predictor model will have a coefficient matrix $\mathbf{b}$ of dimension 4×2. Suppose $n = 6$ observations

$$\mathbf{y} = \begin{bmatrix} 0 & 3 \\ 2 & 8 \\ 1 & 9 \\ 1 & 8 \\ 2 & 7 \\ 0 & 6 \end{bmatrix} \quad \mathbf{x} = \begin{bmatrix} 1 & 3 & 2 & 6 \\ 1 & 4 & 2 & 7 \\ 1 & 5 & 4 & 9 \\ 1 & 2 & 1 & 8 \\ 1 & 3 & 2 & 3 \\ 1 & 6 & 5 & 5 \end{bmatrix} \qquad (10.43)$$

We can calculate $\mathbf{b}$ using the following steps

$$\mathbf{Sx} = \mathbf{x}^T\mathbf{x} = \begin{bmatrix} 1 & 1 & 1 & 1 & 1 & 1 \\ 3 & 4 & 5 & 2 & 3 & 6 \\ 2 & 2 & 4 & 1 & 2 & 5 \\ 6 & 7 & 9 & 8 & 3 & 5 \end{bmatrix} \begin{bmatrix} 1 & 3 & 2 & 6 \\ 1 & 4 & 2 & 7 \\ 1 & 5 & 4 & 9 \\ 1 & 2 & 1 & 8 \\ 1 & 3 & 2 & 3 \\ 1 & 6 & 5 & 5 \end{bmatrix} = \begin{bmatrix} 6 & 23 & 16 & 38 \\ 23 & 99 & 72 & 146 \\ 16 & 72 & 54 & 101 \\ 38 & 146 & 101 & 264 \end{bmatrix} \qquad (10.44)$$

$$\mathbf{Sx}^{-1} = \begin{bmatrix} 4.25 & -1.48 & 1.15 & -0.23 \\ -1.48 & 1.29 & -1.21 & -0.04 \\ 1.15 & -1.21 & 1.23 & 0.03 \\ -0.23 & -0.04 & 0.03 & 0.04 \end{bmatrix}$$

$$\mathbf{Sy} = \mathbf{x}^T\mathbf{y} = \begin{bmatrix} 1 & 1 & 1 & 1 & 1 & 1 \\ 3 & 4 & 5 & 2 & 3 & 6 \\ 2 & 2 & 4 & 1 & 2 & 5 \\ 6 & 7 & 9 & 8 & 3 & 5 \end{bmatrix} \begin{bmatrix} 0 & 3 \\ 2 & 8 \\ 1 & 9 \\ 1 & 8 \\ 2 & 7 \\ 0 & 6 \end{bmatrix} = \begin{bmatrix} 6 & 41 \\ 21 & 159 \\ 13 & 110 \\ 37 & 270 \end{bmatrix}$$

$$\mathbf{b} = \mathbf{Sx}^{-1}\mathbf{Sy} = \begin{bmatrix} 4.25 & -1.48 & 1.15 & -0.23 \\ -1.48 & 1.29 & -1.21 & -0.04 \\ 1.15 & -1.21 & 1.23 & 0.03 \\ -0.23 & -0.04 & 0.03 & 0.04 \end{bmatrix} \begin{bmatrix} 6 & 41 \\ 21 & 159 \\ 13 & 110 \\ 37 & 270 \end{bmatrix} = \begin{bmatrix} 0.74 & 2.47 \\ 1.10 & 1.18 \\ -1.30 & -1.04 \\ -0.08 & 0.41 \end{bmatrix}$$

Multicollinearity problems may occur by correlation among the dependent variables as well. In later chapters when we study principal component analysis, we will study ways of reducing dimensionality and selecting combination of variables to reduce multicollinearity.

10.3 TWO-GROUP DISCRIMINANT ANALYSIS

The objective of this analysis is to separate observations into groups or classes. We assume to have an a priori idea of the groups. What we want is to find a function to produce a new variable (discriminant score) that would separate the groups as much as possible. This is to say, finding the maximum **separability** between groups of multivariate observations. Once we know the discriminant function, other objects or observations can be assigned to a group and the discriminant performs as a predictive tool.

Assume that for each group we have n observations of m variables. These are arranged in a matrix **X** which is rectangular $n \times m$, comprised of n rows for observations and m columns for the variables

$$\mathbf{X} = \begin{bmatrix} x_{11} & x_{21} & \ldots & x_{m1} \\ x_{12} & x_{22} & \ldots & x_{m2} \\ \ldots & \ldots & \ldots & \ldots \\ x_{1n} & x_{2n} & \ldots & x_{mn} \end{bmatrix} = \begin{bmatrix} \mathbf{x}_1 & \mathbf{x}_2 & \ldots & \mathbf{x}_m \end{bmatrix}$$

As an example consider two groups, $n = 20$ observations of $m = 3$ variables. Figure 10.4 shows the box plots for each variable (1, 2, and 3) with the two groups side by side. We can see that each variable shows differences between groups. Therefore, if we combine the three variables, we should be able to distinguish between the groups.

Form a vector **z** of "scores" for the observations as linear combination of the m variables using a set of coefficients or weights

$$\mathbf{Z} = \mathbf{XA} \tag{10.45}$$

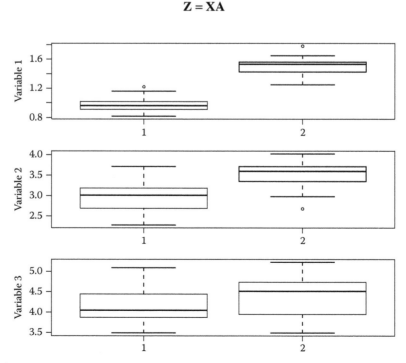

FIGURE 10.4 Box plots of three variables for two groups.

This equation is the linear discriminant function, where matrix $\mathbf{A}$ is $m \times 1$, and is a vector of weights or coefficients.

$$\mathbf{A} = \begin{bmatrix} a_1 \\ a_2 \\ \cdots \\ a_m \end{bmatrix} \tag{10.46}$$

The scores are a measure of **separability**. Multivariate space is collapsed into one variable (score), and the location of each group along this axis is used to separate them. In this sense, linear discriminant analysis is like linear multiple regression where the score $\mathbf{Z}$ is the dependent variable and the elements of X are the independent variables

$$\mathbf{Z} = a_1 \mathbf{x_1} + a_2 \mathbf{x_2} + \cdots + a_m \mathbf{x_m} \tag{10.47}$$

The variables with the largest weights or coefficients are the ones contributing the most to the separation or discrimination of the groups.

For two groups, discriminant analysis is a generalization of the two-sample t test. Recall that for two samples X_1 and X_2 with equal variances, and n_1 and n_2 observations, the difference in sample means $d = \bar{X}_1 - \bar{X}_2$ is tested with a pooled variance given by

$$s_p^2 = \frac{(n_1 - 1)s_1^2 + (n_2 - 1)s_2^2}{n_1 + n_2 - 2} \tag{10.48}$$

where s_1^2 and s_2^2 are the sample variances of the two samples.

Assume two groups of observations $\mathbf{X_1}$ and $\mathbf{X_2}$, with n_1 and n_2 observations, respectively, of m variables.

$$\mathbf{X}_1 = \begin{bmatrix} x_{11} & x_{21} & \cdots & x_{m1} \\ x_{12} & x_{22} & \cdots & x_{m2} \\ \cdots & \cdots & \cdots & \cdots \\ x_{1n1} & x_{2n1} & \cdots & x_{mn1} \end{bmatrix}$$

$$\mathbf{X}_2 = \begin{bmatrix} x_{11} & x_{21} & \cdots & x_{m1} \\ x_{12} & x_{22} & \cdots & x_{m2} \\ \cdots & \cdots & \cdots & \cdots \\ x_{1n2} & x_{2n2} & \cdots & x_{mn2} \end{bmatrix}$$

For each variable we take the sample mean of the observations of each group. These are vectors $1 \times m$ $\mathbf{X} - \bar{\mathbf{X}}_1$ and $\bar{\mathbf{X}}_2$. We can take the difference between sample means to obtain another $1 \times m$ vector

$$\mathbf{D} = \bar{\mathbf{X}}_1 - \bar{\mathbf{X}}_2 \tag{10.49}$$

This is to say, for the ith variable or "attribute," there is difference D_i between the sample mean (average) of that variable for group 1 and the sample mean of that variable for group 2

$$D_i = \overline{X_{1i}} - \overline{X_{2i}} \tag{10.50}$$

where $i = 1,\ldots, m$. Note that the m values of D_i form the entries of the $m \times 1$ matrix $\mathbf{D}$ or vector of differences of the sample means. The matrix $\mathbf{S}_p$ is $m \times m$, or matrix of pooled variances and covariances between groups is a matrix generalization of Equation 10.48

$$\mathbf{S}_p = \frac{(n_1 - 1)\mathbf{S}_1 + (n_2 - 1)\mathbf{S}_2}{n_1 + n_2 - 2} \tag{10.51}$$

where $\mathbf{S}_1$ and $\mathbf{S}_2$ are covariance matrices for groups 1 and 2, respectively. Note that when $n_1 = n_2$, then $\mathbf{S}_p$ is just the average of $\mathbf{S}_1$ and $\mathbf{S}_2$. Now solve for the matrix of weights or coefficients $\mathbf{A}$ from $\mathbf{S}_p\mathbf{A} = \mathbf{D}^T$. Recall that to solve for $\mathbf{A}$, you find the inverse of $\mathbf{S}_p$ and premultiply both sides by this inverse

$$\mathbf{A} = \mathbf{S}_p^{-1}\mathbf{D}^T \tag{10.52}$$

Once we find $\mathbf{A}$, the method is used in the following manner. Calculate the scores for each group using Equation 10.45

$$\mathbf{Z}_1 = \mathbf{X}_1\mathbf{A}$$
$$\mathbf{Z}_2 = \mathbf{X}_2\mathbf{A} \tag{10.53}$$

Also calculate a scalar score for the "centroid" of each group

$$\bar{Z}_1 = \overline{\mathbf{X}_1}\mathbf{A} \tag{10.54}$$

$$\bar{Z}_2 = \overline{\mathbf{X}_2}\mathbf{A} \tag{10.55}$$

and a scalar score for the midpoint between centroids of groups 1 and 2

$$Z_0 = \left(\frac{\overline{\mathbf{X}_1 + \mathbf{X}_2}}{2} \right) \mathbf{A} \tag{10.56}$$

Now we can separate the two groups if $\bar{Z}_1$ and $\bar{Z}_2$ are far apart on the score line and the scores of observations of group 1 do not score on the group 2 side (that is, scores between Z_0 and $\bar{Z}_2$ or beyond $\bar{Z}_2$).

This is easily visualized on a plot of observations on the score axis where Z_0, $\bar{Z}_1$, and $\bar{Z}_2$ are marked (see figure 6-3 Davis, 2002, page 476) or by histograms of the data for each group. For example, Figure 10.5. Please realize that the vertical position of the markers (value of the y-axis) does not mean anything, it is a random value used just for display of the observation markers. The markers labeled with numbers indicate centroids. We can appreciate that there is little overlap in the observations of each group.

Clearly, discrimination is given by distances between the group centroids or absolute value of $\bar{Z}_1 - \bar{Z}_2$; this distance can be calculated from a generalization of the Euclidian formula in the standardized version "divided" by the variances (Davis, 2002, page 477–478). That is, to say

$$\mathbf{DS}_p^{-1}\mathbf{D}^T = \mathbf{DA} \tag{10.57}$$

which is the **Mahalanobis distance** denoted as D^2.

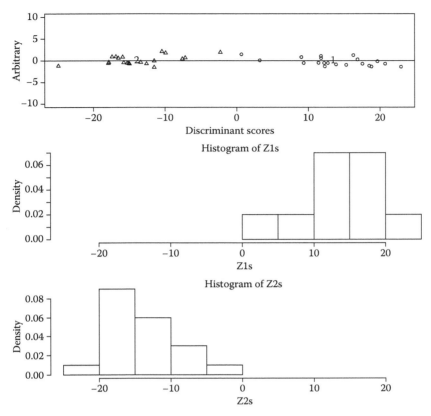

FIGURE 10.5 Discriminant plot and histograms. In the top panel the y-axis does not mean anything, it is just for display of the observation markers. The number markers indicate centroids.

For inferential purposes, we can use a two-sample version of **Hotelling's T² test** (similar to two-sample t test). The Mahalanobis D^2 or difference $\bar{Z}_1 - \bar{Z}_2$ can be used to test the significance of the discrimination with an F test where the F value is

$$F = \frac{(n_1 + n_2 - m - 1)(n_1 n_2)}{m(n_1 + n_2 - 2)(n_1 + n_2)} D^2 \qquad (10.58)$$

with m and $(n_1 + n_2 - m - 1)$ degrees of freedom for numerator and denominator, respectively. The null hypothesis H0 is that the two groups have equal sample means; or in other words that the distance between centroids is zero. Rejecting the null implies that there is enough difference between the sample means to separate the groups.

Let us look at the following numerical example that corresponds to the data displayed in Figure 10.4. These data have $m = 3$ and $n_1 = 20$, $n_2 = 20$, and the following sample means

$$\bar{\mathbf{X}}_1 = \begin{bmatrix} 0.97 & 2.98 & 4.12 \end{bmatrix}$$
$$\bar{\mathbf{X}}_2 = \begin{bmatrix} 1.50 & 3.52 & 4.41 \end{bmatrix} \qquad (10.59)$$

and the following covariance matrices

$$S_1 = \begin{bmatrix} 0.010 & 0.008 & 0.018 \\ 0.008 & 0.160 & -0.020 \\ 0.018 & -0.020 & 0.194 \end{bmatrix} \qquad (10.60)$$

$$S_2 = \begin{bmatrix} 0.013 & -0.011 & -0.019 \\ -0.011 & 0.108 & 0.038 \\ -0.019 & 0.038 & 0.267 \end{bmatrix} \qquad (10.61)$$

We calculate the difference vector subtracting the means given in Equation 10.59

$$D = \overline{X}_1 - \overline{X}_2 = \begin{bmatrix} -0.53 & -0.55 & -0.29 \end{bmatrix} \qquad (10.62)$$

and matrix of pooled variances using covariance matrices given in Equations 10.60 and 10.61 using Equation 10.51

$$S_p = \begin{bmatrix} 0.012 & -0.002 & -0.001 \\ -0.002 & 0.134 & 0.009 \\ -0.001 & 0.009 & 0.230 \end{bmatrix} \qquad (10.63)$$

Now we calculate the inverse

$$S_p^{-1} = \begin{bmatrix} 85.116 & 1.066 & 0.335 \\ 1.066 & 7.491 & -0.281 \\ 0.335 & -0.281 & 4.353 \end{bmatrix} \qquad (10.64)$$

and with this calculate matrix **A** of coefficients using Equation 10.52

$$A = S_p^{-1} D^T = \begin{bmatrix} -46.09 \\ -4.59 \\ -1.30 \end{bmatrix} \qquad (10.65)$$

The centroids

$$\overline{Z}_1 = \overline{X}_1 A = 13.74 \qquad (10.66)$$

$$\overline{Z}_2 = \overline{X}_2 A = -13.74 \qquad (10.67)$$

The Mahalanobis distance

$$D^2 = \text{abs}\,(\overline{Z}_1 - \overline{Z}_2) = 27.48 \qquad (10.68)$$

The degrees of freedom of the numerator = 3 and of the denominator = $(40 - 3 - 1) = 36$. We calculate the F value

$$F = \frac{(40 - 3 - 1) \times 400}{3(40 - 2) \times 40} \times 27.48 = 86.78 \tag{10.69}$$

This is a large F value and thus we would expect a very low p-value, suggesting to reject the null. Therefore, the detected difference between the two groups is highly significant; i.e., the two groups can be "discriminated." The plot shown in Figure 10.5 corresponds to this analysis.

10.4 MULTIPLE ANALYSIS OF VARIANCE (MANOVA)

MANOVA is an extension of ANOVA to detect differences among groups or samples but based on several response variables, not just one response as in ANOVA. Recall that ANOVA tests for the differences of sample means. In MANOVA we test for the differences in the vector of sample means.

Besides detecting differences among samples, MANOVA can help reduce the dimension of a set of variables and thereby simplify models while identifying the variables that differentiate groups the most. In MANOVA, as well as in ANOVA, the categorical variables defining the groups are independent factors and the variables are considered the response or dependent variables.

Let us start by reviewing the one-way ANOVA, but using a slightly different notation and approach. Consider m levels of a factor (or m groups) and the sample means and variances of the response X for each one of the groups. The sample means and variances of the response X for each one of the groups

$$\bar{X}_1, ..., \bar{X}_m$$
$$s_1^2, ..., s_m^2 \tag{10.70}$$

Note that each sample mean is the average of n observations X_{ij} where $j = 1, ..., n$

$$\bar{X}_i = \frac{\sum_{j=1}^{n} X_{ji}}{n} \tag{10.71}$$

The overall variance can be estimated from the average sample variances, that is, to say from the **within** variance.

$$\overline{s_p^2} = \frac{\sum_{i=1}^{m} s_i^2}{m} \tag{10.72}$$

Now note that expanding the variance s_i^2 of each sample in Equation 10.72

$$\overline{s_p^2} = \frac{\sum_{i=1}^{m} \left(\frac{\sum_{j=1}^{ni} (X_{ji} - \bar{X}_i)^2}{(n-1)} \right)}{m} = \frac{\sum_{i=1}^{m} \sum_{j=1}^{ni} (X_{ji} - \bar{X}_i)^2}{m(n-1)} = \frac{\sum_{i=1}^{m} \sum_{j=1}^{ni} (X_{ji} - \bar{X}_i)^2}{N - m} \tag{10.73}$$

where $N = m \times n$ is the total number of observations. Note that this is the sum of squares of the differences or errors (SS_w) divided by the degrees of freedom (df_w) given in the denominator, that is, $N - m$. We refer to the right hand of Equation 10.73 as the mean square of the errors (MS_w). Therefore, the overall within variance is the MS_w.

$$\bar{s}_p^2 = MS_w = \frac{SS_w}{df_w} \tag{10.74}$$

Also the overall sample mean can be estimated from the average of the sample means

$$\bar{\bar{X}} = \frac{\sum_{i=1}^{m} \bar{X}_i}{m}$$

Recall that the mean of the sample means has variance (recall central limit theorem)

$$s_{\bar{X}}^2 = \frac{s_X^2}{n}$$

Therefore, an estimate of the variance from the variance **among** the sample means is

$$s_X^2 = ns_{\bar{X}}^2 = n \left(\frac{\sum_{i=1}^{m} (\bar{X}_i - \bar{\bar{X}})^2}{m-1} \right) = \frac{\sum_{i=1}^{m} n(\bar{X}_i - \bar{\bar{X}})^2}{m-1} \tag{10.75}$$

Thinking of the numerator as a sum of among square differences and the denominator as degrees of freedom, Equation 10.75 is rewritten as the mean square differences among samples

$$s_X^2 = \frac{SS_a}{m-1} = MS_a = \frac{SS_a}{df_a} \tag{10.76}$$

Now when we ratio of "among" variance given in Equation 10.76 to "within" variance given in Equation 10.74, we obtain the F value.

$$F = \frac{MS_a}{MS_w} = \frac{SS_a/df_a}{SS_w/df_w} = \frac{df_w}{df_a} \frac{SS_a}{SS_w} = \frac{N-m}{m-1} \frac{\sum_{i=1}^{m} n(\bar{X}_i - \bar{\bar{X}})^2}{\sum_{i=1}^{m} \sum_{j=1}^{n} (X_{ij} - \bar{X}_i)^2} \tag{10.77}$$

These expressions are used in terms of sum of squares. Now we will employ vectors

$$F = \frac{(N-m)\sum_{i=1}^{m} (\bar{X}_i - \bar{\bar{X}})^2}{(m-1)\sum_{i=1}^{m} \sum_{j=1}^{n} (X_{ij} - \bar{X}_i)^2} = \frac{(N-m)(\mathbf{x} - \bar{\bar{X}})^T (\mathbf{x} - \bar{\bar{X}})}{(m-1)\sum_{i=1}^{m} (\mathbf{x}_i - \bar{X}_i)^T (\mathbf{x}_i - \bar{X}_i)}$$

where the entries of the vector $\bar{\mathbf{X}}$ of dimension m are the group sample means and the vector $\mathbf{x}_i$ for each group is of dimension n.

Now let us derive the multivariate response MANOVA using the same ideas but employing vectors and matrices. Consider m levels of a factor (or m groups) and the vectors of sample means and covariance matrices of the response matrix $\mathbf{X}_i$ of dimension $n \times k$ for each one of the groups

$$\bar{\mathbf{X}}_1, ..., \bar{\mathbf{X}}_m$$
$$\mathbf{S}_1^2, ..., \mathbf{S}_m^2$$
(10.78)

Each one of the $\bar{\mathbf{X}}_i$ entities is a vector of dimension k, and each one of the covariance matrices $\mathbf{S}_i^2$ is square of dimension $k \times k$. The overall covariance matrix is also $k \times k$ and can be estimated from the average of the sample covariance matrices, that is, to say the **within** covariance

$$\overline{\mathbf{S}_p^2} = \frac{\sum_{i=1}^{m} \mathbf{S}_i^2}{m} = \frac{\sum_{i=1}^{m} (\mathbf{X}_i - \bar{\mathbf{X}}_i)^T (\mathbf{X}_i - \bar{\mathbf{X}}_i)}{(N - m)}$$
(10.79)

Also the overall sample mean (it is now a vector of dimension k) can be estimated from the average of the sample means

$$\bar{\bar{\mathbf{X}}} = \frac{\sum_{i=1}^{m} \bar{\mathbf{X}}_i}{m}$$

Now the mean of the sample means has covariance

$$\mathbf{S}_{\bar{X}}^2 = \frac{\mathbf{S}_X^2}{n}$$

And therefore an estimate of the variance from the variance **among** the sample means is

$$\mathbf{S}_X^2 = n\mathbf{S}_{\bar{X}}^2 = n\left(\frac{\sum_{i=1}^{m} (\bar{\mathbf{X}}_i - \bar{\bar{\mathbf{X}}})^2}{m-1}\right) = \left(\frac{n(\bar{\mathbf{X}}_i - \bar{\bar{\mathbf{X}}})^T (\bar{\mathbf{X}}_i - \bar{\bar{\mathbf{X}}})}{m-1}\right)$$
(10.80)

Now we try to find the equivalent of taking the ratio of the **among** variance to the **within** variance to obtain the F value. The equivalent of the division is obtained by using the inverse of the matrix targeted for the denominator. So we use the among covariance matrix of Equation 10.80 and post-multiply it by the inverse of the within covariance given in Equation 10.79

$$\mathbf{S}_X^2 \left(\mathbf{S}_p^2\right)^{-1}$$
(10.81)

However, the matrix used to calculate the statistics is the product of matrices scaled by their degrees of freedom; this is to say

$$\mathbf{S}_w = df_w \left(\mathbf{S}_p^2\right)$$

$$\mathbf{S}_a = df_a \left(\mathbf{S}_X^2\right)$$

And therefore the matrix product is the same as in Equation 10.81, but scaled by the ratio of degrees of freedom

$$\mathbf{S}_a \mathbf{S}_w^{-1} = \frac{df_a}{df_w} \mathbf{S}_X^2 \left(\mathbf{S}_p^2\right)^{-1}$$

Therefore, the multivariate equivalent for F is based not only on the sum of squares among and within groups, as in ANOVA, but also on the sum of cross-products. In this sense, it takes covariance into account as well as group means. Recall that in ANOVA the null hypothesis is that $F = 1$ or no difference in the variances. The equivalent of unity 1 is the identity matrix $\mathbf{I}$; therefore, the null hypothesis is

$$\mathbf{S}_a \mathbf{S}_w^{-1} = \mathbf{I}$$

There are various significance tests based on the F distribution for MANOVA; for example, Hotelling-Lawley, Pillai, Wilks, and Roy.

The **Hotelling-Lawley** statistic is based on the **trace** of matrix given in Equation 10.81 or the "ratio" of among to within

$$H = trace\left(\mathbf{S}_a \mathbf{S}_w^{-1}\right)$$

The **Pillai's statistic** is based on the **trace** of a derived matrix that represents a "ratio" of among to the sum of among + within.

$$P = trace\left(\mathbf{S}_a (\mathbf{S}_w + \mathbf{S}_a)^{-1}\right)$$

The **Wilks' lambda** test is a ratio of **determinants**. It is defined based on a ratio of determinant of matrix of "within" to total (among + within). Therefore, the values vary between 0 and 1. The smaller the lambda, the smaller the within compared to the total.

$$\Lambda = \frac{\left|\mathbf{S}_a\right|}{\left|\mathbf{S}_a + \mathbf{S}_w\right|}$$

All these three statistics are scalars and are converted to F values, which allows for testing the hypothesis. These statistics do not always yield the same F. In many cases Pillai's trace is considered a very robust statistic and is typically preferred. These statistics represent a measure of the difference between sample means of the variables. The t-test, Hotelling's T, and the F test are special cases of Wilks's lambda. Note all three statistics can be expressed as eigenvalues of the matrices given earlier. Another statistic used is the **Roy** statistic, which is the maximum eigenvalue of the matrix $\mathbf{S}_a \mathbf{S}_w^{-1}$. We will cover eigenvalues later in the book, so do not worry about this for now.

10.5 EXERCISES

Exercise 10.1
Assume two uncorrelated variables X_1 and X_2 with the sample means 1.5 and 2.3, respectively, and sample variances 0.2, 0.3, respectively. Assume a dependent variable Y with sample mean of 4.0 and that covariance of Y with X_1 and X_2 are 0.25 and 0.12, respectively. Calculate the coefficients of linear multiple regression.

Exercise 10.2
Repeat the previous exercise but assume that X_1 and X_2 are correlated with covariance of 0.5. Discuss the differences in results with respect to the results of the previous exercise.

Exercise 10.3
Suppose that a multivariate regression model has two dependent variables Y_1 and Y_2 and four independent variables X_1, X_2, X_3, X_4. Determine the dimensions for matrices $\mathbf{x}$, $\mathbf{y}$, $\mathbf{b}$, $\mathbf{x}^T\mathbf{x}$, and $\mathbf{x}^T\mathbf{y}$. Assume $n = 10$ observations.

Exercise 10.4
Suppose

$$\mathbf{X}_1 = \begin{bmatrix} 3 & 1 \\ 1 & 3 \\ 2 & 2 \\ 3 & 3 \\ 2 & 1 \end{bmatrix} \quad \text{and} \quad \mathbf{X}_2 = \begin{bmatrix} 5 & 5 \\ 4 & 5 \\ 6 & 5 \\ 4 & 5 \\ 5 & 4 \\ 4 & 5 \end{bmatrix}$$

Determine, m, n_1, n_2. Show that the group means of X_1 and X_2 are $\bar{\mathbf{X}}_1 = [2.2 \quad 2.0]$ and $\bar{\mathbf{X}}_2 = [4.66 \quad 4.83]$. Calculate the vector $\mathbf{D}$ of differences in group means. Assume that the covariance matrices of X_1 and X_2 are

$$\mathbf{S}_1 = \begin{bmatrix} 0.7 & -0.25 \\ -0.25 & 1 \end{bmatrix} \quad \text{and} \quad \mathbf{S}_2 = \begin{bmatrix} 0.67 & -0.07 \\ -0.07 & 0.17 \end{bmatrix}$$

Show that $\mathbf{S}_p$ and its inverse are

$$\mathbf{S}_p = \begin{bmatrix} 0.68 & -0.15 \\ -0.15 & 0.54 \end{bmatrix} \quad \text{and} \quad \mathbf{S}_p^{-1} = \begin{bmatrix} 1.56 & 0.43 \\ 0.43 & 1.98 \end{bmatrix} \quad \text{and}$$

Show that the vector of weights is $\mathbf{A} = \begin{bmatrix} -5.07 \\ -6.67 \end{bmatrix}$, and that the centroids are $\bar{Z}_1 = 15.71$ and $\bar{Z}_2 = -15.71$. Show that the Mahalanobis distance is $D^2 = 31.4$. Show that $F = 38$ and that the p-value $= 0.000021$. Given that the scores are calculated to be

$$\mathbf{Z}_1 = \begin{bmatrix} 18.33 \\ 15.12 \\ 16.72 \\ 4.98 \\ 23.40 \end{bmatrix} \quad \text{and} \quad \mathbf{Z}_2 = \begin{bmatrix} -18.51 \\ -13.44 \\ -23.58 \\ -13.44 \\ -11.83 \\ -13.44 \end{bmatrix}$$

Draw a sketch of the observations on the Z line. Plot the centroids on the same line and write conclusions.

Exercise 10.5

Suppose we have $m = 3$ groups, $n = 10$ observations for all groups, and $k = 2$ variables, and want to do MANOVA. Calculate degrees of freedom for among and within: df_a and df_w.

Suppose we have calculated the sample means to be

$$\overline{\mathbf{X}}_1 = [1.46 \quad 1.58] \quad \overline{\mathbf{X}}_2 = [2.48 \quad 2.35] \quad \overline{\mathbf{X}}_3 = [3.42 \quad 3.47] \tag{10.82}$$

Assume we have calculated that the covariance matrix of these sample means is

$$\mathbf{S}_{\overline{X}}^2 = \begin{bmatrix} 0.962 & 0.928 \\ 0.928 & 0.910 \end{bmatrix}$$

Show that the among covariance matrix **Sa** is

$$\mathbf{S}_a = \begin{bmatrix} 19.235 & 18.561 \\ 18.560 & 18.200 \end{bmatrix}$$

Assume that the covariance matrices of each group are

$$\mathbf{S}_1^2 = \begin{bmatrix} 0.099 & 0.019 \\ 0.019 & 0.064 \end{bmatrix} \quad \mathbf{S}_2^2 = \begin{bmatrix} 0.103 & -0.031 \\ -0.031 & 0.093 \end{bmatrix} \quad \mathbf{S}_3^2 = \begin{bmatrix} 0.079 & -0.033 \\ -0.033 & 0.093 \end{bmatrix} \tag{10.83}$$

Show that the within covariance matrix

$$\mathbf{S}_w = \begin{bmatrix} 2.531 & -0.411 \\ -0.411 & 2.245 \end{bmatrix}$$

We calculate the inverse to be

$$\mathbf{S}_w^{-1} = \begin{bmatrix} 0.407 & 0.0746 \\ 0.0746 & 0.459 \end{bmatrix}$$

Show that the matrix $\mathbf{S}_a \mathbf{S}_w^{-1}$ is

$$\mathbf{S}_a \mathbf{S}_w^{-1} = \begin{bmatrix} 9.218 & 9.954 \\ 8.917 & 9.739 \end{bmatrix}$$

Show that the values of the Holling-Lawley, Pillai and Wilks statistics are 18.96, 1.00, and 0.048, respectively.

10.6 COMPUTER SESSION: MULTIVARIATE MODELS

10.6.1 MULTIPLE LINEAR REGRESSION

We will use the `airquality` dataset already employed in previous computer sessions when we built contingency tables and performed simple regression. Recall that ozone is an important urban air pollution problem: excessive ozone in the lower troposphere, related to emissions and photo-chemistry, and meteorological variables.

To start load and attach the data frame applying commands

```
>data(airquality)
>attach(airquality)
```

We applied these commands in previous computer sessions; therefore, if you saved your workspace from last session, there should not be a need to apply these commands again.

Then look at the data frame

```
> airquality
    Ozone Solar.R Wind Temp Month Day
1      41     190  7.4   67     5   1
2      36     118  8.0   72     5   2
3      12     149 12.6   74     5   3
4      18     313 11.5   62     5   4
5      NA      NA 14.3   56     5   5
6      28      NA 14.9   66     5   6
...
151    14     191 14.3   75     9  28
152    18     131  8.0   76     9  29
153    20     223 11.5   68     9  30
>
```

which has 153 records. This data set includes NA for some records.

In Chapter 6, we have done a simple regression of ozone vs. temperature and found that temperature only explained ~48% of variation in ozone, and therefore concluded that other variables should be included. First, visually explore by looking at pairwise scatter plots including radiation and wind. Simply, type command

```
>pairs(airquality)
```

Figure 10.1 shows a matrix of scatter plots. From the top row of panels, we see that there seems to be an increasing trend of ozone with increased radiation and temperature and a decreasing trend with wind. Month and Day in this case are not useful. So, let us try to include radiation and wind in a multiple regression model.

There are several ways of proceeding. (1) Include all variables at once, (2) Drop and add variables from a previously built model with fewer variables; (3) An automatic process using the stepwise procedure.

Let us pursue (1) first.

```
> ozone.mlm <- lm(Ozone ~ Solar.R + Temp + Wind)
> summary(ozone.mlm)

Call:
lm(formula = Ozone ~ Solar.R + Temp + Wind)

Residuals:
    Min      1Q  Median      3Q     Max
-40.485 -14.219  -3.551  10.097  95.619

Coefficients:
             Estimate Std. Error t value Pr(>|t|)
(Intercept) -64.34208   23.05472  -2.791  0.00623 **
Solar.R       0.05982    0.02319   2.580  0.01124 *
Temp          1.65209    0.25353   6.516 2.42e-09 ***
Wind         -3.33359    0.65441  -5.094 1.52e-06 ***
---
Signif. codes:  0 '***' 0.001 '**' 0.01 '*' 0.05 '.' 0.1 ' ' 1

Residual standard error: 21.18 on 107 degrees of freedom
  (42 observations deleted due to missingness)
Multiple R-Squared: 0.6059,      Adjusted R-squared: 0.5948
F-statistic: 54.83 on 3 and 107 DF,  p-value: < 2.2e-16
```

The t tests for the coefficients indicate significance for all of them. Note that we increased the explained variation to ~60% (not very good yet) and that the p-value for the variance test F is very low (highly significant). The resulting model is

$$Ozone = -64.34 + 0.05982 \times Solar.R + 1.65 \times Temp - 3.33 \times Wind \qquad (10.84)$$

that is to say, as expected ozone increases with air temperature and radiation and decreases with wind speed. As explained in Chapter 6, the negative value of Ozone forced at zero values for Solar.R, Temp, and Wind due to the negative intercept is an invalid extrapolation considering the range of values for which the relationship is derived.

We can obtain evaluation plots as before with single regression using

```
>par(mfrow=c(2,2))
>plot(ozone.mlm)
```

to obtain Figure 10.2. Here we see a relatively random behavior of the residuals except for the low end of the ozone values. There is not much improvement in the relative spread or influence of outliers. To produce a plot of Ozone vs. fitted values of Ozone (Figure 10.3), we can go through these steps.

```
> airqual <- na.omit(airquality)
> attach(airqual)
> ozone.mlm <- lm(Ozone ~ Solar.R + Temp + Wind, data=airqual)
> > ozone.fit <- matrix(ozone.mlm$fitted.values, ncol=1)
> plot(Ozone, ozone.fit,xlab="Measured Ozone",ylab="Estimated
Ozone",
+ xlim=c(-30,150),ylim=c(-30,150))
```

Often, once you have all independent variables in the model you are interested in seeing whether you can get by with fewer variables. Let's try to drop one of the X's. Use function drop1

```
> drop1(ozone.mlm)
Single term deletions

Model:
Ozone ~ Solar.R + Temp + Wind
        Df Sum of Sq   RSS    AIC
<none>                 48003   682
Solar.R  1      2986 50989   686
Temp     1     19050 67053   717
Wind     1     11642 59644   704
>
```

In this case the AIC is lowest for the current model (line labeled <none>) 682 and therefore you do not want to drop any of the three terms.

Now to illustrate approach (2), recall that we already created an ozone.lm object from airquality that relates ozone to temperature. Use function add1

```
> ozone.add <- add1(ozone.lm, ~ Temp + Solar.R + Wind)
Warning message:
using the 111 / 116 rows from a combined fit in: add1.lm(ozone.lm,
  ~Temp + Solar.R + Wind)
> ozone.add
Single term additions

Model:
Ozone ~ Temp
        Df Sum of Sq   RSS    AIC
<none>                 64110   710
Solar.R  1      4466 59644   704
Wind     1     13121 50989   686
>
```

Based on the lowest AIC for wind (686), we decide to add it as independent variable. You can repeat this by running a new `lm` with just temperature and wind as object `ozone.tw`

```
> ozone.tw <- lm(Ozone ~ Temp + Wind)
> ozone.tw

Call:
lm(formula = Ozone ~ Temp + Wind)

Coefficients:
(Intercept)            Temp            Wind
    -67.322           1.828          -3.295

> summary(ozone.tw)

Call:
lm(formula = Ozone ~ Temp + Wind)

Residuals:
    Min      1Q  Median      3Q     Max
-42.156 -13.216  -3.123  10.598  98.492

Coefficients:
            Estimate Std. Error t value Pr(>|t|)
(Intercept) -67.3220    23.6210  -2.850  0.00524 **
Temp          1.8276     0.2506   7.294 5.29e-11 ***
Wind         -3.2948     0.6711  -4.909 3.26e-06 ***
---
Signif. codes:  0 `***' 0.001 `**' 0.01 `*' 0.05 `.' 0.1 ` ' 1

Residual standard error: 21.73 on 108 degrees of freedom
Multiple R-Squared: 0.5814,      Adjusted R-squared: 0.5736
F-statistic: 74.99 on 2 and 108 DF,  p-value: < 2.2e-16
```

The new R^2 is 0.57, which is only slightly lower than the one for the three variable model (0.59).

When you have many X's the drop/add can be tedious. Now let us try approach (3) and perform stepwise regression. This is an automatic procedure that only requires the simple first two commands. The procedure tries to minimize the AIC and stops when the AIC for the current model (the `none` row) is the lowest among all the other terms.

```
> ozone0.lm <- lm(Ozone ~ 1,)
> step(ozone0.lm, ~ Solar.R + Temp + Wind)
Start: AIC= 779.07
    Ozone ~ 1

          Df Sum of Sq     RSS   AIC
+ Temp     1     59434   62367   707
+ Wind     1     45694   76108   729
+ Solar.R  1     14780  107022   767
<none>                  121802   779
```

```
(continued)

Step:  AIC= 706.77
 Ozone ~ Temp

          Df Sum of Sq     RSS    AIC
+ Wind     1       11378  50989    686
+ Solar.R  1        2723  59644    704
<none>                    62367    707
- Temp     1       59434 121802    779

Step:  AIC= 686.41
 Ozone ~ Temp + Wind

          Df Sum of Sq   RSS    AIC
+ Solar.R  1      2986 48003    682
<none>                 50989    686
- Wind     1     11378 62367    707
- Temp     1     25119 76108    729

Step:  AIC= 681.71
 Ozone ~ Temp + Wind + Solar.R

          Df Sum of Sq   RSS    AIC
<none>                 48003    682
- Solar.R  1      2986 50989    686
- Wind     1     11642 59644    704
- Temp     1     19050 67053    717

Call:
lm(formula = Ozone ~ Temp + Wind + Solar.R)

Coefficients:
(Intercept)         Temp         Wind      Solar.R
  -64.34208      1.65209     -3.33359      0.05982

>
```

The final model is the one we already found using the previous methods. The stepwise procedure pays off when you have many independent variables and you want to see how all the possible models compare to each other.

10.6.2 MULTIVARIATE REGRESSION

Let us consider Ozone and Temp as dependent variables (Y_1 and Y_2) whereas wind and solar radiation as independent variables (X_1 and X_2). We can use the same function lm but now use a response variable composed of the two variables

```
> lm(cbind(Ozone,Temp) ~ Wind+Solar.R)

Call:
lm(formula = cbind(Ozone, Temp) ~ Wind + Solar.R)

Coefficients:
                Ozone        Temp
(Intercept)  77.24604    85.70228
Wind         -5.40180    -1.25187
Solar.R       0.10035     0.02453
```

The function lm evaluates the multivariate regression as a multiple regression of each one of the responses, which is equivalent to using the equation given earlier in the chapter.

$$\mathbf{b} = (\mathbf{x}^T\mathbf{x})^{-1}(\mathbf{x}^T\mathbf{y}) = \mathbf{S}_x^{-1}\mathbf{S}_y$$

In this case matrix **b** is

$$\mathbf{b} = \begin{bmatrix} 77.25 & 85.70 \\ -5.40 & -1.25 \\ 0.10 & 0.03 \end{bmatrix}$$

We can get more info using summary

```
> summary(lm(cbind(Ozone,Temp) ~ Wind+Solar.R))
Response Ozone :

Call:
lm(formula = Ozone ~ Wind + Solar.R)

Residuals:
    Min      1Q  Median      3Q     Max
-45.651 -18.164  -5.959  18.514  85.237

Coefficients:
            Estimate Std. Error t value Pr(>|t|)
(Intercept) 77.24604    9.06751   8.519 1.05e-13 ***
Wind        -5.40180    0.67324  -8.024 1.34e-12 ***
Solar.R      0.10035    0.02628   3.819 0.000224 ***
---
Signif. codes:  0 '***' 0.001 '**' 0.01 '*' 0.05 '.' 0.1 ' ' 1

Residual standard error: 24.92 on 108 degrees of freedom
  (42 observations deleted due to missingness)
Multiple R-Squared: 0.4495,     Adjusted R-squared: 0.4393
F-statistic: 44.09 on 2 and 108 DF,  p-value: 1.003e-14

Response Temp :

Call:
lm(formula = Temp ~ Wind + Solar.R)

Residuals:
     Min      1Q  Median      3Q     Max
-17.2714 -5.0237  0.5837  5.2545 18.4608

Coefficients:
             Estimate Std. Error t value Pr(>|t|)
(Intercept) 85.702275   2.925445  29.295  < 2e-16 ***
Wind        -1.251870   0.217207  -5.763 7.9e-08 ***
Solar.R      0.024533   0.008478   2.894 0.00461 **
---
Signif. codes:  0 '***' 0.001 '**' 0.01 '*' 0.05 '.' 0.1 ' ' 1

Residual standard error: 8.039 on 108 degrees of freedom
  (42 observations deleted due to missingness)
Multiple R-Squared: 0.3014,     Adjusted R-squared: 0.2884
F-statistic: 23.29 on 2 and 108 DF,  p-value: 3.886e-09
```

10.6.3 Two-Group Linear Discriminant Analysis

Use seeg's function `lda2`. This is a function to perform linear discriminant analysis. We will work with data file **X1X2.csv** also available from seeg, and which has $m = 3$ variables, $n_1 = 20$, $n_2 = 20$ observations. These data correspond to the examples presented earlier in the chapter (Figure 10.4 and numerical example given in the linear discriminant Section 10.3).

First, look at the file using a text editor we observe six columns, the first three are for group 1 and the last three for group 2.

```
x11,    x12,    x13,    x21,    x22,    x23
1.01,   3.18,   3.95,   1.57,   3.65,   3.93
1.09,   3.02,   3.51,   1.41,   3.72,   4.26
0.99,   3.18,   4.05,   1.37,   3.34,   4.93
0.89,   2.63,   4.69,   1.54,   2.98,   3.49
   1,   2.28,   4.49,   1.49,   3.53,   4.76
etc
```

we read this file and convert to matrices X1 and X2

```
X1X2 <- read.csv("lab10/X1X2.csv",header=T)
  X1 <- as.matrix(X1X2[,1:3]); X2 <- as.matrix(X1X2[,4:6])
```

Then run `lda2` with X1 and X2 as arguments

```
lda2(X1,X2)
```

The results obtained are given in the following and can be compared to the numerical example given in the discriminant analysis Section 10.3. \$G1 and \$G2 correspond to the sample means $\bar{\mathbf{X}}_1$ and $\bar{\mathbf{X}}_2$, \$S1 and \$S2 to the covariance matrices $\mathbf{S}_1$ and $\mathbf{S}_2$, \$Sp to the pooled variance $\mathbf{S}_p$, \$Spinv to its inverse, \$A to matrix $\mathbf{A}$, \$Z1c.Z2c are centroids, \$D2.F.p.value contain Mahalanobis distance, F statistic, and its p-value, \$Z1s and \$Z2s are scores $\mathbf{Z}_1$ and $\mathbf{Z}_2$.

```
> X.lda2
$m.n1.n2
[1] 3 20 20

$G1
[1] 0.9745 2.9780 4.1240

$G2
[1] 1.5080 3.5255 4.4170
```

(continued)

(*continued*)

$S1
```
           x11          x12          x13
x11 0.010415526  0.008188421   0.01753368
x12 0.008188421  0.159669474  -0.02034421
x13 0.017533684 -0.020344211   0.19423579
```

$S2
```
           x21          x22          x23
x21  0.01313263 -0.01161474 -0.01956947
x22 -0.01161474  0.10847868  0.03791737
x23 -0.01956947  0.03791737  0.26655895
```

$Sp
```
            x11          x12          x13
x11  0.011774079 -0.001713158 -0.001017895
x12 -0.001713158  0.134074079  0.008786579
x13 -0.001017895  0.008786579  0.230397368
```

$Spinv
```
          [,1]        [,2]        [,3]
x11 85.1163790  1.0656104   0.3354051
x12  1.0656104  7.4905920  -0.2809581
x13  0.3354051 -0.2809581   4.3525240
```

$A
```
          [,1]
x11 -46.091284
x12  -4.587282
x13  -1.300404
```

$Z1c.Z2c
```
[1]   13.74113 -13.74113
```

$D2.F.p.value
```
[1] 27.48225 86.78607  0.00000
```

$Z1s
```
[1] 11.4045261  9.0233661 12.1963114 18.4961863 15.2917744
   19.6376784
[7] 20.7141828 22.9319190 16.2798070  0.6557518  3.2175597
   17.5502242
[13] 12.7059666 18.8057204  9.3563646 11.7528371 11.7693240
   13.8756514
[19] 12.2892241 16.8681712
```

$Z2s
```
[1]  -16.536607  -9.912244  -7.196697 -11.508212 -13.378165
  -15.932330
[7]   -2.280115 -14.944910 -11.498039 -10.434962 -24.872461
  -15.082672
[13] -15.248542  -7.597233 -16.864232 -15.780279 -17.851410
  -17.923408
[19] -12.582696 -17.397332
```

The centroid scores Z1c, Z2c and individual scores Z1s, Z2s have been centered on Z0, which is the midpoint or average of the centroids. From A, the weight or coefficient for the first variable is ten times the weight for the second variable, and the latter three to four times the weight of the third. The *F* value is large enough that produces negligible *p*-value allowing us to reject the null, indicating significant difference between the two groups.

Using the individual scores and the centroid scores, the function also produces the plot of Figure 10.5, which displays the difference between the groups in two different manners. The one at the top panel uses markers and the two below use histograms. It indicates little overlap in observations along the score axis and substantial differences between the centroids.

10.6.4 MANOVA

Multiple ANOVA can be performed with the base package function aov. There is a special summary function that applies to manova. Pillai is the default, but Wilks lambda can be requested as argument test = "Wilks". For example, if we apply the test to the X1X2 data of the previous section.

First make the multivariate set, then the factor and then a data frame

```
>X <- rbind(X1,X2)
>grp <- factor(rep(c(1,3), c(20,20,20)))
>X.g <- data.frame(grp, X)
```

The univariate ANOVA can be applied to each variable

```
> summary(aov(X ~grp,data=X.g))
 Response x11 :
            Df    Sum Sq Mean Sq F value     Pr(>F)
grp          1  2.84622 2.84622  241.74 < 2.2e-16 ***
Residuals   38  0.44742 0.01177
---
Signif. codes:  0 '***' 0.001 '**' 0.01 '*' 0.05 '.' 0.1 ' ' 1

 Response x12 :
            Df Sum Sq Mean Sq F value    Pr(>F)
grp          1 2.9976 2.99756  22.358 3.09e-05 ***
Residuals   38 5.0948 0.13407
---
Signif. codes:  0 '***' 0.001 '**' 0.01 '*' 0.05 '.' 0.1 ' ' 1

 Response x13 :
            Df Sum Sq Mean Sq F value  Pr(>F)
grp          1 0.8585 0.85849  3.7261 0.06106 .
Residuals   38 8.7551 0.23040
---
Signif. codes:  0 '***' 0.001 '**' 0.01 '*' 0.05 '.' 0.1 ' ' 1
>
```

We can see that there is significant difference between the two groups for each variable (or response). Now we run MANOVA as to include covariance.

```
> manova(X ~grp,data=X.g)
Call:
    manova(X ~ grp, data = X.g)

Terms:
                        grp      Residuals
resp 1           2.846222       0.447415
resp 2           2.997563       5.094815
resp 3           0.85849        8.75510
Deg. of Freedom        1             38

Residual standard error: 0.1085084 0.3661613 0.4799973
Estimated effects may be unbalanced
>
```

We can use any `test` of the set "Pillai", "Wilks", "Hotelling-Lawley", and "Roy" by adding the argument `test =`.

Let us use Pillai's

```
> summary(manova(X ~grp,data=X.g), test= "Pillai")
          Df Pillai approx  F num Df den Df      Pr(>F)
grp        1 0.87853    86.786      3     36   <2.2e-16 ***
Residuals 38
---
Signif. codes: 0 '***' 0.001 '**' 0.01 '*' 0.05 '.' 0.1 ' ' 1
```

The result is also significant because we have relatively high Pillai 0.87 and high F (86) and very small p-value for this F. Therefore, we conclude that there are significant differences between the groups for the combined responses.

To contrast let us consider the Wilks lambda.

```
> summary(manova(X ~grp,data=X.g), test="Wilks")
          Df   Wilks approx F num Df den Df      Pr(>F)
grp        1 0.12147    86.786      3     36   < 2.2e-16 ***
Residuals 38
---
Signif. codes:  0 '***' 0.001 '**' 0.01 '*' 0.05 '.' 0.1 ' ' 1
>
```

The result is also significant because we have low lambda (0.12), high F (86), and very small p-value for this F. Therefore, we conclude that there are significant differences between the groups for the combined responses. In this case, the Pillai and Wilks offered the same result. In fact, it will result the same with Hotelling-Lawley and Roy.

10.6.5 Computer Exercises

Exercise 10.6

Use `trees` data frame in package **datasets.** Attach this data set. You can get more information about it by using `help(trees)`. Assume that Girth (diameter at breast height) and tree height will be explanatory variables and that Volume will be a dependent variable. Then investigate by linear multiple regression the relationships of Volume to Girth and Height. What is the effect of the correlation between explanatory variables (girth and height) on the results of the multiple regression?

Exercise 10.7

Use `Mandel` data frame of package **car** that has values for two explanatory variables x1,x2 and one dependent variable y. Note: first load package car and then the data set Mandel. Build a predictor. Investigate by linear multiple regression. Are there any collinearity problems? Explain.

Exercise 10.8

Use data from Davis, 2002 Figure 6-2 page 472, and variables are median grain size and sorting coefficient of beach sand samples from Texas. For easy access, seeg includes files **sand1.txt** and **sand2.txt** with these data. Perform linear discriminant analysis to find a discriminant function of `sand1` data from `sand2`.

Exercise 10.9

Use the data from previous exercise. Perform MANOVA on `sand1` and `sand2`.

Exercise 10.10

Use the `airquality` data frame. Form one group of data with `Ozone` and `Temp` (two variables) for all days of the month of May and another group with the same data for the month of August. Develop discriminant analysis and MANOVA for these two groups.

10.6.6 Functions

This is function for two-group linear discriminant analysis.

```
lda2 <- function(X1,X2){

# two-group linear discriminant analysis
# Miguel Acevedo march 2004

# number of variables and observations in each group
m <- dim(X1)[2]
n1 <- dim(X1)[1]
n2 <- dim(X2)[1]

# calculate sample means
G1 <- seq(1,m); G2 <- G1
for(i in 1:m) G1[i] <- mean(X1[,i])
for(i in 1:m) G2[i] <- mean(X2[,i])
# difference of sample means
D <- G1 - G2
# var/cov
S1 <- var(X1)
S2 <- var(X2)
# pooled var-cov
Sp <- (1/(n1+n2-2)) * ((n1-1)*S1 + (n2-1)*S2)
```

(continued)

(continued)

```
# invert
I <- diag(m)
Sinv <- solve(Sp,I)

# coefficients
A <- Sinv%*%D
# scores of means
Z1 <- G1%*%A
Z2 <- G2%*%A
# score midpoint
Z0 <- (1/2)*(G1+G2)%*%A

# Mahalanobis distance
D2 <- t(D)%*%Sinv%*%D
# equivalently
#D2 <- t(D)%*%A
#D2 <- abs(Z1-Z2)

# observation scores centered
Z1s <- c(X1%*%A) - Z0
Z2s <- c(X2%*%A) - Z0

# centroid scores centered
Z1c <- Z1 - Z0
Z2c <- Z2 - Z0

smin <- min(Z1s,Z2s); smax <- max(Z1s,Z2s)
# plot
panel3(size=7)
plot(Z1s,rnorm(n1), xlim=c(smin,smax),ylim=c(-10,10), col=1,
     ylab="Arbitrary",xlab="Discriminant Scores")
points(Z2s,rnorm(n2), pch=2, col=1)
abline(h=0)
points(Z1c,0, pch="1", col=1,cex=2)
points(Z2c,0, pch="2", col=1,cex=2)

hist(Z1s,xlim=c(smin,smax),prob=T)
hist(Z2s,xlim=c(smin,smax),prob=T)

#Two-sample version of Hotelling's T2 test (similar to two sample t
  test)
    dof.num <- m
    dof.den <- n1 + n2 - m + 1
    if(dof.den > 0) {
    F1 <- (n1 +n2 -m -1)/ (m*(n1+n2-2))
    F2 <- (n1*n2)/(n1+n2)
    F <- F1*F2*D2
    p.value <- 1 - pf(F, dof.num, dof.den)
    }
    else {
    D2 <- NA; F <- NA; p.value <- NA
    }
```

```
(continued)
m.n1.n2 <- c(m,n1,n2)
Z1c.Z2c <- c(Z1c, Z2c)
D2.F.p.value <- c(D2, F, p.value)

return(list(m.n1.n2=m.n1.n2, G1=G1, G2=G2, S1=S1, S2=S2, Sp=Sp,
  Spinv=Sinv, A=A,
       Z1c.Z2c=Z1c.Z2c, D2.F.p.value=D2.F.p.value, Z1s=Z1s,
  Z2s=Z2s))
}
```

SUPPLEMENTARY READING

Chapter 6, pp. 462–479 and 572–577 (Davis, 2002); Chapter 7, pp. 124–153 (Rogerson, 2001); Chapter 3, pp. 44–59, especially pp. 47–49, Chapter 4, pp. 77–79 (Carr, 1995); Chapter 7, pp. 7-9–7-24 and 7-29–7-38, Chapter 25, pp. 25-1–25-12, pp. 25-36–25-5 (MathSoft, 1999); Chapter 5 (Qian, 2010).

11 Dependent Stochastic Processes and Time Series

In this chapter, we continue the study of random processes initiated in Chapter 7. As you recall in that chapter, we assumed that the values at any time t are independent of the values at previous times. In this chapter, we will relax this assumption to allow for the value at time t to be dependent on values at previous times. Therefore, the value at time t is conditioned on the history of the process.

11.1 MARKOV

11.1.1 DEPENDENT MODELS: MARKOV CHAIN

In many cases we can assume that the random value depends only on recent past values, instead of the entire history. For example, the rainfall today may depend on the rainfall in the past several days, but we could model it assuming that is only dependent on whether it rained yesterday. This is the basis for using **Markov** models.

Consider time steps separated by Δt and denote a time sequence as $0 \times \Delta t$, $1 \times \Delta t$, $2 \times \Delta t$, ..., $n \times \Delta t$, ... or simply 0, 1, 2, ..., n,... for short. Note that any t is given by $t = n \times \Delta t$.

A Markov chain is a very general type of probabilistic model with a finite number of discrete states. The probability of being in a given state at time t depends on the state in the previous time $t - 1$ and a conditional probability of changing to another state (Keen and Spain, 1992; Swartzman and Kaluzny, 1987). Therefore, the change to the future state depends only on the current state; in other words, the history is summarized in the current state. A more general Markov model is pth order: change to the future state depends only on the states on the past p instants of time. We can define a Markov chain based on sequences other than time, for example, distance, strata, depth (Davis, 2002, pp. 168–178).

These conditional probabilities are **transition probabilities**. The process is homogeneous if the matrix is constant, i.e., does not depend on t. Markov models have several applications in geography and environmental analysis; here we will consider examples of simulating weather and vegetation succession.

A Markov model consists of a transition probability matrix $\mathbf{P}$ that projects a vector of probabilities $\mathbf{X}(t)$ through time. Each entry of the vector $\mathbf{X}(t)$ is a probability of being in a given state and therefore the entries of $\mathbf{X}$ are the values of a pmf. The entries of $\mathbf{P}$ are probabilities of transition from state to state. Therefore,

$$\mathbf{X}(t) = \mathbf{P}\mathbf{X}(t-1) \tag{11.1}$$

Consider two states to make up an $\mathbf{X}$ of dimension 2×1

$$\mathbf{X} = \begin{bmatrix} X_1 \\ X_2 \end{bmatrix} \tag{11.2}$$

The probabilities X_1 and X_2 add up to 1 since the process can only be in one of these states. Then $\mathbf{P}$ is 2×2

$$\mathbf{P} = \begin{pmatrix} p_{11} & p_{12} \\ p_{21} & p_{22} \end{pmatrix} \tag{11.3}$$

An entry p_{ij} is the probability of transitioning from j to i. Probabilities p_{11} and p_{21} have to add to 1 since the system has to transition somewhere from the source state, state 1 in this case. Likewise, p_{12} and p_{22} have to add to 1. In general, for more than two states, then

$$\sum_j p_{ij} = 1 \tag{11.4}$$

After multiplying the probability vector $\mathbf{X}$ (state) by the matrix $\mathbf{P}$ many times, we obtain a stationary or stable probability vector. The stable distribution tells us the probability that the system will be one of the states after a long-term run.

The stable pmf or stationary distribution $\mathbf{X}^*$ can be calculated from Equation 11.1 by making $\mathbf{X}(t) = \mathbf{X}(t-1) = \mathbf{X}^*$ to obtain

$$\mathbf{X}^* = \mathbf{P}\mathbf{X}^* \tag{11.5}$$

or

$$(\mathbf{P} - \mathbf{I})\mathbf{X}^* = 0 \tag{11.6}$$

and then we need to solve for $\mathbf{X}^*$.

Take for example a system with two states

$$\begin{bmatrix} X_1^* \\ X_2^* \end{bmatrix} = \begin{bmatrix} 1 - p_{21} & p_{12} \\ p_{21} & 1 - p_{12} \end{bmatrix} \begin{bmatrix} X_1^* \\ X_2^* \end{bmatrix} \tag{11.7}$$

also $X_2^* = 1 - X_1^*$, therefore

$$X_1^*\left(1 - p_{21}\right) + p_{12}\left(1 - X_1^*\right) = X_1^* \tag{11.8}$$

and solving for X_1^*

$$X_1^* = \frac{p_{12}}{p_{21} + p_{12}} \tag{11.9}$$

Therefore, X_2^* is

$$X_2^* = \frac{p_{21}}{p_{12} + p_{21}} \tag{11.10}$$

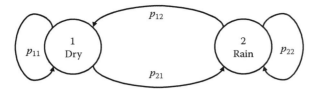

FIGURE 11.1   A Markov chain diagram for the sequence of dry or rainy days.

11.1.2   Two-Step Rainfall Generation: First Step Markov Sequence

Recall from Chapter 7 that a method used often to generate daily rainfall consists of two steps. First, determine the occurrence of rain in a day (say when precipitation exceeds 0.2 mm). We did this in Chapter 7 using a Poisson process. Second, generate the precipitation amount for that day according to a distribution skewed such as to reflect the fact that daily rainfall frequency has higher values toward the low values of rainfall.

In this chapter, the sequence of rain days is generated using a Markov process (Richardson and Nicks, 1990). As an example consider two states for a day, state 1 = dry day (it does not rain) or state 2 = wet day (it rains) (see Figure 11.1). Each day, the system can be in either a dry or a wet state.

$$\mathbf{P} = \begin{bmatrix} 0.4 & 0.2 \\ 0.6 & 0.8 \end{bmatrix} \tag{11.11}$$

Note that entry p_{21} is P(wet|dry) and p_{12} is P(dry|wet). Remember that the bar denotes conditional probability. Also $p_{21} = 1 - p_{11}$ and $p_{12} = 1 - p_{22}$

The stationary distribution is

$$X_1^* = \frac{0.2}{0.2 + 0.6} = \frac{2}{8} = \frac{1}{4} = 0.25 \tag{11.12}$$

and $X_2^* = 3/4 = 0.75$. Therefore, as a long-term average, the probability of a day being dry is ¼ and the probability of being wet is ¾.

As another example, consider that at a site, as a long-term average there are about 50% percent of rainy days in a year, and 20% of the dry days are followed by rainy days. How often are dry days followed by dry days? In this case 50% of rainy days in a year means that, whereas 20% of the dry days are followed by rainy days means that $p_{21} = 0.2$. Using Equation 11.10 we obtain

$$p_2^* = \frac{p_{21}}{p_{21} + p_{12}} = \frac{0.2}{0.2 + p_{12}} = 0.5$$

Rearrange and then solve for p_{12} from $p_{12} + 0.2 = 2 \times 0.2$ to get $p_{12} = 0.2$, then $p_{11} = 1 - p_{12} = 0.8$. The answer is that 80% of the time dry days are followed by dry days.

Figure 11.2 shows examples of realizations of Markov sequences for three cases: (1) rain days are very common $p_{22} = 0.8$, (2) rain occurrence is independent because $p_{21} = p_{12} = 0.5$, (3) dry days are very common $p_{11} = 0.8$.

11.1.3   Combining Dry/Wet Days with Amount on Wet Days

The total rainfall in a month would be the sum of the rainfall amounts for the wet days, which is a random variable R combination of the number of wet days in the month (occurring at random according to the Markov chain model) and the amount on wet days, according to the distribution selected.

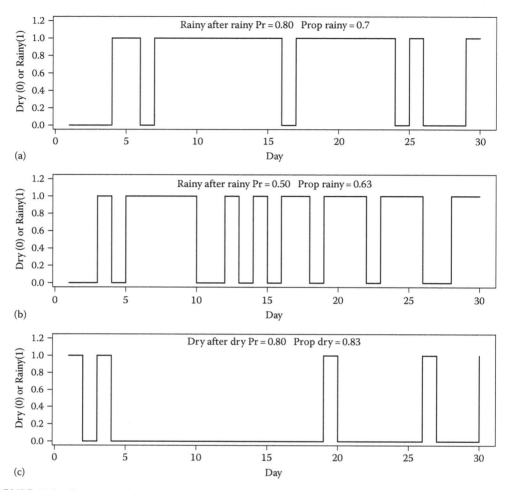

FIGURE 11.2 Examples of realizations of Markov sequences of rain days in a month. Wet days followed by wet days with high probability (a), independent case (b), and dry days followed by dry days with high probability (c).

The expected number of wet days is nX_2^* where n is the total number of days in the month and X_2^* is the probability of getting a wet day given by Equation 11.10. Therefore, the expected rainfall total for the month is the expected number of wet days times the expected rainfall for a wet day

$$\mu_R = nX_2^* \mu_X$$

For example, when using a gamma pdf with for the monthly rainfall amount

$$\mu_R = nX_2^* cb$$

Where c and b are the parameters of the gamma. The variance of R is

$$\sigma_R^2 = nX_2^* \sigma_X^2 \left(1 + cX_1^* \frac{p_{22} + p_{11}}{p_{21} + p_{12}} \right)$$

As a numeric example, take a month with $n = 30$ days and with transition matrix given in Equation 11.11 for which we calculated $X_2^* = 3/4 = 0.75$. For a gamma with $c = 0.8$, $b = 6.25$ we have mean = 5, and therefore the mean daily rainfall is $0.75 \times 5 = 3.75\,$mm and the mean rainfall total is

$$\mu_R = 30 \times 0.75 \times 5 = 112.5 \text{ mm}$$

and variance

$$\sigma_R^2 = 30 \times 0.75 \times 0.8 \times 6.25^2 \times \left(1 + 0.8 \times 0.25 \frac{0.4 + 0.8}{0.6 + 0.2}\right) =$$

$$= 703.13 \times 1.3 = 914.06$$

Figure 11.3 depicts the results of performing the two steps for a high probability of following a rainy day by a rainy day and a Weibull distribution with shape 1.3. These would be conditions for a wet month. Figure 11.4, on the other hand, illustrates the results for a dry month, i.e., when the probability of following a dry day by a dry day is high. In general, many realizations under the same monthly conditions represent samples of the daily behavior for the month. Figure 11.4, on the other

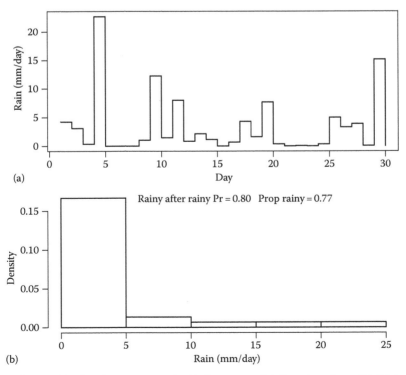

FIGURE 11.3 One realization for a wet month. (a) Sequence of daily rain values. (b) Histogram of rain values.

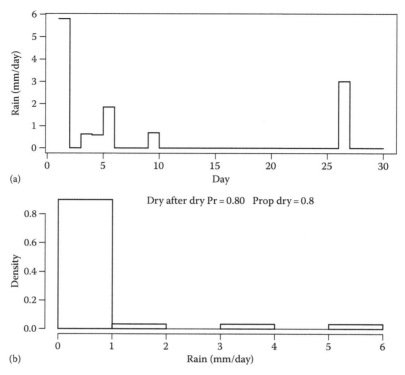

FIGURE 11.4 One realization for a dry month. (a) Sequence of daily rain values. (b) Histogram of rain values.

hand, illustrates the results for a dry month, i.e., when the probability of following a dry day by a dry day is high.

In general, many realizations under the same monthly conditions represent samples of the daily behavior for the month. As examples for wet month, see Figures 11.5 and 11.6.

11.1.4 Forest Succession

Using a Markov approach, each forest stand type is a state and we describe the dynamic changes from type to type by probabilities of transition. $\mathbf{X}(t)$ represents occupancy probabilities or fraction of space occupied by each forest type at time t.

There are many ways of defining forest types as states of a Markov model. As an example, we will combine two properties, canopy gap creating (associated with mortality) and canopy gap requiring (associated with regeneration). Using these properties, we define four ecological roles to classify the tree species. Role 1: gap-creating and gap-requiring, shade-intolerant trees that grow to a large size. Role 2: gap-creating and non-gap-requiring, shade-tolerant trees that grow to a large size. Role 3: non-gap creating and gap-requiring, shade-intolerant trees that grow to relatively small size. Role 4: non-gap-creating and non-gap-requiring, shade-tolerant trees that grow to relatively small size (Acevedo et al., 1996).

We will estimate the long-term patterns of regeneration and succession in a canopy gap. A gap-size forest plot makes transitions among several states defined by dominance of one of the four roles (Figure 11.7). At any particular time t, a tree species belonging to either one of the four roles

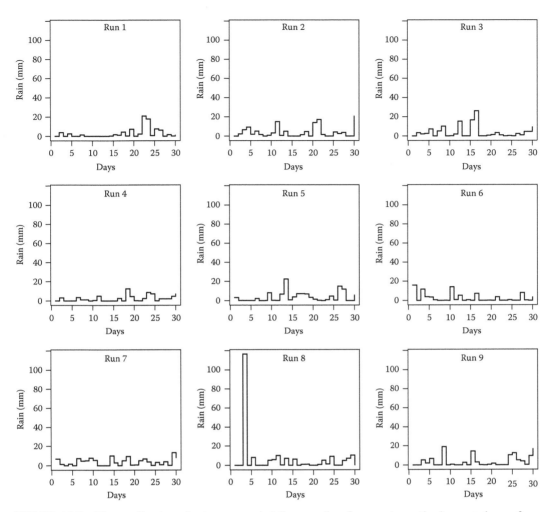

FIGURE 11.5 Nine realizations for two-step rainfall generation for a wet month. Amount drawn from Weibull density.

dominates the plot. The total canopy space covered by a collection of gap-size plots will be distributed among the roles according to proportions X_i. Logically, we will require that

$$\sum_{i=1}^{4} X_i = 1 \tag{11.13}$$

that is, the total canopy cover consists of the sum of proportions X_i of each role i in the canopy. The probability p_{ij} is associated with the transition from role j to role i. Some transitions are unlikely, and therefore their associated transition probabilities have very low values. For example, since role 3 requires gaps for regeneration and roles 3 and 4 do not produce them, we assume that p_{33} and p_{34} have low values. Likewise, p_{14}, p_{13} are negligible because roles 4 and 3 do not produce the gaps required by role 1.

Some other probabilities are also small and depicted as dashed lines because shade-tolerant species are likely to be outcompeted in gaps. Thus, p_{21}, p_{22}, p_{41}, p_{42} have low values. Since roles

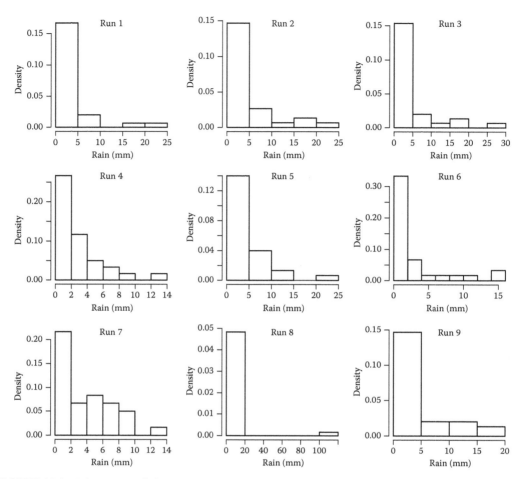

FIGURE 11.6   Histograms of nine realizations shown in Figure 11.5.

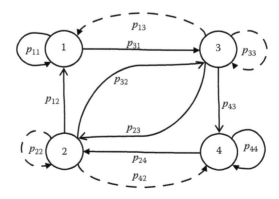

FIGURE 11.7   Markov model of forest succession based on gap creation upon mortality and shade tolerance.

3 and 4 are considered to be low stature relative to role 1 and 2, a two-layer canopy vertical structure is implicitly assumed in the model, even though the states of the model do not explicitly consider the relative proportions in a mixed canopy. A more detailed model can be written to consider the states as composed of role i in the upper canopy and role j in the lower canopy (Acevedo et al., 1995).

Several scenarios can be analyzed using this model; for example, extremes of shade-tolerance and gap-creation characteristics. These conditions will tend to make the previous set of low-value transition probabilities almost negligible. Therefore, the transition matrix $\mathbf{P}$ is

$$\mathbf{P} = \begin{bmatrix} p_{11} & p_{12} & 0 & 0 \\ 0 & 0 & p_{23} & p_{24} \\ 1-p_{11} & 1-p_{12} & 0 & 0 \\ 0 & 0 & 1-p_{23} & 1-p_{24} \end{bmatrix} \tag{11.14}$$

which has only four independent probabilities, e.g., $p_{11}, p_{12}, p_{23}, p_{24}$. These are p_{11}, probability of role 1 reoccupying gaps produced by themselves; p_{12}, probability of role 1 reoccupying gaps produced by mortality of role 2; p_{23}, probability of role 2 species reoccupying space previously occupied by role 3; and p_{24}, probability of role 2 species reoccupying space previously occupied by role 4.

A simulation using Equation 11.1 with matrix

$$\mathbf{P} = \begin{bmatrix} .5 & .4 & 0 & 0 \\ 0 & 0 & .9 & .5 \\ .5 & .6 & 0 & 0 \\ 0 & 0 & .1 & .5 \end{bmatrix}$$

yields results shown in Figure 11.8. As we can note, the fraction of space occupied by each role X_i fluctuates, but settles down to a steady state within 10 years of simulation.

The steady-state values, i.e.,

$$X_i(t) = X_i(t-1) = X_i^* \quad i = 1,\ldots,4 \tag{11.15}$$

are obtained as

$$X_1^* = p_{12}p_{24}D^{-1}$$
$$X_2^* = X_3^* = (1-p_{11})p_{24}D^{-1} \tag{11.16}$$
$$X_4^* = (1-p_{23})(1-1p_{11})D^{-1}$$

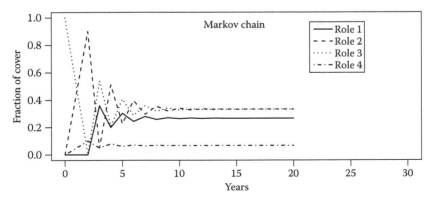

FIGURE 11.8 Markov chain dynamics with transitions among forest cover types.

where D is

$$D = (1 - p_{11})(2p_{24} + 1 - p_{23}) + p_{12}p_{24}$$

We can estimate some general patterns of forest composition. For example, if p_{11} is high, so that $p_{31} = 1 - p_{11} \approx 0$, the equilibrium tends to approximately $[1, 0, 0, 0]^T$, that is, a forest dominated by role 1. On the contrary, if p_{11} is low so that $p_{31} = 1 - p_{11} \approx 1$, the equilibrium contains all four roles in proportions $[p_{12}p_{24}, p_{24}, p_{24}, 1 - p_{23}]^T$. The value of p_{11}, discriminating colonization of gaps created by role 1, controls the dominance of role 1. As another example, a high value of p_{23}, indicating few transitions to role 4, would produce an equilibrium $[p_{11}p_{24}, (1 - p_{11})(2p_{24}), (1 - p_{11})(2 - p_{24}), 0]^T$ with a low proportion of role 4.

Succession modeling using the steady state derived from a Markov chain to infer the long-term distribution does not account for the time spent in a given state before making a transition. In the next section, we address this limitation by using semi-Markov models.

11.2　SEMI-MARKOV PROCESSES

The calculations in the previous section do not consider the different longevities and growth rates of the species. We can extend the previous approach by using semi-Markov models. In these models, transitions depend not only on the source state but also on the time spent in the source state. This is the **holding time** of the transition and it is a random variable with a given probability density function.

The Markov process underlying the semi-Markov process is the "embedded chain" and corresponds to the basic model of Figure 11.7, with associated transition probability matrix **P**. The holding time densities h_{ij}, that is, the probability densities for the time spent in making the transition from role j to i, for every pair of roles i, j will be important in determining the transients as well as the steady state.

To analyze this model, a convenient form to use for the holding time density is an Erlang density we studied in Chapter 7 (Acevedo, 1980; Acevedo et al., 1996; Hennessey, 1980; Lewis, 1977)

$$h_{ij}(\tau) = \frac{d_{ij}^{k_{ij}} \tau^{k_{ij}-1} \exp(-d_{ij}\tau)}{(k_{ij} - 1)!} \tag{11.17}$$

which has two parameters d_{ij}, and k_{ij}. The first one d_{ij} is a first-order rate corresponding to a Poisson process, and the second k_{ij} is an integer representing the order of pdf. In this expression, τ is time to make the transition. For $k_{ij} > 1$ the probability of making the transition in zero time is zero. Recall that a first-order Erlang density $k_{ij} = 1$ is an exponential. Thus, a semi-Markov process with exponential holding time is the same as a continuous time Markov process.

The mean and variance of this pdf are equal to

$$\mu_{ij} = \frac{k_{ij}}{d_{ij}} \quad \sigma_{ij}^2 = \frac{k_{ij}}{(d_{ij})^2} \tag{11.18}$$

Statistics of interest, e.g., occupancy probabilities, are calculated from available semi-Markov results (Howard, 1971) in the following manner.

The waiting time density, that is, the probability density of the time spent in role j, before making a transition to any one of the other roles is given by

$$W_j(t) = \sum_{i=1}^{4} p_{ij} h_{ij}(t) \quad j = 1, 2, \ldots, 4 \tag{11.19}$$

as the sum of all the holding time densities corresponding to transitions out of state j. The mean waiting time in role j is then the mean of $W_j(T)$, and denoted as M_j,

$$M_j = \sum_{i=1}^{4} p_{ij} \mu_{ij} \quad j = 1, 2, \ldots, 4 \tag{11.20}$$

or the sum of all products of the probability p_{ij} and the mean k_{ij}/d_{ij} of the $h_{ij}(t)$ density. In turn, the mean time between transitions M is a weighted sum of the mean waiting times in each state

$$M = \sum_{j=1}^{4} X_j^* M_j \tag{11.21}$$

where the weights used in the average are the stationary proportions of the embedded chain as given in the previous section. The steady-state occupancy probabilities, and therefore the stationary proportions X_i^{**} for all the roles can be calculated as

$$X_j^{**} = X_j^* M_j M^{-1} \quad j = 1, 2, \ldots, 4 \tag{11.22}$$

where X_i^{**} represents the fraction of space occupied by role i after a sufficiently long time has elapsed.

Under the assumption of zero probabilities for the transition shown in dashed lines in Figure 11.7, that is, the extreme-role case already discussed in the previous section, application of the previous equations yields

$$\begin{bmatrix} X_1^{**} \\ X_2^{**} \\ X_3^{**} \\ X_4^{**} \end{bmatrix} = \begin{bmatrix} p_{12}p_{24}(a_{11} + a_{31}) \\ (1 - p_{11})p_{24}(a_{12} + a_{32}) \\ (1 - p_{11})p_{24}(a_{23} + a_{43}) \\ (1 - p_{11})(1 - p_{23})(a_{44} + a_{24}) \end{bmatrix} M^{-1} D^{-1} \tag{11.23}$$

where $a_{ij} = p_{ij}\mu_{ij}$ for short and M is given by

$$M = [p_{12}p_{24}(a_{11} + a_{31}) + (1 - p_{11})p_{24}(a_{12} + a_{32}) +$$

$$+ (1 - p_{11})p_{24}(a_{23} + a_{43}) + (1 - p_{11})(1 - p_{23})(a_{44} + a_{24})]D^{-1} \tag{11.24}$$

Note that for each one of the eight transitions, three parameters require values: the rate and order for each one of eight time lags, and one transition probability. As a hypothetical example, assume the following values for the transition probabilities and the lag orders

$$\mathbf{P} = \begin{bmatrix} .5 & .4 & 0 & 0 \\ 0 & 0 & .9 & .5 \\ .5 & .6 & 0 & 0 \\ 0 & 0 & .1 & .5 \end{bmatrix} \quad \mathbf{k} = \begin{bmatrix} 2 & 2 & 0 & 0 \\ 0 & 0 & 2 & 2 \\ 2 & 2 & 0 & 0 \\ 0 & 0 & 2 & 2 \end{bmatrix} \tag{11.25}$$

Instead of specifying the rates, it is easier to start with the means of the lags, estimated from the longevities of tree species represented by each role

$$
m = \begin{bmatrix} 80 & 130 & 0 & 0 \\ 0 & 0 & 50 & 100 \\ 80 & 130 & 0 & 0 \\ 0 & 0 & 50 & 100 \end{bmatrix}
\tag{11.26}
$$

and then compute the rates as k_{ij}/μ_{ij},

$$
d = \begin{bmatrix} .025 & .0154 & 0 & 0 \\ 0 & 0 & .04 & .02 \\ .025 & .0154 & 0 & 0 \\ 0 & 0 & .04 & .02 \end{bmatrix}
\tag{11.27}
$$

Figure 11.9 illustrates four realizations of a semi-Markov process with these parameter matrices. We can see that all runs start with role 3 and then at various times the plot transitions to other states. Figure 11.10 illustrates the distribution of holding times. Averaging many realizations we can obtain dynamics like the one shown in Figure 11.11.

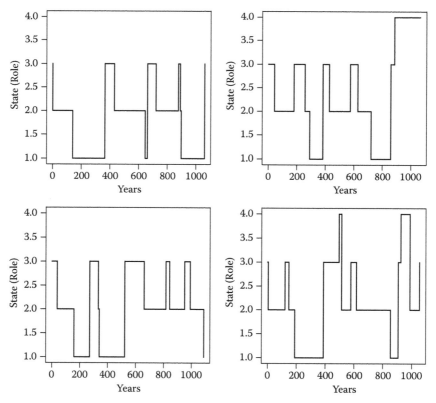

FIGURE 11.9 Four realizations of semi-Markov state transitions.

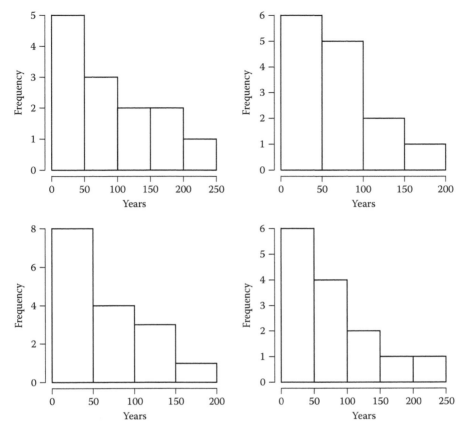

FIGURE 11.10   Histograms of holding times for realizations of Figure 11.9.

In Markov and semi-Markov models, there is the potential for expanding the state description to include environmental and other biotic factors important in ecosystem dynamics. For example, at a landscape level the parameters (probabilities and holding times) would depend on elevation, slope and its aspect, soils, etc.

In the following sections, we turn to time series and allow for dependence on a longer history of the process. A more general Markov model is pth order: change to future state depends on the states on the past p instants of time.

11.3   AUTOREGRESSIVE (AR) PROCESS

An **autoregressive** (AR) process of order p, denoted AR(p), is such that

$$x(t) = a_1 x(t-1) + a_2 x(t-2) + \cdots + a_p x(t-p) + e(t) \tag{11.28}$$

that is, at time t, $x(t)$ is a linear combination of terms x lagged up to p plus some noise (or residual variability) $e(t)$. This noise is Gaussian white noise $N(0, \sigma)$. The $x(t)$ is a process with zero mean. An AR(p) process is in essence a pth order Markov process but with an additional noise term $e(t)$. AR models are linear prediction models.

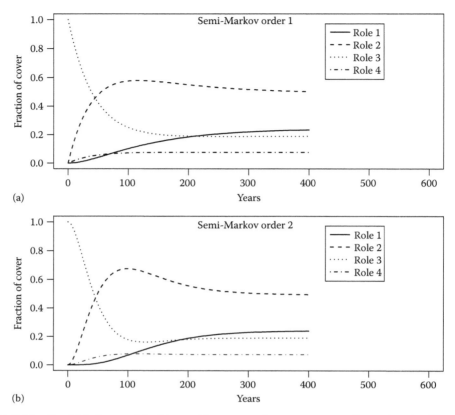

(a)

(b)

FIGURE 11.11 Semi-Markov model with transitions among forest cover types illustrating the effect of holding time in each transition. (a) Erlang order 1 and (b) Erlang order 2.

For example, an AR(1) process is simply

$$x(t) = a_1 x(t-1) + e(t) \tag{11.29}$$

which is similar to a Markov chain plus white noise. Figure 11.12 compares a realization with $a_1 = 0$ and $a_1 = -0.5$. Note that of course for $a_1 = 0$ we have a white noise process and how the ACF for $a_1 = -0.5$ now includes a negative spike at a lag of 1. Scatter plots of lagged values shows that for $a_1 = 0$ (white noise only, Figure 11.13) there is no indication of a relationship between lagged values. However, for $a_1 = -0.5$ we notice the negative relationship for values lagged by one Δt in Figure 11.14, while no relationship is suggested for other lags.

Increasing p from 1 to 2 we now have an AR(2)

$$x(t) = a_1 x(t-1) + a_2 x(t-2) + e(t) \tag{11.30}$$

Figure 11.15 shows two realizations with $a_1 = 0.6$ and $a_2 = -0.5$. Note how the ACF includes a large positive spike at lag of 1 for both realizations; however, a negative spike at lag = 2 is smaller than the spike at lag = 3 in the second realization.

The **Yule-Walker** (YW) equations are relationships in terms of covariance or of correlation. Assume $\rho(h)$ is autocorrelation of an AR(p) process $x(t)$, at lag h, then the a_i coefficients of the AR(p) satisfy the YW equations

$$\rho(h) = a_1 \rho(1-h) + a_2 \rho(2-h) + \cdots + a_p \rho(p-h) \quad \text{where } h = 1, 2, \ldots, p \tag{11.31}$$

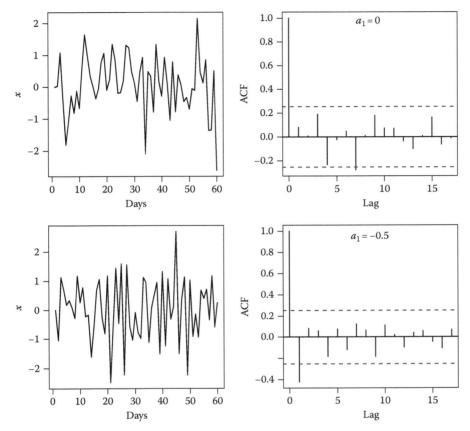

FIGURE 11.12 AR(1) one realization and its ACF for $a_1 = 0$ and $a_1 = -0.5$.

This is a system of p linear equations (recall that autocorrelation is even, $\rho(k) = \rho(-k)$)

$$\rho(1) = a_1\rho(0) + a_2\rho(1) + \cdots + a_p\rho(p-1)$$

$$\rho(2) = a_1\rho(1) + a_2\rho(0) + \cdots + a_p\rho(p-2)$$

... (11.32)

$$\rho(p) = a_1\rho(p-1) + a_2\rho(p-2) + \cdots + a_p\rho(0)$$

which we write in matrix form (recall that $\rho(0) = 1$)

$$\begin{bmatrix} \rho(1) \\ \rho(2) \\ ... \\ \rho(p) \end{bmatrix} = \begin{bmatrix} 1 & \rho(1) & ... & \rho(p-1) \\ \rho(1) & 1 & ... & \rho(p-2) \\ ... & ... & ... & ... \\ \rho(p-1) & \rho(p-2) & ... & 1 \end{bmatrix} \begin{bmatrix} a_1 \\ a_2 \\ ... \\ a_p \end{bmatrix} \quad (11.33)$$

Or for short using vector and matrix notation

$$\mathbf{r} = \mathbf{F}\,\mathbf{a} \quad (11.34)$$

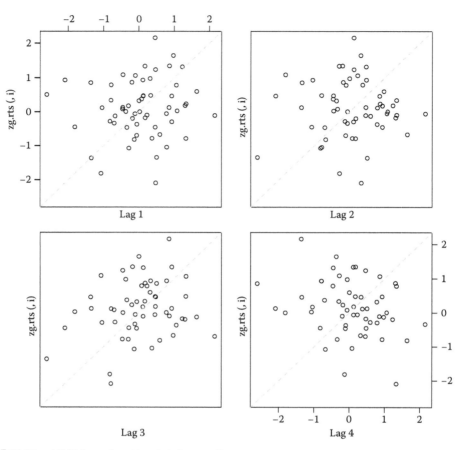

FIGURE 11.13   AR(1) lagged scatter plots for $a_1 = 0$.

Matrix $\boldsymbol{\Phi}$ is symmetric and has an inverse. Therefore, we can solve for coefficients $\mathbf{a} = \boldsymbol{\Phi}^{-1}\boldsymbol{\rho}$. For illustration, when AR(1) there is only one equation with obvious solution

$$\rho(1) = a_1 \tag{11.35}$$

For example, consider the series in Figure 11.16, the autocorrelation at lag 1 is $\rho(1) = -0.48$ and relatively smaller for higher lags. Thus, we could model the series as AR(1), then solving Equation 11.35 we have $a_1 = \rho(1) = -0.48$.

When AR(2) we have two equations

$$\rho(1) = a_1 + a_2\rho(1)$$
$$\rho(2) = a_1\rho(1) + a_2 \tag{11.36}$$

in matrix form

$$\begin{bmatrix} \rho(1) \\ \rho(2) \end{bmatrix} = \begin{bmatrix} 1 & \rho(1) \\ \rho(1) & 1 \end{bmatrix} \begin{bmatrix} a_1 \\ a_2 \end{bmatrix} \tag{11.37}$$

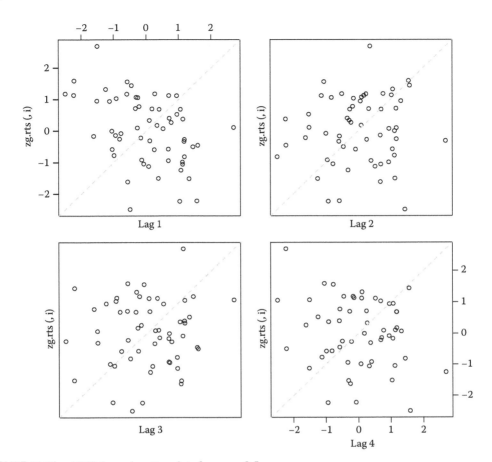

FIGURE 11.14   AR(1) lagged scatter plots for $a_1 = -0.5$.

Consider Figure 11.17, assume AR(2) and estimate two coefficients by solving Equation 11.36.

$$\begin{bmatrix} 0.3923 \\ -0.2735 \end{bmatrix} = \begin{bmatrix} 1 & 0.3923 \\ 0.3923 & 1 \end{bmatrix} \begin{bmatrix} a_1 \\ a_2 \end{bmatrix} \tag{11.38}$$

First, $a_1 = 0.59$ and $a_2 = -0.51$. Note that we arbitrarily decided to ignore the correlation at lag = 3. This series could be modeled as AR(3) by solving one more equation. An important question is then how to identify the order p of the AR(p).

The **partial autocorrelation function** (PACF) is the ACF truncated after lag p (Box and Jenkins, 1976). Partial autocorrelation is obtained by recursion: fit AR(p) models successively from $p = 1$ to the maximum lag solving the YW equations. The structure of the equations makes them solvable by a recursive method. For each step in the recursion, keep the pth coefficient. We try several values of p until the partial autocorrelation value for next p is very low or within a confidence interval. For example, the AR(2) process estimated above yields PACF 0.39 and -0.51, which correspond to the ACF at lag 1 and the a_2 estimated by solving YW shown in Equation 11.38.

The residual time series should behave like Gaussian white noise; red flags to look for deviation from white noise are outliers, trends, and drift. A handy tool to check that the residuals behave like white noise is to do an autocorrelation of the residuals. All spikes except the lag = 0 should be within the confidence interval.

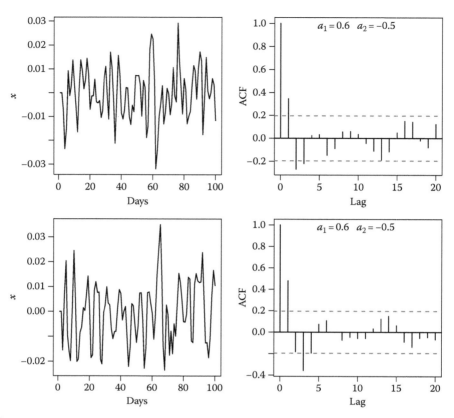

FIGURE 11.15 AR(2) two realizations for the same values of coefficients.

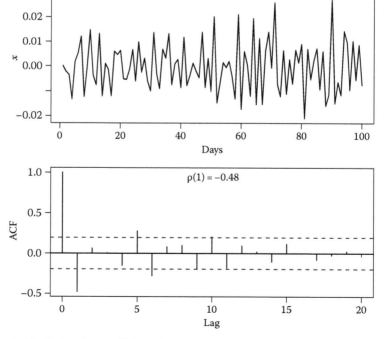

FIGURE 11.16 AR(1) Estimating coefficients from autocorrelation.

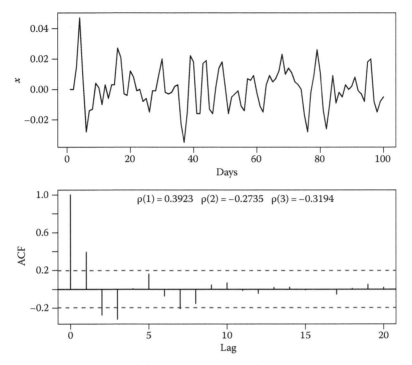

FIGURE 11.17 AR(2) Estimating coefficients from autocorrelation.

To identify the order p we can use the Akaike information criterion (AIC) already studied in previous chapters. Its objective is to balance reduction of estimated error variance with number of estimated parameters; this is accomplished by minimizing

$$AIC(p) \sim \log\left(\sigma_X^2\right) + 2p \tag{11.39}$$

Figure 11.18 illustrates the results for the simple example just discussed. We can see the plot of PACF with only two prominent spikes (top left panel), the AIC plot dropping to zero at $p = 3$ (top right panel), and Gaussian noise behavior of the residuals (bottom panels). In the computer session, we will learn how to employ these techniques for a general value of p.

11.4 ARMA AND ARIMA MODELS

An **autoregressive moving average** (ARMA) time-series model of order p and q, denoted ARMA(p, q) is

$$x(t) = a_1 x(t-1) + a_2 x(t-2) + \cdots + a_p x(t-p)$$

$$+ e(t) + b_1 e(t-1) + b_2 e(t-2) + \cdots + b_q e(t-q) \tag{11.40}$$

composed of two processes, an AR of order p and a moving average (MA) model of order q. Here $e(t)$ is white noise and called the "innovation" process; usually $N(0, \sigma)$ (but is not restricted to

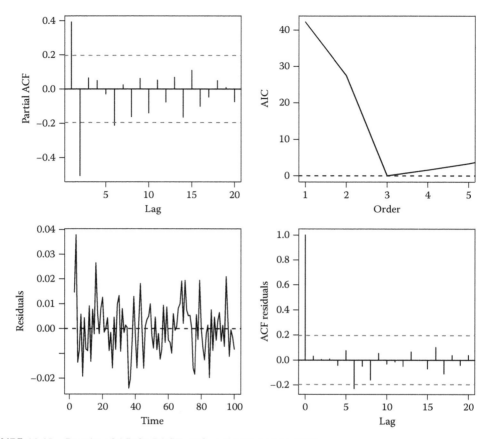

FIGURE 11.18   Results of AR(2): PACF, AIC, residual, and its ACF.

Gaussian). The process $x(t)$ is zero mean. ARMA(p, q) includes linear combination of the lagged white noise (up to q order), a_i are coefficients of an AR(p) model, b_j are coefficients of the MA(q) model. An ARMA model is only valid if the process is stationary; if not, we need to include more terms to form a more complicated model as follows.

An **autoregressive integrated moving-average** (ARIMA) model of order p, d, and q; denoted ARIMA(p, d, q) use the difference in values of $x(t)$. For example, the first-order ($d = 1$) difference $w(t) = x(t) - x(t - 1)$ and the second-order ($d = 2$) difference $w(t) = x(t) - 2x(t)x(t - 1) + x(t - 2)$. Usually, we do not include more than $d = 2$.

The Box–Jenkins method to model ARIMA series consists of three steps. First, perform **Identification**: that is to say, determine order p, d, and q, determine partial autocorrelation, estimate AIC. Then invoke Parsimony principle, i.e., use the minimum value of p, d, q which explain the data. Second, perform **Estimation**, i.e., estimate the coefficients, and for this there are several methods, such as maximum likelihood and least squares. Third, perform **Evaluation**: that is to say, study the residuals and correct model if necessary. The residuals must behave like white noise; check autocorrelation of residuals. We can also use chi-square to check the Gaussian behavior of the sum of the squares of autocorrelation of residuals up to a given lag.

One important application of time series is to use the covariance structure to model the series and then **predict** future values of the process. It is useful to establish confidence interval using the standard error. As we have seen, an underlying process (i.e., the signal) plus some noise can model a time series. It is often of interest to separate the signal from the noise. Several options are moving average filter, running median filter, and exponential smoothing filter with forecast.

11.5   EXERCISES

Exercise 11.1
What is the stationary probability distribution when $p_{21} = 0.6$ and $p_{12} = 0.4$?

Exercise 11.2
Calculate the steady state for the matrix

$$\mathbf{P} = \begin{bmatrix} .5 & .4 & 0 & 0 \\ 0 & 0 & .9 & .5 \\ .5 & .6 & 0 & 0 \\ 0 & 0 & .1 & .5 \end{bmatrix}$$

Demonstrate that it is equal to $X^* = [0.27, 0.33, 0.33, 0.07]^T$. Confirm by examining the steady state in Figure 11.8.

Exercise 11.3
Evaluate the stationary states according to Equation 11.23 and demonstrate that we get $X^{**} = [0.24, 0.49, 0.19, 0.08]^T$, compare to the X^* for the embedded process. This exercise should demonstrate the importance of the holding time in determining the stationary distribution of a semi-Markov process.

Exercise 11.4
Write the matrix of YW equations for AR(3) in a similar manner as we did in Equation 11.37.

11.6   COMPUTER SESSION: MARKOV PROCESSES AND AUTOREGRESSIVE TIME SERIES

11.6.1   WEATHER GENERATION: RAINFALL MODELS

We will use SEEG's function **markov.rain**. As explained at the end of the chapter, this implements a Markov chain of dry–rainy days based on a transition probability matrix P. Also, for rainy days draws a quantity of rain using **rain.day** from one of a set of pdf's: exponential, Weibull, gamma, and skewed normal. Arguments are the transition matrix, the number of days, and the statistics of the daily rainfall amount: mean, standard deviation, skewness, shape, and the model pdf. The function returns the sequence of rain value one for each day, as well as statistics of the sample sequence and expected values for the sequence. In addition, the function produces a graphics window with plots for the sequence and histogram of rain amounts.

As an example, let us use a month with high probability (say 0.80) that rainy days are followed by rainy days, and use mean 5 and Weibull with shape 1.3.

```
> amount.param=list(mu=5,std=8,skew=2, shape=1.3, model.pdf ="w")
> ndays=30
> # rainy followed by rainy
> P <- matrix(c(0.4,0.2,0.6,0.8), ncol=2, byrow=T)
> rainy1 <- markov.rain(P, ndays, amount.param)
> mtext(side=3,line=-1,paste("Rainy after rainy Pr=0.80","Prop
  rainy=",round(rainy1$wet.days/ndays,2)),cex=0.8)
```

The result is as shown in Figure 11.3. We see how there are many wet spells in the month and that most rain values were low. Let us look at the numeric results.

```
> rainy1
$x
 [1]   0.00   0.83   3.88   3.40   0.00   0.00 11.46 26.60   1.99   0.83
   10.53   4.22
[13] 13.11   2.69   0.05 17.98   1.32   7.04   0.00   1.22 18.62   6.04
    8.09   0.00
[25]   1.44   0.00   6.63   5.64   0.88   0.15

$wet.days
[1] 24

$expected.wet.days
[1] 22.5

$dry.days
[1] 6

$expected.dry.days
[1] 7.5

$rain.tot
[1] 154.65

$expec.rain.tot
[1] 112.5

$rain.avg
[1] 5.15

$expected.rain.avg
[1] 3.75

$rain.wet.avg
[1] 6.44

$expected.wet.avg
[1] 5
```

We see that most days were rainy, and the overall rain average of 5.15 mm (counting dry days as well) was below the average of 6.44 mm for only the wet days. We can see that the sample statistics are not too far from the expected results.

Note that every time we run this function we obtain a different realization. We can do many realizations by writing a loop around the function as given in the following script. As an example, we will use nine realizations.

```
amount.param=list(mu=5,std=8,skew=2, shape=1.3, model.pdf ="w")
ndays=30
# rainy followed by rainy
P <- matrix(c(0.4,0.2,0.6,0.8), ncol=2, byrow=T)
nruns <- 9

# define array
z <- matrix(ncol=nruns, nrow=ndays)
```

(continued)

```
# loop realizations
for(j in 1:nruns) {
 rainy <- markov.rain(P, ndays, amount.param,plot.out=F)
 z[,j] <- rainy$x
} # end of realization loop

# plot
 mat<- matrix(1:9,3,3,byrow=T)
 layout(mat,rep(7/3,3),rep(7/3,3),res=TRUE)
 par(mar=c(4,4,1,.5), xaxs="r", yaxs="r")

for(i in 1:nruns) {
 plot(z[,i], type="s", ylab="Rain(mm)", xlab="Days",
   ylim=c(0,max(z)))
 mtext(side=3,line=-1,paste("run",i),cex=0.8)
}

for(i in 1:nruns){
 hist(z[,i],prob=T,main="",xlab="Rain (mm)")
 mtext(side=3,line=-1,paste("run",i),cex=0.8)
}
```

To obtain results given in Figure 11.5 and their histograms in Figure 11.6.

11.6.2 SEMI-MARKOV

SEEG's function **semimarkov** simulates semi-Markov dynamics assuming Erlang densities for the holding times. The function is explained at the end of the chapter. We will apply it here to forest succession based on roles, using the parameter matrices given earlier in the chapter. P, Hk, and Ha are 4×4 matrices corresponding to **P**, **k**, **d**, given in Equations 11.25 and 11.27.

```
> P <- matrix(c(0.5,0.4,0.0,0.0, 0.0,0.0,0.9,0.5, 0.5,0.6,0.0,0.0,
  0.0,0.0,0.1,0.5), ncol=4, byrow=T)
> Ha <- matrix(c(0.025,0.0154,0.0,0.0, 0.0,0.0,0.04,0.02,
  0.025,0.0154,0.0,0.0, 0.0,0.0,0.04,0.02), ncol=4, byrow=T)
> Hk <- matrix(c(2,2,0.0,0.0, 0.0,0.0,2,2, 2,2,0.0,0.0, 0.0,0.0,2,2),
  ncol=4, byrow=T)

> P
     [,1] [,2] [,3] [,4]
[1,]  0.5  0.4  0.0  0.0
[2,]  0.0  0.0  0.9  0.5
[3,]  0.5  0.6  0.0  0.0
[4,]  0.0  0.0  0.1  0.5
> Ha
      [,1]    [,2]  [,3]  [,4]
[1,]  0.025  0.0154  0.00  0.00
[2,]  0.000  0.0000  0.04  0.02
[3,]  0.025  0.0154  0.00  0.00
[4,]  0.000  0.0000  0.04  0.02
```

(continued)

```
(continued)
> Hk
      [,1] [,2] [,3] [,4]
[1,]    2    2    0    0
[2,]    0    0    2    2
[3,]    2    2    0    0
[4,]    0    0    2    2
>
```

The function semi-Markov call includes five arguments: the first three are the matrices and then the simulation time and initial state. We execute four runs and call semi-Markov for each run storing results in a list.

```
nruns=4; y <- list()
for(i in 1:nruns){
y[[i]] <- semimarkov(P, Hk,Ha, tsim=1000, xinit=3)
}
```

Then we plot the realizations to obtain Figure 11.9 presented earlier.

```
panel4(size=7)
for(i in 1:nruns)
plot(y[[i]]$t,y[[i]]$x,type="s",xlab="Years",ylab="State
  (Role)",ylim=c(1,4))
```

and in addition we can plot histograms of each run to obtain Figure 11.10, which was also presented earlier.

```
panel4(size=7)
for(i in 1:nruns)
hist(y[[i]]$tau,xlab="Years",main="Hist of Holding time",cex.
  main=0.7)
```

11.6.3 AR(P) MODELING AND FORECAST

We start simple by using the series given in Figure 11.17. The data are in file **lab11/sample-rts.txt**, which has 100 records, time in days, and two time-series. We will work with the first series (second column of the file). First, we read the data, convert to time series, and then apply function ar.yw, which solves the YW equations and calculates PACF.

```
X <- read.table(file="lab11/sample-rts.txt",header=T)
Xt <- ts(X[,2],start=1,deltat=1)
ar.Xt <- ar.yw(Xt)
```

ar.Xt yields two coefficients a_1 and a_2 as we determined earlier in this chapter solving the YW. The list ar.Xt components include the estimates for the coefficients, the PACF coefficients, AIC values, the residuals, and more information; you could of course query, list and plot all of these. For example,

```
> ar.Xt$order.max
[1] 20
> ar.Xt$order
[1] 2
> ar.Xt$ar
[1]  0.5904834 -0.5051648
> ar.Xt$partialacf
, , 1

                     [,1]
 [1,]   0.392304792
 [2,]  -0.505164768
 [3,]   0.063999246
...etc ...

> ar.Xt$aic
         0            1          2          3          4          5
         6            7
42.174939 27.462806  0.000000  1.589569  3.356292  5.272075
2.686618  4.633660
         8            9         10         11         12         13
        14           15
 3.987635  5.612929  5.629536  7.362118  8.771779 10.314398
  9.572589 10.398995
        16           17         18         19         20
11.380417 13.165393 14.925897 16.920165 18.372494
> ar.Xt$resid
Time Series:
Start = 1
End = 100
Frequency = 1
   [1]            NA              NA  1.467071e-02  3.781346e-02
    -1.350453e-02
   [6] -8.719925e-03  5.740402e-03 -1.920713e-02  4.274692e-03
    -8.258361e-03
  [11] -8.899110e-03  9.080713e-03 -1.315238e-02  7.729109e-03
    -2.131724e-03
etc ......
>
```

The maximum order p explored was 20, finally an order of $p = 2$ was selected, and estimates of a_1 and a_2 are 0.59 and -0.51. The PACF are 0.39 and -0.51. Truncation to order 2 is also related to the AIC that becomes 0 at 3. We can also plot the PACF, the AIC, time plot of the residuals, and the ACF of the residuals to obtain Figure 11.18 already discussed.

```
panel4(size=7)
pacf(Xt)
ts.plot(ar.Xt$aic,ylab="AIC",xlab="Order",xli
  m=c(1,5));abline(h=0,lty=2)
ts.plot(ar.Xt$resid, ylab="Residuals");abline(h=0,lty=2)
acf(ar.Xt$resid,na.action=na.pass,ylab="ACF residuals")
```

You can forecast from the end of the series up to a future time by using the function `predict`. Use this function to forecast for the two cycles (10 days) ahead from day 50 of the data series, calculate upper and lower limit based on the double of the standard error `se`. Usually, the forecast is made for a relatively short time horizon.

```
Xt.pred <- predict(ar.Xt, ts(Xt[1:50]), n.ahead=10)
# plot
up <- Xt.pred$pred + 2*Xt.pred$se
low <- Xt.pred$pred - 2*Xt.pred$se
```

Now we will plot and plot the data together with the forecast. First, we plot the forecast by itself for says 51 to 60, and the series (top panel of Figure 11.19). Second, we plot the forecast together with the observed values in the period 51–60 for evaluation purposes (bottom panel of Figure 11.19).

```
minx<-min(Xt,low)
maxx<-max(Xt,up)
panel2(size=7)
ts.plot(ts(Xt[1:50]),Xt.pred$pred,col=1,lty=c(1,2),ylim=c
   (minx,maxx),ylab="X")
lines(up, col=1, lty=3)
lines(low, col=1, lty=3)
ts.plot(ts(Xt[1:60]),Xt.pred$pred,col=1,lty=c(1,2),ylim=c
   (minx,maxx),ylab="X")
lines(up, col=1, lty=3)
lines(low, col=1, lty=3)
```

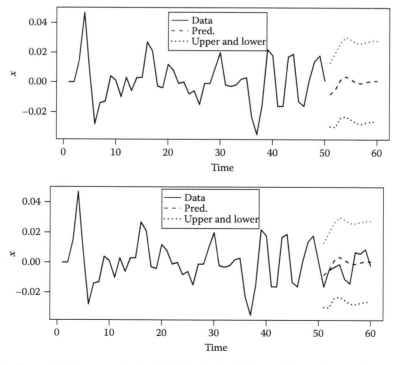

FIGURE 11.19 Forecast 10 days ahead using AR(2) for the first 50 days of data and comparing to observed data.

Note that this forecast overestimates in the period 51–57 and underestimates in the remainder of the forecasting period.

11.6.4   ARIMA(P, D, Q) MODELING AND FORECAST

Now we will learn to use function `arima` to estimate an ARIMA(p, d, q) model. One can use the more general ARIMA for the simpler AR, MA, or ARMA by doing ARIMA(p, 0, 0) for AR(p), ARIMA(0, 0, q) for MA(q), ARIMA(p, 0, q) for ARMA(p, q).

We will apply `arima` to the sunspots example; we will use the **lab7/year-spot1700-2011.txt** data set, which contains yearly number of sunspots in the period 1700–2011 (SIDC-Team, 2012). We started to study this time-series in Chapter 7 and repeat here several steps for easy reference.

First, we read the file as a `matrix` object and check

```
> yrspots <- matrix(scan("lab7/year-spot1700-2011.txt",skip=6),
  ncol=2,byrow=T)
> yrspots
       [,1]  [,2]
 [1,]  1700    5
 [2,]  1701   11
 [3,]  1702   16
```

Next create regular time series with function `ts` applied to second column of `yrspots`, with a time interval of 1 year (`deltat = 1` year) starting in 1700; next do a plot of the time series (top panel Figure 11.20) and its autocorrelation (bottom panel Figure 11.20)

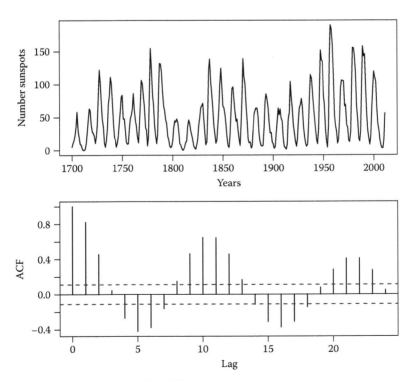

FIGURE 11.20   Sunspots time series and its ACF.

```
yrspots.rts <- ts(yrspots[,2], start=1700, deltat=1)
panel2(size=7)
ts.plot(yrspots.rts, ylab="Number Sunspots",xlab="Years")
acf(yrspots.rts)
```

Before we use `arima`, we will start by using the function `ar.yw` to fit an AR model and have a first approximation to the order p.

```
> ar.yrspots <- ar.yw(yrspots.rts)
> ar.yrspots$order.max
[1] 24
> ar.yrspots$order
[1] 9
```

The maximum order p explored was 24, finally an order of $p = 9$ was selected. For evaluation we produce plots as in the previous section

```
panel4(size=7)
pacf(yrspots.rts)
ts.plot(ar.yrspots$aic,ylab="AIC",xlab="Order",xlim=c(1,10));
  abline(h=0,lty=2)
ts.plot(ar.yrspots$resid, ylab="Residuals");abline(h=0,lty=2)
acf(ar.yrspots$resid,na.action=na.pass,ylab="ACF residuals")
```

The results (Figure 11.21) indicate that PACF values are positive for order 1, then negative for order 2–3, and then positive for orders 6–9. The AIC is zero for order 10, and there is Gaussian noise behavior of the residuals.

Now we will use function `arima` to estimate the parameters given the order $p = 9$ and a simple first approximation of $d = 0$, $q = 0$. This is basically an AR(9)

```
> arima.yrspots <- arima(yrspots.rts, order=c(9,0,0))
```

The AIC value

```
> arima.yrspots$aic
[1] 2596.879
```

Next try ARMA increasing q to 1

```
> arima.yrspots <- arima(yrspots.rts, order=c(9,0,1))
> arima.yrspots$aic
[1] 2598.859
```

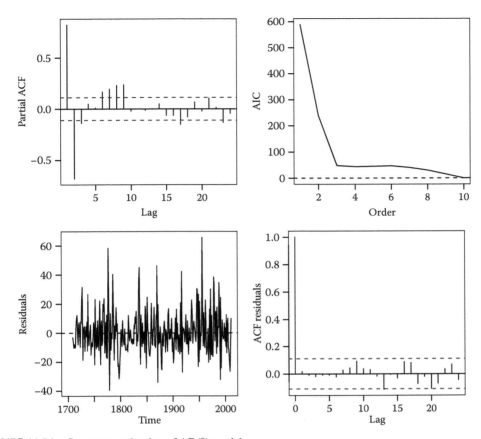

FIGURE 11.21 Sunspots evaluation of AR(9) model.

The AIC increased, thus go back to $q = 0$, but increase d

```
> arima.yrspots <- arima(yrspots.rts, order=c(9,1,0))
> arima.yrspots$aic
[1] 2593.13
```

We now have decreased the AIC and therefore try incrementing d to 2, you will find that the AIC increases. Therefore, we settle in a model ARIMA(9, 1, 0) and move on to evaluation and forecast. Note that if we invoke parsimony we could have just used an AR(9) to model the series.

Some of the results contained in the object `arima.yrspots` are the coefficients

```
> arima.yrspots$coef
        ar1             ar2             ar3             ar4             ar5
        ar6
 0.19600775 -0.20559508 -0.35935858 -0.19868828 -0.28260689
-0.25095088
        ar7             ar8             ar9
-0.19675363 -0.27280480 -0.01552607
>
```

Function `tsdiag` provides diagnostic of ARIMA models; try

```
> tsdiag(arima.yrspots)
```

To obtain Figure 11.22 that allows evaluating the white-noise behavior of the residuals. The bottom panel plot is of the p-values from chi-square tests of the Gaussian behavior of the test statistic that relates to the sum of the squares of the autocorrelation values up to a given lag.

You can forecast from the end of the series up to a future time by using the function `predict` as explained in the previous section. Use this function to forecast for the two cycles (22 years) ahead from the end of the data series (years 2012–2033), and plot the data together with the forecast. Also, for detail, we zoom in last part of the series by limiting the ranges of the x axis (time) using `xlim` to years 1960–2033.

```
yrspots.pred <- predict(arima.yrspots, n.ahead=22)
up <- yrspots.pred$pred + 2*yrspots.pred$se
low <- yrspots.pred$pred - 2*yrspots.pred$se
minx<-min(yrspots.rts,low)
maxx<-max(yrspots.rts,up)

panel2(size=7)
ts.plot(yrspots.rts,yrspots.pred$pred, col=1, lty=c(1,2),xlim=c(1700,
    2033),ylim=c(minx,maxx),ylab="X")
lines(up, col=1, lty=3)
lines(low, col=1, lty=3)
legend("top",leg=c("Data","Pred","Upper & Lower"), lty=c(1,2,3))

ts.plot(yrspots.rts,yrspots.pred$pred, col=1, lty=c(1,2),xlim=c(1960,
    2033),ylim=c(minx,maxx),ylab="X")
lines(up, col=1, lty=3)
lines(low, col=1, lty=3)
legend("top",leg=c("Data","Pred","Upper & Lower"), lty=c(1,2,3))
```

to obtain a plot like the one in Figure 11.23. Note that the prediction is lower maxima and slightly higher minima for the period 2012–2033. However, the upper bound of the confidence interval has higher maxima, whereas the lower bound has smaller minima. The use of 22 years here is for illustration of periodical behavior of the series.

One alternative, as we did in the previous section (Figure 11.19) is to identify and estimate the model with only one part of the data set, and set aside the remaining data for evaluation, e.g., 22 years in this case. Then forecast for the remaining period of 22 years and compare the forecast result with the reserved part of the dataset.

11.6.5 COMPUTER EXERCISES

Exercise 11.5
Study the rainfall values when dry days tend to be followed by dry days. The transition matrix is c(0.8, 0.6, 0.2, 0.4). Provide plots and numeric results. Discuss results. Compare to Figure 11.4.

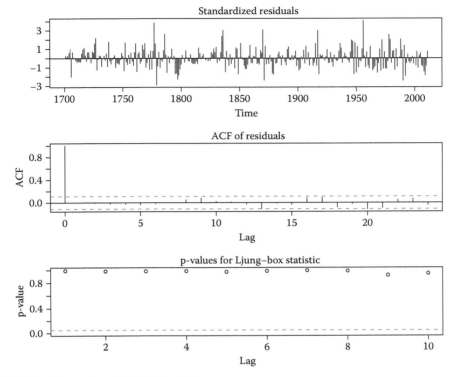

FIGURE 11.22   Diagnostic of ARIMA(9, 1, 0).

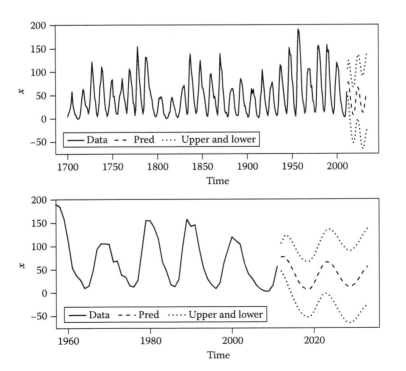

FIGURE 11.23   Sunspots forecast with 22 years horizon.

Exercise 11.6

Add lines of code to the script listed earlier to generate nine realizations (Figures 11.5 and 11.6) to calculate the sample mean of (1) the number of wet days and (2) monthly averages. Use 10 and 100 sample size (realizations). Compare and discuss. Hint: the sample mean of the number of wet days would be the average of the number of wet days for all runs. Likewise, the sample mean of the monthly means would be the average of the mean for all runs (realizations). Also, note that this can be done for the monthly of only wet days or the monthly of all days.

Exercise 11.7

Explore the effect of first-order Erlang pdf in the semi-Markov example. To modify the parameter values, edit the Hk matrix to contain all 1. Compare to the results given in the example (Figures 11.9 and 11.10). Repeat for third-order Erlang.

Exercise 11.8

Develop a semi-Markov model for five states. Four of these are the four roles already studied and the fifth state is a canopy gap or opening. Only those roles that start in gaps should have non-zero transition probabilities from gap state. Only those roles that create gaps should have non-zero transition probabilities to the gap state. Draw a transition graph, write a matrix, assign parameter values, calculate steady-state values, and develop a simulation using function semimarkov.

Exercise 11.9

Predict sunspots 22 years ahead from year 2011 (2012–2033) using a simple AR(9) model. Compare to the results obtained with ARIMA(9, 1, 0). Predict sunspots 22 years ahead from year 1980 (1981–2003). Compare to observed values in that time period.

Exercise 11.10

Consider the flow of the Neches River in Texas at station USGS 08040600 near Town Bluff, TX. File **lab7/TB-flow.csv** contains daily flow data 1952–2010. We used this file for one of the exercises in Chapter 7. Read the file, convert the flow to time series, plot the time series and autocorrelation, and develop an ARIMA model. Hint: when applying ts use freq = 365, start = 1952, end = 2010.

11.6.6 SEEG Functions

Function rain.day uses one argument, which is composed of parameters to generate the amount of rain in a day, and given as a list. As explained in the first few lines of the function, the list contains the mean, standard deviation, skew coefficient, shape, and the distribution to be sampled. The function markov.rain calls function rain.day once it determines that the current day is rainy. The function returns the amount of rain generated for the day to the calling function.

```
rain.day <- function(param){
 # mu, std and skew of daily rain used for skew
 # shape for gamma, exp, and Weibull
 # model.pdf "s" skewed, "w" weibull and "e" exponential, "g" gamma
  mu <- param[[1]]; std <-param[[2]]; skew <- param[[3]]
  shape <- param[[4]]; model.pdf <- param[[5]]
```

```
(continued)
# calc rain
 if(model.pdf=="e"){
   u <- runif(1,0,1) # generate uniform
   scale <- mu
   y <- scale*(-log(u))
 }
if(model.pdf=="w"){
   u <- runif(1,0,1) # generate uniform
   scale <- mu/gamma(shape+1)
   y <- scale*(-log(u))^shape
 }
 if(model.pdf=="g"){
   scale <- mu/shape
   y <- rgamma(1,scale,shape)
 }
 if (model.pdf == "s"){
   z <- rnorm(1,0,1) # generate standard normal
   y <- mu+ 2*(std/skew)*((((skew/6)*(z-skew/6)+1))^3 -1)
 }

 return(y)
 }
```

The function `markov.rain` implements the two-step rainfall generation. Its arguments include the transition probability of the Markov chain and the number of days, which are used for the first step. In addition, one argument is the list of amount parameters, which r are passed internally to the `rain.day` function described earlier.

```
markov.rain <- function(P, ndays, amount.param){

# arguments: markov matrix P & number of days
# and amount stats parameters

mu <- amount.param[[1]]

# define array with all 0
x <- rep(0,ndays); wet <- 0

# start first day with rain at random
y <- runif(1,0,1)

if(y > 0.5) {x[1] <- rain.day(amount.param); wet<- wet+1}

# loop for remaining days
for(i in 2:ndays){
                                                    (continued)
```

```
(continued)
# apply markov
  y <- runif(1,0,1)
  if(x[i-1]==0) {
    if(y > P[1,1]) x[i] <- rain.day(amount.param)
  }
  else {
    if(y > P[1,2]) x[i] <- rain.day(amount.param)
  }
 if(x[i] >0) wet <- wet+1

} # end of days loop
expec.wet.days <- ndays*P[2,1]/(P[1,2]+P[2,1])
expec.dry.days <- ndays - ndays*P[2,1]/(P[1,2]+P[2,1])
dry <- ndays-wet
rain.tot<-round(sum(x),2);expec.rain.tot <- expec.wet.days*mu
rain.avg=round(mean(x),2); expec.avg <- expec.rain.tot/ndays
rain.wet.avg=round(sum(x)/wet,2); expec.wet.avg <- mu

mat<- matrix(1:2,2,1,byrow=T)
layout(mat,c(7,7),c(3.5,3.5),res=TRUE)
par(mar=c(4,4,1,.5), xaxs="r", yaxs="r")
plot(x,type="s",xlab="Day",ylab="Rain (mm/day)")
Rain <- x
hist(Rain,prob=T,main="",xlab="Rain (mm/day)")

return(list(x=round(x,2),wet.days=wet,expected.wet.days=expec.wet.days,
        dry.days=dry,expected.dry.days=expec.dry.days,
        rain.tot=rain.tot, expec.rain.tot=expec.rain.tot,
        rain.avg=rain.avg,expected.rain.avg = expec.avg,
        rain.wet.avg=rain.wet.avg,expected.wet.avg=expec.wet.avg))
}
```

Function `semimarkov` simulates semi-Markov dynamics assuming Erlang densities for the holding times. It uses five arguments: the first three are the matrices and then the simulation time and initial state.

```
semimarkov <- function(P, Hk, Ha, tsim, xinit){

# arguments: markov matrix P
# Ha rates of holding times, Hk order of holding times
# simulation time

# dim state
nstate <- dim(P)[1]
# initial
x <- array(); tau <- array(); t <- array()
i=1;t[i]=0; x[1] <- xinit
```

```
(continued)
while(t[i]< tsim){
 i<- i+1
 y <- runif(1,0,1)
 z[1] <- 0; for(j in 2:(nstate+1)) z[j] <- z[j-1] + P[j-1,x[i-1]]
 for(j in 2:(nstate+1)){
  if(y>z[j-1]&& y<=z[j]) {x[i] <- j-1;tau[i]<- rgamma(1,shape=Hk[j-1,
    x[i-1]],rate=Ha[j-1,x[i-1]])}
 }
 t[i] <- t[i-1]+ tau[i]
 } # end while
return(list(x=round(x,0),tau=round(tau,2),t=t))
 }
```

It returns the state transitions, the holding times, and the simulation time.

SUPPLEMENTARY READING

Chapters 1–2 (Lewis, 1977); Chapter 16, pp. 16-1–16-27, Chapter 17, pp. 17-1–17-30 (MathSoft, 1999); Chapters 1–5, pp. 1–170 (Box and Jenkins, 1976); Chapter 8 (Manly, 2009); (Shumway and Stoffer, 2006); (Cryer and Chan, 2010); (Cowpertwait and Metcalfe, 2009).

12 Geostatistics
Kriging

12.1 KRIGING

In this chapter, we will continue the study of **geostatistics** that we commenced in Chapter 8. We will look at **Kriging** methods, which are techniques to predict values of regionalized variables in nonsampled points of a spatial domain using a collection of sampled points. This set of sampled points form a **marked point pattern**, which can be regular or irregular (Figure 12.1). As explained in Chapter 8, we calculate similarity (using the covariance) or dissimilarity (using the semivariance, Figure 12.2) between points separated at given distances and generate a model of the covariance and semivariance as a function of lag (Figure 12.3).

We then conduct kriging, which consists of using the covariance model to predict the values of the variables at the nonsampled points. A regular grid like the one shown in Figure 12.4 determines the target points for prediction. This way, we have values distributed in the entire domain; some of these are measured values while others are kriging estimates. We visualize the results as a grid or raster image with overlaid contour lines (Figure 12.5). By decreasing step size between the prediction grid points, we produce a higher resolution image (Figure 12.6).

There is a variety of Kriging procedures, including kriging in three-dimensional spatial domains. In this book, we cover only two basic methods and limit ourselves to two-dimensional domains.

12.2 ORDINARY KRIGING

Ordinary kriging helps to interpolate values of the regionalized variable $Z(x, y)$ assuming that there is not a **trend**, that is, the regionalized variable is random with constant mean. This assumes that Z is stationary, that is, its mean and variance do not change with location x, y. Since we now know matrices and vectors, we will use vector notation $\mathbf{x}$ for a generic point at coordinates x, y. Thus, $\mathbf{x}$ is a 2×1 vector with entries x and y.

More specifically, denote $\mathbf{x_0}$ as a point $\mathbf{x}$ where we want to estimate the value of Z. The estimate of $Z(\mathbf{x_0})$ is obtained by linear combination of n known values $Z(\mathbf{x_i})$ around the target location $\mathbf{x_0}$ (Figure 12.7). This procedure requires a set of weights or coefficients λ_i of this linear combination to form the equation

$$\widehat{Z}(\mathbf{x_0}) = \sum_{i=1}^{k} \lambda_i Z(\mathbf{x_i}) \tag{12.1}$$

Note that the **kriging error** is the difference between the estimate and the real value

$$e = Z(\mathbf{x_0}) - \widehat{Z}(\mathbf{x_0})$$

We want to preserve the so-called **intrinsic hypothesis**; that is, that the mean and variance are constant. Thus, the expected value of the estimate should be equal to the mean or expected

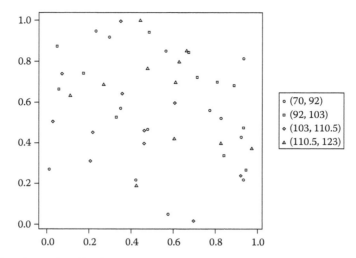

FIGURE 12.1   Sampled regionalized variable as marked point pattern.

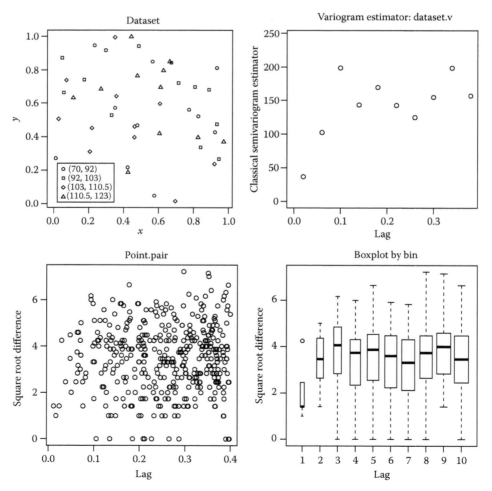

FIGURE 12.2   Semivariogram diagnostics of point pattern of previous figure.

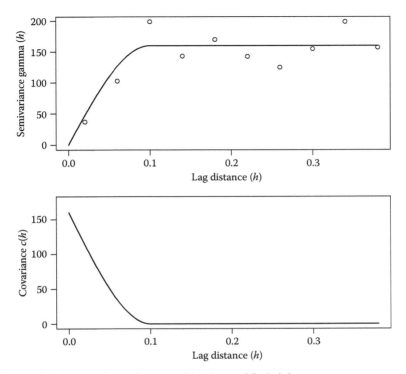

FIGURE 12.3 Semivariance and covariance model to be used for kriging.

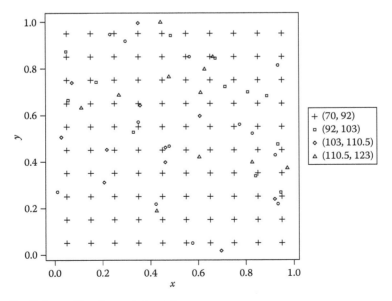

FIGURE 12.4 Prediction grid and sampled points.

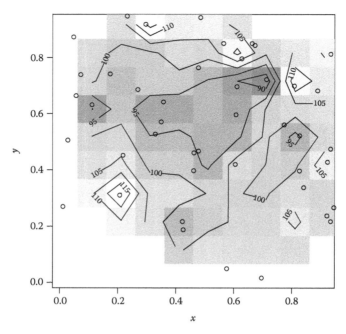

FIGURE 12.5 Kriged regionalized variable using 10 × 10 grid.

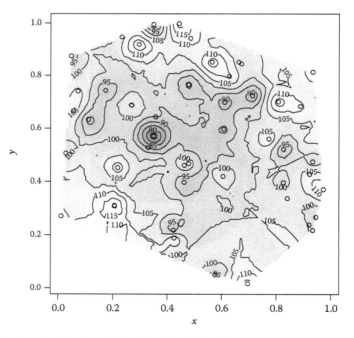

FIGURE 12.6 Kriged regionalized variable using 100 × 100 grid.

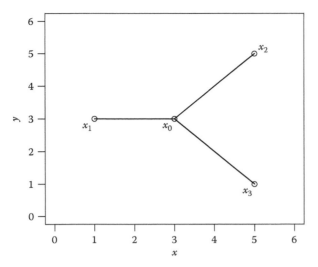

FIGURE 12.7 Calculating distances from three points to target point x0.

value of Z at any point and therefore the expected value of the kriging error should be zero. This is to say

$$E[e] = E\left[Z(\mathbf{x_0}) - \widehat{Z}(\mathbf{x_0})\right] = \mu_Z - E\left[\widehat{Z}(\mathbf{x_0})\right] = 0 \qquad (12.2)$$

The idea is to find the weights λ_i by minimizing the expected value of the square of the error

$$E[e^2] = E\left[\left(Z(\mathbf{x_0}) - \widehat{Z}(\mathbf{x_0})\right)^2\right] \qquad (12.3)$$

which happens to be equal to the **variance of the kriging error** because $E[e] = 0$.

$$\sigma_e^2 = E[e^2] - E[e]^2 = E[e^2] - 0 = E[e^2]$$

It is important to generate an image of this variance to see how the error varies with location. Figure 12.8 shows the variance of the error for the low-resolution grid, whereas Figure 12.9 illustrates this variance for the higher resolution grid.

In other words, the procedure consists of finding the λ_i such that

$$\min_{\lambda_i} \sigma_e^2 = \min_{\lambda_i} E\left[\left(Z(\mathbf{x_0}) - \widehat{Z}(\mathbf{x_0})\right)^2\right] \qquad (12.4)$$

Equation 12.2 means that we want an **unbiased** estimator.

$$E\left[\widehat{Z}(\mathbf{x_0})\right] = \mu_Z \qquad (12.5)$$

Note that

$$E\left[\widehat{Z}(\mathbf{x_0})\right] = E\left[\sum_{i=1}^{k} \lambda_i Z(\mathbf{x_i})\right] = \sum_{i=1}^{k} \lambda_i E[Z(\mathbf{x_i})] = \mu_Z \sum_{i=1}^{k} \lambda_i \qquad (12.6)$$

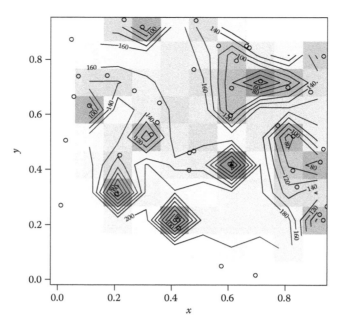

FIGURE 12.8   Variance of kriging error using 10 × 10 grid.

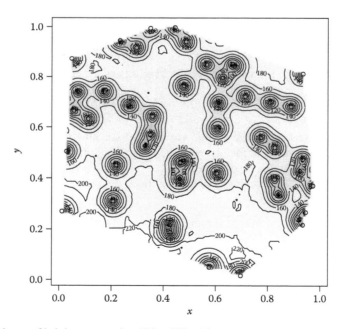

FIGURE 12.9   Variance of kriging error using 100 × 100 grid.

Because the expected value at all points $\mathbf{x_i}$ is equal to the constant mean. Therefore, to obtain Equation 12.5 we need to require that the weights sum up to 1

$$\sum_{i=1}^{k} \lambda_i = 1 \tag{12.7}$$

in order to obtain the unbiased estimates. The estimates at the measured points are the same as the observed values. This means that kriging is an **exact interpolator**.

In essence, this implies solving the optimization problem (Equation 12.4) subject to a constraint (Equation 12.7). This type of problems is solved by taking partial derivatives of the objective function (variance of the kriging error in this case) with respect to the coefficients λ_i and using an extra parameter called a **Lagrange multiplier** μ. Using this method we form the matrix equation

$$\begin{bmatrix} c(h_{11}) & c(h_{12}) & \dots & c(h_{1k}) & 1 \\ c(h_{21}) & c(h_{22}) & \dots & c(h_{2k}) & 1 \\ \dots & \dots & \dots & \dots & \dots \\ c(h_{k1}) & c(h_{k2}) & \dots & c(h_{kk}) & 1 \\ 1 & 1 & \dots & 1 & 0 \end{bmatrix} \begin{bmatrix} \lambda_1 \\ \lambda_2 \\ \dots \\ \lambda_k \\ -\mu \end{bmatrix} = \begin{bmatrix} c(h_{01}) \\ c(h_{02}) \\ \dots \\ c(h_{0k}) \\ 1 \end{bmatrix} \tag{12.8}$$

where $c(h_{ij})$ denotes the covariance model evaluated at lag h_{ij}. This is to say the covariance of $Z(\mathbf{x_i})$ and $Z(\mathbf{x_j})$ for points $\mathbf{x_i}$ and $\mathbf{x_j}$ separated by lag h_{ij}. As you can see, all the known quantities of this equation are obtained by evaluating the covariance model, or equivalently the semivariance model since

$$c(h) = \sigma_Z^2 - \gamma(h) \tag{12.9}$$

The modeled semivariogram is obtained from the empirical semivariogram as explained in Chapter 8 (see Figure 12.3).

As a simple example to facilitate understanding consider Figure 12.7 and let us calculate the estimate Z at a point $\mathbf{x_0}$ with coordinates $(3, 3)$ from measurements at three points $\mathbf{x_1}$ at $(1, 3)$, $\mathbf{x_2}$ at $(5, 5)$, and $\mathbf{x_3}$ at $(5, 1)$. The coordinate information and the values at each point are given in Table 12.1. We calculate Euclidian lag distances among all these points as shown in Table 12.2.

Assume a semivariance spherical model. We calculate semivariance for each distance of the table using the spherical model equation

$$\gamma(h) = \begin{cases} 0 & \text{when } h = 0 \\ \gamma(0^+) + \left[c(0) - \gamma(0^+) \right] \left(\dfrac{3h}{2a} - \dfrac{h^3}{2a^3} \right) & \text{when } 0 < h \le a \\ \gamma(0^+) + \left[c(0) - \gamma(0^+) \right] = c(0) & \text{when } h > a \end{cases} \tag{12.10}$$

TABLE 12.1

Values at Each Point

Point label	x	y	Z
x0	3	3	Unknown
x1	1	3	2
x2	5	1	4
x3	5	5	8

TABLE 12.2

Distances between Pair of Points

	x0	x1	x2	x3
x0	0.00			
x1	2.00	0.00		
x2	2.83	4.47	0.00	
x3	2.83	4.47	4.00	0.00

TABLE 12.3

Semivariance for Each Pair of Points

	x0	x1	x2	x3
x0	0.20			
x1	0.65	0.20		
x2	0.81	0.99	0.20	
x3	0.81	0.99	0.96	0.20

Assume nugget 0.2, sill or variance 1, range 5. We obtain semivariance for each distance of the table as shown in Table 12.3. Now we calculate covariance for each one of the distances subtracting from the sill or variance,

$$c(h) = \sigma_Z^2 - \gamma(h)$$

To obtain the results in Table 12.4. Now arrange the covariance values in matrix form according to Equation 12.8

$$\begin{bmatrix} 0.8 & 0.01 & 0.01 & 1 \\ 0.01 & 0.8 & 0.04 & 1 \\ 0.01 & 0.04 & 0.8 & 1 \\ 1 & 1 & 1 & 0 \end{bmatrix} \begin{bmatrix} \lambda 1 \\ \lambda 2 \\ \lambda 3 \\ -\mu \end{bmatrix} = \begin{bmatrix} 0.35 \\ 0.19 \\ 0.19 \\ 1.00 \end{bmatrix} \tag{12.11}$$

TABLE 12.4

Covariance for Each Pair of Points

	x0	x1	x2	x3
x0	0.80			
x1	0.35	0.80		
x2	0.19	0.01	0.80	
x3	0.19	0.01	0.04	0.80

This equation is solved to find

$$
\begin{bmatrix} \lambda_1 \\ \lambda_2 \\ \lambda_3 \\ \mu \end{bmatrix} = \begin{bmatrix} 0.475 \\ 0.262 \\ 0.262 \\ -0.035 \end{bmatrix} \tag{12.12}
$$

Now finally use the coefficients to make the estimate

$$
Z(\mathbf{x_0}) = \lambda_1 Z(\mathbf{x_1}) + \lambda_2 Z(\mathbf{x_2}) + \lambda_3 Z(\mathbf{x_3})
$$

$$
= 0.475 \times 2 + 0.262 \times 4 + 0.262 \times 8
$$

$$
= 4.1 \tag{12.13}
$$

As discussed earlier, of interest is to make the kriging predictions over many points over the spatial domain. A convenient way is to estimate the variable over a regular grid (Figure 12.4). Then for each point on the grid, we calculate the estimate as just shown. Once we have an estimate for each grid point, we can plot images and contour line maps of the regionalized variable as shown in Figures 12.5 and 12.6.

If the variable is isotropic, we can apply the omnidirectional variogram, but if the variable is anisotropic then when calculating distances and covariance at these distances we must take into account the directional variograms.

12.3   UNIVERSAL KRIGING

For ordinary kriging, we require the mean to be spatially constant so that we can derive an unbiased estimate. When $Z(\mathbf{x})$ is not stationary, we can no longer apply ordinary kriging. We know that a regionalized variable is nonstationary when the semivariogram does not reach a bound or sill. This indicates that there is a spatial trend in the data. In these cases, we use **Universal Kriging**. The regionalized variable $Z(\mathbf{x})$ can be decomposed in two parts,

$$
Z(\mathbf{x}) = m(\mathbf{x}) + R(\mathbf{x}) \tag{12.14}
$$

where
  $m(\mathbf{x}) = E[Z(\mathbf{x})]$ is the "trend"
  $R(\mathbf{x})$ is the "residual"

The residual should be stationary with zero mean.

The mean depends on location x, y. That is the expected value of $Z(x, y)$ at all points is not longer a constant, but a function of location

$$
E\big[Z(x,y)\big] = \mu_Z(x,y) \tag{12.15}
$$

For example, it may vary linearly along the x axis. Universal Kriging can be thought of as filtering and has the analogs of high-pass and low-pass filtering that we studied in Chapter 7. In **high-pass filtering**, we estimate the high-frequency fluctuation or the residual (also called the "variable"), and remove the slowly varying component or the trend. In low-pass filtering, we estimate the trend and remove the residual ("variable"). In Universal kriging parlance, the terms "**kriging with a trend**" or "**estimate the variable**" are used for high pass; whereas "**kriging the trend**" or "**estimate the trend**" for low pass.

The trend $m(\mathbf{x})$ is given as linear combination of deterministic functions of $\mathbf{x}$. Ideally, it is deducted from knowledge of the problem (e.g., topography) and usually estimated in practice from a low-order polynomial, that is, a linear or quadratic trend. One can think of the process in the following way: first find a polynomial trend surface, then obtain residuals and apply ordinary Kriging to the residuals, then add the trend and the kriged residuals. However, universal kriging is the procedure to work with the trend and avoid a sequence of the earlier steps.

A first-order polynomial for the trend is

$$m(\mathbf{x}) = \beta_0 + \beta_1 x + \beta_2 y \tag{12.16}$$

And a second order

$$m(\mathbf{x}) = \beta_0 + \beta_1 x + \beta_2 y + \beta_3 x^2 + \beta_4 y^2 + \beta_5 xy \tag{12.17}$$

The optimization problem (Equation 12.4) is now subject to additional constraints like Equations 12.16 or 12.17 which requires additional Lagrange multipliers.

12.4   DATA TRANSFORMATIONS

An implicit assumption in kriging is that the variable is normally distributed and the spatial pattern uniform. This does not always occur. For example, when the pattern is clustered or the variable is skewed. In these cases, it is necessary to transform the data before applying kriging.

For example, a clustered pattern can be de-clustered using weights. When the data are not normal, we can use a nonparametric transformation. One such transform is an indicator function that assigns the value 1 when the variable is less than a cutoff value Z_c and the value 0 otherwise

$$I(\mathbf{x}) = \begin{cases} 1 & \text{when } Z(\mathbf{x}) \le Z_c \\ 0 & \text{when } Z(\mathbf{x}) > Z_c \end{cases} \tag{12.18}$$

When the semivariogram is very noisy, an indicator transform based on a cutoff value equal to the median can make it spherical. In addition, indicator kriging reduces influence of outliers.

12.5   EXERCISES

Exercise 12.1
Calculate the estimate Z at a point $\mathbf{x_0}$ with coordinates (2, 3) from measurements at three points x1 at (1, 3), x2 at (5, 5), and x3 at (5, 1). Sketch a plot on the plane to show the location of these points. Assume a semivariance spherical model with nugget 0.2, sill or variance 1, range 5. Use the same Z values as in the example.

Exercise 12.2
Consider a waste site of size 1 km by 1 km and that we sampled for toxicant concentration at 50 points randomly distributed on the site. Determine the maximum distance to employ for the empirical semivariance calculation. Design a kriging grid that would produce predictions every 50 m in both spatial directions. Determine the number of grid columns and rows.

Exercise 12.3
Consider the situation of the previous exercise and, in addition, the site has a slope toward the bottom row. Data suggest that concentrations increase as elevation decrease. What type of kriging would you employ? Describe the kriging process.

12.6 COMPUTER SESSION: GEOSTATISTICS, KRIGING

12.6.1 ORDINARY KRIGING

Use the **lab12/example-1x1.txt** file, which is in geoEAS format. It contains x, y coordinates and marks labeled "z"

```
xyz marked point pattern
3
x
y
z
0.4785 0.4668 103
0.7734 0.5599 110
0.934  0.8124 113
0.3497 0.5704  70
0.9335 0.4743  97
0.4852 0.9415 114
0.2065 0.3103 122
… etc …
```

First, read the file, convert to point pattern, and plot the point pattern marking by quartiles using the script

```
xy <- scan.geoeas.ppp("lab12/example-1x1.txt")
xy.ppp <- ppp(xy$x, xy$y, marks=xy$z)
xyz <- point(xy)
plot.point.bw(xyz,v='z',legend.pos=2,pch=c(21:24),main="",cex=0.7)
mtext(side=1,line=2,"x")
mtext(side=2,line=2,"y")
title("Mark by quartiles",cex.main=0.7)
```

which will produce Figure 12.1. Next, calculate the omnidirectional semivariogram and plot it. Use 0.4 for maxdist in the semivariogram calculation.

```
> xyz.v <- vario(xyz,num.lags=10,type='isotropic', maxdist=0.4)
```

This function produced Figure 12.2. We check the variance in order to have a first approximation for the sill.

```
> var(xyz$z)
[1] 144.3876
>
```

Next, we build a spherical model of the semivariogram. From Figure 12.2, the following parameter values seem reasonable: nugget 0, sill 160, and range 0.1.

```
> m.xyz.v <- model.semivar.cov(var=xyz.v, nlags=10, n0=0, c0=160,
  a=0.1)
```

They provide a satisfactory fit as shown in Figure 12.3, which is produced by the function `model.semivar.cov`. Instead of using the sgeostat `fit.variogram` function, we will use SEEG's `make.variogram` function to directly force a model with these parameter values in the following manner:

```
xyz.vsph <- make.variogram(nugget=0, sill=160, range=0.1)
```

Now we use the model presented earlier to perform ordinary kriging on a 10 × 10 grid shown in Figure 12.4. For this, we use SEEG's function `Okriging` given at the end of the chapter. First, we have to select a step for the grid for the prediction. Use minimum and maximum values in each axis to select a distance step. In this case, we will use step = 0.1.

```
> xyz.ok <- Okriging(xyz, xyz.vsph, step=0.1, maxdist=0.25)
```

We obtain a dataset of the kriged values of the variable (marks z) over the prediction grid together with the variance of the kriging error. Examine `xyz.ok` and note that it has the following contents:

```
> xyz.ok
          x       y        zhat      varhat
1    0.0112 0.0163         NA          NA
2    0.1112 0.0163         NA          NA
3    0.2112 0.0163         NA          NA
4    0.3112 0.0163         NA          NA
5    0.4112 0.0163         NA          NA
6    0.5112 0.0163         NA          NA
7    0.6112 0.0163         NA          NA
8    0.7112 0.0163         NA          NA
9    0.8112 0.0163         NA          NA
10   0.9112 0.0163         NA          NA
11   0.0112 0.1163         NA          NA
12   0.1112 0.1163         NA          NA
13   0.2112 0.1163         NA          NA
14   0.3112 0.1163         NA          NA
15   0.4112 0.1163    99.37335   219.39366
16   0.5112 0.1163   103.36147   208.16463
17   0.6112 0.1163   102.52114   200.27077
18   0.7112 0.1163   102.73155   209.75406
19   0.8112 0.1163   101.90547   194.92739
... etc ...
```

Columns x and y contain the x and y coordinates of the predictions, `zhat` is the predicted value, and `varhat` is variance estimate. Also, note that the prediction was made on a grid yielding 100 values.

Now to obtain maps just apply the SEEG function `plot.kriged` given at the end of the chapter. Just do

```
> plot.kriged(xyz, xyz.ok,outpdf="lab12/xyz-kriged.pdf")
```

and obtain two maps in the PDF file just declared. The first Figure 12.5 has a raster image of the kriged values. For additional visualization, we superimpose a contour map and a plot of the original point pattern (measured points). The second map (Figure 12.8) is the variance of the kriging error and provides a visual idea of how the error varies over the domain. The function produces also output that we can use for other purposes.

We can accomplish the same results using the SEEG add on to the Rcmdr. First, select xyz as active dataset. Then go to **Spatial|Ordinary Kriging**. Enter appropriate text and values in dialog box; that is, semivariance model = xyz.vsph, step = 0.1, maxdist = 0.25, dataset to store results = xyz. ok. Press Ok and obtain the same results as given earlier for xyz.ok. Now to do the plots: reselect xyz as active dataset. Then, go to Spatial and then select Plot Kriged and we get the same plots as given earlier in Figures 12.5 and 12.8.

12.6.2 UNIVERSAL KRIGING

We will learn to do universal kriging (UK) using the **lab12/example-trend-1x1.txt** file, which is in geoEAS format. It contains x,y coordinates and marks labeled "z." It is similar to the file we used in the previous example but the marks have a trend.

As before, first read the file, convert to point pattern, and plot the point pattern marking by quartiles using the script

```
xy <- scan.geoeas.ppp("lab12/example-trend-1x1.txt")
xy.ppp <- ppp(xy$x, xy$y, marks=xy$z)
xyz <- point(xy)
plot.point.bw(xyz,v='z',legend.pos=2,pch=c(21:24),main="",cex=0.7)
mtext(side=1,line=2,"x")
mtext(side=2,line=2,"y")
title("Mark by quartiles",cex.main=0.7)
```

The result is in Figure 12.10. The trend is not evident, but a careful scrutiny would reveal that higher values are toward the NW and lower values toward the SE.

We proceed to calculate semivariance and develop a model of it (Figures 12.3 and 12.11).

```
xyz.v <- vario(xyz,num.lags=10,type='isotropic', maxdist=0.4)
m.xyz.v <- model.semivar.cov(var=xyz.v, nlags=10, n0=0, c0=160,
  a=0.1)
xyz.vsph <- make.variogram(nugget=0, sill=160, range=0.1)
```

Now we are ready for UK. In order to accomplish this, we proceed as follows:

1. Find a polynomial trend surface and extract the residuals (the residuals are examined, to find the model of the trend that provides the best fit)
2. Generate models of the semivariance and covariance for these residuals obtained in (1)
3. Apply ordinary kriging to these residuals, using the models developed in (2)
4. Add the trend obtained in (1) to the kriged residuals obtained in (3)

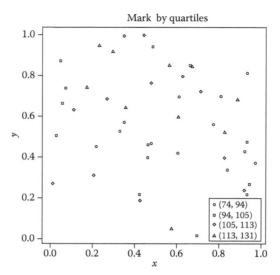

FIGURE 12.10 Point pattern with trend.

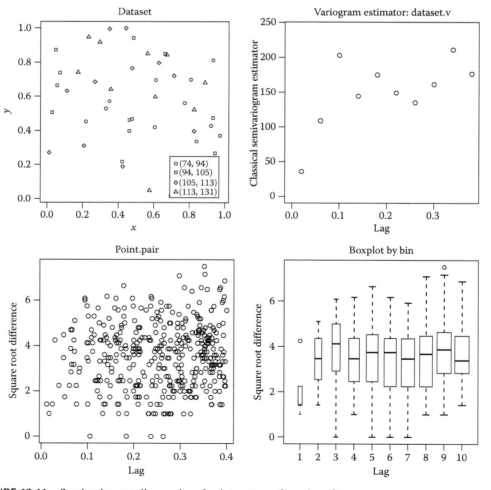

FIGURE 12.11 Semivariogram diagnostics of point pattern of previous figure.

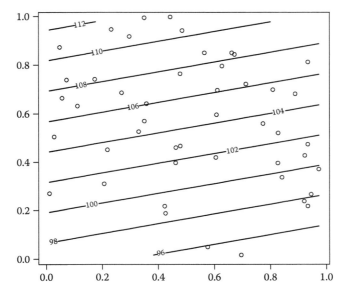

FIGURE 12.12 Extracted trend.

Let us follow this process.

1. First, find the possible trend assuming linear (polynomial order 1)

```
xyz.tr <- fit.trend(xyz,'z',np=1,plot.it=T)
```

We get the coefficients in the beta component, then x and y coordinates and a contour line plot in Figure 12.12.

```
> xyz.tr
$beta
   x^0 y^0    x^1 y^0    x^0 y^1
96.965084  -3.207682  15.952920

$R
           x^0 y^0    x^1 y^0     x^0 y^1
[1,]  -6.855655  -3.592013  -3.9008966
[2,]   0.000000  -1.978792   0.4236878
[3,]   0.000000   0.000000  -1.6719790

$np
[1] 1

$x
  [1] 0.4785 0.7734 0.9340 0.3497 0.9335 0.4852 0.2065 0.9207 0.1734
     0.8408
 [11] 0.6625 0.0112 0.6722 0.9727 0.4618 0.2963 0.4779 0.5757 0.0279
     0.4628
 [21] 0.4255 0.8262 0.8884 0.9226 0.9336 0.9453 0.2182 0.0562 0.7130
     0.4430
```

(continued)

```
(continued)
[31] 0.3572 0.6049 0.6074 0.5644 0.6106 0.4228 0.8085 0.0476 0.0722
   0.3496
[41] 0.6958 0.1109 0.8271 0.2690 0.2323 0.6272 0.3294

$y
 [1] 0.4668 0.5599 0.8124 0.5704 0.4743 0.9415 0.3103 0.2379 0.7415
   0.3373
[11] 0.8498 0.2700 0.8434 0.3711 0.3971 0.9175 0.7639 0.0493 0.5051
   0.4604
[21] 0.1879 0.3964 0.6819 0.4276 0.2175 0.2669 0.4514 0.6643 0.7219
   0.9977
[31] 0.6418 0.4195 0.5961 0.8499 0.6964 0.2173 0.6987 0.8726 0.7388
   0.9944
[41] 0.0163 0.6318 0.5210 0.6859 0.9465 0.7960 0.5268

$z
 [1] 105 111 115  74  95 121 124  95  98 101 108 106 106 112  91 130
  93  88 104
[20] 106 109  96 113 107  93  90 118 117  88 131 100 103  90 125  88
  91 121 102
[39] 115  94 113  95  89 112 112 116  94

$residuals
 [1]   2.1229685   7.5836971   8.0707385 -30.9429036  -6.5371829
  10.5716086
 [7]  22.7471109  -2.8069709 -10.2379627   1.3510150  -0.3967865
   4.7635532
[13]  -2.2635733  12.2348996 -10.8186813  19.3485475 -14.6185688
  -7.9049005
[19]  -0.9334101   3.1747065  10.4022308  -4.6386348   8.0063243
   6.1728547
[25]  -4.4401522  -8.1906966  14.5336837   9.6176624 -18.1944200
  19.5396903
[31]  -6.0578845   1.2829927 -14.5262739  16.2869446 -18.1160872
  -8.0754458
[37]  15.4820214  -8.7329170   6.4804928 -17.7072626  18.0067885
 -11.6884074
[43] -13.6234817   4.9556742   0.6806212   8.3482494 -10.3124722

attr(,"class")
[1] "trend.surface"
>
```

The equation of the trend is

$$m(x, y) = 97 - 3.21x + 15.95y$$

Now we can remove this trend by extracting residuals

```
xyz.res <- data.frame(xyz.tr$x,xyz.tr$y, xyz.tr$residuals)
colnames(xyz.res) <- c("x","y","Res")
```

2. Re-calculate the empirical variogram and generate new model variogram

```
xyz.res.v <- vario(xyz.res,num.lags=10,type='isotropic', maxdist=0.4)
xyz.res.vsph <- make.variogram(nugget=0, sill=160, range=0.1)
```

In this case we can keep the same model
3. Now, we perform kriging of the residuals with this semivariance model.

```
xyz.res.ok <- Okriging(xyz.res, xyz.res.vsph, step=0.1, maxdist=0.25)
```

We map the predicted results and variance of error as before

```
plot.kriged(xyz, xyz.res.ok,outpdf="lab12/xyz-uni-res-kriged.pdf")
```

and obtain Figures 12.13 and 12.14. We can appreciate that now we have positive and negative areas because the estimated values are of the residuals (values after removing the trend).

The same thing can be done using Rcmdr select active dataset xyz.res, go to **Spatial| Ordinary Kriging** enter text and values in dialog box and then reselect xyz.res as active dataset, go to **Spatial|Plot Kriged** and fill in dialog box with xyz.res.vsph, 0.1, 0.25, xyz.res.ok.
4. The final step is to add the trend to the kriged residuals. To do this, first find the values of the trend at each point in the grid

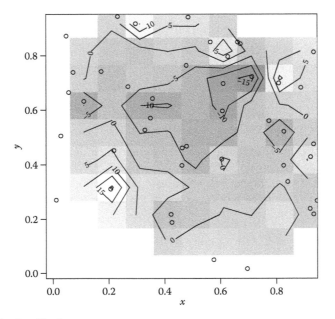

FIGURE 12.13   Kriged residuals.

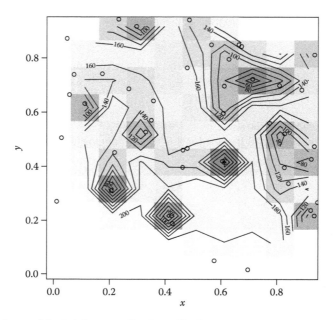

FIGURE 12.14   Variance of the kriging error for the residuals.

```
xyz.trend <- 97 -3.21*xyz.res.ok$x +15.95*xyz.res.ok$y
```

and then create and update the `zhat` using this trend (Figure 12.15)

```
xyz.uk <- xyz.res.ok
xyz.uk$zhat=xyz.trend+xyz.res.ok$zhat
```

Then use function `plot.kriged` to obtain the superimposed trend and fitted residuals

```
plot.kriged(xyz, xyz.uk,outpdf="lab12/xyz-uni-plus-kriged.pdf")
```

This last command can also be executed using the Rcmdr SEEG add on using **Spatial|Plot Kriged** and filling in the dialog box.

We can perform UK on a higher resolution grid (100 × 100) and obtain results as shown in Figure 12.16. You will develop the process as one of the exercises at the end of the chapter.

12.6.3   REGULAR GRID DATA FILES

Using a text editor, browse through the **lab12/grid30x30.txt** file. This file is in a format similar to GeoEAS but with additional information and specific convention for a regular grid. It contains values of one variable z (e.g., it could be toxicant concentration in a waste site). The file has a header with a title, then number of columns and rows for the grid; in this case 30 columns and 30 rows or 30 × 30 cells = 900 data values. Then it specifies the cell size (100 m). Therefore, the waste site is

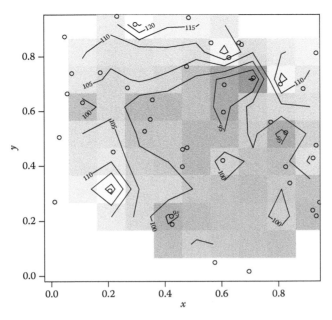

FIGURE 12.15 Kriged values by universal kriging. Trend plus kriged residuals.

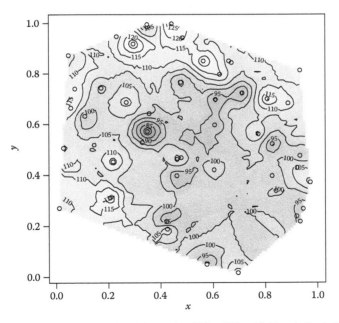

FIGURE 12.16 Kriged values by universal Kriging using 100 × 100 grid. Trend plus kriged residuals.

3000 m × 3000 m. Then the file specifies the number of variables (1) in the file, and labels for the measurements. After that, we have a stream of 900 values. In general, we can have more than one variable and therefore each record can have more than one number. The first 30 records are the southernmost row of the grid; each row goes from west to east. This is a modified GeoEAS raster or regular grid format. This is how the first few lines should look like.

```
Example Grid 30x30
ncols 30
nrows 30
size 100
nvars 1
names z
107.59
98.92
85.84
115.64
105.93
95.96
101.66
```

Let us read this file into a matrix object and then use the `image` and `contour` functions. For this, use SEEG's function `scan.map.ras`. Apply it in the following manner:

```
test.ras <- scan.map.ras("lab12/grid30x30.txt")
```

We get contour lines superimposed on an image plot Figure 12.17.

We can perform this procedure using the Rcmdr. Go to **Spatial|Read and plot geoEAS raster**. Then fill in the dialog box and browse to the file. Open and obtain same results as shown in Figure 12.17.

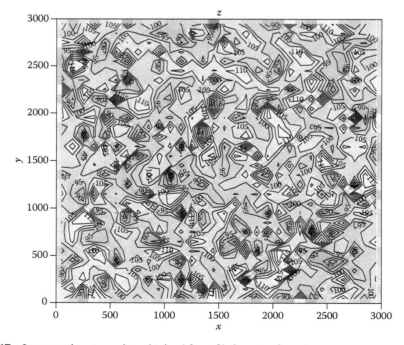

FIGURE 12.17   Image and contour plots obtained from file in raster format.

12.6.4 FUNCTIONS

The following function Okriging.R performs ordinary kriging the function is

```
Okriging <- function (dataset, vario, step, maxdist,border.
sw=F,border.poly="none"){
# dataset: columns 1 and 2 are x,y coordinates, 3 is variable
# vario is variogram model, step is interval for the prediction grid
# maxdist is max distance for prediction

# extract names of coord and variable
 x <- names(dataset)[1]
 y <- names(dataset)[2]
 v <- names(dataset)[3]

# First, select a grid for the prediction.
# Use min and max of the original dataset and a distance step

grid <- list(x=seq(min(dataset[,1]),max(dataset[,1]),by=step),
             y=seq(min(dataset[,2]),max(dataset[,2]),by=step))

# Get the range and span of x for this prediction grid

grid$xr <- range(grid$x)
grid$xs <- grid$xr[2] - grid$xr[1]

#Get the range and span of y for this prediction grid

grid$yr <- range(grid$y)
grid$ys <- grid$yr[2] - grid$yr[1]

# Then build two matrices to arrange grid coordinates for prediction

xmat <- matrix(grid$x, length(grid$x), length(grid$y))
ymat <- matrix(grid$y, length(grid$x), length(grid$y), byrow=TRUE)

#Reconvert to vectors, bind columns and convert to dataframe

grid$xy <- data.frame(cbind(c(xmat), c(ymat)))
colnames(grid$xy) <- c("x", "y")

# convert grid and datset to point patterns

grid$point <- point(grid$xy)
data.point <- point(dataset,x=x,y=y)

# Apply function krige( ) to predict over this grid point pattern
  using measured points in the dataset,
# using the variogram model and over distances from measured points
  not to exceed maxdist.
if(border.sw==F)grid$krige <- krige(grid$point,data.point,v,vario,
  maxdist=maxdist,extrap=F)
else grid$krige <- krige(grid$point,data.point,v,vario,
  maxdist=maxdist,extrap=F,border=border.poly)

result <- data.frame(x=grid$krige$x, y=grid$krige$y,
  zhat=grid$krige$zhat, varhat=grid$krige$sigma2hat)

return(result)
}
```

Note in the krige function extrap = F, limits extrapolation, the prediction area is confined around the measured points. Also note that we can apply an optional border.

Another useful function is `plot.kriged` to plot a dataset and the kriged results.

```
plot.kriged <- function (dataset, kriged, outpdf="dataset-kriged.
  pdf",border.sw=F,border.poly="none"){

# dataset: columns 1 and 2 are x,y coordinates
x <- names(dataset)[1]
y <- names(dataset)[2]

# prediction grid given by kriged dataset
grid <- list(x=unique(kriged$x),y=unique(kriged$y))

# Get the range and span of x and y for the prediction grid
grid$xr <- range(grid$x)
grid$xs <- grid$xr[2] - grid$xr[1]
grid$yr <- range(grid$y)
grid$ys <- grid$yr[2] - grid$yr[1]

# Build two matrices to arrange grid coordinates for prediction
xmat <- matrix(grid$x, length(grid$x), length(grid$y))
ymat <- matrix(grid$y, length(grid$x), length(grid$y), byrow=TRUE)

#Reconvert to vectors, bind columns and convert to dataframe
grid$xy <- data.frame(cbind(c(xmat), c(ymat)))
colnames(grid$xy) <- c("x", "y")

# Convert grid and dataset to point patterns
grid$point <- point(grid$xy)
data.point <- point(dataset,x=x,y=y)

pdf(file=outpdf,7,7)
# Declare a square plot, and find maximum to limit the plot.
par(pty="s")
grid$max <- max(grid$xs, grid$ys)
plot(grid$xy, type="n", xlim=c(grid$xr[1], grid$xr[1]+grid$max),
                   ylim=c(grid$yr[1], grid$yr[1]+grid$max))
# Form matrix of kriged values for the image plot
Zest <- matrix(kriged$zhat, length(grid$x),length(grid$y))

# Plot using image and then overlay a contour line map
image(grid$x, grid$y, Zest, col=grey(seq(0.5,1,0.02)),add=T)
contour(grid$x, grid$y, Zest, col=1,add=T)

#Plot of the original point pattern (measured points)
points(data.point)
title("Kriged values")
if(border.sw==T) lines(border.poly$x,border.poly$y,col=1,lwd=2.5)
# plot variance
par(pty="s")
plot(grid$xy, type="n", xlim=c(grid$xr[1], grid$xr[1]+grid$max),
                   ylim=c(grid$yr[1], grid$yr[1]+grid$max))
Sigmaest <- matrix(kriged$varhat,length(grid$x),length(grid$y))
image(grid$x,grid$y, Sigmaest, col=grey(seq(0.5,1,0.02)),add=T)
contour(grid$x,grid$y, Sigmaest, col=1, add=T)
points(data.point)
title("Variance Kriging error")
if(border.sw==T) lines(border.poly$x,border.poly$y,col=1,lwd=2.5)
dev.off()

}
```

The function `scan.map.ras` is based on reading the file and using `image` and `contour` commands applied to the transpose. When scanning, we take into account that the first row is the southernmost row and the last row is the northernmost row.

```
scan.map.ras <- function(filename){
  # function to scan semi-geoeas files in raster formats
  ncols <- as.numeric(scan(filename, skip=1, what=c("",0),
    nlines=1)[2])

 nrows <- as.numeric(scan(filename, skip=2, what=c("",0),
    nlines=1)[2])
  size <- as.numeric(scan(filename, skip=3, what=c("",0),
    nlines=1)[2])
  nvar <-  as.numeric(scan(filename, skip=4, what=c("",0),
    nlines=1)[2])
  namesvar <- as.character(structure(1:nvar))
  variab.ras <- list()
  for(i in 1:nvar)
  namesvar[i] <- scan(filename, skip=5+(i-1), what=c("",""),
    nlines=1)[2]
  variab <- matrix(scan(filename, skip=5+nvar), ncol=nvar, byrow=T)
  x <- seq(0+size/2,size*ncols-size/2,by=size)
  y <- seq(0+size/2,size*nrows-size/2,by=size)

  for(vcol in 1:nvar){
   variab.ras[[vcol]] <- matrix(variab[,vcol], ncol=ncols,byrow=T)
   z <- matrix(t(variab.ras[[vcol]]),length(x),length(y))
   image(x,y,z,col=grey(seq(0.5,1,0.02)))
   contour(x,y,z,add=T,col=1)
   mtext(side=3,line=2, namesvar[vcol]);
  }
  return(variab.ras)
}
```

An additional SEEG function `make.variogram` allows bypassing the sgeostat's `fit.variogram` function and instead creating a spherical model object that can be input to kriging. This way we can force a variogram model to have the parameter values we estimate from some other function different to `fit.variogram`.

```
make.variogram <- function (nugget=0, sill = 1000, range = 1000){
  spherical.v <- function(h, parameters) ifelse(h == 0, 0,
    ifelse(h <= parameters[3], parameters[1] + parameters[2] *
      (3/2 * h/parameters[3] - 1/2 * (h/parameters[3])^3),
      parameters[1] + parameters[2]))
  parameters <- c(nugget, sill, range)
  v.m.object <- list(parameters = parameters, model = spherical.v)
  names(v.m.object$parameters) <- c("nugget", "sill", "range")
  attr(v.m.object, "class") <- "variogram.model"
  attr(v.m.object, "type") <- "spherical"
  return(v.m.object)
}
```

12.6.5   COMPUTER EXERCISES

Exercise 12.4

Use the **lab12/example-1x1.txt** file. Use the spherical model built in the computer session example. Perform ordinary kriging on a 100×100 grid. Produce maps of the prediction and of the prediction error. Compare to Figures 12.6 and 12.9.

Exercise 12.5

Use the **lab12/example-trend-1x1.txt** file. Use the spherical model built in the computer session example. Perform universal kriging on a 100×100 grid. Produce maps of the prediction. Compare to Figure 12.16.

Exercise 12.6

Work with the `maas` dataset of **sgeostat** package. Recall that we worked with this spatial pattern in Chapter 8. We calculated a model for the semivariogram, and we stored the model variogram in `maas.vsph`. Use this model to perform ordinary Kriging and produce maps of kriged and variance of kriging error. Use step = 100, maxdist = 1000.

Exercise 12.7

Work with the `maas` dataset of **sgeostat** package as in the previous exercise. Perform universal Kriging and produce maps of the trend, residuals and errors, and the prediction.

SUPPLEMENTARY READING

Chapter 5 pp. 416–443 (Davis, 2002); Chapter 6 pp. 150–160, pp. 184–206, (Carr, 1995); Chapter IV pp. 61–65, pp. 91–94, pp. 107–113 (Deutsch and Journel, 1992); Chapter 4 pages 107–119 (Kaluzny et al. 1996); Chapter 12 pp. 279–322, Chapter 16, pp. 369–399 (Isaaks and Srivastava, 1989).

13 Spatial Auto-Correlation and Auto-Regression

13.1 LATTICE DATA: SPATIAL AUTO-CORRELATION AND AUTO-REGRESSION

Lattice spatial data are such that the spatial domain is divided into regions and the observations or variable values are associated with regions. There are two types of lattice data: **regular**, e.g., grid or raster, and **irregular**, e.g., polygons. Variables have a unique value for an entire region. The regions have a **neighborhood structure** given by distances between centroids or by the amount of shared borders.

One important analysis method of lattice data is **spatial auto-correlation**. Its objective is to detect spatial patterns based on correlation of a variable among regions, given the neighborhood structure. This information is useful to understand spatial patterning and to make decisions regarding the applicability of correlation and regression methods among variables. Another important method is **spatial auto-regression** (SAR). Its objective is to predict the outcome or value of a variable in a region based partially on the values of the same variable in neighboring regions and partially on other variables.

13.2 SPATIAL STRUCTURE AND VARIANCE INFLATION

An important reason for performing auto-correlation is to determine whether the assumptions of lack of serial correlation to perform regression are appropriate. You should recall now two important aspects of regression. First, for simple regression: it assumes that values of the independent variable are independent observations, i.e., they are uncorrelated. This is why we checked for auto-correlation in time when doing exploratory data analysis. Second, for multiple regression: we demonstrated that the various independent variables should not be correlated or collinear because this would lead to distorted values of the regression coefficients, giving more importance to some variables and causing variance inflation.

Correlation among values of the independent variable can occur because they have spatial dependence. Therefore, we need to make sure that the spatial structure does not affect the estimation of the coefficients. We investigate the potential for this problem using **spatial auto-correlation**, and the effect of spatial structure is included using **spatial auto-regression**.

13.3 NEIGHBORHOOD STRUCTURE

Neighborhood structure provides the covariance structure needed for spatial auto-correlation and auto-regression. There are several ways of defining neighbor regions: one is by the amount of common borders, and the other is by the distance separating a reference point of each region. For example, Figure 13.1 illustrates nine regions. The label identifies the region. The neighborhood structure is not necessarily symmetric, because it depends on how we define neighbors.

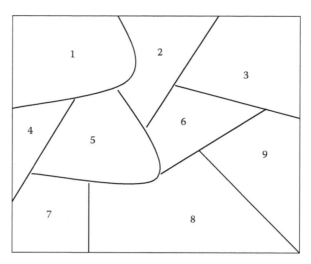

FIGURE 13.1 Lattice data: irregular or polygon.

The neighborhood structure can be stored in a **binary** matrix $\mathbf{W}$: entries w_{ij} are 1 or 0; 1 if the pair of regions are neighbors, and 0 if the pair of regions are not neighbors. For the aforementioned example, defining neighbors as those regions sharing borders, we have

$$
\mathbf{W} = \begin{bmatrix}
0 & 1 & 0 & 1 & 1 & 0 & 0 & 0 & 0 \\
1 & 0 & 1 & 0 & 1 & 1 & 0 & 0 & 0 \\
0 & 1 & 0 & 0 & 0 & 1 & 0 & 0 & 1 \\
1 & 0 & 0 & 0 & 1 & 0 & 1 & 0 & 0 \\
1 & 1 & 0 & 1 & 0 & 1 & 1 & 1 & 0 \\
0 & 1 & 1 & 0 & 1 & 0 & 0 & 1 & 1 \\
0 & 0 & 0 & 1 & 1 & 0 & 0 & 1 & 0 \\
0 & 0 & 0 & 0 & 1 & 1 & 1 & 0 & 1 \\
0 & 0 & 1 & 0 & 0 & 1 & 0 & 1 & 0
\end{bmatrix}
\tag{13.1}
$$

Note that we excluded self-neighbors, i.e., we write 0 in the main diagonal. This is an $n \times n$ matrix where n is the number of regions. Note that the sum of all non-zero intensities is the number of 1s.

By expressing graphically the existence of neighboring relationship with a line, we obtain another interesting diagram. Nodes represent region and the lines are links connecting the nodes when the regions are neighbors (Figure 13.2). The matrix $\mathbf{W}$ corresponds to Figure 13.2.

Another way to define neighbors is by distance separating their centroids. For example, neighbors are those regions with distance between centroids shorter than a threshold or cutoff distance. For example, if we look at distance between region 1 and all other regions (Figure 13.3), we may decide that only regions 1 and 4, 1 and 2 are neighbors. However, increasing the cutoff distance, also regions 1 and 5 would be neighbors.

We can also assign values other than 1 to obtain a **weighted** neighbor matrix. For example, the amount of shared border between 1 and 5 is smaller than the amount shared between 1 and 2 and 1 and 4. We can also assign weights based on distance or lengths of the links; the shorter the link connecting two nodes, the higher the weight. Consider Figure 13.2. The following matrix summarizes approximate weights based on distances

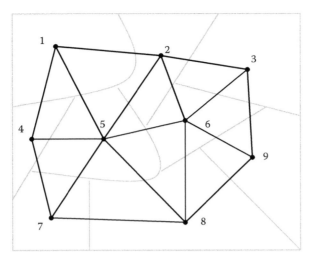

FIGURE 13.2 Neighborhood node-link diagram.

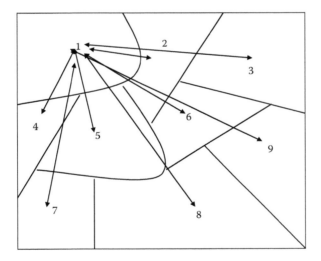

FIGURE 13.3 Distance between centroids from region 1 to all other regions.

$$
\mathbf{W} = \begin{bmatrix}
0 & 1/3 & 0 & 1/3 & 1/3 & 0 & 0 & 0 & 0 \\
0.15 & 0 & 0.35 & 0 & 0.15 & 0.35 & 0 & 0 & 0 \\
0 & 1/3 & 0 & 0 & 0 & 1/3 & 0 & 0 & 1/3 \\
0.25 & 0 & 0 & 0 & 0.5 & 0 & 0.25 & 0 & 0 \\
0.1 & 0.1 & 0 & 0.3 & 0 & 0.3 & 0.1 & 0.1 & 0 \\
0 & 0.22 & 0.22 & 0 & 0.22 & 0 & 0 & 0.12 & 0.22 \\
0 & 0 & 0 & 0.4 & 0.4 & 0 & 0 & 0.2 & 0 \\
0 & 0 & 0 & 0 & 0.2 & 0.3 & 0.2 & 0 & 0.3 \\
0 & 0 & 1/3 & 0 & 0 & 1/3 & 0 & 1/3 & 0
\end{bmatrix} \qquad (13.2)
$$

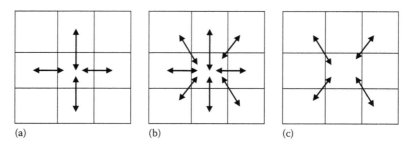

(a)   (b)   (c)

FIGURE 13.4   (a) First order (rook), (b) second order (queen); (c) diagonal (bishop).

These values were not obtained by a precise calculation and are only for illustration purposes. Note that for each node we make all weights add up to 1.

When the regions are cells of a regular lattice (grid or raster), then neighbors can be defined in several ways: first order (rook pattern), second order (queen pattern), and diagonal (bishop pattern). See Figure 13.4. These ideas can also be extended to hexagonal grids.

13.4   SPATIAL AUTO-CORRELATION

Assumes that the data are spatially stationary (i.e., no spatial trends in mean or variance). The auto-correlation depends on the weights given to the neighborhood structure.

13.4.1   Moran's *I*

Moran's *I* statistic is a common way of expressing auto-correlation. Using centered values for the variable $z_i = x_i - \mu_X$ at a region i, the Moran's *I* statistic is

$$I = \frac{\dfrac{\sum_i^n \sum_j^n w_{ij} z_i z_j}{\sum_i^n \sum_j^n w_{ij}}}{\dfrac{\sum_i^n z_i^2}{n}} \tag{13.3}$$

Here n is the number of regions. This is a ratio of standardized sum over the cross products at all pairs of regions to the sum of the squares over all regions. The numerator of the ratio is standardized by dividing over the sum of the weights for the relationship between observations at regions i and j. This equation can be rewritten in the form most commonly employed

$$I = \frac{n \sum_i^n \sum_j^n w_{ij} z_i z_j}{\left[\sum_i^n \sum_j^n w_{ij}\right] \sum_i^n z_i^2} \tag{13.4}$$

Note that if neighborhood structure is binary (weights are 1 or 0), then the sum of weights

$$\sum_i^n \sum_j^n w_{ij} \tag{13.5}$$

reduces to the number w' of connections or neighbors and therefore

$$I = \frac{n \sum_i^n \sum_j^n w_{ij} z_i z_j}{w' \sum_i^n z_i^2} \tag{13.6}$$

We can see that when neighbors have similar values, then Moran's I will tend to produce high values. The Moran's I values are always between 0 and 1 and is interpreted similarly to a correlation coefficient; values near 1 indicate spatial pattern, values near 0 indicate no spatial pattern.

Moran's I can also be calculated as a function of distance since matrix W can be defined at various values of distance d using this value as the cutoff to define the neighborhood. Typically, $I(d)$ will show a decay from positive or negative values to zero as distance increases.

The expected value of I is

$$E(I) = \mu_I = \frac{-1}{n-1} \tag{13.7}$$

which for large n becomes very small. Moran's I for large n is normal, and therefore the mean and variance of I can be used to create a standard normal.

$$Z = \frac{I - \mu_I}{\sigma_I^2} \tag{13.8}$$

Then we can perform a test with H0: there is no spatial correlation. The significance of Moran's I value is given by the p-value.

This is the **normality** assumption and requires the mean and variance to be the same for all regions, i.e., the values of the variable in each region is drawn from the same normal distribution. We can also assume that all permutations of the values of the variable are equally likely; this is the **randomization** assumption.

13.4.2 TRANSFORMATIONS

Sometimes none of these two assumptions may be valid. For example, when regions have strikingly different population densities and we are studying rates by region. These rates depend on population density and therefore variances are different in regions with low density of populations (normality does not hold). In these cases, variance depends on the mean and it is important to apply a variance-stabilization transform. One such method is the Freeman-Tukey (FT) transform.

Let X be the number of events by region, say number of crimes, and P the population by region. The FT transform is

$$y_i = \sqrt{P_i N} \left[\sqrt{x_i/P_i} + \sqrt{(x_i+1)/P_i} \right] \tag{13.9}$$

where N is a scale factor (e.g., 1,000 or 10,000 depending on the units of population) and x_i and P_i are the number of events and population in region i.

13.4.3 GEARY'S C

The Geary statistic c is a measure of dissimilarity, because it uses squared differences between values of the variable instead of the product and is divided by 2.

$$c = \frac{\dfrac{\sum_i^n \sum_j^n w_{ij}(z_i - z_j)^2}{2\left[\sum_i^n \sum_j^n w_{ij}\right]}}{\dfrac{\sum_i^n z_i^2}{n-1}} \tag{13.10}$$

This can be rewritten in the more common form

$$c = \frac{(n-1)\sum_i^n \sum_j^n w_{ij}(z_i - z_j)^2}{2\left[\sum_i^n \sum_j^n w_{ij}\right]\sum_i^n z_i^2} \tag{13.11}$$

Note that for binary weights (1 or 0) this expression reduces to

$$c = \frac{(n-1)\sum_i^n \sum_j^n w_{ij}(z_i - z_j)^2}{2w'\sum_i^n z_i^2} \tag{13.12}$$

where w' is the number of neighbor connections. Geary's c can also be calculated as a function of distance since matrix $\mathbf{W}$ will vary with cutoff distance. In this case, the numerator of Equation 13.10 is similar to the definition of semi-variance. Therefore, Geary's c is a ratio of semi-variance to square values or a standardized semi-variance.

Geary's c is also normally distributed for large number of regions n. The mean is 1, always positive. A value $c = 1$ indicates no correlation. Values of c below the mean indicate positive correlation, whereas values above the mean indicate negative correlation. Low value of c (near 0) indicates strong positive correlation. We can set up a test with null hypothesis of no correlation and the p-value gives the significance of the correlation coefficient.

13.5 SPATIAL AUTO-REGRESSION

This is a method to predict the value of a variable for regions of the lattice, taking into account the spatial structure. The most common model is the **simultaneous auto-regression model (SAR)**. Recall that a linear regression model

$$\mathbf{y} = \mathbf{xb} + \mathbf{e} \tag{13.13}$$

where
    $\mathbf{y}$ is a $n \times 1$ vector of observations of one dependent or response variable Y
    $\mathbf{x}$ is a $n \times (m + 1)$ matrix of observations of m independent or explanatory variables X_i
    $\mathbf{b}$ is $(m + 1) \times 1$ vector of coefficients
    $\mathbf{e}$ is a $n \times 1$ vector of errors

There are two approaches to include the spatial structure. One is to include auto-regression of the errors (the spatial error model), and the other to include auto-regression of the response Y (the spatial lag model).

The lagged response or **spatial lag model** includes an additional term consisting of including the lagged responses as part of the prediction

$$\mathbf{y} = \mathbf{xb} + \mathbf{e} + \rho\mathbf{Wy} \tag{13.14}$$

The lagged responses $\rho\mathbf{Wy}$ are obtained with the weighted neighbor structure matrix $\mathbf{W}$, and the auto-correlation parameter ρ. The spatial correlation structure determines parameter ρ, which then gives the intensity of the correlation of the response.

Note the lagged response at each region is the linear combination of the response at all other regions, scaled by ρ. Note that the weights are the entries of the spatial structure matrix $\mathbf{W}$

$$z_i = \rho \sum_{j=1}^{n} w_{ij} y_j \tag{13.15}$$

We apply this spatial lag model when the spatial effect seems to occur due to structural or process relationships among the variables across the regions. Another way to look at this is to think of the regression error as $\mathbf{e} + \rho\mathbf{Wy}$, which is the residual error plus a spatial effect given by $\rho\mathbf{Wy}$. Therefore, for a spatially uncorrelated response (i.e., $\rho = 0$), the error is just the residual term in a linear regression.

The **spatial error model** consists of including an additional term of lagged error or fluctuation $\mathbf{u}$

$$\mathbf{y} = \mathbf{xb} + \mathbf{e} + \lambda\mathbf{Wu} \tag{13.16}$$

As mentioned earlier, the lagged fluctuations are obtained with the weighted neighbor structure matrix $\mathbf{W}$, and the auto-correlation parameter λ. The spatial correlation structure determines λ. This parameter gives the intensity of the correlation of the fluctuation.

Note the lagged fluctuation at each region is the linear combination of the fluctuation at all other regions, scaled by λ, and where the weights are the entries of the spatial structure matrix $\mathbf{W}$

$$z_i = \lambda \sum_{j=1}^{n} w_{ij} u_j \tag{13.17}$$

We apply this spatial error model when the spatial effect seems to occur due to unknown factors or missing variables. As given earlier, the problem is how to find parameters $\mathbf{b}$ and λ, which minimize the error. Also as given earlier, another way to look at this is to think of the regression error as $\mathbf{e} + \lambda\mathbf{Wu}$, which is the residual error plus a spatial effect given by $\lambda\mathbf{Wu}$. So, for a spatially uncorrelated error, $\lambda = 0$, the error is just the residual term in a linear regression.

The goal in both models, spatial lag or spatial error, is to find parameters $\mathbf{b}$ and ρ (in the case of spatial lag) and λ (in the case of spatial error) to minimize square of residual errors $\mathbf{e}$ or residual variance σ^2. Then once we have the estimated value of the parameters $\mathbf{b}$ and ρ (or λ), we can use these in the predictor equation. For example, using the method of maximum likelihood it can be shown that the estimation of ρ reduces to maximizing

$$\ln\left(|\mathbf{I}-\rho\mathbf{W}|\right) - \left(\left((\mathbf{I}-\rho\mathbf{W})\mathbf{y}\right)^T \left(\mathbf{I}-\mathbf{x}(\mathbf{x}^T\mathbf{x})^{-1}\mathbf{x}^T\right)^T \left(\mathbf{I}-\mathbf{x}(\mathbf{x}^T\mathbf{x})^{-1}\mathbf{x}^T\right)(\mathbf{I}-\rho\mathbf{W})\mathbf{y}\right) \qquad (13.18)$$

which has two major pieces; one is the log of the determinant and the other a sum of squares. Recall that T denotes matrix transpose and $\mathbf{I}$ is the identity matrix.

Once we find an estimate, $\hat{\rho}$, we can find the estimate of the regression coefficient as we did for multiple regression. Note that Equation 13.14 can be rewritten as

$$\mathbf{y}-\hat{\rho}\mathbf{W}\mathbf{y} = (\mathbf{I}-\hat{\rho}\mathbf{W})\mathbf{y} = \mathbf{x}\mathbf{b} \qquad (13.19)$$

as before we need to solve $\mathbf{b}$, and this can be done by pre-multiplying both sides by $\mathbf{x}^T$

$$\mathbf{x}^T(\mathbf{I}-\hat{\rho}\mathbf{W})\mathbf{y} = \mathbf{x}^T\mathbf{x}\mathbf{b} \qquad (13.20)$$

And then pre-multiplying both sides by the inverse of $\mathbf{x}^T\mathbf{x}$

$$\mathbf{b} = (\mathbf{x}^T\mathbf{x})^{-1}\mathbf{x}^T(\mathbf{I}-\hat{\rho}\mathbf{W})\mathbf{y} \qquad (13.21)$$

Using another symmetric matrix $\mathbf{D}$, the covariance matrix $\mathbf{S}$ of the spatially correlated dependent variables can be calculated

$$\mathbf{S} = (\mathbf{I}-\rho\mathbf{W})^T\mathbf{D}^{-1}(\mathbf{I}-\rho\mathbf{W})^{-1}\sigma^2 \qquad (13.22)$$

The diagonal matrix $\mathbf{D}$ is also called a "weight" matrix, but it is **not** the matrix of neighbor weights. Be careful not to confuse them: the same word is used for two different things.

Alternatively, we could calculate it using the conditional auto-regression (CAR) model

$$\mathbf{S} = (\mathbf{I}-\rho\mathbf{W})^{-1}\mathbf{D}\sigma^2 \qquad (13.23)$$

or the moving average spatial regression (MA) model

$$\mathbf{S} = (\mathbf{I}+\rho\mathbf{W})\mathbf{D}(\mathbf{I}+\rho\mathbf{W})^T\sigma^2 \qquad (13.24)$$

We will not discuss CAR and MA in this book.

13.6   EXERCISES

Exercise 13.1
Table 13.1 gives the matrix of distances (in km) between county seats for five counties. Define neighborhood based on distance between county seats less than 30 km. Form a binary neighborhood matrix.

TABLE 13.1
Distances for Pairs of County Seats

0				
55	0			
23	21	0		
60	13	50	0	
80	45	10	70	0

TABLE 13.2
Values of Two Variables for Each Region

Region	X	Y
1	2.6	3.8
2	4.5	4.5
3	2.7	4.3
4	1.2	2.5
5	0.8	2.8

Exercise 13.2

Consider distances of Table 13.1 and the results of the previous exercise. For each pair of neighbors (distance <30 km) calculate the inverse of distance and use it to form a weighted neighborhood matrix. Finally, row-standardize this matrix, i.e., divide the values of inverse distance in each row by the number of connections in the row.

Exercise 13.3

Table 13.2 provides the values of two variables X and Y for the aforementioned regions. First, center each variable with respect to the mean (this is to say, subtract the mean from each value). Use the centered values and the **W** resulting from the previous exercise to calculate Moran's I for both X and Y.

Exercise 13.4

Write the SAR equation (Equation 13.14) to predict Y from X of the previous exercise, using the **W** matrix from the first exercise.

13.7 COMPUTER SESSION: SPATIAL CORRELATION AND REGRESSION

13.7.1 PACKAGES

We will use the **spdep** package for spatial regressions. One way of calculating neighbors in this package requires a triangulation package called **tripack**. In addition, spdep requires package **maptool**. First, install all these packages from the CRAN website and then load these packages: **maptool**, **tripack**, and **spdep**. The packages can be loaded automatically by modifying your **Rprofile.site**.

```
# loading packages automatically
local({
old <- getOption("defaultPackages")
options(defaultPackages = c(old, "spatstat", "sgeostat", "Rcmdr",
        "maptools", "tripack", "spdep"))
})
```

13.7.2 MAPPING REGIONS

As an example, we will use the dataset in **lab13/demo-spreg.txt**. It is on hypothetical disease incidence by region as an irregular lattice. First, read the dataset and examine it

```
> xd <- read.table("lab13/demo-spreg.txt")
> xd
   id X1    X2   X3     x     y
1   1  9  8785 4488 0.476 0.157
2   2  7  6917 3664 0.799 0.299
3   3 10  9883 4863 0.953 0.357
4   4 18 18094 8755 0.028 0.266
5   5 17 17177 8633 0.240 0.523
6   6 19 19060 9505 0.968 0.182
7   7 14 14069 6972 0.199 0.710
8   8 17 17006 8597 0.003 0.465
9   9 16 16036 7957 0.916 0.451
10 10  4  4037 2054 0.490 0.854
11 11  2  1918 1005 0.574 0.995
12 12  3  2871 1618 0.984 0.706
13 13 16 16006 8076 0.055 0.287
14 14  8  7810 3881 0.354 0.737
15 15 16 15999 8050 0.293 0.858
16 16 13 13028 6682 0.362 0.736
17 17 15 15121 7575 0.596 0.862
18 18  3  2900 1457 0.710 0.985
19 19 13 12922 6669 0.155 0.748
20 20 14 14118 7065 0.190 0.265
>
```

We have 20 regions with id, then three variables X1, X2, X3, and coordinates x,y that define the location of the region's main village. Variable X1 is a discrete count of incidence for each region in a given period. Variable X2 is the population per region, and X3 is the old-population suspect to be prone to the disease.

Of course, a first step is to look at a map of the regions we are dealing with. We use function `readShapePoly` from maptools to read a file **lab13/demo-spreg.shp**. This file is in shape format (shp); the function converts to a polygon usable by maptools and spdep. Then function `plot` will plot the map, and we use `text` to write a label on each region (Figure 13.5).

```
xp <- readShapePoly("lab13/demo-spreg.shp")
plot(xp,axes=T,xlim=c(-0.25,1))
lab.xp <- as.character(seq(1,20,1))
text(coordinates(xp), labels=lab.xp, cex=0.6)
```

We can plot a map of regions coded according to the value of a variable, say X1. First, we examine the values of X1 and cut the values of the variable in intervals

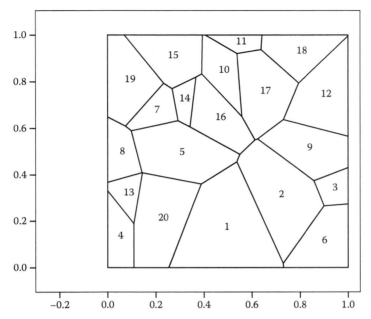

FIGURE 13.5 Map of regions.

```
> xd$X1
 [1]  9  7 10 18 17 19 14 17 16  4  2  3 16  8 16 13 15  3 13 14

> x1 <- as.ordered(cut(xd$X1, breaks=seq(0,20,2), include.lowest=TRUE))
> unclass(x1)
 [1] 4 3 4 7 7 8 5 7 6 2 1 2 6 3 6 5 6 2 5 5
attr(,"levels")
[1] "[0,2]"  "(2,4]"  "(6,8]"  "(8,10]"  "(12,14]"  "(14,16]"  "(16,18]"
[8] "(18,20]"
>
```

We see how each region is assigned a code from the numbers 1–8. Then we assign a gray tone (darker for higher values) to each interval and plot the map of polygons according to the intervals and color codes selected (Figure 13.6)

```
cols <- grey(seq(1,0.3,-0.1))
plot(xp, col=cols[unclass(x1)], border = par("fg"),
axes=T,xlim=c(-0.25,1))
legend("topleft", legend=paste("x1", levels(x1)), fill=cols,
  bty="n")
```

As explained in the help of **spdep**, we can also view the map as probabilities of observing these values if rates were to follow a Poisson distribution. First, compute the parameter for the Poisson with reference to the population X2

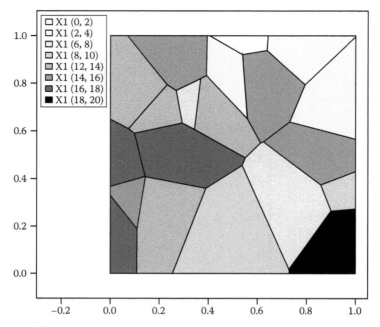

FIGURE 13.6 Map of colored polygons to show categorical values of variables.

```
> xp.phat <- sum(xd$X1) / sum(xd$X2)
```

apply the cumulative of the Poisson (ppois)

```
> pm <- ppois(xd$X1, xp.phat*xd$X2)
```

As given earlier cut, assign gray colors, and map

```
pm.f <- as.ordered(cut(pm, breaks=seq(0.50,0.65,0.005), include.
  lowest=TRUE))
cols <- grey(seq(1,0.3,-0.1))
plot(xp, col=cols[unclass(pm.f)],border = par("fg"),axes=T,xli
  m=c(-0.25,1))
legend("topleft", legend=paste("prob.", levels(pm.f)), fill=cols,
  bty="n")
```

We obtain Figure 13.7.

13.7.3 NEIGHBORHOOD STRUCTURE

For auto-correlation and auto-regression, we need to build a neighbor structure to be stored as object of class nb. Let us find the weighted neighbor matrix. Neighbors can be defined according to distance between region's main villages or amount of shared borders. The weights would

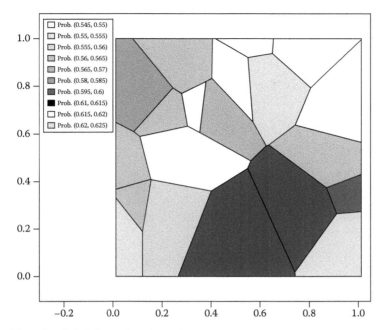

FIGURE 13.7   Map of probabilities assigned to polygons.

represent intensity of neighbor relationship (e.g., extent of common boundary, closeness of centroids). Assigning weights is critical since spatial correlation and spatial regression models will eventually depend on the weights.

13.7.4   Structure Using Distance

First, let us use the distance. Bind the coordinates of region villages as a 20×2 matrix.

```
> coords <- cbind(xd$x, xd$y)
```

Then use function `dnearneigh` to find neighbors (start with villages <0.3 distance of each other). This function will exclude the redundant cases when a region is neighbor with itself.

```
> coords <- cbind(xd$x, xd$y)
> xd.nb <- dnearneigh(coords, 0, 0.3, row.names = rownames(xd))
> xd.nb
Neighbour list object:
Number of regions: 20
Number of nonzero links: 82
Percentage nonzero weights: 20.5
Average number of links: 4.1
1 region with no links:
1
```

You can find the link for each region by addressing the id number, for example, for regions 1 and 2

```
> xd.nb[1:2]
[[1]]
[1] 0
[[2]]
[1] 3 6 9

>
```

This means that region 1 has no neighbors and region 2 has regions 3, 6, and 9 as neighbors. Note that self-neighbors are excluded.

As you go through xd.nb you realize that this distance may be too restrictive. This is more evident if you plot the links

```
> plot(xd.nb, coords)
```

to obtain Figure 13.8 where you see disconnection among regions. Thus, let us increase the neighbor cutoff distance to 0.4 to see if we include these regions.

```
> xd.nb <- dnearneigh(coords, 0, 0.5, row.names = rownames(xd) )
> plot(xd.nb, coords)
```

confirming that indeed we do as seen in Figure 13.9. The links can be overlaid on the region polygons to visualize the geographical relationships (Figure 13.10).

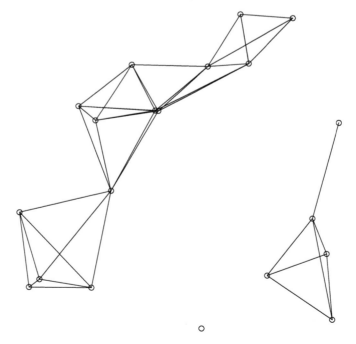

FIGURE 13.8   Neighborhood structure with cutoff distance at 0.3.

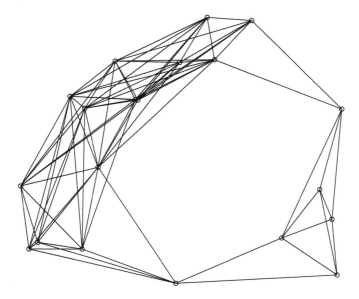

FIGURE 13.9   Neighborhood structure with increased cutoff distance at 0.5.

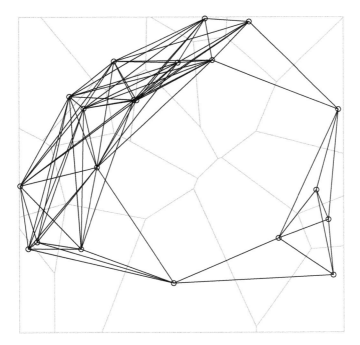

FIGURE 13.10   Regions and links.

```
> plot(xp, border = "grey")
> plot(xd.nb, coords, add=T)
```

Next, we need weights for neighborhood structure. In this case, a simple calculation of weights is the inverse of the distance; i.e., the closest the regions, the larger the weight (the closest two regions are, the more intense the neighbor effect).

We can get distances between each pair

```
> xd.dists <- nbdists(xd.nb, coords)
> summary(unlist(xd.dists))
   Min. 1st Qu. Median Mean 3rd Qu. Max.
0.008062 0.178700 0.286900 0.296100 0.434300 0.499700
>
```

And also we can get an intensity of neighborhood using the inverse of the distance applied to all pairs

```
> inten <- lapply(xd.dists, function(x) 1/x)
> inten
[[1]]
[1] 2.834181 2.168871 2.296261 2.029901 2.269558 3.271049

[[2]]
[1] 2.834181 6.076810 4.865043 5.213356 2.236773

[[3]]
[1] 6.076810 5.693409 9.899049 2.854092

[[4]]
[1] 2.168871 3.001596 2.101764 4.985935 29.235267 2.006217 6.172722

etc
>
```

Next, we produce a neighbor matrix with these weights. This is a 20 × 20 matrix. Style "W" is row-standardized, i.e., we divide by the number of neighbors in a row. For brevity, we only show the first 5 rows and 5 columns

```
> xd.nbmat <- nb2mat(xd.nb, glist=inten, style="W")
> xd.nbmat[1:5,1:5]
        [,1]       [,2]      [,3]       [,4]       [,5]
1 0.00000000 0.1905995 0.0000000 0.14585722 0.15442425
2 0.13352299 0.0000000 0.2862887 0.00000000 0.00000000
3 0.00000000 0.2477968 0.0000000 0.00000000 0.00000000
4 0.04366353 0.0000000 0.0000000 0.00000000 0.06042787
5 0.05531961 0.0000000 0.0000000 0.07231194 0.00000000
>
```

The same result is stored in a list because the function to compute Moran statistics will require a list.

```
> xd.nblsw <- nb2listw(xd.nb, glist=inten, style="W")
> xd.nblsw
Characteristics of weights list object:
Neighbour list object:
Number of regions: 20
Number of nonzero links: 160
Percentage nonzero weights: 40
Average number of links: 8

Weights style: W
Weights constants summary:
   n  nn S0       S1        S2
W 20 400 20 9.194488 82.19478
```

13.7.5 STRUCTURE BASED ON BORDERS

We can build a neighborhood relation using the polygons themselves instead of centroids. Neighboring relations are established based on shared borders instead of distance between region villages.

```
> xd.nb.pol <- poly2nb(xp, row.names = rownames(xd))
```

having similar structure as the xd.nb object we built earlier using distance. We can examine the differences

```
> diffnb(xd.nb, xd.nb.pol, verbose=TRUE)
Neighbour difference for region id: 1 in relation to id: 4 13
Neighbour difference for region id: 2 in relation to id: 5 12 16 17
Neighbour difference for region id: 3 in relation to id: 12
Neighbour difference for region id: 4 in relation to id: 1 5 7 8 19
Neighbour difference for region id: 5 in relation to id: 2 4 10 15
  17 19
Neighbour difference for region id: 6 in relation to id: 9
Neighbour difference for region id: 7 in relation to id: 4 10 11 13
  16 17 20
Neighbour difference for region id: 8 in relation to id: 4 14 15
  16 20
Neighbour difference for region id: 9 in relation to id: 6 17
Neighbour difference for region id: 10 in relation to id: 5 7 14
  18 19
Neighbour difference for region id: 11 in relation to id: 7 14 15
  16 19
Neighbour difference for region id: 12 in relation to id: 2 3
Neighbour difference for region id: 13 in relation to id: 1 7 19
Neighbour difference for region id: 14 in relation to id: 8 10 11 17
  18 19 20
```

(continued)

```
(continued)
Neighbour difference for region id: 15 in relation to id: 5 8 11
  17 18
Neighbour difference for region id: 16 in relation to id: 2 7 8 11
  18 19
Neighbour difference for region id: 17 in relation to id: 2 5 7 9 14
  15 19
Neighbour difference for region id: 18 in relation to id: 10 14
  15 16
Neighbour difference for region id: 19 in relation to id: 4 5 10 11
  13 14 16 17 20
Neighbour difference for region id: 20 in relation to id: 7 8 14 19
Neighbour list object:
Number of regions: 20
Number of nonzero links: 90
Percentage nonzero weights: 22.5
Average number of links: 4.5
>
```

Next, we need weights for neighborhood structure. The weights would represent intensity of neighbor relationship (in this case extent of common boundary). Results are put in a list because later the function to compute Moran's I statistic will require a list. Shown here only the first four rows

```
> xd.nblsw.pol <- nb2listw(xd.nb.pol, glist=NULL, style="W")
> xd.nblsw.pol$weights[1:4]
[[1]]
[1] 0.25 0.25 0.25 0.25

[[2]]
[1] 0.1428571 0.1428571 0.1428571 0.1428571 0.1428571 0.1428571
  0.1428571

[[3]]
[1] 0.3333333 0.3333333 0.3333333

[[4]]
[1] 0.5 0.5

>
```

Now, with these results we are ready to apply spatial auto-correlation (e.g., using the Moran's I statistic) and auto-regression (SAR). We do this in the following sections.

13.7.6 SPATIAL AUTO-CORRELATION

Apply the Moran's I spatial auto-correlation test to the X1 variable with the neighbor structures built in the previous section. The arguments are the variable, the neighborhood structure, and a decision on whether we assume normality or randomization. Let us assume normality.

```
> moran.test(xd$X1, xd.nblsw, randomisation=F)

        Moran's I test under normality

data:  xd$X1
weights: xd.nblsw

Moran I statistic standard deviate = 1.5318, p-value = 0.06278
alternative hypothesis: greater
sample estimates:
Moran I statistic         Expectation              Variance
      0.14996736         -0.05263158            0.01749245

>
```

Here although the value for the Moran statistic is low and would seem to indicate no spatial pattern, the low variance makes the Z value high and consequently the p-value is low enough to conclude that there is spatial pattern.

We have assumed that the mean and variance are the same for all regions. As discussed in Kaluzny et al. (1996), the variance of incidence rate increases for regions with low population. Thus, transform the data using the Freeman-Tukey (FT) transform of the X1 variable and multiply by the square root of population, to achieve constant variance

```
ft.X1 <- sqrt(1000)*(sqrt(xd$X1/xd$X2) + sqrt((xd$X1+1)/xd$X2))
tr.X1 <- ft.X1*sqrt(xd$X2)
names(tr.X1) <- rownames(xd)
```

Now, modify the Moran test as follows:

```
> moran.test(tr.X1, xd.nblsw, randomisation=F)

        Moran's I test under normality

data: tr.X1
weights: xd.nblsw

Moran I statistic standard deviate = 1.7034, p-value = 0.04424
alternative hypothesis: greater
sample estimates:
Moran I statistic     Expectation          Variance
      0.17266079     -0.05263158        0.01749245

>
```

This produces a slight increase in Moran's I, and consequently a slight improvement in the p-value. Let us try randomization

```
> moran.test(xd$X1, xd.nblsw, randomisation=T)

        Moran's I test under randomisation

data: xd$X1
weights: xd.nblsw
```

(continued)

```
(continued)

Moran I statistic standard deviate = 1.4928, p-value = 0.06775
alternative hypothesis: greater
sample estimates:
Moran I statistic       Expectation          Variance
      0.14996736        -0.05263158        0.01841946

>
```

This does not represent an improvement in p-value compared to normality. So far, results indicate that there is spatial auto-correlation of the data in these regions. We can confirm with Monte Carlo simulations (calculating Moran's I many times).

```
> moran.mc(xd$X1, xd.nblsw, nsim=1000)

        Monte-Carlo simulation of Moran's I

data: xd$X1
weights: xd.nblsw
number of simulations + 1: 1001

statistic = 0.15, observed rank = 921, p-value = 0.07992
alternative hypothesis: greater

>
```

We have a relatively larger p-value.
    You can repeat with the Geary statistic, by just changing "moran" for "geary".

```
> geary.test(xd$X1, xd.nblsw, randomisation=T)

        Geary's C test under randomisation

data: xd$X1
weights: xd.nblsw

Geary C statistic standard deviate = 1.9534, p-value = 0.02539
alternative hypothesis: Expectation greater than statistic
sample estimates:
Geary C statistic       Expectation          Variance
      0.73766223        1.00000000        0.01803641
>
```

The results are consistent.

13.7.7 Spatial Auto-Regression Models

The spatial auto-regression (SAR) linear model (SLM) object is sarslm. Package **spdep** provides functions lagsarslm for the spatial lag model and errorsarslm for the spatial error model, to perform SAR auto-regression. We will use lagsarslm to estimate the coefficients of a predictor and the auto-correlation parameter ρ.

First, assume that X3 sub-population is more prone to the disease, and therefore X1 is modeled as a function of X3

```
> xd.lag <- lagsarlm(X1 ~ X3, data=xd, xd.nblsw, tol.solve=1e-12)
> xd.lag

Call:
lagsarlm(formula = X1 ~ X3, data = xd, listw = xd.nblsw, tol.solve =
   1e-12)
Type: lag

Coefficients:
          rho   (Intercept)            X3
  0.012966193  -0.184136343   0.001995207

Log likelihood: 2.532483
>
```

Note that the results include estimates of the coefficients (intercept and slope) and the parameter *rho* for ρ. This model is then used to predict the X1 rate by region. It was necessary to decrease the tolerance to 10^{-12}. By default, it is 10^{-7}. Using summary, we get more information

```
> summary(xd.lag)

Call:lagsarlm(formula = X1 ~ X3, data = xd, listw = xd.nblsw, tol.
   solve = 1e-12)

Residuals:
      Min          1Q      Median          3Q          Max
-0.293580  -0.167328   -0.018474    0.123499     0.516407

Type: lag
Coefficients: (asymptotic standard errors)
              Estimate   Std. Error   z value  Pr(>|z|)
(Intercept) -1.8414e-01  2.3065e-01   -0.7983    0.4247
X3           1.9952e-03  1.8606e-05  107.2326    <2e-16

Rho: 0.012966, LR test value: 0.43271, p-value: 0.51066
Asymptotic standard error: 0.019494
    z-value: 0.66513, p-value: 0.50597
Wald statistic: 0.44239, p-value: 0.50597

Log likelihood: 2.532483 for lag model
ML residual variance (sigma squared): 0.045449, (sigma: 0.21319)
Number of observations: 20
Number of parameters estimated: 4
AIC: 2.935, (AIC for lm: 1.3677)
LM test for residual autocorrelation
test value: 3.9696, p-value: 0.046328

>
```

The coefficient estimates have low p-values. A special function is the likelihood ratio test (LR) to check whether the parameter rho is nonzero. In this case, the estimate is not so good with a p-value of 0.51. This tells us that ρ is not significantly different from zero. The Lagrange multiplier (LM) test for lack of serial correlation of the residuals yields low p-value and therefore the null hypothesis of serially correlated residuals is rejected.

Predicted values and residuals are available as part of the `slm` object type as `fitted` and `resid`. This model can be diagnosed much in the same manner that we evaluated linear regression models: scatter plots, qq plots, and residual-predicted plots. For example

```
lim<- max(xd$X1, fitted(xd.lag))
split.screen(c(2,1))
screen(1)
par(mar=c(4,4,1,.5),xaxs="r", yaxs="r")
plot(fitted(xd.lag), resid(xd.lag),xlab="Estimated",ylab=
  "Residuals")
abline(h=0)
split.screen(c(1,2), screen = 2)
screen(3)
plot(xd$X1, fitted(xd.lag),xlab="Observed",ylab="Estimated",xlim=
  c(0,lim),ylim=c(0,lim))
abline(a=0,b=1)
screen(4)
qqnorm(resid(xd.lag))
qqline(resid(xd.lag))
```

to obtain Figure 13.11 where we see that the residuals behave like a normal distribution.

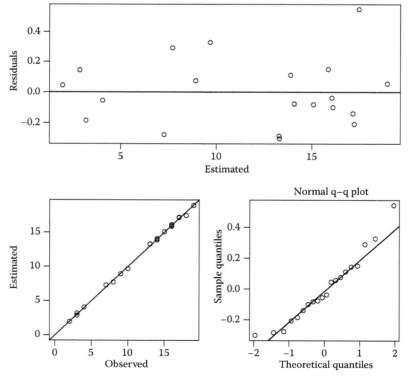

FIGURE 13.11   Evaluation of regression.

13.7.8   NEIGHBORHOOD STRUCTURE USING TRIPACK

This is an alternative method to the generation of the nb object. Use the `tri2nb` function (which requires package tripack) to find neighbors according to distances between region centers.

```
> xd.tri.nb <- tri2nb(coords, row.names=rownames(xd))
> xd.tri.nb
Neighbour list object:
Number of regions: 20
Number of nonzero links: 96
Percentage nonzero weights: 24
Average number of links: 4.8
```

This object excludes self-neighbors. Take a look at the summary

```
> summary (xd.tri.nb, coords)
Neighbour list object:
Number of regions: 20
Number of nonzero links: 96
Percentage nonzero weights: 24
Average number of links: 4.8
Link number distribution:

3  4  5  6  7  8
2  9  4  2  2  1
2 least connected regions:
18 19 with 3 links
1 most connected region:
5 with 8 links
Summary of link distances:
    Min.   1st Qu.    Median      Mean   3rd Qu.      Max.
0.008062  0.163600  0.203100  0.260900  0.328500  0.618000

  The decimal point is 1 digit(s) to the left of the |

  0 | 113366
  1 | 0011334444444466666666677778888888999999
  2 | 000011444455666677
  3 | 0011111122555599
  4 | 22446699
  5 | 2222
  6 | 000022

>
```

We can plot this set of neighbors (Figure 13.12)

```
plot(xd.tri.nb, coords)
```

Now we calculate distances, intensity, apply function `nb2listw` to the `nb` object to create weights

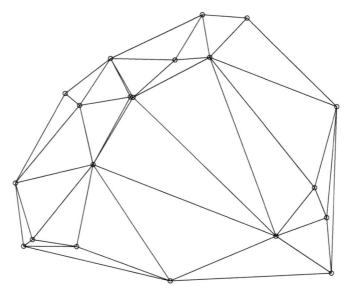

FIGURE 13.12   Neighborhood using tripack.

```
> xd.tri.dists <- nbdists(xd.tri.nb, coords)
> inten <- lapply(xd.tri.dists, function(x) 1/x)
> xd.nblsw <- nb2listw(xd.tri.nb, glist=inten, style="W")
> xd.nblsw$weights[1:4]
[[1]]
[1]  0.2249303 0.1721290 0.1822391 0.1610999 0.2596017

[[2]]
[1]  0.11839210 0.25384632 0.06936608 0.20322721 0.21777730
0.06759254 0.06979845

[[3]]
[1]  0.2477968 0.2321627 0.4036579 0.1163826

[[4]]
[1]  0.05095697 0.11714303 0.68687377 0.14502624

>
```

13.7.9   NEIGHBORHOOD STRUCTURE FOR GRID DATA

The nb spatial neighbor structure is defined by the cell2nb function for regular lattices (grid or raster based); options include the rook and queen moves. In addition, optionally we can employ a torus to remove edge effects. For example, the nb object for a $5 \times 5 = 25$ cells grid can be built with the rook move using

```
> nb5x5 <- cell2nb(5,5, type='rook')
> nb5x5[1:4]
[[1]]
[1]  2 6
```

```
(continued)
[[2]]
[1] 1 3 7

[[3]]
[1] 2 4 8

[[4]]
[1] 3 5 9
```

Then once we have an object of type nb, we can proceed as in the previous exercises to find weights.

```
> nb5x5.w <- nb2listw(nb5x5, style="W")
> nb5x5.w$weights[1:4]
[[1]]
[1] 0.5 0.5

[[2]]
[1] 0.3333333 0.3333333 0.3333333

[[3]]
[1] 0.3333333 0.3333333 0.3333333

[[4]]
[1] 0.3333333 0.3333333 0.3333333

>
```

Finally, using these results, we can apply spatial auto-correlation and auto-regression.

13.7.10 COMPUTER EXERCISES

Exercise 13.5

Use nc.sids dataset of **spdep** package. This dataset is on SIDS (Sudden Infant Death Syndrome) incidence by county in North Carolina (NC). In R go to Help, then Html help, then packages, then look for spdep. Then look for nc.sids. All variables and details are given in the Help. In addition, you can find a description and guidance in Kaluzny et al. (1996). You are required to do the following:

- Map the regions (polygons showing the borders) and produce maps according to levels of SIDS rates.
- Calculate the spatial neighborhood structure. Document and give rationale for your neighbor selection criteria. Provide plots.
- Calculate Moran's I and Geary's c for variables SID79 and NWBIR79. Determine if these variables are auto-correlated for this set of neighborhoods.
- Use SAR to build a predictor of SID79 rates from NWBIR79 birth rates. Evaluate and discuss the results.

Exercise 13.6

Use columbus dataset of **spdep** package. This dataset is for 49 neighborhoods in Columbus, Ohio. It has 49 rows and 22 columns. In R go to Help, then Html help, then packages, then look for spdep. Then look for columbus. All variables and details are given in the Help. In addition, you can find a description and guidance for package spdep that uses columbus as example at http://sal.agecon.uiuc.edu/csiss/pdf/spdepintro.pdf

You are required to do the following:

- Map the regions (polygons showing the borders) and produce maps according to levels of crime rates.
- Calculate the spatial neighborhood structure. Document and give rationale for your neighbor selection criteria. Provide plots.
- Calculate Moran's I and Geary's c for variables Housing value and Income. Determine if these variables are auto-correlated for this set of neighborhoods.
- Use SAR to build a predictor of crime rates from Housing value and Income. Evaluate and discuss the results.

SUPPLEMENTARY READING

Rogerson, 2001 Chapter 8 pp. 167–175, Rogerson, 2001 Chapter 9 pp. 179–190; Kaluzny et al., 1996, Chapter 5 pages 121–160.

14 Multivariate Analysis I
Reducing Dimensionality

14.1 MULTIVARIATE ANALYSIS: EIGEN-DECOMPOSITION

In this chapter and the next, we will work with several or many variables X_i. First, we will not necessarily assume that some of the variables are independent variables X_i, influencing one or more dependent, or response variable Y. Instead, we try to uncover relationships among all the variables and ways of reducing the dimensionality of the dataset. In this chapter, we will cover the methods of **principal component analysis** (PCA), **factor analysis** (FA), and **correspondence analysis** (CA).

All of these methods are based on the eigenvalues and eigenvectors of selected matrices based on the variance–covariance matrix or often called the **covariance matrix**. We first look at these matrices and their eigen-decomposition.

14.2 VECTORS AND LINEAR TRANSFORMATION

In the following, we think of n observations of a variable as a column vector $n \times 1$ in such a way that the entries define the coordinates of a point in n-space. The length of the vector will be the distance from the origin of coordinates to the point (Davis, 2002 pp. 141–152, Carr, 1995 pp. 50–57). Denote the direction of the vector by an arrow pointing from the origin toward the point. For example, vector $\mathbf{v} = \begin{bmatrix} 2 \\ 1 \end{bmatrix}$ in two-dimensional space is as in Figure 14.1. Its length is $\sqrt{2^2 + 1^2} = \sqrt{5} = 2.24$.

Usually when a square matrix is postmultiplied by a column vector, the result is a vector with a different length and different direction than the original vector. For example,

$$\mathbf{Av} = \begin{bmatrix} 1 & 1 \\ -2 & 4 \end{bmatrix} \begin{bmatrix} 2 \\ 1 \end{bmatrix} = \begin{bmatrix} 3 \\ 0 \end{bmatrix}$$

will give the result shown in Figure 14.2 with length $\sqrt{3^2 + 0^2} = \sqrt{9} = 3$

14.3 EIGENVALUES AND EIGENVECTORS

However, there is a particular class of vectors for each square matrix that when premultiplied by the matrix the resulting vector **preserves the direction**, and only changes the length of the original vector by a scalar factor. A vector with this property is an **eigenvector**. The scale factor associated with the transformation of the eigenvector is the **eigenvalue**.

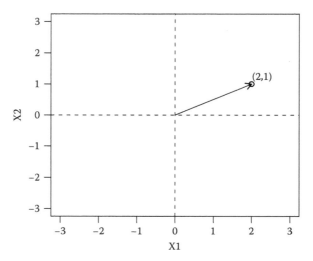

FIGURE 14.1 A vector in 2D.

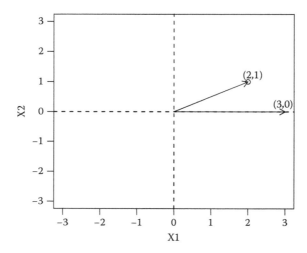

FIGURE 14.2 A vector v transformed to another vector by matrix multiplication.

For example, consider the vector $\mathbf{v} = \begin{bmatrix} -0.707 \\ -0.707 \end{bmatrix}$ shown in Figure 14.3. When premultiplied by matrix $\mathbf{A}$ as earlier, we get

$$\mathbf{Av} = \begin{bmatrix} 1 & 1 \\ -2 & 4 \end{bmatrix} \begin{bmatrix} -0.707 \\ -0.707 \end{bmatrix} = \begin{bmatrix} -1.414 \\ -1.414 \end{bmatrix}$$

We can see that it does not change direction but the length has doubled (Figure 14.4). To check that indeed the length has doubled, calculate the new length

$$\sqrt{1.414^2 + 1.414^2} = \sqrt{(2 \times 0.707)^2 + (2 \times 0.707)^2}$$

$$= \sqrt{4 \times (0.707)^2 + 4 \times (0.707)^2} = 2 \times \sqrt{(0.707)^2 + (0.707)^2}$$

Formally, an eigenvalue, eigenvector pair of $\mathbf{A}$ is any real or complex number-vector pair denoted as $(\lambda, \mathbf{v})$, such that $\mathbf{Av} = \lambda\mathbf{v}$. Even if we multiply this equation by a scalar k, the equality will hold.

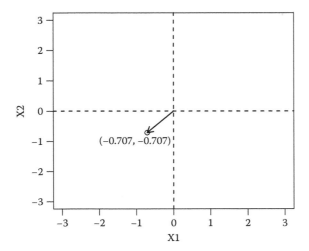

FIGURE 14.3 A special vector for the matrix **A**.

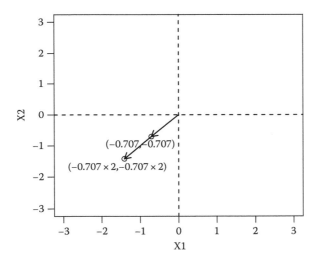

FIGURE 14.4 The special vector preserves direction but has doubled in length.

Therefore, there are an infinite number of eigenvectors associated with a particular eigenvalue. Note that an eigenvalue is a scalar, whereas an eigenvector is a vector.

As an example, consider the vector given earlier

$$\mathbf{Av} = \begin{bmatrix} 1 & 1 \\ -2 & 4 \end{bmatrix} \begin{bmatrix} -0.707 \\ -0.707 \end{bmatrix} = \begin{bmatrix} -1.414 \\ -1.414 \end{bmatrix} = 2 \begin{bmatrix} -0.707 \\ -0.707 \end{bmatrix} = 2\mathbf{v}$$

We can see that $\lambda = 2$ is the eigenvalue.

14.3.1 FINDING EIGENVALUES

The definition provided earlier $\mathbf{Av} = \lambda\mathbf{v}$ is equivalent to $\mathbf{Av} - \lambda\mathbf{v} = \mathbf{0}$ or equivalently $\lambda\mathbf{v} - \mathbf{Av} = \mathbf{0}$. Factoring out the vector $\mathbf{v}$, and inserting the identity matrix $\mathbf{I}$ (remember the identity matrix does not change the value of the vector), we get

$$(\lambda\mathbf{I} - \mathbf{A})\mathbf{v} = 0. \tag{14.1}$$

If the inverse of $\lambda\mathbf{I} - \mathbf{A}$ exists, then the solution will be $\mathbf{v} = 0$, the null eigenvector, which is not very interesting. Therefore, non-null vectors of $\mathbf{v}$ exist only if the $\lambda\mathbf{I} - \mathbf{A}$ does not have an inverse. The inverse does not exist when the matrix $\lambda\mathbf{I} - \mathbf{A}$ is singular, that is to say when the determinant

$$\left| \lambda\mathbf{I} - \mathbf{A} \right| = 0 \tag{14.2}$$

This condition is the **characteristic equation** for $\mathbf{A}$ and is an nth degree polynomial in λ for an $n \times n$ matrix. Therefore, there are at most n distinct values of λ that satisfy the characteristic Equation 14.2.

Let us consider the matrix $\mathbf{A}$:

$$\mathbf{A} = \begin{bmatrix} 1 & 1 \\ -2 & 4 \end{bmatrix} \tag{14.3}$$

The characteristic equation of this matrix is

$$\left| \begin{bmatrix} \lambda & 0 \\ 0 & \lambda \end{bmatrix} - \begin{bmatrix} 1 & 1 \\ -2 & 4 \end{bmatrix} \right| = \begin{vmatrix} \lambda - 1 & -1 \\ +2 & \lambda - 4 \end{vmatrix} = 0 \tag{14.4}$$

Calculating the determinant

$$(\lambda - 1)(\lambda - 4) + 2 = \lambda^2 - 5\lambda + 6 = 0 \tag{14.5}$$

and factoring

$$(\lambda - 3)(\lambda - 2) = 0 \tag{14.6}$$

From this we can conclude that $\lambda_1 = 3$ and $\lambda_2 = 2$ are two distinct solutions to the quadratic equation (Equation 14.5). Each one of these eigenvalues is associated with an eigenvector.

14.3.2 Finding Eigenvectors

The eigenvectors can be found by solving $\mathbf{Av} = \lambda\mathbf{v}$ for the eigenvalues given earlier. Take the $\lambda_1 = 3$ eigenvalue, the equation would be

$$\begin{bmatrix} 1 & 1 \\ -2 & 4 \end{bmatrix} \cdot \begin{bmatrix} v_1 \\ v_2 \end{bmatrix} = 3 \begin{bmatrix} v_1 \\ v_2 \end{bmatrix} \tag{14.7}$$

yielding

$$v_1 + v_2 = 3 \times v_1$$
$$-2 \times v_1 + 4 \times v_2 = 3 \times v_2 \tag{14.8}$$

We can see that $v_2 = 2v_1$ will satisfy both Equations 14.8.

This result emphasizes the point that each eigenvector is a member of a class of eigenvectors that satisfy the condition $v_2 = 2v_1$; any eigenvector that satisfies this condition will work. Arbitrarily choose

$$\begin{bmatrix} v_1 \\ v_2 \end{bmatrix} = \begin{bmatrix} 1 \\ 2 \end{bmatrix} \tag{14.9}$$

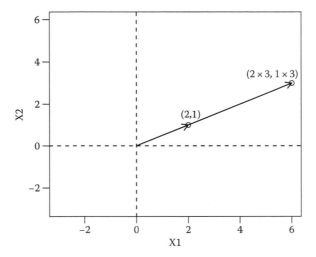

FIGURE 14.5 Transformed eigenvector preserves direction and scales in length.

as the representative eigenvector. Geometrically we can see that if we transform premultiplying by **A**, we get

$$3\begin{bmatrix} v_1 \\ v_2 \end{bmatrix} = 3\begin{bmatrix} 1 \\ 2 \end{bmatrix} = \begin{bmatrix} 3 \\ 6 \end{bmatrix}$$

which has the same direction but is 3 times longer than **v** (Figure 14.5). The length is

$$\sqrt{6^2 + 3^2} = \sqrt{3^2(2^2 + 1^2)} = 3 \times 2.27 = 6.71$$

14.4 EIGEN-DECOMPOSITION OF A COVARIANCE MATRIX

Most multivariate analysis methods are based on **eigen-decomposition of the covariance matrix**. Therefore, understanding eigenvalues and eigenvectors of a symmetric matrix is very important because the **covariance matrix is symmetric**. Thus, we need a good grasp of eigen-analysis of symmetric matrices.

14.4.1 COVARIANCE MATRIX

Assume the multivariate data set consists of n observations and m variables, and therefore we have a matrix **X** which is rectangular and of dimension $n \times m$, where n is the number of rows for observations (individuals, samples, objects) and m columns for m variables (attributes, characteristics, responses)

$$\mathbf{X} = \begin{bmatrix} x_{11} & x_{21} & x_{31} & \cdots & x_{m1} \\ x_{12} & x_{22} & x_{32} & \cdots & x_{m2} \\ \cdots & \cdots & \cdots & \cdots & \cdots \\ x_{1n} & x_{2n} & x_{3n} & \cdots & x_{mn} \end{bmatrix} \tag{14.10}$$

There are relations between individuals and attributes. The aim is to understand these relations and possibly reduce the dimensionality of the dataset.

In general, for m variables, the trace will be the sum of m variances and it is equal to the sum of the m eigenvalues. Therefore, by eigen-decomposition, we allocate the total variance to the eigenvalues in such a way that the amount is sorted from the largest to smallest eigenvalue.

First, center each column vector on its sample mean. This is to say, subtract the sample mean of Xi from all observations x_{ij}, in the ith column of $\mathbf{X}$,

$$\mathbf{x} = \begin{bmatrix} x_{11} - \overline{X}_1 & x_{21} - \overline{X}_2 & x_{31} - \overline{X}_3 & \cdots & x_{m1} - \overline{X}_m \\ x_{12} - \overline{X}_1 & x_{22} - \overline{X}_2 & x_{32} - \overline{X}_3 & \cdots & x_{m2} - \overline{X}_m \\ \cdots & \cdots & \cdots & \cdots & \cdots \\ x_{1n} - \overline{X}_1 & x_{2n} - \overline{X}_2 & x_{3n} - \overline{X}_3 & \cdots & x_{mn} - \overline{X}_m \end{bmatrix} \tag{14.11}$$

We have used lowercase $\mathbf{x}$ to denote this centered data matrix. Now, premultiply $\mathbf{x}$ by $\mathbf{x}^T$ to obtain the major product matrix (we covered this concept in Chapter 9)

$$\mathbf{x}^T\mathbf{x} = \begin{bmatrix} \sum(x_{1i} - \overline{X}_1)^2 & \sum(x_{2i} - \overline{X}_1) \times(x_{2i} - \overline{X}_2) & \sum(x_{1i} - \overline{X}_1) \times(x_{3i} - \overline{X}_3) & \cdots & \sum(x_{1i} - \overline{X}_1) \times(x_{mi} - \overline{X}_m) \\ \sum(x_{1i} - \overline{X}_1) \times(x_{2i} - \overline{X}_2) & \sum(x_{2i} - \overline{X}_2)^2 & \sum(x_{2i} - \overline{X}_2) \times(x_{3i} - \overline{X}_3) & \cdots & \sum(x_{2i} - \overline{X}_2) \times(x_{mi} - \overline{X}_m) \\ \sum(x_{1i} - \overline{X}_1) \times(x_{3i} - \overline{X}_3) & \sum(x_{2i} - \overline{X}_2) \times(x_{3i} - \overline{X}_3) & \sum(x_{3i} - \overline{X}_3)^2 & \cdots & \sum(x_{3i} - \overline{X}_3) \times(x_{mi} - \overline{X}_m) \\ \vdots & \vdots & \vdots & & \\ \sum(x_{1i} - \overline{X}_1) \times(x_{mi} - \overline{X}_m) & \sum(x_{2i} - \overline{X}_2) \times(x_{mi} - \overline{X}_m) & \sum(x_{3i} - \overline{X}_3) \times(x_{mi} - \overline{X}_m) & \cdots & \sum(x_{mi} - \overline{X}_m)^2 \end{bmatrix} \tag{14.12}$$

The entries are the sum of squares and cross-products of x's. Note that $\mathbf{x}^T\mathbf{x}$ is a square $m \times m$ symmetric matrix.

Now if we multiply by the scalar $1/(n - 1)$, we can identify the diagonal terms as variances of each variable and the off-diagonal terms as covariance of each pair of variables. Therefore, this new matrix is the "**variance-covariance**" **matrix** or **covariance matrix** for short (see also pages 147, 510, 518 in Davis, 2002). This matrix is symmetric. For brevity, we can use the notation $\mathbf{C} = (n - 1)^{-1} \mathbf{x}^T\mathbf{x}$. Thus,

$$\mathbf{C} = \begin{bmatrix} s_{X_1}^2 & s_{\text{cov}\,X_1,X_2} & s_{\text{cov}\,X_1,X_3} & \cdots & s_{\text{cov}\,X_1,X_m} \\ s_{\text{cov}\,X_1,X_2} & s_{X_2}^2 & s_{\text{cov}\,X_2,X_3} & \cdots & s_{\text{cov}\,X_2,X_m} \\ s_{\text{cov}\,X_1,X_3} & s_{\text{cov}\,X_2,X_3} & s_{X_3}^2 & \cdots & s_{\text{cov}\,X_3,X_m} \\ \vdots & \vdots & \vdots & \ddots & \\ s_{\text{cov}\,X_1,X_m} & s_{\text{cov}\,X_m,X_2} & s_{\text{cov}\,X_m,X_3} & \cdots & s_{X_m}^2 \end{bmatrix} \tag{14.13}$$

14.4.2 BIVARIATE CASE

To gain some insight with a simple case, let us develop the sample covariance matrix for $m = 2$,

$$\mathbf{X} = \begin{bmatrix} x_{11} & x_{21} \\ x_{12} & x_{22} \\ \dots & \dots \\ x_{1n} & x_{2n} \end{bmatrix} \tag{14.14}$$

We first center the columns by subtracting the mean

$$\mathbf{x} = \begin{bmatrix} x_{11} - \bar{X}_1 & x_{21} - \bar{X}_2 \\ x_{12} - \bar{X}_1 & x_{22} - \bar{X}_2 \\ \dots & \dots \\ x_{1n} - \bar{X}_1 & x_{2n} - \bar{X}_2 \end{bmatrix} \tag{14.15}$$

then calculate the major product

$$\mathbf{x}^T\mathbf{x} = \begin{bmatrix} \sum(x_{1i} - \bar{X}_1)^2 & \sum(x_{1i} - \bar{X}_1)(x_{2i} - \bar{X}_2) \\ \sum(x_{1i} - \bar{X}_1)(x_{2i} - \bar{X}_2) & \sum(x_{2i} - \bar{X}_2)^2 \end{bmatrix} \tag{14.16}$$

Divide each entry by $n - 1$ and use sample variance notation

$$\mathbf{C} = \begin{bmatrix} s_{X_1}^2 & s_{\mathrm{cov}(X_1,X_2)} \\ s_{\mathrm{cov}(X_1,X_2)} & s_{X_2}^2 \end{bmatrix} \tag{14.17}$$

The trace of $\mathbf{C}$ is the sum of the diagonal terms

$$trace(\mathbf{C}) = (s_{X_1})^2 + (s_{X_2})^2 \tag{14.18}$$

The determinant of $\mathbf{C}$ is calculated to be

$$|\mathbf{C}| = (s_{X_1})^2 (s_{X_2})^2 - (s_{\mathrm{cov}(X_1,X_2)})^2 \tag{14.19}$$

Let us see the use of eigen-decomposition using this simple 2×2 matrix. First, we calculate the eigenvalues of $\mathbf{C}$ solving

$$|\lambda\mathbf{I} - \mathbf{C}| = \lambda^2 - \lambda(s_{X_1}^2 + s_{X_1}^2) - (\mathrm{cov}_{X_1X_2}^2 - s_{X_1}^2 \ s_{X_1}^2) = 0 \tag{14.20}$$

and then the eigenvectors for each one of these eigenvalues. These vectors can be drawn to be directions of semi-axis of an ellipse. See Figures 3.1, 3.2 and 3.3 of Davis, 2002 and Figure 3.5 in Carr, 1995, for geometric interpretation based on orientation (eigenvectors) and lengths (eigenvalues) of ellipse semi-axis.

Let us work out an example. Suppose the covariance matrix is

$$\mathbf{C} = \begin{bmatrix} 0.5 & 0.3 \\ 0.3 & 0.4 \end{bmatrix}$$

with trace = 0.9, which is also the total variance. The characteristic equation is

$$|\lambda\mathbf{I} - \mathbf{C}| = \begin{bmatrix} \lambda & 0 \\ 0 & \lambda \end{bmatrix} - \begin{bmatrix} 0.5 & 0.3 \\ 0.3 & 0.4 \end{bmatrix} = \begin{vmatrix} \lambda - 0.5 & -0.3 \\ -0.3 & \lambda - 0.4 \end{vmatrix} = 0$$

Or expanding the determinant

$$\lambda^2 - (0.4 + 0.5)\lambda + 0.2 - 0.09 = \lambda^2 - 0.9\lambda + 0.11 = 0$$

The solution of a quadratic equation

$$a\lambda^2 + b\lambda + c = 0 \quad \text{is} \quad \lambda = \frac{-b \pm \sqrt{b^2 - 4ac}}{2a}$$

Therefore,

$$\lambda = \frac{0.9 \pm \sqrt{.81 - 4 \times 0.11}}{2} = \frac{0.9 \pm 0.6}{2}$$

We have two eigenvalues, one for each sign in the numerator of this ratio

$$\lambda_1 = 0.75$$

$$\lambda_2 = 0.15$$

Note that the sum of the eigenvalues is equal to the trace of $\mathbf{C}$. That is to say to the total variance 0.9. The trace is the total variance and it is equal to the sum of the eigenvalues. Therefore, by eigendecomposition, we allocate the total variance to the eigenvalues in such a way that the amount is sorted from the largest to smallest eigenvalue.

The eigenvectors can be calculated using

$$\mathbf{C}\mathbf{v}_1 = \lambda_1\mathbf{v}_1$$

$$\mathbf{C}\mathbf{v}_2 = \lambda_2\mathbf{v}_2$$

Let us start with $\mathbf{v}_1$ with entries v_{11}, v_{12} corresponding to $\lambda_1 = 0.75$

$$\begin{bmatrix} 0.5 & 0.3 \\ 0.3 & 0.4 \end{bmatrix}\begin{bmatrix} v_{11} \\ v_{12} \end{bmatrix} = \lambda_1 \begin{bmatrix} v_{11} \\ v_{12} \end{bmatrix}$$

This can be expanded as

$$0.5 \times v_{11} + 0.3 \times v_{12} = \lambda_1 \times v_{11} = 0.75 \times v_{11}$$

$$0.3 \times v_{11} + 0.5 \times v_{12} = \lambda_1 \times v_{12} = 0.75 \times v_{12}$$

using the first one of these two equations

$$v_{12} = \frac{(0.75-0.5)\times v_{11}}{0.3} = \frac{0.25\times v_{11}}{0.3} = 0.833\times v_{11}$$

Next, we work with the second eigenvector $\mathbf{v}_2$ with entries v_{21}, v_{22} corresponding to $\lambda_2 = 0.15$

$$\begin{bmatrix} 0.5 & 0.3 \\ 0.3 & 0.4 \end{bmatrix} \begin{bmatrix} v_{21} \\ v_{22} \end{bmatrix} = \lambda_2 \begin{bmatrix} v_{21} \\ v_{22} \end{bmatrix}$$

Expand

$$0.5\times v_{21} + 0.3\times v_{22} = \lambda_2 \times v_{11} = 0.15\times v_{21}$$

$$0.3\times v_{21} + 0.5\times v_{22} = \lambda_1 \times v_{22} = 0.15\times v_{22}$$

using the first one of these two equations

$$v_{22} = \frac{(0.15-0.5)\times v_{21}}{0.3} = \frac{-0.35\times v_{21}}{0.3} = -1.166\times v_{21}$$

or

$$v_{21} = \frac{v_{22}}{-1.16} = -0.833\times v_{22}$$

Therefore, the eigenvectors are

$$\mathbf{v}_1 = \begin{bmatrix} 1 \\ 0.833 \end{bmatrix} \quad \mathbf{v}_2 = \begin{bmatrix} -0.833 \\ 1 \end{bmatrix}$$

Recall that there are infinitely many ways of scaling the eigenvectors. For the sake of visualization, we can represent how these vectors can span the two-dimensional space as in Figure 14.6. They can serve as semi-axis of an ellipse (Davis, 2002).

When standardizing the entries in the covariance matrix, we obtain a correlation matrix

$$\mathbf{R} = \begin{bmatrix} 1 & r_{X1,X2} \\ r_{X1,X2} & 1 \end{bmatrix}$$

We can calculate eigenvalues and eigenvectors of $\mathbf{R}$ in the same manner. Note that the trace of $\mathbf{R}$ is always 2 in the bivariate case. Now let us calculate a numerical example using the covariance matrix $\mathbf{C} = \begin{bmatrix} 0.5 & 0.3 \\ 0.3 & 0.4 \end{bmatrix}$. Divide the diagonal terms by the variances 0.5 and 0.4 and the off-diagonal terms by the product of standard deviations $\sqrt{5} = 0.7$ and $\sqrt{4} = 0.63$ to get

$$\mathbf{R} = \begin{bmatrix} 1 & 0.3/(0.7\times 0.63) \\ 0.3/(0.7\times 0.63) & 1 \end{bmatrix} = \begin{bmatrix} 1 & 0.22 \\ 0.22 & 1 \end{bmatrix}$$

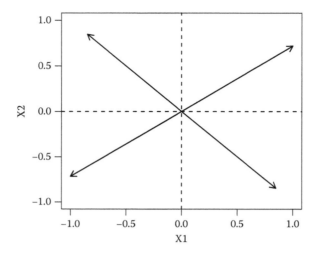

FIGURE 14.6 Graphical representation of eigenvectors of a covariance matrix.

Note that the trace is $1 + 1 = 2$ or the number of variables. Calculate the eigenvalues

$$\lambda_1 = 1.22 \quad \text{and} \quad \lambda_2 = 0.78$$

and note that the sum of the eigenvalues is also 2.

$$\mathbf{v}_1 = \begin{bmatrix} 0.707 \\ 0.707 \end{bmatrix} \quad \text{and} \quad \mathbf{v}_2 = \begin{bmatrix} 0.707 \\ -0.707 \end{bmatrix}$$

In this case, it can be scaled to

$$\mathbf{v}_1 = \begin{bmatrix} 1 \\ 1 \end{bmatrix} \quad \text{and} \quad \mathbf{v}_2 = \begin{bmatrix} 1 \\ -1 \end{bmatrix} \tag{14.21}$$

Note that for the correlation matrix the eigenvectors will always have the values given in Equation 14.21 and that they bisect the quadrants as in Figure 14.7. Again, see geometric interpretation in Davis, 2002.

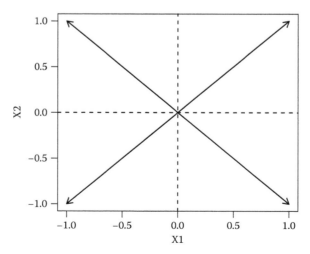

FIGURE 14.7 Graphical representation of eigenvectors of a correlation matrix.

14.5   PRINCIPAL COMPONENTS ANALYSIS (PCA)

The objective is to explain the data from new variables called **principal components**. These are linear combination of the original variables. We want to generate new and fewer variables capable of explaining most of the variance. The new variables (components) are the eigenvectors of the covariance matrix. These new variables are orthogonal (perpendicular in 2D). All new variables explain data variance and just a few of them explain most of the variance (Davis, 2002 pp. 509–526).

Consider $\mathbf{X}$ to be an $n \times m$ data matrix as in the previous section. Its columns correspond to values of the m variables $[X_1, X_2, ..., X_m]$; the rows to the n individual observations. We assume that $n \gg m$, that is to say that we have many more observations than attributes.

The covariance matrix is $m \times m$, $\mathbf{C} = \text{cov}(\mathbf{X})$. The trace of $\mathbf{C}$ is the sum of diagonal terms, which is equal to the total variance and in turn equal to the sum of m eigenvalues. Therefore, each eigenvalue is a % of the total variance. When selecting components, a typical approach is to select those components that explain 90% of total variance. Alternatively, there are other criteria, such as the Kaiser rule, that retain a component as long as it explains as much as one of the original variables. Instead of $\mathbf{C}$ we can also use correlation matrix $\mathbf{R} = \text{cor}(\mathbf{X})$ to scale uniformly.

The new variables $[Z_1, Z_2, ... Z_m]$ make up the new matrix $\mathbf{Z}$, which is also $n \times m$. Note that $\mathbf{Z}$ is linear function of $\mathbf{X}$, that is to say $\mathbf{Z} = \mathbf{XA}$, where $\mathbf{A}$ is a matrix $m \times m$ composed of the eigenvectors of the covariance matrix.

$$\mathbf{A} = \begin{bmatrix} v_{11} & v_{21} & ... & v_{31} \\ v_{12} & v_{22} & ... & v_{32} \\ ... & ... & ... & ... \\ v_{1m} & v_{2m} & ... & v_{mm} \end{bmatrix} \tag{14.22}$$

The entries of $\mathbf{A}$ are the **loadings**. The entries of $\mathbf{Z}$ are **scores** or new coordinate values. Observations are plotted as points in this new set of coordinates. Points close to each other are similar. See the simple examples (Davis, 2002 page 510–516), for two variables, $\mathbf{X} =$ is $n \times 2$, $\mathbf{C}$ is 2×2, $\mathbf{A}$ is 2×2.

The matrix $\mathbf{A}$ satisfies

$$\mathbf{C} = \mathbf{ALA}^T \tag{14.23}$$

where $\mathbf{L}$ is a diagonal matrix made up with the eigenvalues of $\mathbf{C}$ arranged in descending order.

$$\mathbf{L} = \begin{bmatrix} \lambda_1 & 0 & ... & 0 \\ 0 & \lambda_2 & ... & 0 \\ ... & ... & ... & 0 \\ 0 & 0 & ... & \lambda_m \end{bmatrix} \tag{14.24}$$

Here, $\lambda_1 \geq \lambda_2 \geq ... \geq \lambda_m$. The trace of $\mathbf{L}$ is equal to the total variance.

For example in a similar manner to the example in Carr (1995, p. 96), consider the data matrix of $n = 5$ observations of two variables

$$\mathbf{X} = \begin{bmatrix} 3 & 14 \\ 4 & 13 \\ 5 & 6 \\ 6 & 5 \\ 7 & 0 \end{bmatrix} \tag{14.25}$$

First, center the columns by subtracting the means

$$
\mathbf{x} = \begin{bmatrix} -2 & 6.4 \\ -1 & 5.4 \\ 0 & -1.6 \\ 1 & -2.6 \\ 2 & -7.6 \end{bmatrix}
\tag{14.26}
$$

Now the covariance matrix is

$$
\mathbf{C} = \frac{1}{n-1}\mathbf{x}^T\mathbf{x} = \frac{1}{5-1}\begin{bmatrix} 10 & -36 \\ -36 & 137.2 \end{bmatrix} = \begin{bmatrix} 2.5 & -9 \\ -9 & 34.3 \end{bmatrix}
\tag{14.27}
$$

The total variance is the trace of $\mathbf{C} = 34.3 + 2.5 = 36.8$. The eigenvalues are

$$
\lambda_1 = 36.67 \quad \text{and} \quad \lambda_2 = 0.13
\tag{14.28}
$$

Their sum is also equal to the total variance ($36.67 + 0.13 = 36.8$). The first eigenvalue 36.67 is a large fraction of total variance 36.8. Calculate % of total variance explained by first eigenvalue $100 \times (36.67/36.8) = 99.6\%$. Therefore, only one component explains most of the variance. The eigenvectors are

$$
\mathbf{v}_1 = \begin{bmatrix} -0.25 \\ 0.97 \end{bmatrix} \quad \text{and} \quad \mathbf{v}_2 = \begin{bmatrix} -0.97 \\ -0.25 \end{bmatrix}
$$

and are arranged as columns in a matrix $\mathbf{A}$ to form the **loadings**

$$
\mathbf{A} = \begin{bmatrix} -0.25 & -0.97 \\ 0.97 & -0.25 \end{bmatrix}
$$

The new or transformed data matrix $\mathbf{Z}$ is obtained by postmultiplying the data by the loadings

$$
\mathbf{Z} = \mathbf{XA} = \begin{bmatrix} 3 & 14 \\ 4 & 13 \\ 5 & 6 \\ 6 & 5 \\ 7 & 0 \end{bmatrix}\begin{bmatrix} -0.25 & -0.97 \\ 0.97 & -0.25 \end{bmatrix} = \begin{bmatrix} 12.77 & -6.47 \\ 11.55 & -7.18 \\ 4.52 & -6.36 \\ 3.30 & -7.08 \\ -1.78 & -6.77 \end{bmatrix}
$$

We can verify that $\mathbf{L}$ is diagonal with eigenvalues in main diagonal

$$
\mathbf{L} = \mathbf{A}^T\mathbf{CA} = \begin{bmatrix} -0.25 & -0.97 \\ 0.97 & -0.25 \end{bmatrix}\begin{bmatrix} 2.5 & -9 \\ -9 & 34.3 \end{bmatrix}\begin{bmatrix} -0.25 & -0.97 \\ 0.97 & -0.25 \end{bmatrix} = \begin{bmatrix} 36.67 & 0 \\ 0 & 0.13 \end{bmatrix}
$$

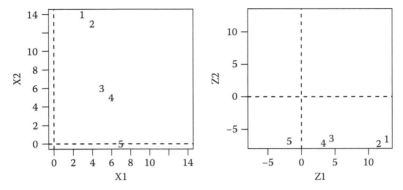

FIGURE 14.8 Original data X (left panel) and transformed data (scores) Z (right panel) using observation number for labels.

Note that in this case the transpose of $\mathbf{A}$ is the same matrix $\mathbf{A}$. Graphically we can compare the original data $\mathbf{X}$ to the transformed scores $\mathbf{Z}$ as shown in Figure 14.8. Here, we label the points with observation numbers. Proximity of pairs of observations indicates similarity between those observations. The first principal component (horizontal axis) explains most of the variance and observations sort themselves along this axis; observation 1 has the highest vale and observation 5 the lowest (Figure 14.8). However, all observation pairs differ little along the vertical axis (second principal component).

When the observations are not all on the same scale, which could happen for example when the variables are not in the same units, it is important to first standardize the observations or equivalently to use the correlation matrix. In the example at hand, the second column has much larger values than the first; therefore, it is important to standardize or use the correlation matrix.

To standardize the observations, divide the centered columns by the standard deviation of the column. For example, matrix X is standardized

$$\mathbf{x}_s = \begin{bmatrix} -1.26 & 1.09 \\ -0.63 & 0.93 \\ 0 & -0.27 \\ 0.63 & -0.44 \\ 1.26 & -1.30 \end{bmatrix}$$

We can either calculate the correlation matrix from the covariance matrix

$$\mathbf{R} = \begin{bmatrix} 2.5/2.5 & \dfrac{-9}{\sqrt{2.5}\sqrt{34.3}} \\ \dfrac{-9}{\sqrt{2.5}\sqrt{34.3}} & 34.3/34.3 \end{bmatrix} = \begin{bmatrix} 1 & -0.97 \\ -0.97 & 1 \end{bmatrix}$$

or calculate it directly from the standardized observations themselves

$$\mathbf{R} = \frac{1}{n-1}\mathbf{x}_s^{\mathrm{T}}\mathbf{x}_s = \begin{bmatrix} 1 & -0.97 \\ -0.97 & 1 \end{bmatrix}$$

The total correlation (scaled variance) is the trace of $\mathbf{R} = 1 + 1 = 2$. The eigenvalues are

$$\lambda_1 = 1.97 \quad \text{and} \quad \lambda_2 = 0.03$$

Their sum is also equal to the total correlation ($1.97 + 0.03 = 2$). The first eigenvalue 1.97 is a large fraction of total correlation. Calculate % of total correlation explained by the first eigenvalue $100 \times (1.97/2) = 98.5\%$. Therefore, only one component explains most of the variance, but note that the standardization has made the value smaller, indicating inflation before. The eigenvectors

$$\mathbf{v}_1 = \begin{bmatrix} -0.71 \\ 0.71 \end{bmatrix} \quad \text{and} \quad \mathbf{v}_2 = \begin{bmatrix} -0.71 \\ -0.71 \end{bmatrix}$$

have all the same magnitude due to the standardization. These are arranged as columns in a matrix $\mathbf{A}_s$ to form the **loadings**

$$\mathbf{A}_s = \begin{bmatrix} -0.71 & -0.71 \\ 0.71 & -0.71 \end{bmatrix}$$

The new or transformed data matrix $\mathbf{Z}_s$ is obtained by postmultiplying the data by the loadings.

$$\mathbf{Z}_s = \mathbf{x}_s \mathbf{A}_s = \begin{bmatrix} -1.26 & 1.09 \\ -0.63 & 0.93 \\ 0 & -0.27 \\ 0.63 & -0.44 \\ 1.26 & -1.30 \end{bmatrix} \begin{bmatrix} -0.71 & -0.71 \\ 0.71 & -0.71 \end{bmatrix} = \begin{bmatrix} 1.67 & 0.12 \\ 1.10 & -0.20 \\ -0.19 & 0.19 \\ -0.76 & -0.13 \\ -1.81 & 0.02 \end{bmatrix}$$

Plots of the data and scores allow comparison as before (Figure 14.9). On the right-hand-side panel, sorting along the horizontal axis is nearly the same, but now we can appreciate differences among observation pairs along the vertical axes (second component). Observations 2 and 4 are relatively similar, and so are 1 and 3. However, these two pairs differ among themselves; 2 and 4 are negative, whereas 1 and 3 are positive.

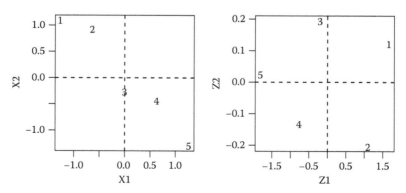

FIGURE 14.9 Original data (standardized) Xs (left panel) and transformed data (scores) Zs (right panel) using observation number for labels.

14.6 SINGULAR VALUE DECOMPOSITION AND BIPLOTS

Singular value decomposition (SVD) is a method to obtain a diagonal form for the covariance matrix **C**, similar to Equation 14.23, but it is more general and applied directly to the rectangular data matrix **X**. SVD relates to the Eckart–Young theorem. SVD applies to many multivariate techniques: PCA, CA, and canonical correlation analysis.

We can work directly with the **X** data matrix or the centered observations **x** or the standardized observations $\mathbf{x_s}$. In the following, we work with the centered column data matrix **x**. The SVD of matrix **x** is

$$\mathbf{x} = \mathbf{U\Gamma V^T} \tag{14.29}$$

where
   U is $n \times m$
   Γ is $m \times m$
   V is $m \times m$

U and **V** are matrices of singular vectors, the columns of **U** are eigenvectors of $\mathbf{xx^T}$ (referred to as **left** eigenvectors) and the columns of **V** are eigenvectors of $\mathbf{x^Tx}$ (referred to as **right** eigenvectors). Note that $\mathbf{xx^T}$ can be a large matrix because it is $n \times n$ and that $\mathbf{x^Tx}$ is smaller because it is $m \times m$. Matrix **Γ** is a diagonal matrix of singular values, the square root of eigenvalues of $\mathbf{x^Tx}$, arranged in non-decreasing order. The nonzero eigenvalues of $\mathbf{xx^T}$ are the same as the eigenvalues of $\mathbf{x^Tx}$.

$$\Gamma = diag\left(\sqrt{\lambda_i} \quad i = 1, ..., m\right) \tag{14.30}$$

A least-squares approximation of the matrix **x** is obtained using the first few (say k) dominant singular values and vectors

$$\hat{\mathbf{x}}_k = \mathbf{U_k \Gamma_k V_k^T} \tag{14.31}$$

In this equation, the subscript k means we take the first k columns of **U**, **Γ** and $\mathbf{V^T}$. A useful graphical visualization of the reduced dimensionality is called a **biplot** (Gabriel, 1971). This plot shows both the observations and the variables simultaneously. The prefix **bi** refers to the simultaneous display of **both** rows and columns of the transformed data matrix given by Equation 14.31, and not to the fact that the plots are bi-dimensional. Biplots are usually drawn in 2D ($k = 2$) for ease of interpretation, but conceptually they can be done in 3D and multi-dimensional space ($k > 2$).

A factorization of Equation 14.31 is

$$\hat{\mathbf{x}}_k = \mathbf{U_k \Gamma_k^a \Gamma_k^{1-a} V_k^T} = \mathbf{GH^T} \tag{14.32}$$

where **G** and $\mathbf{H^T}$ are

$$\mathbf{G} = \mathbf{U_k \Gamma_k^a}$$
$$\mathbf{H^T} = \mathbf{\Gamma_k^{1-a} V_k^T} \tag{14.33}$$

and the value of the scale parameter α $(0 \leq \alpha \leq 1)$ determines whether emphasis is placed on the rows or columns of $\mathbf{x}$. The biplot display in 2D is the plot of the row "markers" $\mathbf{G}$ and column "markers" $\mathbf{H}$ given in Equation 14.33 for $k = 2$

$$\mathbf{G} = \mathbf{U}_2 \boldsymbol{\Gamma}_2^{\alpha}$$

$$\mathbf{H} = \mathbf{V}_2 \boldsymbol{\Gamma}_2^{1-\alpha}$$

(14.34)

In other words, the biplot is a plot of the coordinates associated with $\mathbf{G}$ or columns of $\mathbf{G}$, superimposed over the coordinates associated with $\mathbf{H}$ or columns of $\mathbf{H}$.

Although any values of α are possible to accomplish the factorization, three are most commonly used—1, ½, and 0. When $\alpha = 1$ is selected (the original value used in Gabriel, 1971), the result is a **row metric preserving** biplot. This display is useful for studying relationships among the observations. When the value 0 is selected, the result is a **column metric preserving** biplot. This display is useful for interpreting relationships between variables (e.g., interpreting a covariance or correlation matrix). The other value of α, ½, gives equal scaling or weight to the rows and columns. It is useful for interpreting interaction in two factor experiments (Gower and Hand, 1996).

We can see that PCA is a special case of SVD, since in PCA $\mathbf{X}^T\mathbf{X}$ is transformed (centered and scaled) to be the covariance or correlation matrix. This assumption is not generally made in SVD. Therefore, the singular values of PCA are eigenvalues of a covariance or a correlation matrix. The singular values of an SVD of the centered observations $\mathbf{x}$ must be squared and divided by $n - 1$ to obtain variances. The $\mathbf{V}$ matrix is equivalent to the loadings matrix $\mathbf{A}$.

Let us look at the numerical example given in the previous section and apply the SVD to the centered observations given in Equation 14.26 and the $\mathbf{x}^T\mathbf{x}$ matrix given in Equation 14.27

$$\mathbf{x}^T\mathbf{x} = \begin{bmatrix} 10 & -36 \\ -36 & 137.2 \end{bmatrix}$$

with eigenvalues

$$\lambda_1 = 146.68 \quad \lambda_2 = 0.52$$

take the square roots to obtain the singular values

$$s_1 = \sqrt{146.68} = 12.11$$

$$s_2 = \sqrt{0.52} = 0.72$$

and the singular values matrix has the diagonal elements equal to the singular values

$$\boldsymbol{\Gamma} = \begin{bmatrix} 12.11 & 0 \\ 0 & 0.72 \end{bmatrix}$$

In this case, the matrix $\mathbf{x}$ is of rank 2 (only two variables) and therefore the least-squares approximation for $k = 2$ is exact. The singular values matrix is just 2×2 and have no zeros in the main diagonal. Note that $12.11^2/4 = 36.67$ and $0.72^2/4 = 0.13$ are the eigenvalues of the covariance matrix.

The eigenvectors of the $\mathbf{x}^T\mathbf{x}$ matrix are arranged as columns in a matrix $\mathbf{V}$

$$\mathbf{V} = \begin{bmatrix} -0.25 & -0.96 \\ 0.96 & -0.25 \end{bmatrix}$$

Now calculate the $\mathbf{xx}^T$ matrix, note that $n = 5$ and therefore this matrix will be 5×5

$$\mathbf{xx}^T = \begin{bmatrix} 44.96 & 36.56 & -10.24 & -18.64 & -52.64 \\ 36.56 & 30.16 & -8.64 & -15.04 & -43.04 \\ -10.24 & -8.64 & 2.56 & 4.16 & 12.16 \\ -18.64 & -15.04 & 4.16 & 7.76 & 21.76 \\ -52.64 & -43.04 & 12.16 & 21.76 & 61.76 \end{bmatrix}$$

with eigenvalues

$$\lambda_1 = 146.68$$

$$\lambda_2 = 0.52$$

$$\lambda_3 = 0.0$$

$$\lambda_4 = 0.0$$

$$\lambda_5 = 0.0$$

Note that the first two are nonzero eigenvalues and are the same as the eigenvalues of $\mathbf{x}^T\mathbf{x}$. This fact is part of the Eckart–Young theorem and SVD.

The eigenvectors of the $\mathbf{xx}^T$ matrix can be arranged as columns of matrix $\mathbf{U}$

$$\mathbf{U} = \begin{bmatrix} -0.55 & -0.42 & 0.72 & 0.00 & 0.00 \\ -0.45 & 0.56 & -0.01 & -0.68 & -0.06 \\ 0.12 & -0.56 & -0.23 & -0.49 & -0.60 \\ 0.22 & 0.42 & 0.43 & 0.25 & -0.72 \\ 0.64 & -0.00 & 0.50 & -0.47 & 0.33 \end{bmatrix}$$

We can verify that the SVD $\mathbf{x} = \mathbf{U}\Gamma\mathbf{V}^T$ yields the correct results. Let us calculate the projection matrices $\mathbf{G}$ and $\mathbf{H}$ for the biplot in 2D ($k = 2$). Use scale equal to 1.

$$\mathbf{G} = \mathbf{U}_2\Gamma_2^\alpha = \mathbf{U}_2\Gamma = \begin{bmatrix} -0.55 & -0.42 \\ -0.45 & 0.56 \\ 0.12 & -0.56 \\ 0.22 & 0.42 \\ 0.64 & 0.00 \end{bmatrix} \begin{bmatrix} 12.11 & 0 \\ 0 & 0.72 \end{bmatrix} = \begin{bmatrix} -6.70 & -0.30 \\ -5.47 & 0.41 \\ 1.54 & -0.41 \\ 2.76 & 0.30 \\ 7.86 & -0.00 \end{bmatrix}$$

$$\mathbf{H} = \mathbf{V}_2\Gamma_2^{1-\alpha} = \mathbf{V}_2 = \mathbf{V} = \begin{bmatrix} -0.25 & -0.96 \\ 0.96 & -0.25 \end{bmatrix}$$

Now we can construct the biplot in 2D using $\mathbf{G}$ and $\mathbf{H}$ as in Figure 14.10. In this plot, the observations are labeled 1, 2, 3, 4, 5 and the variables V1 and V2. The bottom- and left-axis labels and scale correspond to the observations (coordinates related to U), whereas the top- and right-axis labels and scale correspond to the variables (coordinates of V). Proximity between pairs of observations denotes degree of similarity between those observations, whereas the cosine of the angle in between

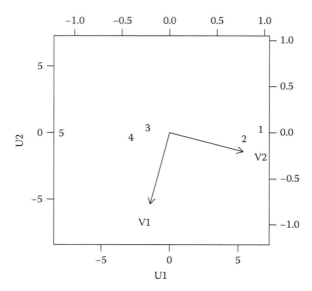

FIGURE 14.10   Biplot for the example.

variables represents the correlation between those variables. In the example, we see that observation pairs 1, 2 are similar and 3, 4 are similar.

The biplot concept can be applied to the PCA, the observations are plotted on the two principal components PC1, PC2. Their coordinates given by **G** are the scores. The variables are given by coordinates of **H**, which are equivalent to the loadings or matrix **A**.

14.7   FACTOR ANALYSIS

In PCA and SVD, we use the entire dataset and explore to see if we can reduce dimension. In FA we have some a priori idea of the underlying "factors" that can summarize the observable data and then we search for the unknown linear combinations of the known variables that would yield good predictors from these factors (Davis, 2002 pp. 526–548). The factors are mutually uncorrelated.

For example, suppose we have the answers from 100 individuals on a survey of 10 questions. Each question is one of the 10 observed variables, and we use FA to find the two factors, say "environmental awareness" and "recreational drive," that the survey was designed to measure. These factors will each be a linear combination of the 100 questions.

In FA, we build a predictor of the m variables from the k factors; k is less than m

$$\mathbf{z} = \mathbf{Af} + \mathbf{u} \tag{14.35}$$

where
   $\mathbf{z}$ is the predicted $m \times 1$ vector
   $\mathbf{f}$ is $k \times 1$ vector of "common factors"
   $\mathbf{u}$ is $m \times 1$ vector of "unique factors"
   $\mathbf{A}$ is the projection $m \times k$ matrix or "loadings." The "common factors" are supposed to be the
      underlying factors.

The $m \times m$ covariance matrix $\mathbf{C}$ is decomposed in two matrices, the common factor covariance and the error covariance or unique factor covariance

$$\mathbf{C} = \mathbf{AA}^{\mathrm{T}} + \mathrm{var}(\mathbf{u}) \tag{14.36}$$

Typically, FA is performed with the correlation matrix $\mathbf{R}$ instead of the covariance matrix. Therefore,

$$\mathbf{R} = \mathbf{A}\mathbf{A}^T + cor(\mathbf{u}) \tag{14.37}$$

The diagonal of $\mathbf{A}\mathbf{A}^T$ contains the "communalities" h_j^2, which are the sum (by column) of the squares of entries of $\mathbf{A}$. That is to say for the jth column

$$h_j^2 = \sum_{i=1}^{k} a_{ij}^2 \tag{14.38}$$

The communalities add up to the first eigenvalue. The diagonal terms of the error covariance are also called the "uniquenesses." Note that Equation 14.36 can be written as

$$\text{var}(\mathbf{u}) = \mathbf{C} - \mathbf{A}\mathbf{A}^T \tag{14.39}$$

to emphasize that the error covariance is the residual variance from the estimation of the covariance by the common factor covariance.

The goal of FA is to estimate the loadings $\mathbf{A}$ and the unique variance var($\mathbf{u}$). There are two approaches to do this: (1) optimization method, such as maximum likelihood, and (2) by SVD methods. Whatever approach is used, once $\mathbf{A}$ and var($\mathbf{u}$) are found, $\mathbf{A}$ can be transformed or rotated to achieve more understanding of the meaning of the factors. The reason for this is that there is more than one matrix, $\mathbf{A}$, which can satisfy this relation to $\mathbf{C}$. To see this note that if $\mathbf{A}$ is multiplied by a rotation matrix $\mathbf{G}$

$$\mathbf{C} = \mathbf{A}\mathbf{G}(\mathbf{A}\mathbf{G})^T + \text{var}(\mathbf{u}) = \mathbf{A}\mathbf{G}\mathbf{G}^T\mathbf{A}^T + \text{var}(\mathbf{u}) = \mathbf{A}\mathbf{A}^T + \text{var}(\mathbf{u}) \tag{14.40}$$

Therefore, any $\mathbf{G}$ would do; we have to select one that yields a desired structure. There are several rotation schemes; for example, the **varimax** rotation.

The number of observable variables m restricts the number of factors k. There have to be at least m variables to select k factors such that the degrees of freedom exceed or is equal to zero ($df \geq 0$). The degrees of freedom are calculated from

$$df = 0.5\left[(m-k)^2 - (m+k)\right] \tag{14.41}$$

For example, $m = 2$, $k = 1$ would not have enough degrees of freedom.

$$df = 0.5\left[(2-1)^2 - (2+1)\right] = 0.5[1-3] = -2$$

And $m = 3$, $k = 1$ would have zero degrees of freedom

$$df = 0.5\left[(3-1)^2 - (3+1)\right] = 0.5[4-4] = 0$$

So for one factor we must have at least three variables.

The **maximum likelihood estimation** (MLE) method makes some assumptions about the structure of the common and unique factors. The common factors are mutually independent normally distributed with mean 0 and variance 1, whereas the unique factors are mutually inde-pendent normally distributed with mean 0 and variance var($\mathbf{u}$). Columns of matrix $\mathbf{A}$ of loadings and var($\mathbf{u}$) are found iteratively. The initial estimate for the first column of $\mathbf{A}$ is based on the first eigenvector of the covariance matrix $\mathbf{C}$ scaled such that the sum of their squares is the equal to the first eigenvalue. The initial estimate for the var(u) is calculated using Equation 14.39 and the first estimate of $\mathbf{A}$. Then a new matrix is formed based on this var($\mathbf{u}$) estimate and the first eigen-value and eigenvector (again scaled such that the sum is equal to the first eigenvalue) is used to update the estimate of first column of $\mathbf{A}$, which in turn is used to update the estimate of var($\mathbf{u}$). Iteration continues until the first column of A converges. At this point, we have an MLE estimate of A and the need for additional factors is checked with a chi-square procedure. If a second factor is needed, then the residual from Equation 14.39 is used for eigen-decomposition. The second eigenvalue and eigenvector are used to find the second column of A using the iterative method until it converges.

The other approach to estimate $\mathbf{A}$ and var($\mathbf{u}$) is to perform SVD or PCA. There are two modes of FA: **R-mode** and **Q-mode**. For R-mode FA uses the $m \times m$ correlation matrix $\mathbf{R} = cor(\mathbf{X})$ of m attri-butes or variables; this matrix is proportional to $\mathbf{X}^T\mathbf{X}$. Recall that $\mathbf{C} = (n-1)^{-1}\mathbf{X}^T\mathbf{X}$. The matrix $\mathbf{R}$ is eigen-decomposed and its eigenvectors are used to estimate the columns of $\mathbf{A}$. The eigenvectors are converted to columns of $\mathbf{A}$ or factor loadings by multiplying by the square root of the corresponding eigenvalue, so that the sum of the squares of the elements of each factor loading is equal to the cor-responding eigenvalue. Only the first k loadings are selected. The residual variance after retaining the first k components is the uniqueness.

Q-mode FA uses the $n \times n$ "similarity" (cosine value) matrix between pairs of observations; this matrix is proportional to $\mathbf{X}\mathbf{X}^T$. As we know this matrix can be eigen-decomposed, and has only rank m, so it has a maximum of m nonzero eigenvalues. The eigenvectors form matrix $\mathbf{V}$.

To illustrate numerically we want at least three variables because $m = 2$ will not provide enough degrees of freedom. We will work with the same example as before but one more column; in a simi-lar fashion as in Carr (1995 page 111).

$$\mathbf{X} = \begin{bmatrix} 3 & 14 & 8 \\ 4 & 13 & 6 \\ 5 & 6 & 4 \\ 6 & 5 & 2 \\ 7 & 0 & 1 \end{bmatrix} \quad (14.42)$$

It is not easy to demonstrate the hand calculation of the MLE of $\mathbf{A}$ and var(u). The results for load-ings and uniquenesses obtained by computer are using correlation

$$\mathbf{A} = \begin{bmatrix} -0.998 \\ 0.971 \\ 0.996 \end{bmatrix} \quad \text{and} \quad cor(\mathbf{u}) = \begin{bmatrix} 0.005 \\ 0.058 \\ 0.009 \end{bmatrix}$$

Note that

$$\mathbf{A}\mathbf{A}^T = \begin{bmatrix} 0.996 & -0.969 & -0.994 \\ -0.969 & 0.943 & 0.967 \\ -0.994 & 0.967 & 0.992 \end{bmatrix}$$

The communalities are the squares of the elements of **A** or the diagonal terms of **AA**^{T}

$$\mathbf{h}^2 = \begin{bmatrix} 0.996 \\ 0.942 \\ 0.992 \end{bmatrix}$$

Adding up all communalities we get 2.93, which is equal to the first eigenvalue of **AA**^{T}, which is the only nonzero eigenvalues. The total correlation for three variables would be 3.00. Therefore 2.93/3 = 0.976. Therefore, this FA explains 97.6% of the variance.

Write the **cor(u)** as diagonal matrix

$$\mathbf{cor(u)} = \begin{bmatrix} 0.005 & 0 & 0 \\ 0 & 0.058 & 0 \\ 0 & 0 & 0.009 \end{bmatrix}$$

and now we can see that

$$\mathbf{AA}^T + \mathbf{cor(u)} = \begin{bmatrix} 1.00 & -0.97 & -0.99 \\ -0.97 & 1.00 & 0.97 \\ -0.99 & 0.97 & 1.00 \end{bmatrix} \tag{14.43}$$

which is equal to the correlation matrix **R** of X.

14.8 CORRESPONDENCE ANALYSIS

Correspondence analysis (CA) was originally derived for categorical data; however, it can be used for a variety of data including counts, continuous and mixed data. Therefore, it expands the capabilities of PCA, which applies only to continuous data. Each row and column value is transformed so that it has an associated "mass"; i.e., proportion related to the row or column totals. This is the same as a joint probability when the data are counts. Thus, the data table becomes a contingency table. One advantage is that the observations and the variables are represented in the same space. Therefore, relationships among observations, among variables, and between observations and variables are visualized in the same space. The chi-squared distance takes the place of the Euclidean distance. The idea is finding a new orthogonal coordinate system as in PCA that maximizes the representation of the "inertia," which now plays the role of the variance.

As before the $n \times m$ matrix **X** is the data matrix of n observations and m variables. To convert the data matrix in the form of a contingency table, let us calculate the grand total X_{tot} as the sum of all entries in **X**

$$X_{tot} = \sum_j \sum_i X_{ij} \tag{14.44}$$

A new matrix **Y** of proportions of the grand total is built based on **X** scaled by X_{tot}

$$\mathbf{Y} = \frac{\mathbf{X}}{X_{tot}} \tag{14.45}$$

The row totals are obtained summing over the columns (over the variables)

$$wc_i = \sum_j Y_{ij} \tag{14.46}$$

Note that $i = 1, \ldots, n$, then $\mathbf{wc}$ is an $n \times 1$ vector. We can think of wc_i as the "mass" of the ith row. Similarly, the column totals are obtained summing over the rows (over the observations)

$$wr_j = \sum_i Y_{ij} \tag{14.47}$$

Note that $j = 1, \ldots, m$, then $\mathbf{wr}$ is an $m \times 1$ vector. We can think of wr_j as the "mass" of the jth column. $\mathbf{wc}$ and $\mathbf{wr}$ are marginal probabilities (recall contingency tables).

We build an $n \times m$ matrix $\mathbf{S_x}$ such that its elements are like the square root of a chi-square calculation

$$\mathbf{S_x} = diag\left(1/\sqrt{wr_i}\right) \mathbf{Y} \, diag\left(1/\sqrt{wc_i}\right) = \mathbf{WrYWc} \tag{14.48}$$

Here, $\mathbf{Wr}$ is the diagonal matrix with entries equal to the inverse of the square root of entries of vector $\mathbf{wr}$, and $\mathbf{Wc}$ is the diagonal matrix with entries equal to the inverse square root of entries of vector $\mathbf{wc}$.

Now proceed with R-mode and Q-mode decomposition or equivalently SVD of matrix $\mathbf{S_x}$. In R-mode, $\mathbf{R} = \mathbf{S_x^T S_x}$ is an $m \times m$ **similarity** matrix that plays the role of the correlation matrix. Each element of this matrix is analogous to the chi-square statistic. Now we do an eigen-decomposition of $\mathbf{R}$; the loadings matrix $\mathbf{A}$ is formed by the eigenvectors of $\mathbf{R}$ multiplied by square root of corresponding eigenvalues λ_i (i.e., singular value s_i)

$$\mathbf{A} = \mathbf{U} \, diag\left(\sqrt{\lambda_i}\right) = \mathbf{U} \, diag(s_i) = \mathbf{UL}$$

where
 $\mathbf{U}$ is matrix formed by eigenvectors of $\mathbf{R}$
 $\mathbf{L}$ is the diagonal matrix with singular values s_i of $\mathbf{R}$

Now the scores are

$$\mathbf{Z}_R = \mathbf{S_x A} \tag{14.49}$$

In Q-mode, $\mathbf{Q} = \mathbf{S_x S_x^T}$ is an $n \times n$ matrix, which we proceed to decompose. The loadings are the eigenvectors of $\mathbf{Q}$ multiplied by square root of corresponding eigenvalues (singular value) $\mathbf{B} = \mathbf{V} diag\left(\sqrt{a_i}\right)$ where $\mathbf{V}$ is matrix formed by eigenvectors of $\mathbf{Q}$ and a_i are the eigenvalues of $\mathbf{Q}$. These are the same as λ_i of $\mathbf{R}$ but with two additional zero value eigenvalues. Now the scores are $\mathbf{Z}_Q = \mathbf{S_x^T B}$.

We can alternatively use SVD and reduce dimensionality by selecting matrices $\mathbf{G}$ and $\mathbf{H}$ as in biplots and scaling differently. Because the rows and columns add up to 1.00, one of the eigenvalues is zero and therefore the dimensionality is reduced to $m - 1$. A biplot will show observations and variables on the same space and can be compared simultaneously.

To illustrate numerically with a simple example, we want at least three variables because $m = 2$ will lead to only one coordinate. We will work with the same example as in the FA section. The data matrix is $\mathbf{X}$ given in Equation 14.42. The grand total is $X_{tot} = 84$. The $\mathbf{Y}$ matrix is $\mathbf{X}/X_{tot}$

$$\mathbf{Y} = \begin{bmatrix} 0.04 & 0.17 & 0.10 \\ 0.05 & 0.15 & 0.07 \\ 0.06 & 0.07 & 0.05 \\ 0.07 & 0.06 & 0.02 \\ 0.08 & 0.00 & 0.01 \end{bmatrix} \tag{14.50}$$

The column and row totals are

$$\mathbf{wc} = \begin{bmatrix} 0.30 \\ 0.27 \\ 0.18 \\ 0.15 \\ 0.10 \end{bmatrix} \quad \text{and} \quad \mathbf{wr} = \begin{bmatrix} 0.30 \\ 0.45 \\ 0.25 \end{bmatrix}$$

The inverse of the square root of each element yields

$$\begin{bmatrix} 1/\sqrt{0.30} \\ 1/\sqrt{0.27} \\ 1/\sqrt{0.18} \\ 1/\sqrt{0.15} \\ 1/\sqrt{0.10} \end{bmatrix} = \begin{bmatrix} 1.83 \\ 1.91 \\ 2.37 \\ 2.54 \\ 3.24 \end{bmatrix} \quad \text{and} \quad \begin{bmatrix} 1/\sqrt{0.30} \\ 1/\sqrt{0.45} \\ 1/\sqrt{0.25} \end{bmatrix} = \begin{bmatrix} 1.83 \\ 1.49 \\ 2 \end{bmatrix}$$

Therefore, the diagonal matrices formed with the inverse of the square root of these vectors are

$$\mathbf{Wc} = \begin{bmatrix} 1.83 & 0 & 0 & 0 & 0 \\ 0 & 1.91 & 0 & 0 & 0 \\ 0 & 0 & 2.37 & 0 & 0 \\ 0 & 0 & 0 & 2.54 & 0 \\ 0 & 0 & 0 & 0 & 3.24 \end{bmatrix} \tag{14.51}$$

$$\mathbf{Wr} = \begin{bmatrix} 1.83 & 0 & 0 \\ 0 & 1.49 & 0 \\ 0 & 0 & 2 \end{bmatrix} \tag{14.52}$$

And now we calculate the $\mathbf{S_x}$ matrix by Equation 14.48

$$\mathbf{S_x} = \mathbf{WrYWc} = \begin{bmatrix} 0.12 & 0.45 & 0.35 \\ 0.16 & 0.44 & 0.27 \\ 0.25 & 0.25 & 0.23 \\ 0.33 & 0.22 & 0.12 \\ 0.49 & 0.00 & 0.08 \end{bmatrix}$$

For illustration proceed with R-mode. The $\mathbf{R}$ similarity matrix

$$\mathbf{R} = \mathbf{S_x^T S_x} = \begin{bmatrix} 0.46 & 0.27 & 0.22 \\ 0.27 & 0.51 & 0.36 \\ 0.22 & 0.36 & 0.26 \end{bmatrix}$$

Its entries are equivalent to chi-square values. The eigenvalues are $\lambda 1 = 1.000$, $\lambda 2 = 0.240$, $\lambda 3 = 0.006$. The first eigenvalue is trivial with value 1.00, because we converted $\mathbf{X}$ to $\mathbf{Y}$ such that the rows and columns add up to 1.00. The singular values are the square root of the eigenvalues 1, 0.49, 0.075.

The loadings are the eigenvectors of $\mathbf{R}$ multiplied by singular values.

$$\mathbf{A} = \begin{bmatrix} -0.55 & 0.83 & -0.08 \\ -0.67 & -0.50 & -0.55 \\ -0.50 & -0.24 & 0.83 \end{bmatrix}\begin{bmatrix} 1 & 0 & 0 \\ 0 & 0.490 & 0 \\ 0 & 0 & 0.075 \end{bmatrix} = \begin{bmatrix} -0.55 & 0.41 & -0.01 \\ -0.67 & -0.24 & -0.04 \\ -0.50 & -0.11 & 0.06 \end{bmatrix}$$

The scores are

$$\mathbf{Z_R} = \mathbf{S_x A} = \begin{bmatrix} 0.12 & 0.45 & 0.35 \\ 0.16 & 0.44 & 0.27 \\ 0.25 & 0.25 & 0.23 \\ 0.33 & 0.22 & 0.12 \\ 0.49 & 0.00 & 0.08 \end{bmatrix}\begin{bmatrix} -0.55 & 0.41 & -0.01 \\ -0.67 & -0.24 & -0.04 \\ -0.50 & -0.11 & 0.06 \end{bmatrix} = \begin{bmatrix} -0.55 & -0.10 & 0.00 \\ -0.52 & -0.07 & -0.00 \\ -0.42 & 0.02 & 0.00 \\ -0.39 & 0.07 & -0.00 \\ -0.31 & 0.19 & 0.00 \end{bmatrix}$$

You could develop the same process for Q-mode to find, matrix $\mathbf{Q} = \mathbf{S_x S_x^T}$, the matrix $\mathbf{V}$ of eigenvectors of $\mathbf{Q}$, the loadings matrix $\mathbf{B} = \mathbf{V}diag\left(\sqrt{a_i}\right)$, using eigenvalues of $\mathbf{Q}$, and the scores

$$\mathbf{Z_Q} = \mathbf{S_x^T B} = \begin{bmatrix} -0.55 & -0.20 & -0.00 & 0 & 0 \\ -0.67 & 0.12 & -0.00 & 0 & 0 \\ -0.50 & 0.06 & 0.00 & 0 & 0 \end{bmatrix} \tag{14.53}$$

We can use the $\mathbf{Z_R}$ and $\mathbf{Z_Q}$ scores to make a plot as in Figure 14.11, where markers indicate points from $\mathbf{Z_R}$ (observations) and arrows indicate points from $\mathbf{Z_Q}$ (variables). However, more conveniently, we can

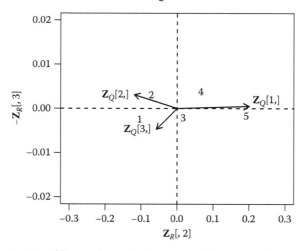

FIGURE 14.11 Plotting scores of R- and Q-modes of correspondence analysis.

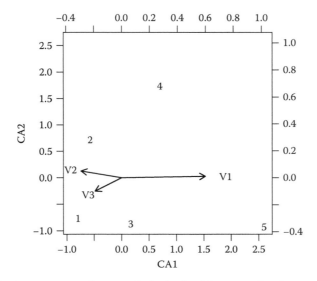

FIGURE 14.12   Biplot from correspondence analysis obtained from cca of package vegan.

scale as a biplot such that proximity of observations and variables indicates similarity. Figure 14.12 shows an example obtained from a function in an R package applied to this simple dataset.

14.9   EXERCISES

Exercise 14.1

Show that the vector $\begin{bmatrix} 1 \\ 1 \end{bmatrix}$ is an eigenvector associated with the eigenvalue $\lambda = 2$ of matrix $\mathbf{A} = \begin{bmatrix} 1 & 1 \\ -2 & 4 \end{bmatrix}$ Plot it together with its transformation by this matrix.

Exercise 14.2

Suppose we have a covariance matrix $\mathbf{C} = \begin{bmatrix} 0.6 & 0.2 \\ 0.2 & 0.5 \end{bmatrix}$, the two eigenvalues are $\lambda 1 = 0.76$, $\lambda 2 = 0.34$. The eigenvectors are $\mathbf{v}_1 = \begin{bmatrix} 0.78 \\ 0.61 \end{bmatrix}$ $\mathbf{v}_2 = \begin{bmatrix} -0.61 \\ 0.78 \end{bmatrix}$. Calculate total variance and discuss its distribution among the eigenvalues. Would one component suffice or would you keep both? Draw a diagram showing the eigenvectors. Write the matrix of loadings.

Exercise 14.3

Suppose we have a covariance matrix

$$\mathbf{C} = \begin{bmatrix} 0.6 & 0.2 & 0.1 \\ 0.2 & 0.5 & 0.2 \\ 0.1 & 0.2 & 0.7 \end{bmatrix}$$

The eigenvalues are

$$\lambda 1 = 0.94 \quad \lambda 2 = 0.55 \quad \lambda 3 = 0.31$$

The eigenvectors are

$$\mathbf{v}_1 = \begin{bmatrix} -0.51 \\ -0.54 \\ -0.67 \end{bmatrix} \quad \mathbf{v}_2 = \begin{bmatrix} 0.72 \\ 0.15 \\ -0.68 \end{bmatrix} \quad \mathbf{v}_3 = \begin{bmatrix} 0.47 \\ -0.83 \\ 0.31 \end{bmatrix}$$

Calculate total variance and discuss its distribution among the eigenvalues. Would you keep all three components? Or would two suffice? Or would one component suffice? Draw a diagram showing the first two eigenvectors. Write the matrix of loadings.

Exercise 14.4
Complete the calculations of Q-mode CA to arrive at the scores Z_Q shown in Equation 14.53.

14.10　COMPUTER SESSION: MULTIVARIATE ANALYSIS, PCA

14.10.1　Eigenvalues and Eigenvectors of Covariance Matrices

Recall that covariance matrices are symmetric. As an example, consider the matrix $\mathbf{C} = \begin{pmatrix} 2 & 4 \\ 4 & 10 \end{pmatrix}$. First, create the matrix

```
> C <- matrix(c(2,4,4,10), byrow=T, ncol=2)
> C
      [,1]    [,2]
[1,]    2      4
[2,]    4     10
```

Calculate the eigenvalues and eigenvectors using **eigen(C)**

```
> eigen(C)
$values
[1] 11.6568542  0.3431458

$vectors
          [,1]        [,2]
[1,] 0.3826834 -0.9238795
[2,] 0.9238795  0.3826834
```

Note that the trace is the sum of the eigenvalues

```
> sum(diag(C))
[1] 12
> sum(eigen(C)$values)
[1] 12
>
```

The trace of a covariance matrix is the total variance. Therefore, the eigen-decomposition is one way of decomposing the variance.

14.10.2   PCA: A SIMPLE 2 × 2 EXAMPLE USING EIGENVALUES AND EIGENVECTORS

Before we use the PCA functions and for better understanding of the fundamentals of PCA, we will develop an example from scratch based on eigen-decomposition for the 2 × 2 example given in the previous chapter. Of course, you do not really need to do PCA on a bivariate dataset; this is only to facilitate your understanding.

```
> X <- cbind(c(3,4,5,6,7), c(14,13,6,5,0))
> X
     [,1] [,2]
[1,]    3   14
[2,]    4   13
[3,]    5    6
[4,]    6    5
[5,]    7    0
```

First, standardize each variable

```
> x <- cbind((X[,1]-mean(X[,1]))/sd(X[,1]), (X[,2]-mean(X[,2]))/
  sd(X[,2]))
> x
             [,1]          [,2]
[1,] -1.2649111    1.0927804
[2,] -0.6324555    0.9220335
[3,]  0.0000000   -0.2731951
[4,]  0.6324555   -0.4439421
[5,]  1.2649111   -1.2976768
>
```

Calculate the covariance matrix, eigenvalues and eigenvectors

```
> # covariance matrix
> C <-  var(x)
> # same as cor(X)
> C
             [,1]          [,2]
[1,]  1.0000000   -0.9719086
[2,] -0.9719086    1.0000000
> # eigen decomp
> eigen(C)
$values
[1] 1.97190864 0.02809136

$vectors
             [,1]          [,2]
[1,] -0.7071068   -0.7071068
[2,]  0.7071068   -0.7071068

>
```

The first eigenvalue 1.97 is a very large fraction of total correlation (2.00), then the variance explained by the first eigenvalue is $100 \times (1.97/2) = 98.5\%$. Therefore, only one component explains most of the variance. The loadings are the eigenvectors. Then, project new coordinates (scores)

```
> A <- eigen(C)$vectors
> # scores
> z <- x%*%A
> z
              [,1]            [,2]
[1,]    1.6671397   0.12171473
[2,]    1.0991897  -0.20476254
[3,]   -0.1931781   0.19317812
[4,]   -0.7611280  -0.13329916
[5,]   -1.8120232   0.02316886
>
```

Compare the original data versus new data (scores)

```
#plot data and transformed data
par(mfrow=c(1,2),pty="s")
plot(x[,1],x[,2], xlab="X1", ylab="X2", pch=as.character(seq(1:n)))
abline(h=0,v=0,lty=2)
plot(z[,1],z[,2], xlab="Z1", ylab="Z2", pch=as.character(seq(1:n)))
abline(h=0,v=0,lty=2)
```

(Figure 14.8). Note that we can always scale the eigenvectors; for example to get exactly same results as in the previous section

```
> lambda <- eigen(C)
> lambda$vectors[,1]
[1] -0.7071068   0.7071068
> pca1 <- lambda$vectors[,1]/lambda$vectors[2,1]
> pca2 <- lambda$vectors[,2]/lambda$vectors[2,2]
> A <- cbind(pca1,pca2)
> A
      pca1 pca2
[1,]    -1    1
[2,]     1    1
> zA <- x %*% A
> zA
              pca1            pca2
[1,]    2.3576915  -0.17213062
[2,]    1.5544890   0.28957797
[3,]   -0.2731951  -0.27319511
[4,]   -1.0763976   0.18851348
[5,]   -2.5625878  -0.03276571
>
```

14.10.3   PCA: A 2 × 2 EXAMPLE

We will do the same exercise but using R functions. We apply function **princomp** to matrix **x** of standardized variables

```
> pca2x2 <- princomp(x)
> pca2x2
Call:
princomp(x = x)

Standard deviations:
    Comp.1    Comp.2
1.2559964 0.1499103

 2  variables and  5 observations.
> summary(pca2x2)
Importance of components:
                             Comp.1      Comp.2
Standard deviation      1.2559964 0.14991025
Proportion of Variance  0.9859543 0.01404568
Cumulative Proportion   0.9859543 1.00000000
>
```

When variables are given in disparate units, it is important to standardize the variables before performing PCA as we have just done or to perform PCA using the correlation matrix instead of the covariance matrix; for this we could have applied function `princomp` with argument `cor = T`.

```
> pca2x2 <- princomp(X,cor=T)
```

Using either approach, we obtain that one component explains most (98.5%) of the variance and the bivariate dataset can be reduced to a univariate one.

Note that the loadings are the same as the eigenvectors obtained in the previous section.

```
> pca2x2$loadings

Loadings:
      Comp.1 Comp.2
[1,] -0.707 -0.707
[2,]  0.707 -0.707

                Comp.1 Comp.2
SS loadings        1.0    1.0
Proportion Var     0.5    0.5
Cumulative Var     0.5    1.0
>
```

You could also look at the scores using

```
> pca2x2$scores
          Comp.1      Comp.2
[1,]  1.6671397  0.12171473
[2,]  1.0991897 -0.20476254
[3,] -0.1931781  0.19317812
[4,] -0.7611280 -0.13329916
[5,] -1.8120232  0.02316886
>
```

Now note that these scores given by `princomp` are the scores obtained in the previous section.

We can produce several useful plots. First, the loadings can be visualized as a barplot. Second, we can produce a plot of variances for each component, and finally a "biplot" (Figure 14.13).

```
panel4(size=7)
barplot(loadings(pca2x2), beside=T)
plot(pca2x2)
plot(pca2x2, type="l")
biplot(pca2x2)
```

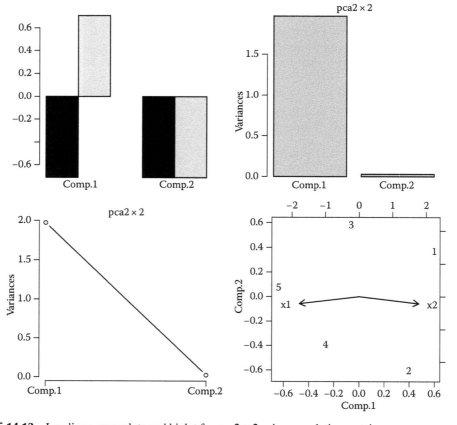

FIGURE 14.13 Loadings, screeplots and biplot for pca2 × 2 using correlation matrix.

14.10.4 PCA HIGHER-DIMENSIONAL EXAMPLE

Let us work with data in file **lab14/canopy.csv**. The dataset is from Acevedo et al. (2001). It refers to fragmentation metrics of canopy images of a tropical cloud forest taken at different locations along two transects. Each image comes from a photograph of the canopy taken from the ground. A canopy opening is defined as intensity of color in a pixel of the picture. Contiguous gaps make up patches of different sizes. First row are variable labels, and first column is sampling points ID label.

```
ID,POC,PRL,LAI,NP,PD,MPS,LPI,MSI,MPFD
1n,2.98,8.02,2.29,1915,7.37,0.14,15.73,1.3,1.55
2n,1.72,8.76,2.94,2207,12.29,0.08,4.23,1.3,1.55
3n,1.21,5.33,2.78,1602,13.08,0.08,8.35,1.29,1.56
4n,1.71,7.27,2.53,2342,13.45,0.07,15.3,1.28,1.56
5n,5.19,16.57,3.02,2953,6.51,0.15,5.08,1.34,1.54
And so on
```

The columns are

```
POC= percent of open area
PRL= percent of area occupied by reflecting leaves
LAI= Leaf area index
NP= number of patches
PD= patch density
MPS= mean patch size
LPI= largest patch index
MSI= mean shape index
MPFD = mean patch fractal dimension
```

First, we read the file as data frame such that rows have names equal to the ID of the point

```
frag <- read.table("lab14/canopy.csv", header=T,row.names=1,sep=",")
```

By examining the data, we see that the variables have disparate ranges. It is important to standardize the variables before applying PCA or to perform the PCA using the correlation matrix instead of the covariance matrix.

We standardize by

```
frag.s <- frag
 for(j in 1:9){
   muf <- mean(frag[,j]); sdf <- sd(frag[,j])
   for(i in 1:22) frag.s[i,j] <- (frag[i,j]-muf)/sdf
 }
```

For simplicity, instead of standardizing, we will perform PCA using function `princomp` and setting argument `cor = T`.

```
> frag.pca <- princomp(frag,cor=T)
```

Check the summary

```
> summary(frag.pca)
Importance of components:
                            Comp.1      Comp.2      Comp.3      Comp.4
   Comp.5
Standard deviation       2.0772715   1.7037539  0.84735546  0.81178307
   0.52856526
Proportion of Variance   0.4794508   0.3225308  0.07977903  0.07322131
   0.03104236
Cumulative Proportion    0.4794508   0.8019816  0.88176061  0.95498192
   0.98602428
                            Comp.6      Comp.7      Comp.8
Comp.9
Standard deviation       0.25556321  0.169020175  0.155242159
   0.0883232141
Proportion of Variance   0.00725695  0.003174202  0.002677792
   0.0008667767
Cumulative Proportion    0.99328123  0.996455431  0.999133223
   1.0000000000
>
```

We can see that it takes three components to explain almost 90% of the variance (88%) and that the first four components explain 95% of variance. Now we plot loadings, variances, and biplots. We draw several biplots to include different pairwise combinations of components; for example, Comp2 vs. Comp 1 and Comp3 vs. Comp2.

```
panel2(size=7)
barplot(loadings(frag.pca), beside=T)
plot(frag.pca)
par(mfrow=c(1,1))
win.graph(); biplot(frag.pca, choices=c(1,2))
win.graph();biplot(frag.pca, choices=c(1,3))
```

which would yield plots shown in Figures 14.14 through 14.16.

14.10.5 PCA USING THE RCMDR

First, make the dataset active. Let us work with the canopy data. We have already built a data frame frag. In the R Commander, go to **Data |Active Dataset|Select Active dataset** and pick

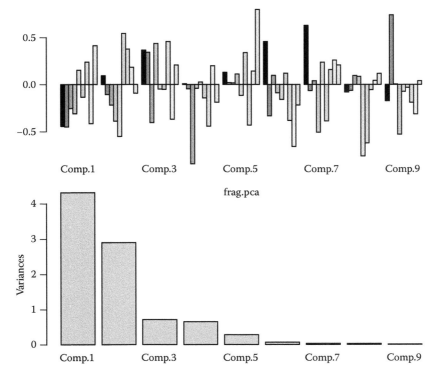

FIGURE 14.14　Plots for PCA of frag data using correlation.

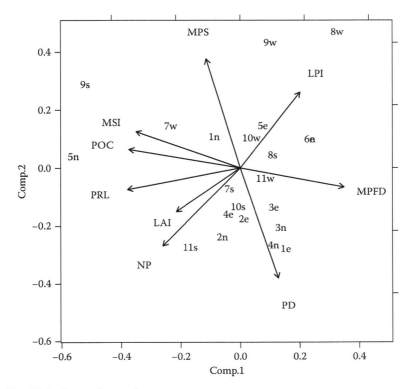

FIGURE 14.15　Biplot first and second components.

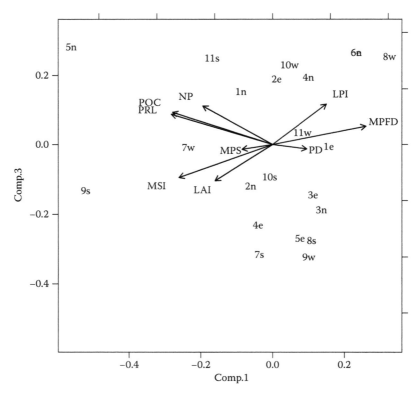

FIGURE 14.16 Biplot first and third component.

frag from the list. Confirm that frag shows in the box for Active dataset and to reconfirm use **View dataset** button (Figure 14.17). Then go to **Statistics|Dimensional Analysis|Principal Component Analysis** and select all variables in the dialog box (Figure 14.18). View the results in the output window:

```
> .PC <- princomp(~LAI+LPI+MPFD+MPS+MSI+NP+PD+POC+PRL, cor=TRUE,
  data=frag)

> unclass(loadings(.PC))  # component loadings
          Comp.1      Comp.2       Comp.3       Comp.4       Comp.5
  Comp.6
LAI    0.2547197 -0.21755149 -0.40525709  0.838928196  0.01614738
  0.09555842
LPI   -0.2384965  0.37703785  0.45740749  0.442962293 -0.43055062
  -0.38161386
MPFD -0.4134113 -0.09432869  0.20760463  0.188078783  0.79767246
  -0.21806475
MPS    0.1351543  0.54409040 -0.05261879  0.141972252  0.33927647
  0.11834354
MSI    0.4148974  0.18227589 -0.36925140 -0.200067213  0.14199172
  -0.65609528
```

```
(continued)

NP     0.3090014 -0.38770939  0.43673267  0.040930393  0.11070335
  -0.08914598
PD    -0.1520914 -0.54973797 -0.04871851 -0.025182320 -0.11586042
  -0.15845612
POC    0.4435619  0.09250391  0.36567621 -0.007517188  0.12935857
  0.45817063
PRL    0.4489555 -0.10615869  0.34278573  0.045401498  0.02003502
  -0.33203213
              Comp.7        Comp.8         Comp.9
LAI   -0.03830865 -0.09421151  0.004457308
LPI   -0.15848931  0.05599095 -0.190441212
MPFD  -0.20685433 -0.11668407  0.038502592
MPS    0.38641960  0.61912758 -0.031957843
MSI   -0.25992605 -0.04261214 -0.313467962
NP     0.50557004 -0.08542145 -0.527909765
PD    -0.23711780  0.75587439 -0.074470172
POC   -0.63056266  0.08081357 -0.173568986
PRL    0.06563551  0.06376156  0.740665674


> .PC$sd^2  # component variances
     Comp.1     Comp.2     Comp.3     Comp.4     Comp.5     Comp.6
  Comp.7
4.31505686 2.90277740 0.71801127 0.65899175 0.27938124 0.06531255
  0.02856782
     Comp.8     Comp.9
0.02410013 0.00780099


> summary(.PC) # proportions of variance
Importance of components:
                        Comp.1     Comp.2     Comp.3     Comp.4
  Comp.5
Standard deviation      2.0772715 1.7037539 0.84735546 0.81178307
  0.52856526
Proportion of Variance 0.4794508 0.3225308 0.07977903 0.07322131
  0.03104236
Cumulative Proportion  0.4794508 0.8019816 0.88176061 0.95498192
  0.98602428
                        Comp.6     Comp.7     Comp.8
  Comp.9
Standard deviation      0.25556321 0.169020175 0.155242159
  0.0883232141
Proportion of Variance 0.00725695 0.003174202 0.002677792
  0.0008667767
Cumulative Proportion  0.99328123 0.996455431 0.999133223
  1.0000000000

> screeplot(.PC)

> remove(.PC)
```

frag

	POC	PRL	LAI	NP	PD	MPS	LPI	MSI	MPFD
1n	2.98	8.02	2.29	1915	7.37	0.14	15.73	1.30	1.55
2n	1.72	8.76	2.94	2207	12.29	0.08	4.23	1.30	1.55
3n	1.21	5.33	2.78	1602	13.08	0.08	8.35	1.29	1.56
4n	1.71	7.27	2.53	2342	13.45	0.07	15.30	1.28	1.56
5n	5.19	16.57	3.02	2953	6.51	0.15	5.08	1.34	1.54
6n	1.57	4.87	1.35	1308	9.23	0.11	25.52	1.28	1.56
7s	1.65	6.64	3.07	1601	10.11	0.10	8.95	1.31	1.55
8s	1.20	4.00	2.09	1111	9.56	0.11	8.64	1.30	1.55
9s	4.36	13.16	2.68	1955	5.26	0.19	7.77	1.36	1.53
10s	1.59	7.88	2.69	1883	11.61	0.09	12.25	1.30	1.55
11s	2.54	11.74	2.65	3030	11.84	0.08	5.49	1.30	1.55
1e	1.23	6.14	2.42	1921	14.29	0.07	6.85	1.28	1.56
2e	1.99	8.97	2.07	2338	11.70	0.09	5.97	1.30	1.56
3e	1.23	5.07	2.19	1552	12.43	0.08	8.71	1.29	1.55
4e	1.67	7.49	3.21	1867	11.18	0.09	8.77	1.30	1.55
5e	1.26	4.56	1.78	1003	8.54	0.12	12.13	1.32	1.55
6e	1.57	4.87	1.35	1308	9.23	0.11	25.52	1.28	1.56
7w	2.81	10.95	2.24	1834	7.29	0.14	13.37	1.33	1.54
8w	1.13	2.99	2.22	508	5.56	0.18	67.22	1.27	1.56
9w	1.39	3.67	2.22	548	4.74	0.21	20.67	1.32	1.56
10w	1.98	7.58	1.85	1660	8.94	0.11	31.19	1.30	1.55
11w	1.60	6.05	1.79	1665	10.57	0.10	10.05	1.29	1.55

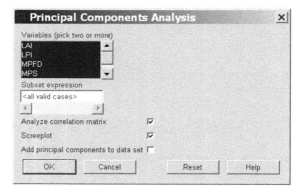

FIGURE 14.17 Make frag the active dataset and view it.

FIGURE 14.18 Select all variables in dialog box.

which at the end produces a screeplot (Figure 14.19). Note that the object. PC is removed at the end. If we re-execute the commands shown in the script window except this remove, then we can use `biplot(.PC)` from the console to obtain a biplot.

14.10.6 Factor Analysis

To gain insight, work with the simple data matrix **X** given in Equation 14.42. Apply the function `factanal` that uses the maximum likelihood method for FA.

```
>Xf <- factanal(X, factors =1)
> Xf

Call:
factanal(x = X, factors = 1)
```

```
(continued)
Uniquenesses:
    x1      x2      x3
0.005 0.058 0.009

Loadings:
    Factor1
x1 -0.998
x2  0.971
x3  0.996

                Factor1
SS loadings      2.929
Proportion Var   0.976

The degrees of freedom for the model is 0 and the fit was 0.1591
>
```

This provides information on uniqueness and loadings. We account for 97.6% of the variance with one factor and zero degrees of freedom. We can use the loadings and uniqueness to form Equation 14.43

```
# use loadings to form A
A <- c(-0.998,0.971,0.996)
# use uniquenesses to form var(u)
Vu <- c(0.005,0.058,0.009)
A%*%t(A)+diag(Vu)
```

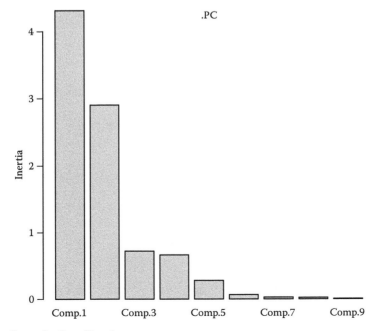

FIGURE 14.19 Screeplot from Rcmdr.

Now, for a more real example, let us work with the same frag data as in the previous section:

```
> frag.fa <- factanal(frag, factors =2)
> frag.fa
```

This provides information on uniqueness and loadings and ends with the following information:

```
SS loadings          3.964    2.832
Proportion Var       0.440    0.315
Cumulative Var       0.440    0.755

Test of the hypothesis that 2 factors are sufficient.
The chi square statistic is 50.27 on 19 degrees of freedom.
The p-value is 0.000119
```

We account for 75.5% of the variance with these two factors. The low p-value suggests to reject the null that two factors are sufficient. Let us try three factors

```
frag.fa <- factanal(frag, factors =3)
```

We can see that

```
SS loadings          3.002    2.877    1.828
Proportion Var       0.334    0.320    0.203
Cumulative Var       0.334    0.653    0.856

Test of the hypothesis that 3 factors are sufficient.
The chi square statistic is 17.89 on 12 degrees of freedom.
The p-value is 0.119
>
```

Now the proportion of variance explained is 85.6% and the higher p-value suggests not rejecting the null that three factors are sufficient.

We obtain uniquenesses from the previous call to the function

```
Uniquenesses:
  POC    PRL    LAI     NP     PD    MPS    LPI    MSI   MPFD
0.054  0.007  0.676  0.025  0.014  0.060  0.180  0.005  0.272
```

We can calculate the communalities simply as 1-uniquenesses

```
> round((1- frag.fa$uniquenesses),3)
  POC    PRL    LAI     NP     PD    MPS    LPI    MSI   MPFD
0.946  0.993  0.324  0.975  0.986  0.940  0.820  0.995  0.728
>
```

Their sum should be equal to the sum of the first three eigenvalues or correlation explained. Therefore, we confirm the fraction explained.

```
> sum(1- frag.fa$uniquenesses)
[1] 7.706775
> 7.70/9
[1] 0.855
```

To produce scores, this calculation should be specified as an argument `bartlett` or `regression`. Both provide estimates of the scores.

```
> frag.fa <- factanal(frag, factors =3, scores="Bart")
> frag.fa$scores

> frag.fa$scores
         Factor1      Factor2      Factor3
1n    0.70768305   0.60986319  -0.95931514
2n   -0.91378470   0.35865535   0.43876934
3n   -1.07003356  -0.70852715   0.43850069
4n   -1.34533014   0.42236143  -0.55831409
5n    1.10812004   2.76150176   0.20604776
6n    0.09628633  -0.33282430  -1.29443801
7s   -0.12004278  -0.55334364   0.88231261
.... and so on
```

To perform rotation use argument rotation = "name of rotation function", by default rotation = varimax.

14.10.7 FACTOR ANALYSIS USING RCMDR

As in PCA, first make the dataset active. Go to **Data |ActiveDatset|Select Active dataset** and pick frag from the list. Confirm that frag shows in the box for Active dataset and to reconfirm use View dataset. See Figure 14.17. Then go to **Statistics|Dimensional Analysis|Factor Analysis** to obtain dialog box as shown in Figure 14.20. Here, we selected all variables and varimax rotation and Bartlett's method. Then in the next pop-up window move the slider to select number of factors = 3 (Figure 14.21). Look at the results in the output window

```
> .FA <- factanal(~LAI+LPI+MPFD+MPS+MSI+NP+PD+POC+PRL, factors=3,
+     rotation="varimax", scores="Bartlett", data=frag)

> .FA

Call:
factanal(x = ~LAI + LPI + MPFD + MPS + MSI + NP + PD + POC +
  PRL, factors = 3, data = frag, scores = "Bartlett", rotation =
  "varimax")

Uniquenesses:
  LAI   LPI  MPFD   MPS   MSI    NP    PD   POC   PRL
0.676 0.180 0.272 0.060 0.005 0.025 0.014 0.054 0.007
```

(continued)

```
(continued)
Loadings:
      Factor1 Factor2 Factor3
LAI  -0.152   0.390   0.385
LPI   0.422  -0.266  -0.756
MPFD -0.405  -0.482  -0.576
MPS   0.969
MSI   0.577   0.359   0.731
NP   -0.414   0.869   0.219
PD   -0.991
POC   0.443   0.828   0.255
PRL   0.126   0.922   0.355

                Factor1 Factor2 Factor3
SS loadings       3.002   2.877   1.828
Proportion Var    0.334   0.320   0.203
Cumulative Var    0.334   0.653   0.856

Test of the hypothesis that 3 factors are sufficient.
The chi square statistic is 17.89 on 12 degrees of freedom.
The p-value is 0.119

> frag$F1 <- .FA$scores[,1]

> frag$F2 <- .FA$scores[,2]

> frag$F3 <- .FA$scores[,3]

> remove(.FA)
```

The three extracted factors (scores) have been added to the dataset as F1, F2, and F3. Because the dataset has been modified, we can save frag as frag.dat.fa to save these scores and reconstruct the original frag from reading the file again for the forthcoming work on CA.

```
frag.dat.fa <- frag
frag <- read.table("lab14/canopy.csv", header=T,row.names=1,sep=",")
```

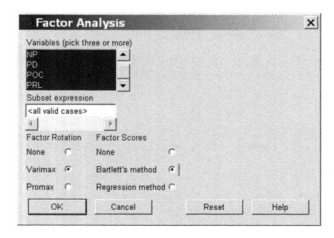

FIGURE 14.20 Dialog window for FA.

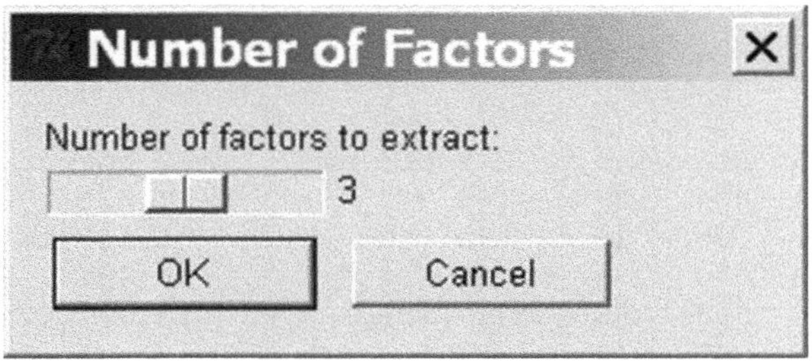

FIGURE 14.21 Slider to select number of factors.

14.10.8 CORRESPONDENCE ANALYSIS

There are functions to perform CA in several packages. We will use function `ca` of package **ca** and function `cca` in package **vegan**. Thus, please install and load packages **ca** and **vegan** if you have not done this yet or it is not in your **Rprofile.site** file. Recall the function `library` offers another way of loading an installed package.

```
>library(ca);library(vegan)
```

Also, recall you can add the following segment to your **etc/Rprofile.site** file.

```
# loading packages automatically copy this in etc/Rprofile.site
local({
 old <- getOption("defaultPackages")
 options(defaultPackages = c(old, "spdep", "maptools", "tripack",
 "MASS", "spatstat", "sgeostat", "ca", "vegan","Rcmdr"))
})
```

First, apply function ca in package ca to the data matrix X given in Equation 14.42. To accomplish this, first make a data frame and declare names that will become labels of the biplot. Then apply the function, and use `plot`, with argument labels set to 2 to indicate that we will label all observations and variables, `col = 1` so that all colors be black, and also `arrows is T` only for variables, not for observations.

```
X.df <- data.frame(X); names(X.df) <- c("V1","V2","V3"); row.
  names(X.df)<-1:5
X.ca <- ca(X.df)
plot(X.ca, labels=c(2,2), col=c(1,1), arrows=c(F,T))
```

The result is in Figure 14.22, which is similar to the one shown in Figure 14.12. Now, extract a summary

```
> summary(X.ca)

Principal inertias (eigenvalues):

 dim    value      %   cum%   scree plot
 1     0.240324  97.7  97.7   ************************
 2     0.005620   2.3 100.0
       --------  -----
 Total: 0.245944 100.0
```

(continued)

```
(continued)
```

```
Rows:
      name    mass  qlt   inr      k=1 cor  ctr      k=2 cor  ctr
1 |      1 |   298 1000   183 |   -385 979  183 |    -57  21  171 |
2 |      2 |   274 1000    89 |   -278 963   88 |     54  37  144 |
3 |      3 |   179 1000     8 |     85 634    5 |    -65 366  133 |
4 |      4 |   155 1000    86 |    347 876   77 |    131 124  471 |
5 |      5 |    95 1000   633 |   1277 997  646 |    -69   3   81 |

Columns:
      name    mass  qlt   inr      k=1    cor ctr      k=2 cor  ctr
1 |     V1 |   298 1000   679 |    749 1000 695 |     12   0    7 |
2 |     V2 |   452 1000   248 |   -362  972 247 |     61  28  301 |
3 |     V3 |   250 1000    73 |   -236  782  58 |   -125 218  692 |

>
```

Next, for a more real example, apply function ca in package ca to the frag data frame we employed in the previous section. Because of high disparity of scales, we will use the standardized dataset frag.s and furthermore make all values positive by adding an offset proportional to the minimum.

```
frag.ss <- frag.s
for(j in 1:9){
  for(i in 1:22)  frag.ss[i,j] <- frag.s[i,j] +
  1.1*abs(min(frag.s[,j]))
}
```

Use plot as before but explore its argument map. The first try is with map = symbiplot, and the second with argument map = colgreen. In the previous exercise, the argument was symmetric by default.

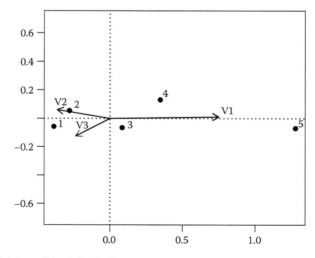

FIGURE 14.22   Biplot from CA of simple data set.

```
frag.ca <- ca(frag.ss)
plot(frag.ca, col=c(1,1), map= "symbiplot", arrows=c(F,T))
plot(frag.ca, col=c(1,1), map= "colgreen", arrows=c(F,T))
```

The results are in Figures 14.23 and 14.24.

The values for argument map are explained in Nenadic and Greenacre (2007). Symbiplot: row and column coordinates are scaled to have variances equal to the singular values; colgreen: columns in

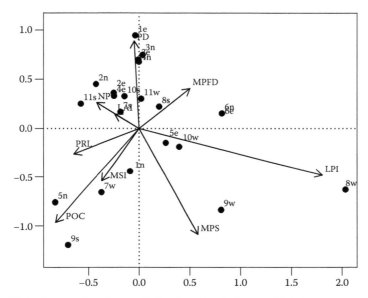

FIGURE 14.23 Biplot from ca function applied to frag data (using symbiplot).

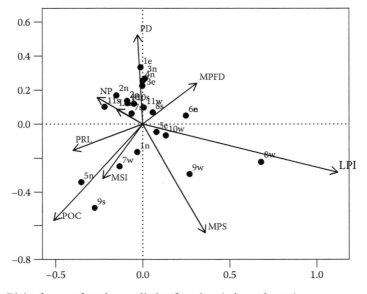

FIGURE 14.24 Biplot from ca function applied to frag data (using colgreen).

principal coordinates and rows in standard coordinates times the square root of the mass. Several other values are available. Symmetric is the default and produces rows and columns in principal coordinates.

We now apply function cca of package vegan to frag.ss data set. Because of its application to ecology, cca assumes that the first argument corresponds to species at each site. A second argument allows for constraints or environmental variables (we will discuss this aspect in the next chapter). The sites would be observations and the species are variables.

Apply function **cca** and examine the results using summary. The results include lambda (eigenvalues or inertia), sites scores (scores of rows or observations), and species scores (scores of columns or variables).

```
> frag.cca <- cca(frag.ss)
> summary(frag.cca)
```

And we can plot

```
>plot(frag.cca)
```

yielding Figure 14.25. To interpret this diagram look for areas where specific sites and variables are close together, indicating relationships. Also see how sites are sorted along gradients, one given by CA1 (the horizontal axes) and another one CA2 (the vertical axes). We can also access scores to employ the biplot function (Figure 14.26). Here we scale the vertical axes differently from the horizontal and employ smaller font in order to improve readability.

```
biplot(scores(frag.cca)$sites, scores(frag.cca)
    $species,ylim=c(-2,1.8),cex=0.7,col=1)
```

As we will see in the next chapter, a commonly used convention will be not to mark variables locations with arrows and instead reserve these for the constraint variables.

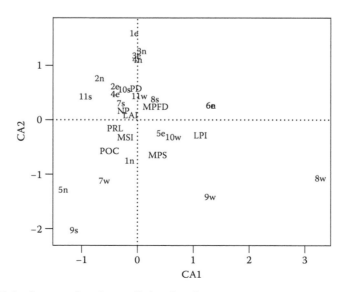

FIGURE 14.25   Biplot from cca function applied to frag data.

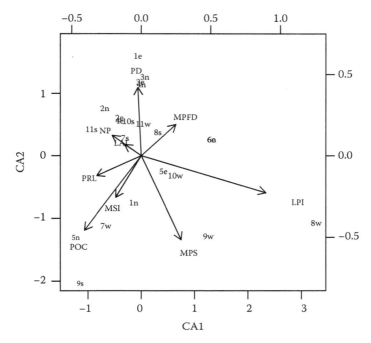

FIGURE 14.26 Biplot from cca function using arrows.

14.10.9 COMPUTER EXERCISES

Exercise 14.5

Use the eigen function to find the eigenvalues and eigenvectors of the following covariance matrix **C**

$$C = \begin{bmatrix} 2 & 1 & 4 \\ 1 & 2 & 3 \\ 4 & 3 & 2 \end{bmatrix}$$

Show that the trace is equal to the sum of eigenvalues. Draw a sketch of the eigenvectors on a plane given by the first and second coordinates. Then draw a sketch of eigenvectors on a plane given by the second and third coordinates. Then draw a sketch of eigenvectors on a plane given by the first and third coordinates.

Exercise 14.6

Use data in file **lab14/watsheds.txt**. It has data for 28 watersheds in Canada. Data from Griffith and Armhein (1991, p. 455). Variables are ID watershed label, AREA (km^2), Flow or discharge (m^3/s), flow/area (flow in 2 months/area), Latitude, longitude, snowfall, precipitation, temp (°C). Perform PCA on **watsheds.txt.** Use correlation matrix if necessary. Select required components to account for more than 90% of the variance. Provide biplots for pairwise combinations of those selected components. Interpret and discuss.

Exercise 14.7

Perform factor analysis on the dataset **watsheds.txt** of the previous exercise. Start with two factors and discuss uniquenesses, communalities and proportion of variance explained. Increase the number of factors to three and repeat. Increase the number of factors to four and repeat. Interpret and discuss. How many factors would you select?

Exercise 14.8

Perform CA on dataset `varspec` of package vegan. Select required components. Provide biplots for pairwise combinations of those selected components. Interpret and discuss.

Exercise 14.9

Perform PCA, FA, and CA on the dataset in file **lab14/blocks.txt**. Dataset is block geometry and dimensions of 25 blocks labeled a–y (Davis, 2002 Figure 6-16, page 507–508). The variable definition is

```
x1= long axis
x2= intermediate axis
x3= short axis
x4= longest diagonal
x5= ratio of radii = (of smallest circumscribed circle)
      over (largest inscribed circle)
x6= ratio of axis = (long + intermediate) over (short)
x7= ratio of (surface area) over (volume)
```

SUPPLEMENTARY READING

Chapter 2 pp. 146–156, Chapter 6 pp. 500–548 (Davis, 2002); Chapter 10 (Rogerson, 2001); Chapter 5 pp. 91–111 (Carr, 1995); Chapters 13 and 14 (MathSoft, 1999); (Nenadic and Greenacre, 2007). More insights into the concept of biplots can be gained from (Gabriel, 1971) and (Gower and Hand, 1996).

15 Multivariate Analysis II
Identifying and Developing Relationships among Observations and Variables

15.1 INTRODUCTION

In this chapter, we continue the study of multivariate methods by covering **Multigroup Discriminant Analysis** (MDA), **Canonical Correlation** (CANCOR), **Constrained Correspondence Analysis** (CCA), **Cluster Analysis**, and **Multidimensional Scaling** (MDS). Of these, CANCOR, MDA, and CCA help find relations between linear combinations of multiple independent and multiple dependent variables. Using MDA we can employ these relationships to separate groups of observations. MDS and cluster analysis deal with the issue of identifying relationships among groups of observations. For these purpose these two methods employ measures of "distance" or dissimilarity between observations.

15.2 MULTIGROUP DISCRIMINANT ANALYSIS (MDA)

To perform linear discriminant analysis for more than two groups, it is necessary to obtain several **discriminant functions**. The number of functions is equal to the number of groups minus one. For example, for three groups we could have one function to differentiate between groups 1 and 2 and 3 combined, another to differentiate between groups 2 and 3. The first function maximizes separability and the second one is orthogonal and will maximize separation once we control for variation along the first axis defined by the first function.

Following Davis (2002), denote by $\mathbf{B}$ and $\mathbf{W}$ the matrices of between-groups and within-groups sums of products. The total sum of products S is $\mathbf{B} + \mathbf{W}$, and we would like the ratio of between to within to be as large as possible. Thus, we form successive optimization problems such that we find the vector $\mathbf{A}_i$ of coefficients of each discriminant function that maximizes $\mathbf{A}_i^T \mathbf{B} \mathbf{A}_i$ subject to $\mathbf{A}_i^T \mathbf{W} \mathbf{A}_i = 1$. The solution $\mathbf{A}_1$ is the eigenvector of the largest eigenvalue of $\mathbf{W}^{-1}\mathbf{B}$, then $\mathbf{A}_2$ is the eigenvector of the second largest eigenvalue, and so forth.

The ratio of the eigenvalues indicates the relative discriminating power of the discriminant functions. For example, if the ratio of two eigenvalues is 1.5, then the first discriminant function accounts for 50% more between-group variance than does the second discriminant function. The eigenvalues also represent proportion of total trace of $\mathbf{W}^{-1}\mathbf{B}$ accounted by each function. The eigenvalues (square of the singular values) provide the values of F for each discriminant axis.

Finally, form matrix $\mathbf{A}$ with columns $\mathbf{A}_1$, $\mathbf{A}_2$,..., $\mathbf{A}_k$ where k is the number of discriminant functions. The scores are then projected from the original data matrix $\mathbf{X}$ by

$$\mathbf{Z} = \mathbf{A}^T \mathbf{X} \tag{15.1}$$

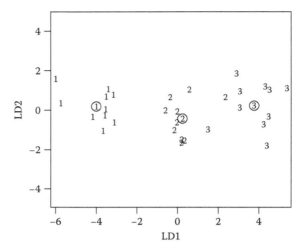

FIGURE 15.1   Two-dimensional (2D) discriminant space.

Group centroids are the mean discriminant scores for each of the categories for each of the discriminant functions. One can create **canonical plots** in which the two axes are two of the discriminant functions, and markers in the plot locate the centroids of each category under analysis. These plots depict a **discriminant function space**. For example, Figure 15.1 shows three groups G1, G2, G3 clearly separated along the first discriminant axes. This example corresponds to three responses (variables) X1, X2, X3. The group means for X1 are 60, 62, 63; for X2 are 90, 92, 94; and for X3 19, 21, 24. ANOVA of each response yields F values of 33, 39, 62. Because these values are sufficiently high, we would expect that these groups can be separated. We calculate two discriminant functions that have eigenvalues 151.3, 1.33. Thus the proportions of total trace are 0.99, 0.01. So, $k = 2$ is sufficient. The matrix **A** is formed by two eigenvectors

$$\mathbf{A} = \begin{bmatrix} 0.33 & -0.84 \\ 0.62 & 0.58 \\ 0.74 & 0.13 \end{bmatrix} \tag{15.2}$$

The scores for the ith observation or ith row of **X** would be

$$\begin{bmatrix} z_{i1} \\ z_{i2} \end{bmatrix} = \begin{bmatrix} 0.33 & 0.62 & 0.74 \\ -0.84 & 0.58 & 0.13 \end{bmatrix} \begin{bmatrix} x_{i1} \\ x_{i2} \\ x_{i3} \end{bmatrix} \tag{15.3}$$

Actually, one discriminant axis would be sufficient to separate the groups as suggested by the proportion of trace explained and as observed in Figure 15.1.

15.3   CANONICAL CORRELATION

This method has similar basis to PCA and to factor analysis (i.e., eigen-decomposition), but the goal is to make predictors as in multiple regression. However, it differs from multiple regression analysis in that the dependent variable is not a single variable but a set of variables (Davis, 2002 Chapter 6 pp. 577–584).

In its predictive sense, canonical correlation is an extension of multiple regression analysis. We now have a set of dependent variables (not just one Y) and a set of independent variables X. The objective is to find linear combinations of X_i that yields the highest correlation with linear combinations of the Y_j.

Each linear combination of the X_i is a new single variable, a **canonical X variable**, whereas each linear combination of the Y_j is a new single variable, a **canonical Y variable**. We then determine the **canonical correlation** between canonical variables X and Y.

Applications include, for example, explaining biotic variables (Y) from environmental conditions (X). In water quality, for example, explaining biotic community indicators (e.g., diversity) from water quality data.

We define the following dimensions: n number of objects or observations, p number of Y variables; q number of X variables. Thus, X matrix is $n \times q$ and the Y matrix is $n \times p$, and $m = (p + q)$ is the total number of variables. Therefore, the dimension of the entire data matrix is $n \times (p + q)$, and we assume that $p < q$ (more Xs variables than Y variables).

The covariance matrix S of the entire data set has dimension as $m \times m$ or $(p + q) \times (p + q)$ and is partitioned into four sub-matrices in the following manner.

$$S = \begin{bmatrix} Syy & Sxy \\ Syx & Sxx \end{bmatrix} \tag{15.4}$$

where

$X^T X = Sxx$ is the covariance matrix of X and its dimension is $q \times q$

$Y^T Y = Syy$ is the covariance matrix of Y and its dimension is $p \times p$

$X^T Y = Sxy$ is covariance matrix of X and Y and its dimension is $p \times q$; its transpose is $Y^T X = Syx$ of dimension $q \times p$. Recall that $Syx = Sxy^T$

We want to find linear transformations of data matrices Y and X such that we maximize covariance. Start by defining the linear transformations as follows. First, YA is linear transformation for Y; here A is $p \times 1$, the vector of weights to convert the Ys into the canonical Y. Second XB is linear transformation for X; here B is $q \times 1$, the vector of weights to convert the X into the canonical X. Do not confuse with the B matrix of discriminant analysis.

The variance of the canonical Y is

$$(YA)^T (YA) = A^T Y^T YA = A^T SyyA \tag{15.5}$$

This is a scalar because $(1 \times p) \times (p \times p) \times (p \times 1)$ yields 1×1. And the variance of the canonical X is

$$(XB)^T (XB) = B^T X^T XB = B^T SxxB \tag{15.6}$$

This is also a scalar because $(1 \times q) \times (q \times q) \times (q \times 1)$ yields 1×1.

The covariance of the canonical X and Y is

$$(YA)^T (XB) = A^T Y^T XB = A^T SyxB \tag{15.7}$$

This is a scalar because $(1 \times q) \times (q \times p) \times (p \times 1)$ yields 1×1.

The objective is to select A and B to maximize covariance subject to the constraint that variances are equal to one. A and B are found by eigen-decomposition of the $q \times q$ matrix L of pooled variances

$$L = Sxx^{-1} Sxy^T Syy^{-1} Sxy$$

This matrix has q eigenvalues and eigenvectors because it is $q \times q$. For each eigenvalue, we build **B** from the corresponding eigenvector and then **A** is constructed as

$$\mathbf{A} = \frac{\mathbf{Syy^{-1}Sxy\ B}}{\sqrt{\lambda}} \tag{15.8}$$

where

    λ is the corresponding eigenvalue. There are q different vectors

    A, **B** are found by repeating this process, the one corresponding to the maximum eigenvalue maximizes the correlation coefficient, which is equal to $\sqrt{\lambda}$

For an interpretation think of **A** and **B** as weights used to transform original variables into canonical variables, and consider canonical variables as factors in factor analysis. It is usually difficult to attach physical meaning to the weights.

Consider the example in Figure 15.2; this is a set of three canonical variables. There is very high canonical correlation between the first canonical variables.

Tests of significance: for an overall test use a null or H0 of all canonical correlations are zero. In this case, the statistic is chi-squared with $p \times q$ degrees of freedom

$$\chi^2 = \left(\frac{p+q+1}{2} - n + 1 \right) \ln \left[(1-\lambda_1)(1-\lambda_2) \ldots (1-\lambda_q) \right] \tag{15.9}$$

where λ_i is ith eigenvalue of the **L** matrix. To test for largest canonical correlation, define null H0 to be that the largest canonical correlation is zero. The statistic is chi-squared

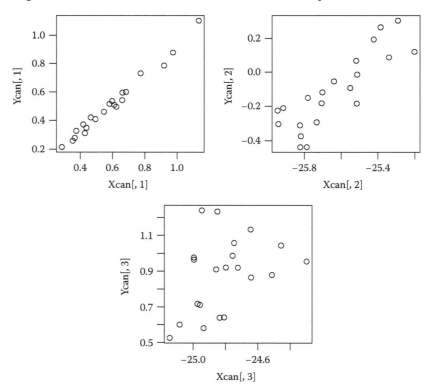

FIGURE 15.2 Canonical correlation plots.

$$\chi^2 = \left(\frac{p+q+1}{2} - n + 1 \right) \ln\left[(1-\lambda_1) \right] \qquad (15.10)$$

where λ_1 is the largest eigenvalue of the **L** matrix. The number of degrees of freedom is given by rounding the following calculation to the nearest integer

$$p + q + 1 + 0.5 \times \left[(p-1)(q-1) \right]^{2/3} \qquad (15.11)$$

15.4 CONSTRAINED (OR CANONICAL) CORRESPONDENCE ANALYSIS (CCA)

This is a very popular multivariate tool in ecology and used often to analyze how various species respond to environmental conditions. Couched this way, this problem is one of constrained ordination.

One has a matrix **Y** of p species abundance at multiple (say n) sites and another matrix **X** of values of q environmental variables at the n sites. Typically, the variables of **Y** are counts or abundance of species and we can define response functions for each species k at each site i

$$\lambda_{ik} = f_k(\mathbf{x}i) \qquad (15.12)$$

where the λ_{ik} is a Poisson rate parameter. The larger λ_{ik} the more abundant the species k is at site i. A direction or **environmental gradient** is a vector $\boldsymbol{\alpha}$ that assigns a weight to each environmental variable. In other words, a vector defining a linear combination of the environmental variables

$$z_i = \boldsymbol{\alpha}^T \mathbf{x}_i \qquad (15.13)$$

Thus, the score $\mathbf{z}_i$ for site i is a combination of the q environmental variables. The Gaussian response model assumes

$$\log\left(f_k(z_i) \right) = a_k - \frac{(z_i - u_k)^2}{2t_k^2} \qquad (15.14)$$

where
   a_k is the optimal log response
   u_k is the optimal value of the environmental score (the response is largest if the species receives score $z_i = u_k$)
   t_k is the tolerance or spread of the response

These parameters are estimated by maximizing the log-likelihood

$$\sum_{k=1}^{p} \sum_{i=1}^{n} -f_k(\mathbf{x}_i) + y_{ik} \log(f_k(\mathbf{x}_i)) \qquad (15.15)$$

Assuming that all tolerances are the same $t_k = t$ for all species. This is accomplished using SVD as in CA.

This method could be applied to situations where instead of species abundances we have other variables that respond to independent forcing functions, as long as the assumptions enumerated

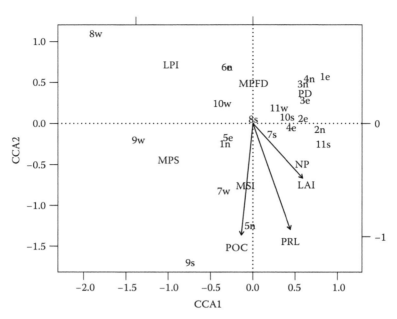

FIGURE 15.3 The CCA plot is similar to correspondence analysis, but it adds the arrows for the environmental variables.

earlier are reasonable. As an example, Figure 15.3 shows values of canopy fragmentation metrics, instead of species abundances, and canopy variables as environmental constraints.

15.5 CLUSTER ANALYSIS

Cluster analysis is a method to classify objects or observations into groups; this analysis helps to identify the possible groups and reveal relations among groups (Davis, 2002 Chapter 6 pages 487–500). It has many applications including remote sensing image classification and numerical taxonomy.

There are several types of cluster analysis; for example, **hierarchical agglomerative clustering**, which successively joins or merges the most similar observations to form clusters, and **model-based clustering**, which uses a statistical model for the clusters, such as a Gaussian density. For this purpose, a dissimilarity matrix is calculated for each iteration.

Assume we have n objects, and m attributes or variables. As always, the data matrix is $n \times m$ matrix. First, the distance (dissimilarity) matrix $\mathbf{D}$ is $m \times m$. Entries can be computed in various ways, for example, calculating the Euclidian distance in m-space between objects i and j

$$d_{ij} = \sqrt{\frac{\sum_{k=1}^{m} (x_{ik} - x_{jk})^2}{m}} \qquad (15.16)$$

The sum is over $k = 1, \ldots, m$ attributes. Here x_{ik} is kth variable measured on object i, and x_{jk} is kth variable measured on object j.

A small distance between two objects implies that the objects are similar. Matrix $\mathbf{D}$ is symmetrical. It is usual to standardize the $n \times m$ data matrix before computing the distance matrix.

The hierarchical agglomerative cluster analysis process consists of the following steps:

- Initially set $i = n$ clusters, this is to say only one object in each cluster, compute $\mathbf{D}$
- Join two clusters to form a new cluster; at this point, we have $i - 1$ clusters. To decide what clusters to join there are several options; for example, (a) the sum of squares method:

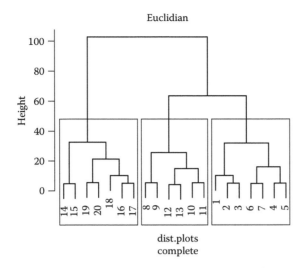

FIGURE 15.4 Dendrogram generated by cluster analysis.

merges those two with the smallest increase in within-cluster sum of squares, (b) the link method: merges two with the smallest distance. There are several options to decide this distance between groups; for example, (a) nearest neighbor, (b) farthest neighbor, (c) centroid, and (d) average.

- Recompute **D**, which is now $(i-1) \times (i-1)$; the new cluster is represented by a "centroid."
- Repeat steps 2 and 3 until $i = 2$, that is to say the set reduces to only 2 clusters.
- Use the distance determined for joining clusters as a vertical axis (height) to draw a tree or dendrogram.

An example is in Figure 15.4 where three clusters are identified at a distance (height) of 50. Increasing the cutoff to a height to 60 we have only two clusters.

Model-based clustering uses a statistical model for the clusters; for example, Gaussian density for each cluster with mean μ and covariance C as parameters. The eigenvectors of **C** determine orientation of the cluster, and the eigenvalues of **C** determine the shape. Recall the geometric interpretation of PCA studied in the previous chapter. If we make all clusters have same ratios of eigenvalues to the dominant eigenvalue, then all clusters have same shape (S*), which is parameterized by the ratios; for example, when ratio = 1 we have hyper spherical clusters, but small ratio values lead to highly elongated hyper ellipsoids.

One difficult part of cluster analysis is to decide how many clusters to include at the end. This typically becomes a judgment call made by looking at the dendrogram. However, some help is available as a Bayes factor

$$B_k = \frac{P(mc_k)}{P(mc_1)}$$

where
mc_k is the event that the model has k clusters
mc_1 that model has 1 cluster

Then $B_1 = 1$ and larger B_k indicates more evidence for the existence of k clusters. The Bayes factor can be converted into the approximate weight of evidence (AWE) for k clusters AWE_k as

$$AWE_k = 2 \times \log (B_k) \tag{15.17}$$

Now $AWE_1 = 0$. The rates of change of AWE and max (AWE) are used to decide on the number of clusters.

15.6 MULTIDIMENSIONAL SCALING (MDS)

This is a collection of techniques to explore similarities and dissimilarities in data consisting of m attributes or variables of n objects (Borg and Groenen, 2005). It is a general method because the distance to be analyzed is not just correlation but any other similarity measure.

First, a matrix of similarities among objects is calculated, then objects are placed in a space with low pre-specified dimension, say $N = 2$ or 3. Then the results can be visualized in this lower dimensional space.

As always, data are an $n \times m$ matrix. First, the distance (dissimilarity) matrix $\mathbf{D}$ is $n \times n$ with entries d_{ij}. The idea is to find n vectors in the N-dimensional space such that a norm or metric (e.g., Euclidian distance) measuring differences between all pairs of these vectors is approximately the same as entries of the dissimilarity matrix measuring the distance between objects in m-space. In other words,

$$\left\| x_i - x_j \right\| \sim d_{ij}, \quad \text{for all} \quad i, j \tag{15.18}$$

Said another way we try to find a representation of the objects in N-space such that the distance between them is preserved. These new vectors are not unique and are found by optimization. Say find the set of vectors x_i, $i = 1, \ldots, n$ such that a cost function is minimized. For example,

$$\min_{x_1, \ldots, x_n} \sum_{i < j} \left(\left\| x_i - x_j \right\| - d_{ij} \right)^2 \tag{15.19}$$

This cost function is called the **stress** and is given more generally by

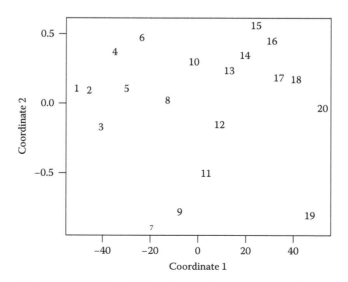

FIGURE 15.5   Example of Metric MDS applied to the same data set of the previous figure.

$$\sum_{i<j}\left(\|x_i - x_j\| - f(d_{ij})\right)^2 \qquad\qquad (15.20)$$

where $f(d_{ij})$ is a function of the original distances and can be non-metric. Other loss functions include the **strain**. In some cases, the optimization problem can be solved by eigen-decomposition.

The example in Figure 15.5 illustrates MDS applied to the same data set of the example illustrating cluster analysis. We can see that many neighboring plots in the coordinate space correspond to those identified by cluster analysis.

There are several alternative MDS methods: classical MDS, metric MDS, and non-metric MDS. The classical MDS uses as input a dissimilarity matrix and produces a coordinate system that minimizes the strain. Metric MDS uses a variety of distance matrices and minimizes the stress loss function. Non-metric MDS uses non-parametric relation of dissimilarities and employs regression.

15.7   EXERCISES

Exercise 15.1
Consider the example leading to Equation 15.2. Calculate the singular values and the ratio of eigenvalues. Calculate the scores of the centroids from the group means using Equation 15.3 applied to the group means instead of the individual x_{ij}. Draw a sketch showing location of the centroids in discriminant space.

Exercise 15.2
Consider the example of the previous exercise. Suppose a new object measures X1, X2, X3 = 60, 90, 20. Apply the discriminant functions to decide to what group it most likely belongs to.

Exercise 15.3
Consider 4 environmental variables X (abiotic), 2 response variables Y (biotic), and 10 observations of all variables. What are the dimensions of matrices **A** and **B** of CANCOR? Assume the largest eigenvalue is 0.7. Calculate chi-square and degrees of freedom to test the H0 that the largest eigenvalue is zero.

Exercise 15.4
Consider Figure 15.4, how many clusters are generated at a height = 30. Compare how members of these clusters relate to each other on the MDS diagram of Figure 15.5.

15.8   COMPUTER SESSION: MULTIVARIATE ANALYSIS II

15.8.1   MULTIGROUP LINEAR DISCRIMINANT ANALYSIS

We will use function lda in package MASS. First, load the package if you have not yet. The function lda performs linear discriminant analysis. We will work with data file **lab15/testgrp3.txt**. It has $m = 3$ variables, and three groups $n_1 = n_2 = n_3 = 10$ observations.

We generated these hypothetical data at random from a normal distribution. The first 10 rows are for group 1, the next 10 rows are for group 2, and the last 10 are for group 3.

First look at file **testgrp3.txt** using a text editor, then scan the file and convert to matrix, also create factor for groups and assemble a data frame

```
x <- matrix(scan(file="lab15/testgrp3.txt"),ncol=3,byrow=T)
grp <- factor(rep(c(1:3), c(10,10,10)))
xg <- data.frame(grp,x)
```

Then run univariate ANOVA for each group and a MANOVA

```
> summary(aov(x ~ grp))
 Response 1 :
            Df Sum Sq Mean Sq F value    Pr(>F)
grp          2 71.610  35.805  33.298 5.146e-08 ***
Residuals   27 29.033   1.075
---
Signif. codes:  0 `***' 0.001 `**' 0.01 `*' 0.05 `.' 0.1 ` ' 1

 Response 2 :
            Df Sum Sq Mean Sq F value    Pr(>F)
grp          2 91.573  45.786  39.085 1.066e-08 ***
Residuals   27 31.629   1.171
---
Signif. codes:  0 `***' 0.001 `**' 0.01 `*' 0.05 `.' 0.1 ` ' 1

 Response 3 :
            Df  Sum Sq Mean Sq F value    Pr(>F)
grp          2 139.035  69.518  61.699 8.525e-11 ***
Residuals   27  30.422   1.127
---
Signif. codes:  0 `***' 0.001 `**' 0.01 `*' 0.05 `.' 0.1 ` ' 1

>
> summary(manova(x ~ grp), test="Wilks")
          Df   Wilks approx F num Df den Df    Pr(>F)
grp        2  0.0746  22.1784      6     50 1.547e-12 ***
Residuals 27
---
Signif. codes:  0 `***' 0.001 `**' 0.01 `*' 0.05 `.' 0.1 ` ' 1
```

We can conclude that there are significant differences among the three groups, so we should be able to find linear discriminant functions. These would hold for all three variables (and their combination). For this purpose now we will run lda with arguments grp ~x1 + x2 + x3. Because the scope are all the variables, it can be abbreviated as a dot for the scope grp ~.

```
xgda <- lda(grp ~., data=xg)
```

The results obtained are

```
> xgda
Call:
lda.formula(grp ~ ., data = xg)

Prior probabilities of groups:
        1         2         3
0.3333333 0.3333333 0.3333333
```

```
> xgda
Call:
lda.formula(grp ~ ., data = xg)

Prior probabilities of groups:
        1         2         3
0.3333333 0.3333333 0.3333333

Group means:
      X1     X2     X3
1 60.036 90.161 19.041
2 62.578 92.148 21.970
3 63.735 94.437 24.303

Coefficients of linear discriminants:
          LD1         LD2
X1 0.3322028 -0.8372882
X2 0.6195081  0.5792344
X3 0.7393298  0.1252196

Proportion of trace:
   LD1    LD2
0.9913 0.0087
>
```

The coefficients would form matrix **A**. We can see that the first variable explains 99% of variance. To obtain the individual scores we can use **predict**(). For brevity we only show a few individual scores.

```
> predict(xgda)$x

            LD1         LD2
1 -3.11922510 -0.63601796
2 -3.52320100  0.67056904
3 -3.48353496  0.87497021
4 -3.54162117  0.02934717
5 -6.00262026  1.56685838
6 -4.18920109 -0.35221570
...etc
>
```

These are centered in such a way that the weighted average of the centroids is at zero. The weights for the groups are given by the prior probabilities. In this example, they are equally weighted (1/3 = 0.3333 each one).

Apply function `plot` to the `lda` object

```
> plot(lda(grp ~., data=xg))
```

It uses the centered individual scores to produce the plot of Figure 15.6, which displays the difference between the groups in 2D discriminant score space. It indicates little overlap in observations in the score space given by these axes.

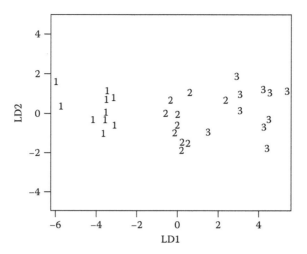

FIGURE 15.6 Discriminant plot.

To this plot, one could add the centroids, calculated by the scores of the group means. Using the $prior component, Z_0 is calculated as a weighted average. The centroid scores are then centered on Z_0.

```
# centroids
  xgc <- matrix(xgda$means%*%coef(xgda), ncol=2, byrow=F)
# midpoint weighted average Z0
  xgcmid <- xgda$prior%*%xgc
# centered centroids
  xgcc <- xgc
  for(i in 1:3) xgcc[i,] <- xgc[i,] - xgcmid
> xgcc
              [,1]          [,2]
[1,]    -4.0030356    0.1906989
[2,]     0.2378833   -0.4199807
[3,]     3.7651523    0.2292818
>
```

Alternatively, we can calculate the centered centroids as means of the centered scores.

```
>      xgcc <- matrix(nrow=3,ncol=2,byrow=T)
>      for(j in 1:2) xgcc[1,j] <- mean(predict(xgda)$x[1:10,j])
>      for(j in 1:2) xgcc[2,j] <- mean(predict(xgda)$x[11:20,j])
>      for(j in 1:2) xgcc[3,j] <- mean(predict(xgda)$x[21:30,j])
```

Then after the command `plot`, use functions `points` to draw circles and labels at centroids as in Figure 15.1.

```
plot(lda(grp ~., data=xg))
points(xgcc,pch=c("1","2","3"),cex=1.2)
points(xgcc,pch=1,cex=2)
```

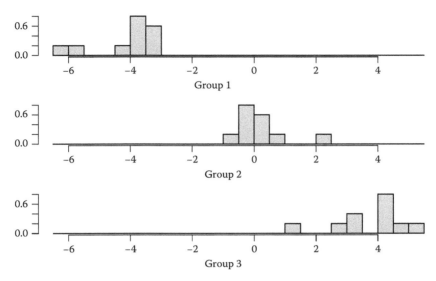

FIGURE 15.7 Discriminant histograms.

We can see substantial differences among centroids in this plot. If we use argument `dimen = 1` of function `plot`, that is only one discriminant axis, then the plot will produce histograms along that one axis (Figure 15.7).

```
plot(lda(grp ~., data=xg),dimen=1)
```

Significance can be evaluated using the singular values of each axis. The singular values are in component `$svd`. The square of each singular value is the corresponding eigenvalue. It also corresponds to the *F* value

```
# significance
> xgda$svd
[1] 12.299848  1.151781
```

The square of each singular value is the corresponding eigenvalue. It also corresponds to the *F* value

```
> xgda$svd^2
[1] 151.2863  1.3266
```

We can see that for the first axis the value of *F* is very large and will have very low p-value. This indicates high discriminating power for the first axis. The second axis has much lower *F*, not too different from 1.00, and will have a higher p-value, indicating much lower discriminating power. The ratio of eigenvalues is 114, indicating that the first discriminant function is much more significant than the second one.

15.8.2 CANONICAL CORRELATION

We will work with the **frag** data set we used in the last chapter. If you have not read the file yet, please do so. We will try canonical correlation of a set of Y (first three variables, POC, PRL, LAI) and X (last six variables, or fragmentation metrics NP, PD, MPS, LPI, MSI, MPFD).

First, check the correlation matrix

```
> round(cor(frag),2)
       POC   PRL   LAI    NP    PD   MPS   LPI   MSI  MPFD
POC   1.00  0.91  0.32  0.59 -0.45  0.40 -0.26  0.74 -0.74
PRL   0.91  1.00  0.48  0.83 -0.13  0.09 -0.45  0.66 -0.71
LAI   0.32  0.48  1.00  0.48  0.18 -0.10 -0.39  0.33 -0.35
NP    0.59  0.83  0.48  1.00  0.39 -0.43 -0.60  0.23 -0.35
PD   -0.45 -0.13  0.18  0.39  1.00 -0.96 -0.45 -0.54  0.39
MPS   0.40  0.09 -0.10 -0.43 -0.96  1.00  0.44  0.53 -0.31
LPI  -0.26 -0.45 -0.39 -0.60 -0.45  0.44  1.00 -0.41  0.36
MSI   0.74  0.66  0.33  0.23 -0.54  0.53 -0.41  1.00 -0.83
MPFD -0.74 -0.71 -0.35 -0.35  0.39 -0.31  0.36 -0.83  1.00
>
```

Next, apply function `cancor` for canonical correlation.

```
X <- as.matrix(frag)
frag.cc <- cancor(X[,4:9],X[,1:3])
> frag.cc
$cor
[1] 0.9908868 0.8531676 0.4626585

$xcoef
                 [,1]          [,2]          [,3]          [,4]
    [,5]
NP      0.0003062606  1.080944e-06 -1.031664e-04 -0.0002582732
    2.274904e-04
PD     -0.0016642643 -1.093119e-01  2.941511e-01 -0.1112145596
    -2.232828e-02
MPS     0.6795669302  2.076802e+00  1.951373e+01 -6.8004561194
    9.748822e+00
LPI     0.0033679803 -1.140155e-02  6.148861e-04 -0.0204329299
    -2.688249e-02
MSI     3.7947516654 -1.089631e+01 -4.070545e-01  0.4285565926
    -2.393664e+01
MPFD -3.2231040952 -6.731858e+00 -1.882487e+01  2.5387766045
    -2.510537e+00
                 [,6]
NP     -9.663245e-05
PD      5.030819e-03
MPS     2.536775e+00
LPI    -7.910227e-03
MSI    -2.141121e+01
MPFD -5.363583e+01
```

```
(continued)

$ycoef
               [,1]             [,2]            [,3]
POC  0.0007522287   0.49317206   0.19636343
PRL  0.0645277373  -0.13808488  -0.09531828
LAI  0.0096847706  -0.02574766   0.50483518

$xcenter
             NP              PD             MPS             LPI             MSI
MPFD
1732.3181818       9.7627273       0.1136364      14.8981818       1.3018182
   1.5518182

$ycenter
      POC          PRL          LAI
1.980909   7.390000   2.337727

>
```

Component `$cor` are the canonical correlations. The results show high values for the first and second canonical correlations. Component `$xcoef` contain canonical vectors **B** for each eigenvalue, whereas component `$ycoef` contains canonical vectors **A** for each eigenvalue. The `$xcenter` and `$ycenter` are the means of X and Y, respectively. These means are subtracted from the variables before the analysis.

We can plot canonical Y as a function of canonical X, one for each canonical correlation.

```
Ycan <- X[,1:3]%*%frag.cc$ycoef
Xcan <- X[,4:9]%*%frag.cc$xcoef
par(mfrow=c(2,2))
plot(Xcan[,1], Ycan[,1])
plot(Xcan[,2], Ycan[,2])
plot(Xcan[,3], Ycan[,3])
```

The first pair shows high correlation (Figure 15.2).

15.8.3 CANONICAL CORRESPONDENCE ANALYSIS

We will apply the `cca` function from package **vegan**, which we already used in the previous chapter to run correspondence analysis. To make it constrained we write

```
>cca(X,Y)
```

Recall that because of its use in ecology, the function refers to X as the matrix of species response and Y the environmental (constraint) matrix. We will use `frag` as in the previous section, but standardized and shifted positive, i.e., `frag.ss` that we generated in the previous chapter. Select for Y the first three variables (POC, PRL, LAI) and for X the last six variables, or fragmentation metrics (NP, PD, MPS, LPI, MSI, MPFD).

Apply `cca` function

```
> X <- frag.ss[,4:9]; Y<-frag.ss[,1:3]
> XY.cca <- cca(X,Y)
> XY.cca
Call: cca(X = X, Y = Y)

              Inertia Proportion Rank
Total          0.3183     1.0000
Constrained    0.1818     0.5713     3
Unconstrained  0.1365     0.4287     5
Inertia is mean squared contingency coefficient

Eigenvalues for constrained axes:
    CCA1      CCA2      CCA3
0.095851 0.084667 0.001307

Eigenvalues for unconstrained axes:
      CA1       CA2       CA3       CA4       CA5
0.0716000 0.0541160 0.0076854 0.0020665 0.0009956
```

We can plot using scaling = 3 to make it symmetric

```
> plot(XY.cca,scaling=3)
>
```

to obtain Figure 15.3. In this type of biplot we use numbers as markers for the sites (observations), symbols as the markers for the species, and arrows ending at the values of the environmental variables. To interpret this diagram look for areas where specific sites, species, and environmental variables are close together, indicating relationships. Also see how sites, species, and constraints are sorted along gradients, one given by CA1 (the horizontal axes) and another one CA2 (the vertical axes).

15.8.4 CLUSTER ANALYSIS

You can perform cluster analysis using functions `dist` and `hclust`. Function `dist` is used to compute a distance matrix, then the result is used in `hclust`. As an example we will use data set in **lab15/forest-20-model-plots.csv** file. This file contains 20 modeled forest plots at an early successional stage (25 years) in a montane coniferous forest with three dominant tree species. Each line has six integer numbers: tree count in each one of two diameter classes for species 1, tree count in each one of 2 diameter classes for species 2, tree count in each one of 2 diameter classes for species 3.

Read the file and calculate distance as Euclidian

```
X <- read.table(file="lab15/forest-20-model-plots.csv",
  header=T,sep=",")
dist.plots <- dist(X, method="euclidean")
> dist.plots
           1          2          3          4          5          6
  7
2   5.291503
3  10.148892  5.000000
4  16.000000 10.770330  5.916080
5  20.904545 15.716234 10.770330  5.000000
6  27.331301 22.113344 17.204651 11.357817  6.480741
7  31.937439 26.720778 21.794495 16.000000 11.090537  4.795832
8  38.065733 32.848135 27.928480 22.113344 17.204651 10.770330
  6.244998
9  43.278170 38.065733 33.136083 27.33130 22.405357 16.000000
  11.357817
10 48.815981 43.600459 38.678159 32.848135 27.928480 21.494185
  16.941074>
```

The segment shown is only a small part of the distance matrix. Note that the object dist.plots is the lower half of the symmetrical 20×20 matrix and excludes the diagonal.

Now we will perform hierarchical clustering, plot a dendrogram, and identify clusters. To do this run the hclust cluster function using dist.plots with default method, plclust, and rect.hclust

```
h.euclid <- hclust(dist.plots)
plclust(h.euclid,main="Euclidian",sub="Complete")
rect.hclust(h.euclid, k=3)
```

to obtain the dendrogram of Figure 15.4. The plot or observation numbers are at the bottom of the dendrogram. The vertical axis is distance. Note that as you go up the number of clusters decrease. At about a distance of 25 we have six clusters; at about 50 you can distinguish three different clusters. One cluster includes plots 1–7, another plots 8–13, and a third cluster includes plots 14–20.

A useful tool to visualize the clusters at a given level is the rect.hclust function. It was applied here to draw rectangles around clusters, given the number of clusters as argument k; here k = 3. Alternatively, we can use an argument to select the distance at which to cut the tree. For example, the following will draw six clusters.

```
> rect.hclust(h.euclid, h=25)
```

One can go back to the nature of the data set to reach conclusions. For example, in this example, the plots were arranged east to west and the resulting clusters may be a response to this arrangement. However, many times, interpreting a dendrogram is difficult.

As another example, use cluster analysis on frags data set

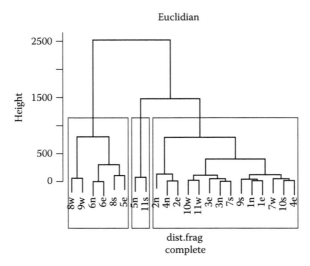

FIGURE 15.8   Dendrogram for frag data set.

```
dist.frag <- dist(frag, method="euclidean")
h.euclid <- hclust(dist.frag)
plclust(h.euclid,main="Euclidian",sub="Complete")
rect.hclust(h.euclid, k=3)
```

As we can see in Figure 15.8, at about distance 600 we have three clusters: one with sites 8w, 9w, 6n, 6e, 8s, 5e, another with sites 5n, 11s, and another with the remaining sites. This result is consistent with the other methods applied to this data set.

15.8.5   MULTIDIMENSIONAL SCALING (MDS)

For classical MDS we can apply function cmdscale in the basic stats functions to the forest plots data set. As input, we use the same Euclidian distance matrix dist.plots calculated for cluster analysis.

```
# k is the number of dim
fit <- cmdscale(dist.plots, eig=T, k=2)
fit
$points
              [,1]         [,2]
 [1,]  -51.158823   0.10182905
 [2,]  -45.950698   0.08897626
 [3,]  -41.022197  -0.17568878
 [4,]  -35.201985   0.36589575
 [5,]  -30.273484   0.10123071
...
...
[18,]   40.453651   0.16721070
[19,]   46.266812  -0.80804779
[20,]   51.532955  -0.03969504
```

```
(continued)

$eig
[1]   1.949072e+04   3.574401e+00   2.256171e+00   1.521291e+00
  1.107309e+00
[6]   6.241452e-01   7.522403e-13   2.427453e-13   1.447484e-13
  1.074462e-13
[11]   8.706662e-14   6.938887e-14   6.175188e-14   1.279484e-14
  -4.403246e-15
[16]   -3.772396e-14  -1.117383e-13  -2.150750e-13  -3.999238e-13
  -1.681675e-12

$x
NULL

$ac
[1]  0

$GOF
[1]  0.9997175 0.9997175
```

We can now plot using component $points

```
# plot solution
x <- fit$points[,1]; y <- fit$points[,2]
plot(x, y, xlab="Coordinate 1", ylab="Coordinate 2",
  main="Metric MDS Plots", type="n")
text(x, y, labels = row.names(X), cex=.7)
```

which produces Figure 15.5. Similarly, we can apply non-metric MDS using function isoMDS of package MASS.

```
fit <- isoMDS(dist.plots, k=2) # k is the number of dim
fit
# plot solution
x <- fit$points[,1]
y <- fit$points[,2]
plot(x, y, xlab="Coordinate 1", ylab="Coordinate 2",
  main="NonMetric MDS Plots", type="n")
text(x, y, labels = row.names(X), cex=.7)
```

As another example, apply MDS to frag data set

```
fit <- cmdscale(dist.frag,eig=T, k=2) # k is the number of dim
fit # view results
# plot solution
x <- fit$points[,1];y <- fit$points[,2]
plot(x, y, xlab="Coordinate 1", ylab="Coordinate 2",
  main="Metric MDS Frag", type="n")
text(x, y, labels = row.names(frag), cex=.7)
```

which produces Figure 15.9.

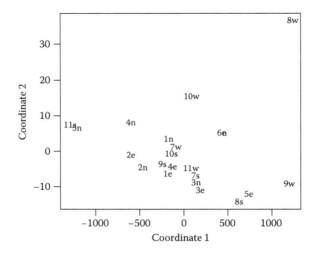

FIGURE 15.9 MDS for `frag` data set.

15.8.6 COMPUTER EXERCISES

Exercise 15.5

Data set of macro-invertebrate diversity in streams (McCuen, 1985). Data reproduced in Carr (1995). The data set is in file **lab15/streams-macroinv.txt**. How well do stream physical variables explain biotic variables? Perform canonical correlation analysis and CCA. Interpret and discuss.

Exercise 15.6

Apply `cca` of vegan to demo data sets `varespec` and `varechem` of vegan to study the relationship of species abundance to environmental variates `Al`, `P`, `K`, and `baresoil`. Study the capabilities of `cca` to construct a formula based on of `Al`, `P`, `K`, and `baresoil`.

Exercise 15.7

Data set in file **lab15/census-tract3.txt**. Data for 15 tracts and 3 variables: median age, % no family, % average income (in thousand dollars). Perform MDS and cluster analysis to explore potential clusters or groups. Select these clusters. Confirm that there may be difference among these groups using MANOVA and then find linear discriminant functions to separate these groups. Interpret and discuss.

Exercise 15.8

Perform canonical correlation, cluster analysis, and metric MDS on the data set in file **lab14/blocks.txt**. Data set is block geometry and dimensions of 25 blocks labeled a-y (Davis, 2002 Figure 6-16, page 507–508). See computer exercises of previous chapter for variable definition.

SUPPLEMENTARY READING

Chapter 6 (Davis, 2002); Chapter 10 (Rogerson, 2001); Chapter 5 (Carr, 1995); Chapter 15 (MathSoft 1999); (Borg and Groenen, 2005).

Bibliography

Acevedo, M.F., 1980. Electrical network simulation of tropical forests successional dynamics, in: Dubois, D. (ed.), *Progress in Ecological Engineering and Management by Mathematical Models*. Centre Belge d'etudes et de documentation (Belgian Center of Studies and Documentation), Liège, Belgium, pp. 883–892.

Acevedo, M.F., 2012. *Simulation of Ecological and Environmental Systems*. CRC Press, Boca Raton, FL.

Acevedo, M.F., McGregor, K., Andressen, R., Ramírez, H., Ablan, M., 1999. Relations of climate variability in Venezuela to tropical Pacific SST, in: *10th Symposium on Global Change Studies, AMS Annual Meeting*. American Meteorological Society, Dallas, TX, pp. 81–84.

Acevedo, M.F., Monteleone, S., Ataroff, M., Estrada, C.A., 2001. Aberturas del dosel y espectro de la luz en el sotobosque de una selva nublada andina de Venezuela. *Ciencia* 9(2), 165–183.

Acevedo, M.F., Urban, D.L., Ablan, M., 1995. Transition and gap models of forest dynamics. *Ecological Applications* 5, 1040–1055.

Acevedo, M.F., Urban, D.L., Shugart, H.H., 1996. Models of forest dynamics based on roles of tree species. *Ecological Modelling* 87, 267–284.

Allen, J.H., Waller, W.T., Acevedo, M.F., Morgan, E.L., Dickson, K.L., Kennedy, J.H., 1996. A minimally-invasive technique to monitor valve movement behavior in bivalves. *Environmental Technology* 17, 501–507.

Becker, R.A., Chambers, J.M., Wilks, A.R., 1988. *The New S Language*. Wadsworth, Pacific Groove, CA.

Bolker, B.M., 2008. *Ecological Models in R*. Princeton University Press, Princeton, NJ, 396pp.

Borg, I., Groenen, P., 2005. *Modern Multidimensional Scaling: Theory and Applications* (2nd edn.). Springer-Verlag, New York.

Box, G.E.P., Jenkins, G.M., 1976. *Time Series Analysis: Forecast and Control*. Holden-Day, San Francisco, CA.

Bréard, A., 2006. Where shall the history of statistics in China begin?, in: Wolfschmidt, G. (ed.), *Algorismus. Es gibt für Könige keinen besonderen Weg zur Geometrie* (Vol. 59). Festschrift für Karin Reich, Augsburg, Germany, pp. 93–100.

Burt, J.E., Barber, G.M., Rigby, D.L., 2009. *Elementary Statistics for Geographers* (3rd edn.). The Guilford Press, New York.

Carr, J.R., 1995. *Numerical Analysis for the Geological Sciences*. Prentice Hall, Englewood Cliffs, NJ.

Carr, J.R., 2002. *Data Visualization in the Geological Sciences*. Prentice Hall, Upper Saddle River, NJ.

Chambers, J.M., Hastie, J.T. (eds.), 1993. *Statistical Models in S*. Chapman & Hall, London, U.K.

Clark, J.S., 2007. *Statistical Computation for Environmental Sciences in R: Lab Manual for Models for Ecological Data*. Princeton University Press, Princeton, NJ.

Cowpertwait, P.S.P., Metcalfe, A.V., 2009. *Introductory Time Series with R*. Springer, New York.

Crawley, M.J., 2002. *Statistical Computing. An Introduction to Data Analysis Using S-Plus*. Wiley, Chichester, U.K.

Crawley, M.J., 2005. *Statistics. An Introduction Using R*. Wiley, Chichester, U.K.

Cryer, J.D., Chan, K.S., 2010. *Time Series Analysis: With Applications in R* (2nd edn.). Springer, New York.

Dalgaard, P., 2008. *Introductory Statistics with R* (2nd edn.). Springer, New York.

Davis, J.C., 2002. *Statistics and Data Analysis in Geology* (3rd edn.). Wiley, New York.

DeCoursey, W.J., 2003. *Statistics and Probability for Engineering Applications with Microsoft® Excel*. Newnes, Burlington, MA.

Deutsch, C.V., Journel, A.G., 1992. *GSLIB Geostatistical Software Library and User's Guide*. Oxford University Press, New York.

Drake, A., 1967. *Fundamentals of Applied Probability Theory*. McGraw-Hill, New York.

Eisenhauer, J.G., 2003. Regression through the origin. *Teaching Statistics* 25, 76–80.

Englund, E., Sparks, A., 1991. *GEO—EAS 1.2.1 Geostatistical Environmental Assessment Software User's Guide*. United States Environmental Protection Agency, Environmental Monitoring Systems Laboratory, Las Vegas, NV.

Everitt, B.S., Hothorn, T., 2010. *Handbook of Statistical Analyses Using R* (2nd edn.). Taylor & Francis Group/ CRC Press, Boca Raton, FL.

Fotheringham, A.S., Brunsdon, C., Charlton, M., 2000. *Quantitative Geography. Perspectives on Spatial Data Analysis.* Sage Publications, Thousand Oaks, CA.

Fox, J., 2005. The R commander: A basic-statistics graphical user interface to R. *Journal of Statistical Software* 14, 1–42.

Gabriel, K.R., 1971. The biplot-graphic display of matrices with application to principal component analysis. *Biometrika* 58, 453–467.

Glass, D.V., 1964. John Graunt and his natural and political observations. *Notes and Records of the Royal Society of London* 19, 63–100.

Gonick, L., Smith, W., 1993. *The Cartoon Guide to Statistics.* Harper Perennial, New York.

Gordon, F., Gordon, S. (eds.), 1992. *Statistics for the Twenty-First Century.* Mathematical Association of America, Washington, DC.

Gotelli, N.J., Ellison, A.M., 2004. *A Primer of Ecological Statistics.* Sinauer, Sunderland, MA.

Gower, C., Hand, D.J., 1996. *Biplots.* Chapman & Hall, London, U.K.

Griffith, D.A., Amrhein, C.G., 1991. *Statistical Analysis for Geographers.* Prentice Hall, Englewood Cliffs, NJ.

Hald, A., 2003. *A History of Probability and Statistics and Their Applications Before 1750.* Wiley, Hoboken, NJ.

Hennessey, J.C., 1980. An age dependent, absorbing semi-Markov model of work histories of the disabled. *Mathematical Biosciences* 51, 283–304.

Herrington, R., 2002. Interactive graphics in R. *Benchmarks Online.* Research and Statistical Support (RSS), University of North Texas, Denton, TX.

Herrington, R., 2003. Interactive graphics in R (Part II—cont.): Kernel density estimation in one and two dimensions. *Benchmarks Online.* Research and Statistical Support (RSS), University of North Texas, Denton, TX.

Howard, R.A., 1971. *Dynamic Probabilistic Systems (Vol. II: Semi-Markov and Decision Processes).* Wiley, New York.

Isaaks, E.H., Srivastava, R.M., 1989. *An Introduction to Applied Geostatistics.* Oxford University Press, New York.

Jensen, J.L., Lake, L.W., Corbett, P.W.M., Goggin, D.J., 1997. *Statistics for Petroleum Engineers and Geoscientists.* Prentice Hall, Upper Saddle River, NJ.

Jones, O., Maillardet, R., Robinson, A., 2009. *Introduction to Scientific Programming and Simulation Using R.* CRC Press/Taylor & Francis Group, Boca Raton, FL.

Kaluzny, S.P., Vega, S.C., Cardoso, T.P., Shelly, A.A., 1996. *S+SPATIALSTATS User's Manual. Version 1.0.* MathSoft, Inc., Seattle, WA.

Keen, R.E., Spain, J.D., 1992. *Computer Simulation in Biology. A BASIC Introduction.* Wiley-Liss, New York, 498pp.

Landsea, C.W., 2012. Atlantic basin tropical cyclone data, "FAQ: Hurricanes, Typhoons, and Tropical Cyclones," Version 2.10, Part E: Tropical Cyclone Records, available online at: http://www.aoml.noaa.gov/hrd/tcfaq/tcfaqE.html

Ledolter, J., Hogg, R.B., 2010. *Applied Statistics for Engineers and Physical Scientists* (3rd edn.). Prentice Hall, Upper Saddle River, NJ.

Lewis, E.R., 1977. *Network Models in Population Biology.* Springer-Verlag, Berlin, Germany.

Manly, B.F.J., 2009. *Statistics for Environmental Science and Management* (2nd edn.). Chapman & Hall/CRC Press, Boca Raton, FL.

Mann Prem, S., 1998. *Introductory Statistics* (3rd edn.). Wiley, New York.

MathSoft, 1999. *S-PLUS 2000 Guide to Statistics* (Vols. 1 and 2). MathSoft Inc., Seattle, WA.

McCuen, R.H., 1985. *Statistical Methods for Engineers.* Prentice Hall, Englewood Cliffs, NJ.

Middleton, G.V., 2000. *Data Analysis in the Earth Sciences Using MATLAB.* Prentice Hall, Upper Saddle River, NJ.

Neitsch, S.L., Arnold, J.G., Kiniry, J.R., Williams, J.R., King, K.W., 2002. Soil and water assessment tool. Theoretical Documentation Version 2000. Grassland, Soil and Water Research Laboratory, Agricultural Research Service, College Station, TX.

Nenadic, O., Greenacre, M., 2007. Correspondence analysis in R, with two- and three-dimensional graphics: The ca package. *Journal of Statistical Software* 20, 1–13.

Oksanen, J., 2011. Multivariate Analysis of Ecological Communities in R: vegan tutorial. Available online at http://cc.oulu.fi/~jarioksa/opetus/metodi/vegantutor.pdf.

Petruccelli, J.D., Nandram, B.P., Chen, M., 1999. *Applied Statistics for Engineers and Scientists.* Prentice Hall, Upper Saddle River, NJ.

Qian, S.S., 2010. *Environmental and Ecological Statistics with R.* Chapman & Hall/CRC Press, Boca Raton, FL.

Quinn, G., Keogh, M., 2002. *Experimental Design & Data Analysis for Biologists*. Cambridge University Press, Cambridge, U.K.

Raftery, A.E., Tanner, M.A., Wells, M.T. (eds.), 2002. *Statistics in the 21st Century*. Chapman & Hall/CRC Press, Boca Raton, FL.

Reimann, C., Filzmoser, P., Garret, R., Dutter, R., 2008. *Statistical Data Analysis Explained: Applied Environmental Statistics with R*. Wiley, Chichester, U.K.

Richardson, C.W., Nicks, A.D., 1990. Weather generator description, in: Sharpley, A.N., Williams, J.R. (eds.), *EPIC-Erosion/Productivity Impact Calculator: 1 Model Documentation*, pp. 93–104.

Rogerson, P.A., 2001. *Statistical Methods for Geography* (3rd edn.). Sage Publications, London, U.K.

Schiff, D., D'Agostino, R.B., 1996. *Practical Engineering Statistics*. Wiley, New York.

Shumway, R.H., Stoffer, D.S., 2006. *Time Series Analysis and Its Applications: With R Examples* (2nd edn.). Springer, New York.

SIDC-Team, 2012. *Monthly Report on the International Sunspot Number*, World Data Center for the Sunspot Index, Royal Observatory of Belgium, Brussel, Belgium. Online catalogue of the sunspot index: http://www.sidc.be/sunspot-data/1700-2011 (last accessed March 25, 2012).

Soetaert, K., Herman, P.M.J., 2009. *A Practical Guide to Ecological Modelling Using R as a Simulation Platform*. Springer, New York.

Sprinthall, R.C., 1990. *Basic Statistical Analysis* (3rd edn.). Prentice Hall, Englewood Cliffs, NJ.

Stevens, M.H.H., 2009. *A Primer of Ecology with R*. Springer, New York.

Sullivan, M., 2004. *Statistics, Informed Decisions Using Data*. Prentice Hall, Upper Saddle River, NJ.

Swartzman, G.L., Kaluzny, S., 1987. *Ecological Simulation Primer*. MacMillan, New York.

Venables, W.N., Smith, D.M. and the R Core Team. 2012. An Introduction to R. Notes on R: A Programming Environment for Data Analysis and Graphics. Version 2.15.1 (2012-06-22). CRAN. (Available Online http://www.r-project.org).

Wadsworth, H.M. (ed.), 1998. *Handbook of Statistical Methods for Engineers and Scientists* (2nd edn.). McGraw-Hill, New York.

Zuur, A.F., Ieno, E.N., Walker, N.J., Saveliev, A.A., Smith, G.M., 2009. *Mixed Effects Models and Extensions in Ecology with R*. Springer, New York.

Index

A

Akaike information criterion (AIC), 341, 387–388
Analysis of variance (ANOVA), 7, 144
  fictional data
    boxplots, 173–174
    interaction plots, 174–175
    invent.mxn function, 172–173, 175
    p-value, 174
  nonparametric
    Friedman test, 151, 172
    Kruskal–Wallis test (*see* Kruskal–Wallis test)
  one-way
    add statistics to datasets, 209–210
    among samples, 146, 148
    anova function, 165
    aov function, 164–165
    boxplots, 146–147, 164
    column vector, 162
    data.frame function, 163–164
    dissolved oxygen (DO), 162
    factor function, 163
    F-test, 146–148
    maxDO, 163–164, 208–209
    mean square of errors, 145–146, 148
    plot means, 209–210
    pooled sample, 147
    residual error, 146, 164
    response variable, 145
    within samples, 145–146, 148, 164
    Shapiro–Wilk test, 209
    table selection, 165
    using Rcmdr, 165
    zone.maxDO, 163–164
  predictive tool, 195–196
  two-way
    among samples, 149
    block and treatment, 148
    boxplots, 149, 167
    data.frame, 166–167
    effects plots, 169
    factor interaction, 149–151, 167
    interaction plot, 167
    nonreplicates and replicates, 148
    residual error, 149, 168
    within samples, 149
    sources of variations, 149
    using Rcmdr, 168
    yield array, 166
    yield.tv data frame, 167–168
ANOVA, *see* Analysis of variance
Autocorrelation function, 101–102
Autocovariance function, 101–102
Autoregressive integrated moving-average (ARIMA) model
  Box–Jenkins method, 388
  estimation, 388
  evaluation, 388
  identification, 388
  modeling and forecast
    arima function, 396
    arima.yrspots object, 397
    ar.yw function, 396
    diagnostic of, 398–399
    lab7/year-spot1700-2011.txt data set, 395
    predict function, 398
    sunspots evaluation, 396–397
    sunspots forecast, 398–399
    time series, 395
    tsdiag function, 398
Autoregressive moving average (ARMA), 387–388
Autoregressive (AR) process
  ACF, 382–383, 387–388
  AIC, 387–388
  coefficients estimation, 384, 386–387
  modeling and forecast, 392–395
  PACF, 385, 387–388
  *p*th order Markov process, 381
  realizations, 382, 386
  residuals, 387–388
  scatter plots, 382, 384–385
  YW equation, 382

B

Bayesian analysis, 109
Bayes' theorem
  many events, 36
  Rcmdr, 54–55
  two events, 35–36
Binomial distribution, 71
  calculation, 85–86
  cmf, 83
  pmf, 82–83
  vector object, 82
Bivariate analysis
  definition, 98
  LLS regression, 177
  marginal boxplots, 135
  Pearson's test, 134
  p-value, 135
  ranked observation results, 133
  rnorm, 132
  uncorrelated sample results, 133–134
Box–Jenkins method, 388

C

Cell count methods
  clustered pattern, 263–264
  spatial randomness testing, 261–263
  uniform patterns testing, 260–261